Systems & Control: Foundations & Applications

Series Editor
Tamer Başar, University of Illinois at Urbana-Champaign, Urbana, IL, USA

Editorial Board
Karl Johan Åström, Lund University of Technology, Lund, Sweden
Han-Fu Chen, Academia Sinica, Beijing, China
Bill Helton, University of California, San Diego, CA, USA
Alberto Isidori, Sapienza University of Rome, Rome, Italy
Miroslav Krstic, University of California, San Diego, CA, USA
H. Vincent Poor, Princeton University, Princeton, NJ, USA
Mete Soner, ETH Zürich, Zürich, Switzerland;
    Swiss Finance Institute, Zürich, Switzerland
Roberto Tempo, CNR-IEIIT, Politecnico di Torino, Italy

More information about this series at http://www.springer.com/series/4895

Alexander B. Kurzhanski • Pravin Varaiya

Dynamics and Control of Trajectory Tubes

Theory and Computation

Alexander B. Kurzhanski
Faculty of Computational Mathematics
  and Cybernetics
Moscow State (Lomonosov)
  University
Moscow, Russia

Electrical Engineering
  and Computer Sciences
University of California, Berkeley
Berkeley, CA, USA

Pravin Varaiya
Electrical Engineering
  and Computer Sciences
University of California, Berkeley
Berkeley, CA, USA

ISSN 2324-9749 ISSN 2324-9757 (electronic)
ISBN 978-3-319-10276-4 ISBN 978-3-319-10277-1 (eBook)
DOI 10.1007/978-3-319-10277-1
Springer Cham Heidelberg New York Dordrecht London

Library of Congress Control Number: 2014948336

Mathematics Subject Classification (2010): 34A60, 34K35, 35E10, 35K25, 37N35

© Springer International Publishing Switzerland 2014
This work is subject to copyright. All rights are reserved by the Publisher, whether the whole or part of the material is concerned, specifically the rights of translation, reprinting, reuse of illustrations, recitation, broadcasting, reproduction on microfilms or in any other physical way, and transmission or information storage and retrieval, electronic adaptation, computer software, or by similar or dissimilar methodology now known or hereafter developed. Exempted from this legal reservation are brief excerpts in connection with reviews or scholarly analysis or material supplied specifically for the purpose of being entered and executed on a computer system, for exclusive use by the purchaser of the work. Duplication of this publication or parts thereof is permitted only under the provisions of the Copyright Law of the Publisher's location, in its current version, and permission for use must always be obtained from Springer. Permissions for use may be obtained through RightsLink at the Copyright Clearance Center. Violations are liable to prosecution under the respective Copyright Law.
The use of general descriptive names, registered names, trademarks, service marks, etc. in this publication does not imply, even in the absence of a specific statement, that such names are exempt from the relevant protective laws and regulations and therefore free for general use.
While the advice and information in this book are believed to be true and accurate at the date of publication, neither the authors nor the editors nor the publisher can accept any legal responsibility for any errors or omissions that may be made. The publisher makes no warranty, express or implied, with respect to the material contained herein.

Printed on acid-free paper

Springer is part of Springer Science+Business Media (www.birkhauser-science.com)

Preface

This book presents a systematic treatment of theoretical methods and computation schemes for solving problems in dynamics and control. The mathematical models investigated here are motivated by processes that emerge in many applied areas. The book is aimed at graduate students, researchers, and practitioners in control theory with applications and computational realizations.

The emphasis is on issues of reachability, feedback control synthesis under complex state constraints, hard or double bounds on controls, and performance in finite continuous time. Also given is a concise description of problems in guaranteed state estimation, output feedback control, and hybrid dynamics using methods of this book. Although its focus is on systems with linear structure, the text also indicates how the suggested approaches apply to systems with nonlinearity and nonconvexity.

Of primary concern is the problem of *system reachability*, complemented by the problem of *solvability*, within a given class of controls. This leads to two basic questions: find the terminal states that the system can reach from a given initial state within a specified time, using all possible controls, and find the initial states from which the system can reach a given terminal ("target") set in specified time, using all possible controls. The answer to these questions takes the form of bundles of controlled trajectories—the reachability tubes—that emanate either forward in time from a given starting set (*the forward reachability tube*) or backward in time from a given target set (the solvability or *backward reachability tube*). The cross-sections ("cuts") of these tubes are the reachability ("reach") sets. Backward reach sets are important in designing *feedback (closed-loop) controls*.

We are thus led to dealing with *trajectory tubes*—set-valued functions that are the main elements of the required solutions. These tubes are also key in describing systems operating under uncertainty, in the form of unknown but bounded parameters in the system model, inputs, and measurement errors. The mathematical properties of the solutions, such as non-differentiability, may also lead to set-valued functions. Hence we have to conduct the investigation of feedback control system dynamics taking into account the set-valued nature of the functions involved. This is achieved using the Hamilton–Jacobi formalism in conjunction with the Dynamic Programming approach and studying value functions for appropriate problems of dynamic optimization. These value functions turn out to be the solutions to related types of the HJB (Hamilton–Jacobi–Bellman) equation with appropriate boundary conditions. The level sets of such value functions, evolving in time, describe the tubes we need.

In addition to theoretical rigor, the solutions to modern control problems must be effectively computable. That is, the results must be available within prescribed time, perhaps on-line, the set-valued trajectory must be calculated with a guaranteed accuracy, and the computation procedures should be able to handle high-dimensional systems. This may require parallelization of solutions and distributed computation. The computation approaches presented here meet such needs. They are based on ellipsoidal calculus introduced in [174], successfully extended and applied in [181, 182], and complemented by appropriate software tools [132].

The book is divided into eleven chapters. Chapter 1 provides an exposition of problems in control theory for linear systems with solutions given in the specific forms needed for transition to computable operations over trajectory tubes.

Chapter 2 is focused on *how to solve*. The main theoretical tools—Hamiltonian methods and Dynamic Programming techniques—as applied to problems of reachability and target control synthesis are described. These are reduced to optimization problems, whose solution is given by an equation of the HJB type. In its turn, the solution to the HJB equation produces value functions whose level sets are the desired reach sets. Usually, attention is needed in interpreting the solution to the HJB equation. However, for "linear-convex" systems (with convex constraints on the controls and convex starting or terminal sets) the value functions are convex in the state variables and usually unique. These functions are directionally differentiable along any direction in the state space. Hence there is no need for subtle definitions of generalized viscosity solutions since for equations used here they may be expressed through conventional classical arguments. Moreover, direct integration of the HJB equation is avoided by calculating the *exact* value function through duality methods of convex analysis. Such a procedure is also used to apply a verification theorem to the concrete HJB equation, confirming that the calculated value function is its solution. This also allows us to derive the feedback control directly from the HJB equation. The derived control is in general a set-valued function and its substitution in the system equation produces a *differential inclusion*. Effective computation of trajectories that steer the system from point to point arrives through an intersection of colliding tubes—the forward reach tube from a given starting point and the backward reach tube from a given terminal point. The final section describes a solution to the problem of *time-optimal* target control.

Chapter 3 indicates *how to calculate* the solutions given as set-valued functions. The approach develops external and internal ellipsoidal-valued functions that approximate the reachability tubes. Parametrized families of ellipsoids are designed, whose intersection produces an external bound and whose union produces an internal bound. Increasing the number of approximating ellipsoids in the limit yields the exact reach set. The next move is to calculate the synthesizing feedback target control. As already mentioned, such a feedback strategy may be calculated directly from the HJB equation. However, this control strategy, being a nonlinear function of time and state, would require an appropriate existence theorem for the synthesized nonlinear differential inclusion. To avoid this possible difficulty, it is convenient to use the "aiming" rule of Krasovski. Namely, calculating the backward reachability set in advance, one should design the feedback control such that it keeps the trajectory within the backward reach tube, following this "bridge" until it reaches the terminal target set. The computation scheme for the aiming rule is as follows: with given starting point in "time-space" that lies within or beyond the backward reach set, one selects an internal ellipsoidal approximation of the backward set that either contains this point or is the closest to it. This aiming strategy may be calculated explicitly or with minimal computational burden.

Chapter 4 offers examples of problems solved by ellipsoidal methods and illustrated graphically. Though attractive and effective, ellipsoidal solutions may run into some degenerate cases that are more common for internal approximations. These potential computational problems are especially evident in large systems where the dimension of controls is much smaller than the system dimension. Ways of *regularizing* such situations are discussed in a section on high-dimensional systems. The described tools have proved effective in computational experiments. The treatment of high-dimensional systems indicated here is reached through a parallelization of solutions, which is a natural extension of the suggested schemes.

The first part of Chap. 5 is devoted to *nonlinearity* and *nonconvexity*. The general approach formulated here is applicable to nonlinear systems and presented in the form of a *comparison principle*. The idea is to approximate the available HJB equation from above or below through relations that ensure guaranteed upper or lower solution estimates by operating with functions simpler than in the original equation. This approach does *not* depend on the type of exact solution to the HJB equation, whether classic or generalized. At the same time, the comparison principle allows one to develop a *deductive approach* to the ellipsoidal calculus of this book in contrast with the *inductive approach* of Chap. 3. These topics are followed by examples of reachability for nonlinear systems and calculation of the set of points *reachable within a time interval*, which is a nonconvex union of convex sets. The second part of the chapter deals with the application of ellipsoidal methods to systems with *non-ellipsoidal constraints*, such as boxes and zonotopes (symmetrical polyhedra).

The chapters that follow cover a variety of useful properties and specific problems related to implementing the proposed approach. Chapter 6 emphasizes the role of *double constraints*—the simultaneous presence of both a hard bound and an integral bound on the controls. Solutions to such problems are needed for treating *impulse controls* investigated in this chapter. Impulse inputs were previously studied as open-loop controls with few special results on feedback solutions [23,36]. A theory of feedback impulse control is presented here for systems of arbitrary dimension, with solutions in terms of *quasi-variational inequalities* of the HJB type. However, impulse inputs are ideal elements and their physical realization by "ordinary" bounded functions may be achieved through approximation by double-bounded controls which are functions whose bound tends to infinity.

Chapter 7 is concerned with dynamics and control of systems under *state constraints*. Previously discussed problems are now subject to additional "viability" restrictions on the state variables. The description covers solution approaches to problems of reachability (forward and backward) and emphasizes specifics of related mathematical techniques including ellipsoidal approximations. The case of linear systems with convex hard bounds on controls and state coordinates are worked out in detail.

The contents of Chaps. 1–7 indicate that one of the main items in treating considered problems are trajectory tubes and the means of their calculation. The same is true for the rest of the chapters. So Chap. 8 begins with fundamentals of a general vision—*the theory of trajectory tubes* with models of their evolutionary dynamics. Indicated results, together with considerations of the previous chapter, are further applied to *closed-loop control under state constraints*, with techniques borrowed from both Hamiltonian approach and duality theory of nonlinear analysis. This brings forward the discussion to complex state constraints in the form of *obstacle problems*, wherein constrained trajectories must simultaneously lie *within* one set and *outside* another.

The next two chapters consider *uncertainty*, which is inherent in realistic problems of control. These chapters may serve as an introduction to a more thorough description of uncertainty.

Chapter 9 is a concise explanation of the *theory of guaranteed state estimation*, also known as the *set-membership bounding approach* to external disturbances in estimation models. In contrast with conventional descriptions, the present exposition involves Hamiltonian methods and is applicable to nonlinear systems. Dynamic estimation of system trajectories under unknown but bounded errors is also formulated as a problem with state constraints, which now are not known in advance, but arrive online, in real time. In the linear case the proposed deterministic "filtering" equations demonstrate connections and differences when compared with stochastic (Kalman) filtering. Both continuous and discrete measurements are considered.

The results of Chap. 9 have a natural application to problems of *output feedback control* under unknown but bounded disturbances in system and measurement inputs. These problems are addressed in Chap. 10. The solutions introduced there are based on the notions of *generalized state* and *information tubes* which describe the overall system dynamics. For the linear case with convex constraints on controls and uncertainties items, computation schemes based on ellipsoidal approximations are presented. Several examples are worked out.

Finally, Chap. 11 is confined to *verification* problems and *hybrid systems*, with a description of exact solutions and ellipsoidal schemes for their computation. The discussion is accompanied by some examples. Special attention is given to possible involvement of impulse inputs in formal mathematical models of hybrid systems. The aim of this chapter is to emphasize the applicability of techniques presented in this book to the investigation of hybrid systems.

References to prior literature and research results related to this book are given in the introduction to each chapter.

The mathematical level of the book presumes reasonable knowledge of advanced calculus, linear algebra and differential equations, as well as basics of variational analysis with optimization methods and computational mathematics.

Throughout the previous years we had the pleasant opportunity for useful discussions on the topics arranged in this book with K. Åström, J.-P. Aubin, J. Baras, T. Başar, F. Chernousko, P. Kokotovic, A. Krener, A.A. Kurzhanskiy, Yu. Ledyayev, G. Leitmann, A. Lindquist, J. Lygeros, M. Milanese, I. Mitchell, S. Mitter, A. Rantzer, J. Sousa, C. Tomlin, I. Vályi, and V. Veliov. Their valuable comments helped to crystallize the contents.

We thank A. Daryin, M. Gusev, T. Filippova, I. Rublev, and P. Tochilin for reading parts of the manuscript, their useful comments, and contributed illustrations.

Our thanks surely goes to the authors of illustrations for the examples of this book. Their names are indicated throughout the text.

The authors are grateful to the US national Science Foundation, the first author is also grateful to the Russian Foundation for Basic Research for the support of their work.

Moscow, Russia	Alexander B. Kurzhanski
Berkeley, CA, USA	Pravin Varaiya

Contents

1	**Linear Control Systems**		1
	1.1	The Controlled System	1
	1.2	Control and State Constraints: Open-Loop and Closed-Loop Control	5
		1.2.1 Constraints on Control and State	5
		1.2.2 Open-Loop and Closed-Loop Control	8
	1.3	Optimal Control with Norm-Minimal Cost: The Controllability Property	10
		1.3.1 Minimum Energy Control	10
		1.3.2 Minimum Magnitude Control	12
		1.3.3 Controllability	16
		1.3.4 The Finite-Dimensional Moment Problem	17
	1.4	The Reachability Problem: Time-Optimal Control	18
		1.4.1 Reachability	18
		1.4.2 Calculating Controls for the Boundary of Reach Set	20
		1.4.3 Time-Optimal Control	23
	1.5	Optimal Open-Loop Control for Linear-Convex Systems: The Maximum Principle	24
		1.5.1 The Necessary Conditions of Optimality	25
		1.5.2 Degenerate Cases	29
		1.5.3 Sufficiency Conditions for Optimality	32
		1.5.4 A Geometrical Interpretation	36
		1.5.5 Strictly Convex Constraints on Control	38
	1.6	Duality Methods of Convex Analysis in Problems of Optimal Control	40
		1.6.1 The Primal and Dual Optimization Problems	41
		1.6.2 General Remark: Feedforward Controls	46

2 The Dynamic Programming Approach 47
2.1 The Dynamic Programming Equation............................. 48
2.2 The Linear-Quadratic Problem 54
2.3 Reachability Through the HJB Equation: Hard Bounds............ 58
2.3.1 Forward Reachability.. 58
2.3.2 Backward Reachability or the Solvability Problem 61
2.4 Reachability for Linear-Convex Systems 64
2.4.1 Forward Reachability: Calculating the Value Functions.. 64
2.4.2 The Conjugate of the Value Function V_0 67
2.4.3 The Value Function (Backward) 68
2.5 Colliding Tubes: Calculating All Reachable Points 70
2.6 The Closed-Loop Target Control 75
2.7 Reachability Within an Interval...................................... 78
2.8 Dynamic Programming: Time-Optimal Control 83

3 Ellipsoidal Techniques: Reachability and Control Synthesis 87
3.1 Linear Systems Under Ellipsoidal Constraints...................... 88
3.2 Ellipsoidal Approximation of Reach Sets 91
3.3 Recurrent Relations: External Approximations 94
3.4 The Evolution of Approximating Ellipsoids 102
3.5 The Ellipsoidal Maximum Principle and the Reachability Tube... 107
3.6 Example 3.6 111
3.7 Reachability Sets: Internal Approximations 114
3.8 Example 3.8 118
3.9 Reachability Tubes: Recurrent Relations—Internal
    Approximations 122
3.10 Backward Reachability: Ellipsoidal Approximations 128
3.11 The Problem of Control Synthesis: Solution Through
    Internal Ellipsoids 131
3.12 Internal Approximations: The Second Scheme 141

4 Solution Examples on Ellipsoidal Methods: Computation
in High Dimensions 147
4.1 The Multiple Integrator 147
4.1.1 Computational Results 151
4.2 A Planar Motion Under Newton's Law 152
4.2.1 Computational Results 158
4.3 Damping Oscillations 168
4.3.1 Calming Down a Chain of Springs in Finite Time 168
4.4 Computation in High-Dimensional Systems.
    Degeneracy and Regularization..................................... 182
4.4.1 Computation: The Problem of Degeneracy............... 182
4.4.2 Regularizing the Ellipsoidal Estimates 186
4.4.3 Regularizing the Estimate for the Reachability Tube 189
4.4.4 Efficient Computation of Orthogonal Matrix S 193
4.4.5 Parallel Computation 195

5 The Comparison Principle: Nonlinearity and Nonconvexity 197
- 5.1 The Comparison Principle for the HJB Equation 198
  - 5.1.1 Principal Propositions for Comparison Principle 198
  - 5.1.2 A Deductive Approach to Ellipsoidal Calculus 202
- 5.2 Calculation of Nonconvex Reachability Sets 208
- 5.3 Applications of Comparison Principle 213
  - 5.3.1 Forward Reachability 213
  - 5.3.2 Systems with Hamiltonians Independent of the State 215
  - 5.3.3 A Bilinear System 218
  - 5.3.4 External Ellipsoids for the Unicycle: Reachability 222
- 5.4 Ellipsoidal Methods for Non-ellipsoidal Constraints 229
  - 5.4.1 Degenerate Ellipsoids: Box-Valued Constraints 229
  - 5.4.2 Integrals of Box-Valued Functions 231
  - 5.4.3 Reachability Tubes for Box-Valued Constraints: External Approximations 235
  - 5.4.4 Reach Tubes for Box-Valued Constraints: Internal Approximations 239
- 5.5 Ellipsoidal Methods for Zonotopes 241
  - 5.5.1 Zonotopes 241
  - 5.5.2 Internal Ellipsoidal Tubes for a Zonotope 243
  - 5.5.3 External Ellipsoidal Tubes for a Zonotope 246

6 Impulse Controls and Double Constraints 253
- 6.1 The Problem of Impulse Controls 254
  - 6.1.1 Open-Loop Impulse Control: The Value Function 255
  - 6.1.2 Closed-Loop Impulse Control: The HJB Variational Inequality 260
- 6.2 Realizable Approximation of Impulse Controls 266
  - 6.2.1 The Realistic Approximation Problem 266
  - 6.2.2 The Approximating Motions 269

7 Dynamics and Control Under State Constraints 275
- 7.1 State-Constrained Reachability and Feedback 275
  - 7.1.1 The Reachability Problem Under State Constraints 276
  - 7.1.2 Comparison Principle Under State Constraints 281
  - 7.1.3 Linear-Convex Systems 285
- 7.2 State-Constrained Control: Computation 289
  - 7.2.1 The Modified Maximum Principle 289
  - 7.2.2 External Ellipsoids 294
  - 7.2.3 Generalized Multipliers 301
  - 7.2.4 An Example 302
  - 7.2.5 Specifics: Helpful Facts for State-Constrained Control Design 307

8 Trajectory Tubes State-Constrained Feedback Control ... 311

8.1 The Theory of Trajectory Tubes: Set-Valued Evolution Equations ... 311
- 8.1.1 Trajectory Tubes and the Generalized Dynamic System ... 312
- 8.1.2 Some Basic Assumptions ... 314
- 8.1.3 The Set-Valued Evolution Equation ... 315
- 8.1.4 The Funnel Equations: Specific Cases ... 316
- 8.1.5 Evolution Equation Under Relaxed Conditions ... 319

8.2 Viability Tubes and Their Calculation ... 319
- 8.2.1 The Evolution Equation as a Generalized Partial Differential Equation ... 319
- 8.2.2 Viability Through Parameterization ... 322

8.3 Control Synthesis Under State Constraints: Viable Solutions ... 325
- 8.3.1 The Problem of State-Constrained Closed-Loop Control: Backward Reachability Under Viability Constraints ... 326
- 8.3.2 State-Constrained Closed-Loop Control ... 328
- 8.3.3 Example ... 331

8.4 Obstacle Problems ... 333
- 8.4.1 Complementary Convex State Constraints: The Obstacle Problem ... 333
- 8.4.2 The Obstacle Problem and the Reach-Evasion Set ... 335
- 8.4.3 Obstacle Problem: An Example ... 336
- 8.4.4 Closed-Loop Control for the Obstacle Problem ... 338

9 Guaranteed State Estimation ... 341

9.1 Set-Membership State Estimation: The Problem. The Information Tube ... 342

9.2 Hamiltonian Techniques for Set-Membership State Estimation ... 344
- 9.2.1 Calculating Information Tubes: The HJB Equation ... 344
- 9.2.2 Comparison Principle for HJB Equations ... 346
- 9.2.3 Linear Systems ... 348

9.3 Ellipsoidal Approximations ... 350
- 9.3.1 Version AE ... 350
- 9.3.2 Version-BE ... 353
- 9.3.3 Example: Information Set for a Linear System ... 354

9.4 Discrete Observations ... 355
- 9.4.1 Continuous Dynamics Under Discrete Observations ... 355
- 9.4.2 Discrete Dynamics and Observations ... 358

9.5 Viability Tubes: The Linear-Quadratic Approximation ... 363

	9.6	Information Tubes vs Stochastic Filtering Equations. Discontinuous Measurements.......	367
		9.6.1 Set-Valued Tubes Through Stochastic Approximations..	367
		9.6.2 The Singular Perturbation Approach: Discontinuous Measurements in Continuous Time	370
10	**Uncertain Systems: Output Feedback Control**..................		**371**
	10.1	The Problem of OFC	372
	10.2	The System: The Generalized State and the Rigorous Problem Formulation..................	373
	10.3	The Overall Solution Scheme: General Case	375
	10.4	Guaranteed State Estimation Revisited: Information Sets and Information States	377
	10.5	Feedback Control in the Information Space	381
		10.5.1 Control of Information Tubes	381
		10.5.2 The Solution Scheme for Problem **C**	383
		10.5.3 From Problem **C** to Problem $\mathbf{C}_{opt}$	385
	10.6	More Detailed Solution: Linear Systems...........	386
		10.6.1 The "Linear-Convex" Solution	386
		10.6.2 The Computable Solution: Ellipsoidal Approximations..	388
	10.7	Separation of Estimation and Control	391
	10.8	Some Examples	392
11	**Verification: Hybrid Systems**.................		**395**
	11.1	Verification Problems	396
		11.1.1 The Problems and the Solution Schemes	396
		11.1.2 Ellipsoidal Techniques for Verification Problems......	398
	11.2	Hybrid Dynamics and Control............	401
		11.2.1 The Hybrid System and the Reachability Problem	401
		11.2.2 Value Functions: Ellipsoidal Approximation.........	407
	11.3	Verification of Hybrid Systems	417
		11.3.1 Verification: Problems and Solutions	417
		11.3.2 Ellipsoidal Methods for Verification.............	418
	11.4	Impulse Controls in Hybrid System Models	419
		11.4.1 Hybrid Systems with Resets in Both Model and System States	420
		11.4.2 Two Simple Examples................	421
		11.4.3 Impulse Controls in Hybrid Systems	423
		11.4.4 More Complicated Example: The Three-Dimensional Bouncing Ball	425

References... 431

Index... 443

Notations

$\mathbb{R}^n$—the n-dimensional Euclidean vector space, $\mathbb{R}^1 = \mathbb{R}$

$\mathbb{R}^{n \times m}$—the linear space of $n \times m$-matrices

$\langle x, y \rangle = x'y$—the scalar (inner) product of vectors $x, y \in \mathbb{R}^n$, with prime as the transpose

$\|x\|_M = \langle x, My \rangle^{1/2}$, $M = M' > 0$

$\|x\| = \|x\|_I$, I—the identity (unit) matrix

$\mathrm{co}\, Q$, $\mathrm{conv}\, Q$—the convex hull of set Q

$\mathrm{comp}\,\mathbb{R}^n$—the variety of all compact subsets $Q \subset \mathbb{R}^n$

$\mathrm{conv}\,\mathbb{R}^n$—the variety of all convex compact subsets $Q \subset \mathbb{R}^n$

$d(x, Q) = \inf\{\|x - y\| \mid y \in Q.\}$—the distance of point x from set Q

$h(P, Q) = \max\{h^+(P, Q), h^+(Q, P).\}$—the Hausdorff distance between sets P, Q

$h^+(P, Q) = \max\{d(x, Q) \mid x \in P\}$—the Hausdorff semi-distance

$G(t, s)$—the fundamental transition matrix for a linear homogeneous system $\dot{x} = Ax$

int X—interior of set X

$I(l|B)$—indicator function of set B: $I(l|B) = 0$ if $l \in B$, $I(l|B) = \infty$ if $x \bar{\in} B$

$\rho(l \mid Q) = \sup\{\langle l, x \rangle \mid x \in Q\}$—the support function of set Q at point $l \in \mathbb{R}^n$

graph $Y(t) = \{\{t, y\} \in T \times \mathbb{R}^n \mid t \in T = [t, \vartheta], y \in Y(t)\}$

epi $f(x) = \{(x, y) \in \mathbb{R}^n \times \mathbb{R} \mid y > f(x)\}$—epigraph of function f

Dom $f = \{x : f(x) < \infty,\}$—effective domain of f

co$\varphi(l)$—closed convex hull of φ

$\partial \varphi(l)$—subdifferential of convex function $\varphi : \mathbb{R}^n \to \mathbb{R}$ at point l

$f^*(l) = \sup_x \{\langle l, x \rangle - f(x)\}$—Fenchel conjugate of function $f(x)$

V_x—partial derivative in x of function V

$C^m[T]$—the space of m-dimensional continuous vector functions $f : T \to \mathbb{R}^m$

$\mathcal{L}_p^m[T]$—the space of m-dimensional vector functions $f : T \to \mathbb{R}^m$ integrable with power p

$\mathbf{V}^m[T]$—space of m-dimensional vector functions of bounded variation on the interval $T = [t_0, \vartheta]$

$\text{Var} U(t), t \in T$—total variation of function $U \in \mathbf{V}^m[T]$, over the interval $t \in T$

$\|f(\cdot)\|_{\mathcal{L}_p}$—the norm of function $f(\cdot)$ in infinite dimensional space $\mathcal{L}[T]$

Chapter 1
Linear Control Systems

Abstract This chapter gives an exposition of control theory for linear systems with emphasis on items and techniques given in a form appropriate for topics in forthcoming chapters. It introduces problems of reachability and optimal target control under constraints, as well as time-optimal control. Indicated are solution approaches to open-loop control that involve the moment problem, the maximum principle, and the duality methods of convex analysis.

Keywords Control theory • Linear systems • Open-loop control • Reachability • Controllability • Maximum principle • Convex analysis

Among the mathematical models of controlled systems considered in this book priority is given to those with *linear structure* [38, 40, 105, 120, 191, 245, 274]. This class of systems allows detailed analysis with solutions in a form appropriate for computational schemes that yield *complete* solutions to specific problems. In this chapter we describe standard models and indicate the difference between the design of *open-loop* control and *closed-loop* control strategies that use on-line measurements. We treat problems of *dynamic optimization*, that is of *optimal control*. We indicate several basic open-loop control problems with solution techniques centered around *necessary and sufficient conditions of optimality* in two main forms: the Maximum Principle and duality methods of convex analysis. We start with the system model.

1.1 The Controlled System

A standard mathematical representation of a *controlled system* is an ordinary differential equation (ODE) of the form

$$\dot{x}(t) = f(t, x, u), \qquad (1.1)$$

in which t is *time*, the vector $x \in \mathbb{R}^n$ is the *state*, and the vector $u \in \mathbb{R}^p$ is the *control*. The function $f(t, x, u)$ is defined in a domain

$$\mathbf{D} = T \times \mathcal{D} \times \mathcal{P}, \; t \in T, \; x \in \mathcal{D}, \; u \in \mathcal{P},$$

in which it is assumed to be continuous. The time interval T may be bounded as $T = [t_0, t_1]$, or unbounded as $T = [t_0, +\infty)$ or $T = (-\infty, t_1]$. The control set $\mathcal{P}$ may be a compact subset of $\mathbb{R}^p$ or it may coincide with $\mathbb{R}^p$. $\mathcal{D}$ is an open set that may coincide with $\mathbb{R}^n$.

The function $f(t, x, u)$ is assumed to satisfy a condition that guarantees uniqueness and extendibility of solutions of (1.1) within the domain $\mathbf{D}$. One such condition is: the partial derivatives $f_{x_i}(t, x, u)$, $i = 1, \ldots, n$, are continuous in x uniformly for $t \in T$, $u \in \mathcal{P}$ and the inequality

$$\langle x, f(t, x, u) \rangle \leq k(1 + \langle x, x \rangle)$$

holds for some $k > 0$ for all $(t, x, u) \in \mathbf{D}$.

The book concentrates on systems with *linear* structure of the form

$$\dot{x}(t) = A(t)x(t) + B(t)u(t) + v(t), \tag{1.2}$$

with $x(t) \in \mathbb{R}^n$, $u(t) \in \mathbb{R}^p$ as before, and with $v(t) \in \mathbb{R}^n$ as a *disturbance* or external forcing term. The $n \times n$ matrix function $A(t) \in \mathbb{R}^{n \times n}$ and the $n \times p$ matrix function $B(t) \in \mathbb{R}^{n \times p}$ are assumed to be continuous in t, whereas $u(t), v(t)$ are assumed to be Lebesgue-integrable or piecewise right-continuous in $t \in T = [t_0, +\infty)$.[1] Under these conditions the solution to (1.1) exists, is unique, and is extendible within $\mathbf{D} = T \times \mathbb{R}^n \times \mathcal{P}$ for any initial position $\{t_0, x^0\}$, $x(t_0) = x^0$.

Given an initial position $\{t_0, x^0\}$, the unique solution $x(t) = x(t, t_0, x^0)$ is given by the integral formula

$$x(t) = G(t, t_0)x^0 + \int_{t_0}^t G(t, s)(B(s)u(s) + v(s))ds. \tag{1.3}$$

Here $G(t, s)$ is the fundamental *transition matrix* solution of the homogeneous equation (1.2). That is, the matrix function $G(t, s)$ satisfies the equation

$$\frac{\partial}{\partial t} G(t, s) = A(t)G(t, s), \quad G(s, s) = I \tag{1.4}$$

---

[1] We use the terms measurable and integrable for Lebesgue-measurable and Lebesgue-integrable functions.

1.1 The Controlled System

in t, as well as the matrix *adjoint* equation

$$\frac{\partial}{\partial s} G(t,s) = -G(t,s)A(s), \quad G(t,t) = I, \tag{1.5}$$

in s.

$G(t,s)$ is nonsingular for all t,s and has the following properties:

$G^{-1}(t,s) = G(s,t)$, so that $G(t,s)G(s,t) = I$ (invertibility formula); and

$G(t,\tau)G(\tau,s) = G(t,s)$ (usually $s \leq \tau \leq t$, superposition formula).

An integral formula for the adjoint equation, similar to (1.3), is given in (1.80).

When $A(t) \equiv A$ is constant, $G(t,s) = \exp(A(t-s))$, in which $\exp A$ is the matrix exponential.

Formula (1.3) may be checked by direct substitution. The existence of a solution to system (1.2) is thus a standard property of linear differential equations (see [38, 105, 120, 274]).

Exercise 1.1.1. Check formula (1.3).

System (1.2) can be put into a simpler form. Transforming the state by

$$z(t) = G(t_0, t)x(t), \tag{1.6}$$

differentiating, and substituting $x(t)$ for $z(t)$ using (1.1), we come to

$$\dot{z}(t) = G(t_0, t)B(t)u(t) + G(t_0, t)v(t), \tag{1.7}$$

with

$$z^0 = z(t_0) = G(t_0, t_0)x^0 = x^0. \tag{1.8}$$

Thus there is a one-to-one correspondence (1.6) between the solutions $x(t)$ and $z(t)$ to Eqs. (1.2) and (1.7), respectively. Their initial values are related through (1.8). The state $z(t)$ satisfies a particularly simple version of (1.2),

$$\dot{z}(t) = B_0(t)u(t) + v_0(t), \quad z(t_0) = x^0, \tag{1.9}$$

with

$$B_0(t) = G(t_0, t)B(t), \quad v_0(t) = G(t_0, t)v(t).$$

Thus we may consider system (1.9) rather than (1.2). In other words, in the notation of (1.2) we may take $A(t) \equiv 0$ with no loss of generality. Note however that the matrix function $B(t)$ is now time-variant (even when (1.2) is time-invariant).

One should realize, however, that the transformation (1.6) allows us to take $A(t) \equiv 0$ only within the time range $\{t \geq t_0\}$, a different t_0 leads to a different transformation (1.6). A similar result is obtained by the substitution

$$z(t) = G(t_1, t) x(t).$$

With this choice of state the original system again gives $A(t) \equiv 0$, but this is correct only for $\{t \leq t_1\}$.

We shall sometimes make use of these transformations to demonstrate some basic techniques with a simpler notation. The reader will always be able to return to $A(t) \neq 0$ as an exercise.

A shrewd reader may have now realized that there should exist a transformation of the state that takes a given linear homogeneous equation

$$\dot{x} = A(t)x, \tag{1.10}$$

into the linear system

$$\dot{z} = A^0(t)z, \tag{1.11}$$

for any preassigned matrix $A^0(t)$ with the same initial condition, $x(t_0) = z(t_0)$.

Exercise 1.1.2. Indicate the transformation that converts Eq. (1.10) into (1.11) and vice versa.

Among the equations of interest are linear matrix differential equations of the form

$$\dot{X}(t) = A(t)X(t) + X(t)A_1(t) + V(t), \tag{1.12}$$

in which the state $X(t)$ and input $V(t)$ are $n \times n$ square matrices.

Introduce two transition matrices $\mathbf{G}(t, s)$ and $\mathbf{G}_1(t, s)$

$$\frac{\partial}{\partial t}\mathbf{G}(t,s) = A(t)\mathbf{G}(t,s), \ \mathbf{G}(s,s) = I; \ \frac{\partial}{\partial t}\mathbf{G}_1(t,s) = -A_1(t)\mathbf{G}_1(t,s),$$

$$\mathbf{G}_1(s,s) = I. \tag{1.13}$$

Assume $X(t_0) = X^0$. Then an integral representation formula (similar to (1.3)) for Eq. (1.12) is true, namely,

$$X(t) = \mathbf{G}(t, t_0) X^0 \mathbf{G}_1(t_0, t) + \int_{t_0}^{t} \mathbf{G}(t, s) V(s) \mathbf{G}_1(s, t) ds. \tag{1.14}$$

For $A_1(t) = A'(t)$ we have $\mathbf{G}_1(t_0, t) = \mathbf{G}'(t, t_0)$ and

$$X(t) = \mathbf{G}(t, t_0) X^0 \mathbf{G}'(t, t_0) + \int_{t_0}^{t} \mathbf{G}(t, s) V(s) \mathbf{G}'(t, s) ds. \tag{1.15}$$

Exercise 1.1.3. (a) Check formula (1.14) by substitution.
(b) Solve Exercise 1.1.2 for Eq. (1.12).

Also within the scope of this book are *quasilinear* systems of the form

$$\dot{x}(t) = A(t)x(t) + B(t)\varphi(t, u) + v(t), \tag{1.16}$$

with $\varphi(t, u)$ continuous in $\{t, u\} \in T \times \mathbb{R}^p$ and convex in u.

More complicated systems that can be treated with the techniques developed in this book are *bilinear* systems of the form

$$\dot{x}(t) = A(t, u)x(t) + v(t), \quad u \in \mathcal{U}, \tag{1.17}$$

with $A(t, u) = A(t)U + D(t)$, with matrix of controls $U \in \mathbb{R}^{n \times n}$ and continuous matrices of coefficients $A(t), D(t) \in \mathbb{R}^{n \times n}$. In a simple case we may have $A(t, u) = A(t)u + D(t)$ with a scalar u.

1.2 Control and State Constraints: Open-Loop and Closed-Loop Control

One objective of this book is to develop solutions for systems subject to constraints on the control and state. The values u of the control may be restricted by various types of bounds. Here are some typical types.

1.2.1 Constraints on Control and State

Hard bounds also known as geometrical or magnitude bounds are constraints of the form

$$u \in \mathcal{P}(t), \tag{1.18}$$

in which $\mathcal{P}(t)$ is a Hausdorff-continuous set-valued function $\mathcal{P} : T \to \text{conv}\mathbb{R}^p$. The symbol $\text{conv}\mathbb{R}^n$ denotes the class of *closed, convex* sets in $\mathbb{R}^n$, while $\text{comp}\mathbb{R}^n$ denotes the class of *compact, convex* sets in $\mathbb{R}^n$. In the sequel, unless specially remarked, we shall suppose that $\mathcal{P}(t)$ has *a nonempty interior* in $\mathbb{R}^p$ for all t.

Hard bounds may also be specified as the *level set* (at level μ) of a function $\varphi(t, u)$:

$$\mathcal{U}_\mu(t) = \{u : \varphi(t, u) \leq \mu\}.$$

Suppose $\varphi(t, u)$ is continuous in t, u and convex in u. Then its level set $\mathcal{P}_\mu(t)$ will be closed and convex. As indicated in [237], for $\mathcal{U}_\mu(t) \neq \emptyset$ to be bounded, hence compact, for any μ, it is necessary and sufficient that

$$0 \in \text{int}\left(\text{Dom}\varphi^*(t, \cdot)\right). \tag{1.19}$$

Here

$$\varphi^*(t, l) = \sup\{\langle l, u \rangle - \varphi(t, u) \mid u \in \mathbb{R}^p\}$$

is the *("Fenchel") conjugate* of $\varphi(t, u)$ in the variable u; and

$$\text{Dom}\varphi^*(t, \cdot) = \{l : \varphi^*(t, l) < \infty\},$$

is the *effective domain* of the function $\varphi^*(t, l)$ for fixed t, see [237] (interior Q denotes the collection of all interior points of a set Q). Under condition (1.19) the level sets $\mathcal{P}_\mu(t)$ will be bounded, $\varphi(t, u) \to \infty$ as $\langle u, u \rangle \to \infty$, and in the definition of $\varphi^*(t, l)$ the operation of "sup" may be replaced by "max," see [72]. Under the given conditions the convex compact set-valued function $\mathcal{P}_\mu(t)$ will be Hausdorff-continuous.

In practice, hard bounds represent limits imposed by equipment or considerations of safety on a control variable such as voltage, force, or torque. Linear control synthesis techniques, like those based on frequency domain methods, which formulate the control as a linear function of the state, cannot satisfy hard bounds.

Integral bounds, also known as soft bounds, are constraints of the form

$$\int_{t_0}^{t_1} \|u(t)\|^q dt \leq \mu, \ \mu > 0, \ q \geq 1, \tag{1.20}$$

in which $\|u\|$ is a norm and q is usually an integer. A more general integral bound is expressed as

$$\int_{t_0}^{t_1} \varphi(t, u(t)) dt \leq \mu, \tag{1.21}$$

in which $\varphi(t, u)$ is continuous in t, u and convex in u, with $\text{int}\left(\text{Dom}\varphi^*(t, \cdot)\right) \neq \emptyset$.

1.2 Control and State Constraints: Open-Loop and Closed-Loop Control

Among the integral bounds one may single out

$$\int_{t_0}^{t_1} \|u(t)\| dt \leq \mu, \ \mu > 0, \tag{1.22}$$

which coincides with (1.20) for $q = 1$, and

$$\int_{t_0}^{t_1} \left(\sum_{i=0}^{k} \|d^i u(t)/dt^i\| \right) dt \leq \mu, \ \mu > 0. \tag{1.23}$$

Constraints (1.22), (1.23) may be satisfied by controls u that weakly converge to generalized functions: delta functions $u = \delta(t - \tau^*)$ and their derivatives such as $u = \delta^{(i)}(t - \tau^*)$.

Integral bounds can represent limits on energy consumption, using (1.20) with $q = 2$. A delta-function control, $u = \delta(t - \tau^*)$, is an *impulse control* that causes an instantaneous change in the state, $x(\tau^* - 0) \neq x(\tau^* + 0)$, as can be seen from (1.3). The constraint (1.23) (with $k = 1$) is used to restrict the instantaneous change in state.

Joint bounds (or *double bounds*) combine hard and soft bounds. For example, u must simultaneously satisfy two constraints (hard and soft):

$$u(t) \in \mathcal{P}(t), \ t \in [t_0, t_1], \ \int_{t_0}^{t_1} \|u(s)\|^2 ds \leq \mu^2, \text{ or} \tag{1.24}$$

$$\langle u(t), u(t) \rangle \leq \mu^2, \ t \in [t_0, t_1], \ \int_{t_0}^{t_1} \langle u(s), u(s) \rangle^{1/2} ds \leq \nu^2. \tag{1.25}$$

Constraint (1.25) can be used to approximate an impulse control by increasing μ^2 while keeping ν fixed.

Constraints may also be imposed on the state space variables. Taking $z(t) = H(t)x(t)$, with $H(t) \in \mathbb{R}^{m \times n}$ continuous, the constraint may be

$$z(t) \in \mathcal{Z}(t), \ t \in T_z,$$

in which T_z may comprise a discrete set of points, $T_z = \{\tau_1, \ldots \tau_k\} \subset T$, or it may be an interval such as $T = T_z$.

The set-valued function $\mathcal{Z}(t) \in \text{conv}\mathbb{R}^m$ is assumed to be piecewise absolutely continuous in t in the Hausdorff metric (see [76]). Note that the solutions to system (1.2) are piecewise absolutely continuous functions (with discontinuities corresponding to impulse controls), hence differentiable almost everywhere. Thus the properties of the bounding function $\mathcal{Z}(t)$ match those of the system solutions.

Also common are integral bounds on the state trajectories of the form

$$\int_{t_0}^{t_1} \varphi(t, x) dt \leq \mu, \tag{1.26}$$

with $\varphi(t, x)$ being convex in x; or mixed integral bounds on both control and state of the form

$$\int_{t_0}^{t_1} \varphi(t, x, u) dt \leq \mu, \qquad (1.27)$$

in which $\varphi(t, x, u)$ is continuous in t and convex in x, u.

In addition to these common constraints we will also meet in this book some nonstandard constraints that are less common, but which are very relevant for our problems. Such, for example, are bounds that define *obstacle problems* of Chap. 8, Sect. 8.4, when the system trajectory must lie beyond one convex set and within another at the same time, or state constraints that are due to on-line measurements under unknown but bounded noise in Chap. 9.

1.2.2 Open-Loop and Closed-Loop Control

In addition to constraints, one needs to specify the class of functions from which the controls are to be selected. The appropriate class of functions should be considered from both the mathematical and control design points of view.

Mathematically, we will consider classes that are functions of time only, $u = u(t)$, and those that are functions of both time and state, $u = u(t, x)$. The two classes are called *open-loop* and *closed-loop* controls, respectively.

An open-loop control, $u = u(t)$, is required to be measurable. For each open-loop control, the solution to (1.1) reduces to solving the equation

$$\dot{x} = f(t, x, u(t)),$$

which, in the linear case, is immediate through the integral formula (1.3). Thus, substituting an open-loop control $u(t)$ in (1.1) or (1.2) gives a unique solution $x[t] = x(t, t_0, x^0)$ for any starting position $\{t_0, x^0\}$, $x(t_0) = x^0$. From the point of view of control design, taking $u = u(t)$ as a function of time only means that the selection of the control action for each time t is fixed, implanted in the system design, and cannot be changed throughout the process, which lasts from t_0 to t_1. The class of open-loop controls is denoted by $\mathcal{U}_O$.

A closed-loop control, $u = u(t, x)$, depends on both t and x. The function $u(t, x)$ may be nonlinear, even discontinuous, in x. The class of admissible closed-loop controls $u(t, x)$ must however be so restricted that nonlinear differential equations of type

$$\dot{x}(t) = f(t, x, u(t, x(t))),$$

1.2 Control and State Constraints: Open-Loop and Closed-Loop Control

or

$$\dot{x}(t) = A(t)x(t) + B(t)u(t, x(t)) + v(t), \quad (1.28)$$

would have a solution in some reasonable sense. That is, there should be an existence theorem ensuring that Eq. (1.28) can be solved.

From the point of view of design and implementation, a closed-loop control selects the control action $u(t, x)$ at each time t depending on the value of the state $x(t)$. For on-line control (that is, when the value of the control action $u(t, x(t))$ is calculated in real time), the design presupposes that the state $x(t)$ is exactly measured for all t and continuously communicated to the control device. A control of this form leads to a *feedback loop* (also "*closed loop*"), hence the name closed-loop or *feedback control*. They are also referred to as *synthesized* controls in contrast with controls of class $\mathcal{U}_O$ which are known as *control programs* or *open-loop controls*. The class of closed-loop controls to be used, complemented with an existence theorem, is denoted by $\mathcal{U}_C$.

If the disturbance term $v(t)$ is fixed, there is no essential difference between $\mathcal{U}_O$ and $\mathcal{U}_C$, as the following exercise indicates.

Exercise 1.2.1. Suppose the disturbance $v(t)$ in (1.28) is fixed. Let $x[t] = x(t, t_0, x^0)$ be a given solution of (1.28) under a closed-loop control $g(t, x)$, i.e.,

$$\dot{x}(t) = A(t)x(t) + B(t)g(t, x(t)) + v(t).$$

Show that there is a corresponding open-loop control $u(t)$ such that $x[t]$ is also the solution of

$$\dot{x}(t) = A(t)x(t) + B(t)u(t) + v(t).$$

Show also that the corresponding open-loop control depends on the disturbance, i.e., if the disturbance input $v(t)$ changes, so does the open-loop control in order to keep the same solution $x[t]$.

Thus the difference in system performance achieved by using open-loop and closed-loop controls becomes apparent for systems subject to unknown disturbances. The control of systems with unknown disturbances is mentioned in Chap. 10.

In practice, exact measurement of the state vector x may not be possible. The measurement may be incomplete because only a part of the vector x or a function of x can be measured, or because the measurements can be taken only at discrete time instants, or because the measurements may be corrupted by unknown disturbances ("noise"). Consideration of such situations will bring us to new classes of feedback control.

In many cases treated later we allow the closed-loop control $u = \mathcal{U}(t, x)$ to be a *set-valued* map, with values $\mathcal{U}(t, x) \in \text{comp}\mathbb{R}^p$, measurable in t and upper-semicontinuous in x [7, 8, 48, 238]. Upon substituting $u = \mathcal{U}(t, x)$, Eq. (1.2) then becomes a nonlinear differential *inclusion*,

$$\dot{x}(t) \in A(t)x(t) + B(t)\mathcal{U}(t, x) + v(t), \qquad (1.29)$$

whose solution exists and in general is set-valued [75, 76]. The class of such set-valued controls is denoted by $\mathcal{U}_{CS}$.

In the next section we consider several problems of *optimal control* within the class of open-loop controls $\mathcal{U}_O$. These are the "norm-minimal" controls that can be obtained through simple considerations.

1.3 Optimal Control with Norm-Minimal Cost: The Controllability Property

The moment problem is a simple way of presenting the two-point boundary problem of control. It naturally allows to present the optimality criterion as one of minimizing the norm of the control $u(t)$ in an appropriate functional space.

Consider system (1.2) on a finite time interval $T = [t, \vartheta]$. We start with one of the simplest problems of optimal control.

1.3.1 Minimum Energy Control

Problem 1.3.1. Given system (1.2) and two points, $x[t] = x$, $x[\vartheta] = x^{(1)}$, find the optimal control that moves the system trajectory $x[s] = x(s, t, x)$ from x to $x^{(1)}$ with minimum cost

$$\mathcal{J}(t, x) = \min \left\{ \int_t^\vartheta \langle u(s), N(s)u(s) \rangle ds \mid u(\cdot) \in \mathcal{L}_2^{(p)}(T) \right\}, \qquad (1.30)$$

in which the continuous $p \times p$ matrix $N(s)$ is positive definite, $N(s) = N'(s) > 0$.

We shall solve this problem applying basic Hilbert space techniques. Using formula (1.3) the *boundary constraints* $x[t] = x$, $x[\vartheta] = x^{(1)}$ may be rewritten in the form

$$\int_t^\vartheta G(\vartheta, s)B(s)u(s)ds = x^{(1)} - G(\vartheta, t)x - \int_t^\vartheta G(\vartheta, s)v(s)ds = c. \qquad (1.31)$$

1.3 Optimal Control with Norm-Minimal Cost: The Controllability Property

Define the $n \times p$ matrix $D(s) = G(\vartheta, s)B(s)$ with rows $d^{(i)}$, $i = 1, \ldots, n$. For a p-dimensional row $d(\cdot)$ denote

$$\langle d'(\cdot), u(\cdot) \rangle_N = \int_t^\vartheta d(s) N(s) u(s) ds.$$

to be an inner product in $\mathcal{L}_2^{(p)}(T)$—the space of square-integrable $\mathbb{R}^p$-valued functions. Consider both $d^{(i)}(\cdot)$ and $u(\cdot)$ as elements of the Hilbert space $\mathcal{L}_2^{(p)}(T)$ space, with norm

$$\|u(\cdot)\|_N = \left(\int_t^\vartheta \langle u(s), N(s) u(s) \rangle ds \right)^{1/2}.$$

The cost of a control u is the square of its norm and Problem 1.3.1 is to find the element of a Hilbert space with minimum norm that satisfies certain linear constraints.[2] We can thus reformulate Problem 1.3.1.

Problem 1.3.1-A: *Minimize $\langle u(\cdot), u(\cdot) \rangle_N$ under the constraints*

$$\langle D'(\cdot), u(\cdot) \rangle = \langle N^{-1}(\cdot) D'(\cdot), u(\cdot) \rangle_N = c. \tag{1.32}$$

We assume that the n functions $d^{(i)}(\cdot)$ are linearly independent, which means that for any n-dimensional row-vector λ, $\lambda D(t) = 0$, a.e. for $t \in T$, only if $\lambda = 0$.

Now consider controls of the form $u(s) = N^{-1}(s) D'(s) l$, for some $l \in \mathbb{R}^n$, and substitute in (1.32), to get

$$W(\vartheta, t) l = c, \quad W(\vartheta, t) = \int_t^\vartheta D(s) N^{-1}(s) D'(s) ds.$$

The Gramian or Gram matrix $W(\vartheta, t)$ is nonsingular if and only if the functions $d^{(i)}$ are linearly independent [82]. As we shall observe later in this section the determinant $|W(\vartheta, t)| \neq 0$ (for any matrix $N(s) = N'(s) > 0$) ensures that system (1.2) is *controllable* (see below, Lemma 1.3.1).

The particular control

$$u^0(s) = N^{-1}(s) D'(s) l^0, \quad l^0 = W^{-1}(\vartheta, t) c, \tag{1.33}$$

satisfies the constraint (1.32). We now show that $u^0(s)$ minimizes the cost (1.30).

---

[2] We could also regard functions $d^{(i)}(s)$ as elements of $\mathcal{L}_{q^*}^{(p)}[T]$, $q^* \geq 1$. Later, while dealing with impulse controls, it will be more convenient to treat functions $d^{(i)}(s)$ as elements of the space $C^{(p)}[T]$.

Indeed, suppose $u(s)$ satisfies (1.32), but $u(s) \neq u^0(s)$. Then $u_e(s) = u(s) - u^0(s) \neq 0$ satisfies

$$\langle N^{-1}(\cdot)D'(\cdot), u_e(\cdot)\rangle_N = \langle N^{-1}(\cdot)D'(\cdot), N(\cdot)u_e(\cdot)\rangle = \langle D'(\cdot), u(s) - u^0(s)\rangle = 0,$$

and

$$\langle u(\cdot), N(\cdot)u(\cdot)\rangle = \langle u^0(\cdot) + u_e(\cdot), N(\cdot)(u^0(\cdot) + u_e(\cdot))\rangle$$
$$= \langle u^0(\cdot), N(\cdot)u^0(\cdot)\rangle + \langle u_e(\cdot), N(\cdot)u_e(\cdot)\rangle$$
$$\geq \langle u^0(\cdot), N(\cdot)u^0(\cdot)\rangle,$$

since $\langle u^0(\cdot), N(\cdot)u_e(\cdot)\rangle \geq 0$ due to (1.33). This gives the next result.

Theorem 1.3.1. *The optimal control for Problem 1.3.1 is given by the continuous function*

$$u^0(s) = N^{-1}(s)D'(s)W^{-1}(\vartheta, t)c.$$

Remark 1.3.1. If $x^{(1)} = 0, v(s) \equiv 0$, then $c = -G(\vartheta, t)x$, so that **the optimal control**

$$u^0(s) = -N^{-1}D'(s)W^{-1}(\vartheta, t)G(\vartheta, t)x, \qquad (1.34)$$

is linear in x. This property will be important later in treating problems of closed-loop control.

We have just solved the simplest case of the *moment problem* for $\mathcal{L}_q^{(p)}$ with $q = 2$ (see Theorem 1.3.4 below).

1.3.2 Minimum Magnitude Control

Problem 1.3.2. Given system (1.2), starting position $x[t] = x$, and convex compact terminal set $\mathcal{M}$ and terminal time ϑ, find the optimal control that moves the system trajectory $x[s] = x(s, t, x)$ from x to some point $x[\vartheta] \in \mathcal{M}$ with minimum cost

$$\mathcal{J}(t, x) = \min \operatorname{ess\,sup} \{\|u(s)\|, \ s \in T\}, \qquad (1.35)$$

in which $\|u\|$ is a norm in $\mathbb{R}^p$.

If $u(s)$ is a continuous or even a piecewise right- or left-continuous function, we may substitute "esssup" by "sup."

1.3 Optimal Control with Norm-Minimal Cost: The Controllability Property

From formula (1.3) the terminal constraint is

$$\int_t^\vartheta G(\vartheta,s)B(s)u(s)ds + G(\vartheta,t)x + \int_t^\vartheta G(\vartheta,s)v(s)ds = x[\vartheta] \in \mathcal{M}. \quad (1.36)$$

Again take $D(s) = G(\vartheta,s)B(s)$. We treat its rows—the p-dimensional functions $d^{(i)}(\cdot)$—as elements of $\mathcal{L}_1 = L_1^{(p)}[T]$ and treat the control $u(\cdot)$ as an element of its conjugate space $\mathcal{L}_\infty = L_\infty^{(p)}[T]$. Denote the bilinear functional

$$\langle u(\cdot), d(\cdot) \rangle = \int_t^\vartheta \langle d(s), u(s) \rangle ds$$

and norms

$$\|d(\cdot)\|_{\mathcal{L}_1} = \int_t^\vartheta \|d(s)\| ds, \quad \|u(\cdot)\|_{\mathcal{L}_\infty} = \text{ess sup}\{\|u(s)\| \mid s \in T\}.$$

Here the finite-dimensional norms $\|u(s)\|$ and $\|d(s)\|$ are conjugate.

Let us now take

$$\mathcal{B}_\infty = \{u(\cdot) : \|u(\cdot)\|_{\mathcal{L}_\infty} \leq 1\},$$

and find the smallest $\mu = \mu^0$ for which there exists $u(\cdot) \in \mu \mathcal{B}_\infty$ that satisfies (1.36).

Since $\mathcal{M}$ is convex and compact, $x(\vartheta) \in \mathcal{M}$ is equivalent to the system of inequalities

$$\langle l, x(\vartheta) \rangle \leq \rho(l \mid \mathcal{M}) = \max\{\langle l, x \rangle \mid x \in \mathcal{M}\}, \forall l \in \mathbb{R}^n.$$

So the necessary and sufficient condition for $u(\cdot)$ to satisfy (1.36) is

$$\int_t^\vartheta l'G(\vartheta,s)B(s)u(s)ds + \langle l, c^* \rangle \leq \rho(l \mid \mathcal{M}), \forall l \in \mathbb{R}^n, \quad (1.37)$$

in which

$$c^* = G(\vartheta,t)x + \int_t^\vartheta G(\vartheta,s)v(s)ds.$$

Hence, for *some* $u(\cdot) \in \mu \mathcal{B}_\infty$ to satisfy (1.36), it is necessary and sufficient that

$$\min_u \left\{ \int_t^\vartheta l'D(s)u(s)ds \mid u(\cdot) \in \mu \mathcal{B}_\infty \right\} + \langle l, c^* \rangle \leq \rho(l \mid \mathcal{M}), \forall l \in \mathbb{R}^n,$$

or

$$\mu \int_t^\vartheta \min\{l'D(s)u(s) \mid \|u(s)\| \leq 1\}ds + \langle l, c^*\rangle \leq \rho(l \mid \mathcal{M})\}, \quad \forall l \in \mathbb{R}^n,$$

from which it follows that

$$\mu \geq \mu^0 = \max_l \left\{ \frac{\langle l, c^*\rangle - \rho(l \mid \mathcal{M})}{\|l'D(\cdot)\|_{L_1}} \right\}. \tag{1.38}$$

Under the assumption $|W(\vartheta, s)| \neq 0$, (with $N(s) \equiv I$)—the controllability property—the maximizer in (1.38) is $l^0 \neq 0$.

Because the numerator and denominator in (1.38) are homogeneous, if l^0 is a maximizer, so is αl^0, $\forall \alpha > 0$. Hence μ^0 may be found by solving

$$\mu^0 = \max_l \left\{ \langle l, c^*\rangle - \rho(l \mid \mathcal{M}) \, \Big| \int_t^\vartheta \|l'D(s)\|ds = 1 \right\}, \tag{1.39}$$

or

$$(\mu^0)^{-1} = \min_l \left\{ \int_t^\vartheta \|l'D(s)\|ds \, \Big| \, \langle l, c^*\rangle - \rho(l \mid \mathcal{M}) = 1 \right\}. \tag{1.40}$$

We are thus led to the next result whose proof follows from (1.40) and the definition of the norms involved.

Theorem 1.3.2. *The minimum magnitude control u^0 that solves Problem 1.3.2 satisfies the maximum condition*

$$\max\{l^{0'}D(s)u \mid \|u\| \leq \mu^0\} = l^{0'}D(s)u^0(s) = \mu^0\|l^{0'}D(s)\|, \quad s \in T, \tag{1.41}$$

in which l^0 is the maximizer in (1.39) or the minimizer in (1.40).

The maximum condition is a particular case of *the maximum principle* discussed further in more detail in Sect. 1.5.

A not uninteresting problem is the following:

Problem 1.3.2-A *Find largest $\mu = \mu^*$ for which the inclusion (1.36) holds for **all** $u(\cdot) \in \mu \mathcal{B}_\infty$.*

In order to fulfill the requirement of Problem 1.3.2 it is necessary and sufficient that

$$\max_u \left\{ \int_t^\vartheta l'D(s)u(s)ds \mid u(\cdot) \in \mu \mathcal{B}_\infty \right\} + \langle l, c^*\rangle \leq \rho(l \mid \mathcal{M}), \quad \forall l \in \mathbb{R}^n,$$

1.3 Optimal Control with Norm-Minimal Cost: The Controllability Property

or

$$\mu \int_t^\vartheta \max\{l'D(s)u \mid \|u\| \leq 1\}ds + \langle l, c^*\rangle \leq \rho(l \mid \mathcal{M}), \quad \forall l \in \mathbb{R}^n,$$

from which

$$\mu^* = \min_l \left\{ \frac{\rho(l \mid \mathcal{M}) - \langle l, c^*\rangle}{\|l'D(\cdot)\|_{L_1}} \right\}. \tag{1.42}$$

(As before, $\|u\|$ is the $\mathbb{R}^p$-dimensional norm conjugate to the norm used in $\|l'D(s)\|$.)

Theorem 1.3.3. *The solution to Problem 1.3.2-A is given by formula (1.42).*

Exercise 1.3.1. Solve Problem 1.3.1 with minimum cost

$$\mathcal{J}(t, x) = \min \left\{ \int_t^\vartheta (u(s), N(s)u(s))^{q/2} ds \,\bigg|\, u(\cdot) \in L_q^{(p)}(T) \right\}, \quad 1 < q < \infty, \, q \neq 2. \tag{1.43}$$

instead of (1.30).

Observe that the existence of solutions to Problems 1.3.1, 1.3.2 (and as one may check, also to (1.43)) requires the solvability of Eq. (1.32) for any vector $c = (c_1, \ldots, c_n)'$. This is guaranteed if and only if the functions $d^{(i)}(\cdot)$, $i = 1, \ldots, n$, which are the rows of the $n \times p$ matrix

$$D(s) = G(\vartheta, s)B(s)$$

are *linearly independent* over $T = [t, \vartheta]$.

Let $1/q + 1/q^* = 1$.

Definition 1.3.1. An array of elements $d^{(i)}(\cdot) \in L_{q^*}^{(p)}[T]$, $q^* \in [1, \infty]$ is said to be **linearly independent** if $l'D(s) = \sum_i l_i d^{(i)}(s) = 0$ a.e. implies $l = 0$.

Lemma 1.3.1. *The functions $d^{(i)}(\cdot)$ are linearly independent if and only if the symmetric matrix $W(\vartheta, t) = \int_t^\vartheta D(s)D'(s)ds$ with elements*

$$W_{ij} = \int_t^\vartheta \langle d^{(i)}(s), d^{(j)}(s)\rangle ds$$

is nonsingular, i.e., its determinant $|W(\vartheta, t)| \neq 0$.

$W(\vartheta, t)$ is known as the *Gram matrix*.

Exercise 1.3.2. Prove Lemma 1.3.1.

Lemma 1.3.1 leads to the following conclusion.

Lemma 1.3.2. *Equation (1.32) is solvable for any vector c within the class of functions $u(\cdot) \in \mathcal{L}_q^{(p)}[T]$, $q \in [1, \infty]$ if and only if $|W(\vartheta, t)| \neq 0$.*

1.3.3 Controllability

The notion of controllability which characterizes the solvability of the two-point boundary control problem *for any pair of points* was introduced by Kalman [107, 109]. Having stimulated intensive research on the structure of time-invariant linear systems [274], it became one of the basic concepts in control theory for any type of system.

The solvability of (1.32) for *any* c means that there exists a control that transfers the system (1.2) from *any* given starting position $\{t, x\}$, $x[t] = x$ to *any* given terminal position $\{\vartheta, x^1\}$, $x[\vartheta] = x^1$. (Since c is defined by (1.31), by choosing appropriate vectors x, x^1, we can get any c.)

Definition 1.3.2. System (1.2) is said to be **controllable over** $[t, \vartheta]$ if given any positions $\{t, x\}$ and $\{\vartheta, x^1\}$, there exists a control $u(s)$, $s \in [t, \vartheta]$ which steers the system from $\{t, x\}$ to $\{\vartheta, x^1\}$. The system is **completely controllable** over $[t, \vartheta]$ if it is completely controllable over any subinterval $[t_1, t_2] \subseteq [t, \vartheta]$, $t_1 < t_2$.

The notion of controllability deals with the existence of controls that solve certain boundary problems [109]. It is applicable to a broad class of systems, allowing many versions.

Corollary 1.3.1. *System (1.2) is controllable over $[t, \vartheta]$ iff $|W(\vartheta, t)| \neq 0$ and it is completely controllable iff $|W(\vartheta, t)| \neq 0$, $\forall [t_1, t_2] \subseteq [t, \vartheta]$, $t_1 < t_2$.*

The Gram matrix $W(\vartheta, t)$ is also known as the *controllability* matrix for system (1.2).

Remark 1.3.2. (i) In the case of constant matrices $A(t) \equiv A$, $B(t) = B$ the property of complete controllability is equivalent to the condition that the set of column vectors in the array

$$B, AB, \ldots, A^{n-1}B$$

contains a subset of n linearly independent vectors. In other words, the matrix $[B, AB, \ldots, A^{n-1}B]$ has full rank n.

1.3 Optimal Control with Norm-Minimal Cost: The Controllability Property

(ii) A stronger version of condition (i) requires

$$b^{(i)}, Ab^{(i)}, \ldots, A^{n-1}b^{(i)}$$

to be linearly independent for each $i = 1, \ldots, p,$. This is the case of **strong complete controllability**.

In the time-varying case the stronger version requires complete controllability of (1.2) with $B(t)$ replaced by $b^{(i)}(t)$.

Exercise 1.3.3. Prove the assertion of Remark 1.3.2(i).

Remark 1.3.3. From Corollary 1.3.1 it follows that the controllability condition for controls in $L_q^{(p)}[T]$, $q \in [1, \infty]$ is the same for all q. However, when solving an optimization problem of type Exercise 1.3.1 within the class of controls from $L_1^{(p)}[T]$, with $q = 1$, the minimum may not be attained in this class but is attained in a broader class that includes generalized delta functions.

Exercise 1.3.4. For the one-dimensional system

$$\dot{x} = ax + u,$$

consider the problem of finding a control that moves the state from $x[t] = x^0 \neq 0$ to $x(\vartheta) = x^{(1)} = 0$ with minimum cost

$$\min\{\int_t^{\vartheta} |u(s)| ds \mid u(\cdot) \in L_1^{(1)}[T]\}.$$

Show that if $a \neq 0$ the minimum is not attained, but is attained in the broader class that includes delta functions.

An interesting issue is therefore to solve an optimization problem like Exercise 1.3.1, but with $u(\cdot) = dU(\cdot)/dt$ being the generalized derivative of U (in the sense of the theory of distributions, [242]) and with cost as the total variation $\mathrm{Var}\, U(\cdot)$ of the function $U(t)$ over the interval T. The solution, which may include delta functions and its derivatives, is considered later in a separate section.

1.3.4 The Finite-Dimensional Moment Problem

Consider Eq. (1.32) with $D(s) = G(\vartheta, s) B(s)$ having rows $d^{(i)}(\cdot) \in L_q^{(p)}[T]$, $T = [t, \vartheta]$. Let $\|d(\cdot)\|_q^{(p)}$ denote the norm in the $L_q^{(p)}[T]$. The classical finite-dimensional **moment problem** seeks solutions $u(\cdot) \in L_q^{(p)}[T]$ of (1.32) with minimum norm [1].

Theorem 1.3.4. *Suppose the $d^{(i)}(\cdot) \in L_{q^*}^{(p)}[T]$, $q^* \in [1, \infty)$, $i = 1, \ldots, n$, are linearly independent. Then (1.32) is solvable by some $u(\cdot) \in L_q^{(p)}[T]$, $\|u(\cdot)\|_q^{(p)} \le \mu$, if and only if*

$$\mu \left(\int_t^\vartheta \langle D'(s)l, D'(s)l \rangle^{q^*/2} ds \right)^{1/q^*} \ge \langle l, c \rangle, \quad \forall l \in \mathbb{R}^n.$$

Here $1/q + 1/q^* = 1$.

Theorem 1.3.4 suggests a scheme to calculate norm-minimal solutions to the two-point boundary control problem for linear systems by reducing it to the moment problem. This approach, which also applies to impulse controls, was brought into linear control theory by Krasovski [120]. Equivalent results are also obtained through the techniques of convex analysis that are used in this book.

So far we considered the problem of reaching a particular state. We now pass to the problem of finding all the states reachable from a given starting point through all possible controls, restricted by a given bound.

1.4 The Reachability Problem: Time-Optimal Control

1.4.1 Reachability

This is one of the important elements in analyzing system performance, namely, what states may be reached in given time by available controls?

Consider system (1.2) with $v(s) \equiv 0$,

$$\dot{x}(t) = A(t)x(t) + B(t)u, \tag{1.44}$$

and control constraint (1.18): $u(s) \in \mathcal{P}(s)$, $s \in [t, \infty)$.

Definition 1.4.1. *The **reach set** $X[\vartheta] = X(\vartheta, t_0, x^0)$ of system (1.44) **at given time** ϑ, **from position** $\{t_0, x^0\}$, is the set of **all points** x for each of which there exists a trajectory $x[s] = x(s, t_0, x^0)$, generated by a control subject to the given constraint (1.18), that transfers the system from position $\{t_0, x^0\}$ to position $\{\vartheta, x\}$, $x = x[\vartheta]$:*

$$X(\vartheta, t_0, x^0) = \{x : \exists u(\cdot) \in \mathcal{P}(\cdot) \text{ such that } x(\vartheta, t, x^0) = x\}.$$

The reach set $X[\vartheta] = X(\vartheta, t_0, X^0)$ **from set-valued position** $\{t_0, X^0\}$ at given time ϑ is the union

$$X(\vartheta, t_0, X^0) = \bigcup \{X(\vartheta, t_0, x^0) \mid x^0 \in X^0\}.$$

The reach set is also known as the *attainability* set.

1.4 The Reachability Problem: Time-Optimal Control

Lemma 1.4.1. *With X^0 and $P(s), s \in T = [t_0, \vartheta]$, convex and compact, the reach set $X[\vartheta]$ is also convex and compact.*

Proof. Consider two points $x' \in X[\vartheta]$ reached through control $u'(s) \in P(s)$, and $x'' \in X[\vartheta]$, reached through $u''(s) \in P(s), s \in [t_0, \vartheta]$. The convexity of X^0 and $P(s)$ implies that $x^{(\alpha)} = \alpha x' + (1 - \alpha x'')$, $\alpha \in [0, 1]$ is reached through $u^{(\alpha)}(s) = \alpha u'(s) + (1-\alpha)u''(s) \in P(s)$ with initial condition $x^{(\alpha)}[t_0] = \alpha x'[t_0] + (1-\alpha)x''[t_0]$. This shows convexity. Boundedness of $X[\vartheta]$ is obvious. That $X[\vartheta]$ is closed follows from the fact that the set of functions $u(\cdot) \in P(\cdot)$ is *weakly compact* in $L_2^{(p)}$, so if x^n is reached through $u^{(n)}(\cdot)$ and $x^n \to x$, then x is reached by a weak limit $u(\cdot)$ of $\{u^{(n)}(\cdot)\}$. □

Problem 1.4.1. Find the reach set $X[\vartheta]$.

Since $X[\vartheta] = X(\vartheta, t_0, x^0)$ is convex and compact, we shall describe it through its support function

$$\rho(l \mid X[\vartheta]) = \max\{\langle l, x \rangle \mid x \in X[\vartheta]\}$$
$$= \max\{\langle l, x[\vartheta] \rangle \mid u(\cdot) \in P(\cdot), x[t_0] = x^0 \in X^0\}. \quad (1.45)$$

Using formula (1.3), rewrite (1.45) as

$$\rho(l \mid X[\vartheta]) = \max \left\{ \int_{t_0}^{\vartheta} l'G(\vartheta, s)B(s)u(s)ds + l'G(\vartheta, t)x^0 \mid u \in P(s), x^0 \in X^0 \right\}$$
$$= \int_{t_0}^{\vartheta} \max\{l'G(\vartheta, s)B(s)u \mid u \in P(s)\}ds + \max\{l'G(\vartheta, t_0)x^0 \mid x^0 \in X^0\}$$
$$= \int_{t_0}^{\vartheta} \rho(B'(s)G'(\vartheta, s)l \mid P(s))ds + \rho(G'(\vartheta, t_0)l \mid X^0). \quad (1.46)$$

Introducing the *column* vector $\psi[t] = \psi(t, \vartheta, l) = G'(\vartheta, t)l$ as the solution to the adjoint equation

$$\dot{\psi}(t) = -A'(t)\psi(t), \quad \psi[\vartheta] = l, \quad (1.47)$$

we may express the support function of $X(\vartheta, t_0, X^0)$ in the following form.

Lemma 1.4.2. *The reach set from set-valued position $\{t_0, X^0\}$ is given by the formula*

$$\rho(l \mid X(\vartheta, t_0, X^0)) = \int_{t_0}^{\vartheta} \rho(B'(s)\psi[s] \mid P(s))ds + \rho(\psi[t_0] \mid X^0). \quad (1.48)$$

The reach set $X[t] = X(t, t_0, X^0)$ regarded as a *set-valued* function of t, has the following interpretation. Consider the differential *inclusion*

$$\dot{x} \in A(t)x + B(t)P(t),$$

which is a differential equation with set-valued right-hand side [7, 75]. From a starting position $\{t_0, X^0\}$ it has a set-valued solution $X[t] = X(t, t_0, X^0)$ which is nothing else than the reach set $X[t]$. This set $X[t]$ may be represented as the set-valued integral

$$X[t] = G(t, t_0)X^0 + \int_{t_0}^{t} G(t, s) B(s) \mathcal{P}(s) ds.$$

The set-valued integral above may be interpreted in the sense of either Riemann or Lebesgue.

1.4.2 Calculating Controls for the Boundary of Reach Set

The support functions in (1.46), (1.48) yield the boundary $\partial X[\vartheta]$ of the set $X[\vartheta]$. We may also need to find a control that leads to each point of this boundary. We will do that by calculating, for each direction l, the points of the boundary that are the *points of support* x_l for $X[\vartheta]$ in the direction l. For any l, x_l is the solution to the optimization problem

$$x_l = \arg\max\{\langle l, x \rangle \mid x \in X[\vartheta]\},$$

$$\max\{\langle l, x \rangle \mid x \in X[\vartheta]\} = \rho(l \mid X[\vartheta])$$

$$= \int_{t_0}^{\vartheta} \rho(B'(s)G'(\vartheta, s)l \mid \mathcal{P}(s)) ds + \rho(G'(\vartheta, t_0)l \mid X^0).$$
(1.49)

Remark 1.4.1. By considering all directions $l \in \mathbb{R}^n$, one gets all the points of support of $X[\vartheta]$. Since, with l fixed, all the vectors αl, $\alpha > 0$, give the same point of support x_l, it suffices to deal only with vectors l of unit norm, $\langle l, l \rangle = 1$.

The calculation of x_l decomposes into separate calculations of the two terms in (1.49). The second term, which gives the appropriate initial condition, is obtained through a convex optimization problem. The first term, which gives the control $u_l(t)$ corresponding to x_l from formula (1.46) is considered next.

Lemma 1.4.3. *The point of support x_l for $X[\vartheta]$ in direction l satisfies relation (1.49). The control $u_l(t)$ that steers the system to point x_l satisfies **the maximum condition***

$$l'G(\vartheta, s)B(s)u_l(s) = \max\{l'G(\vartheta, s)B(s)u \mid u \in \mathcal{P}(s)\} = \rho(l \mid G(\vartheta, s)B(s)\mathcal{P}(s)).$$
(1.50)

1.4 The Reachability Problem: Time-Optimal Control

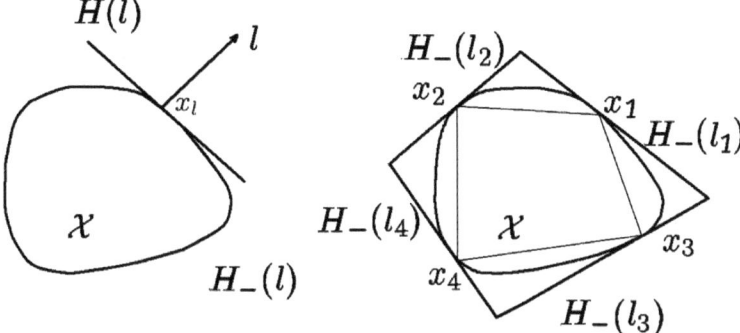

Fig. 1.1 The hyperplane $H(l)$ supports X at x_l and $X \subset H_-(l)$ (*left*). Inner and outer approximation to X (*right*)

Exercise 1.4.1. Prove formula (1.50) and also the assertions stated in Remark 1.4.1.

Definition (1.49) of x_l has the following geometric interpretation. Since

$$\rho_l = \rho(l \mid X[\vartheta]) = \langle l, x_l \rangle \geq \langle l, x \rangle, \quad x \in X = X[\vartheta],$$

the hyperplane $H(l) = \{x \mid \langle l, x \rangle = \rho_l\}$ with normal l *supports* X at the point x_l. See Fig. 1.1.

Observe that the reach set X is contained in the closed half-space

$$H_-(l) = \{x \mid \langle l, x \rangle \leq \rho_l\}.$$

Thus we have the inner and outer approximation to $X[\vartheta]$ of the following exercise.

Exercise 1.4.2. Let $l_1, \cdots, l_k$ be nonzero vectors. Let $\rho_i = \rho(l_i \mid X[\vartheta])$, and $x_i = x_{l_i}$. Show that

$$\text{co}\{x_1, \cdots, x_k\} \subseteq X[\vartheta] \subseteq \cap_{i=1}^k H_-(l_i).$$

Here $\text{co}\{x_1, \cdots, x_k\}$ denotes the convex hull of $\{x_1, \cdots, x_k\}$.

The reach sets may be of course calculated under constraints other than (1.18).

Problem 1.4.2. Find reach set $X[\vartheta] = X(\vartheta, t_0, x^0)$ of Problem 1.4.1 under the integral constraint

$$\|u(\cdot)\|_{L_q} = \left(\int_{t_0}^{\vartheta} \langle u(s), u(s) \rangle^{q/2} ds \right)^{\frac{1}{q}} \leq \mu, \quad \mu > 0, \ q \in (1, \infty), \quad (1.51)$$

with q integer, instead of constraint (1.18).

Since
$$\rho(l \mid X[\vartheta]) = \max_u \{\langle l, x[\vartheta]\rangle \mid \|u(\cdot)\|_{L_q} \leq \mu, x[t_0] = x^0\},$$
we have
$$\rho(l \mid X[\vartheta]) = \mu \left(\int_{t_0}^{\vartheta} (\psi'[t]B(t)B'(t)\psi[t])^{q^*/2} dt\right)^{\frac{1}{q^*}} + \psi'[t_0]x^0, \quad (1.52)$$

where $1/q^* + 1/q = 1$, so that $q^* = q(q-1)^{-1}$. Recall that $\psi[t] = (G^{-1}(t, \vartheta))'l = G'(\vartheta, t)l$.

To obtain (1.52) we used Hoelder's inequality with $d(\cdot) \in \mathcal{L}_{q^*}$, $u(\cdot) \in \mathcal{L}_q$ [3]

$$\int_{t_0}^{\vartheta} \langle u(s), d(s)\rangle ds \leq \|u(\cdot)\|_{\mathcal{L}_q^{(p)}} \|d(\cdot)\|_{\mathcal{L}_{q^*}^{(p)}},$$

with equality if $\|u(s)\|^q \equiv \alpha \|d(s)\|^{q^*}$, $u(s) = \left(\alpha \|d(s)\|^{q^*-q}\right)^{1/q} d(s)$, for some constant $\alpha > 0$.

Lemma 1.4.4. *The reach set $X[\vartheta]$ under constraint (1.51) is given by formula (1.52).*

In particular, if $q = 2$, then also $q^* = 2$, we come to the next result.

Corollary 1.4.1. *With $q = 2$ one has ($\psi[t]$ is a column vector)*

$$\rho(l \mid X[\vartheta]) = \mu \left(\int_{t_0}^{\vartheta} \psi'[t]B(t)B'(t)\psi[t]dt\right)^{1/2} + \psi'[t_0]x^0 = \mu\langle l, W(\vartheta, t_0)l\rangle^{1/2} + l'c, \quad (1.53)$$

with $c = G(\vartheta, t_0)x^0$.

Exercise 1.4.3. For the support function
$$\rho(l \mid X[\vartheta]) = \mu\langle l, W(\vartheta, t_0)l\rangle^{1/2} + l'c, \quad W(\vartheta, t_0) = W(\vartheta, t_0)' > 0,$$

prove that
$$X[\vartheta] = \{x : \langle (x-c), W^{-1}(\vartheta, s)(x-c)\rangle \leq \mu\}, \quad (1.54)$$

which is a nondegenerate ellipsoid.

---

[3] Instead of the traditional letter p paired with q, we use q^*, since p is used to denote for the dimension of u.

1.4 The Reachability Problem: Time-Optimal Control

Calculation of the reach set is more difficult for the bilinear system (1.17), which with $v(s) = 0$ becomes

$$\dot{x} = A(t, u)x. \tag{1.55}$$

The difficulty is due to the fact that even if u is constrained to a convex compact set with $x^0 \neq 0$ the reach set $X[\vartheta]$ may be *nonconvex*. This is seen in the next example.

Exercise 1.4.4. Show that the differential inclusion for $x \in \mathbb{R}^2$,

$$\dot{x}_1 \in [-1, 1]x_2, \quad \dot{x}_2 = 0, \quad t \in [0, 1],$$

with $\mathcal{X}^0 = \{x : x_1^0 = x_1(0) = 0, \; x_2^0 = x_2(0) \in [-\alpha, \alpha], \; \alpha > 0\}$, has a nonconvex reach set $X[\vartheta] = X(\vartheta, 0, \mathcal{X}^0)$, $\vartheta > 0$.

Remark 1.4.2. A more general problem than finding the reach sets $X[t]$ at time t is to find reach sets within a given time interval $t \in T$. The calculation of such sets, $\bigcup\{X[t] \mid t \in T\}$, requires more complex operations and is treated in Sect. 2.6 of Chap. 2.

Exercise 1.4.5. Find a two-dimensional autonomous system $\dot{x}(t) = Ax(t)$ and a convex, compact initial set $\mathcal{X}^0$ so that $\bigcup\{X[t] \mid t \in [0, 1]\}$ is not convex.

The next problem is basic. It initiated research in optimal control.

1.4.3 Time-Optimal Control

Problem 1.4.3. Given starting position $\{t_0, x^0\}$ and final point x^F, find control $u(s) \in \mathcal{P}(s)$, $s \in [t_0, \vartheta]$ that steers the system from x^0 to x^F **in minimum time**, i.e., $\vartheta - t_0$ is minimized.

We shall solve this problem as follows. Consider the reach set $X[\vartheta] = X(\vartheta, t_0, x^0)$ with $\vartheta > t_0$. Assuming $x^F \neq x^0$, let us look for the first instant of time ϑ^0 when $x^F \in X[\vartheta]$. Namely, denoting the (Euclidean) distance as $\epsilon(\vartheta) = d(x^F, X[\vartheta])$, and assuming $\epsilon(t_0) > 0$, we will find the smallest root ϑ^0 of the equation $\epsilon(\vartheta) = 0$. Then $\vartheta^0 - t_0$ will be the minimum time.

Taking $\mathcal{B}(0) = \{x : \langle x, x \rangle \leq 1\}$ and expressing $x^F \in X[\vartheta] + \epsilon \mathcal{B}(0)$ in terms of its support function,

$$\langle l, x^F \rangle \leq \rho(l \mid X[\vartheta] + \varepsilon \mathcal{B}(0)) = \rho(l \mid X[\vartheta]) + \varepsilon \langle l, l \rangle^{1/2}, \quad \forall \{l : \langle l, l \rangle \leq 1\},$$

we find that the distance $d(x^F, X[\vartheta]) = \epsilon(\vartheta)$ is the smallest ε satisfying this inequality,

$$\epsilon(\vartheta) = \max\{\langle l, x^F \rangle - \rho(l \mid X[\vartheta]) \mid \langle l, l \rangle \leq 1\}. \tag{1.56}$$

Hence

$$\vartheta^0 = \min\{\vartheta \mid \epsilon(\vartheta) = 0\}.$$

Recall that the support function $\rho(l \mid X[\vartheta])$ was calculated earlier (see Lemma 1.4.2).

Theorem 1.4.1. *The minimal time for transferring system (1.44) from starting point $x[t_0]$ to final point $x[\vartheta^0] = x^F$ is $\vartheta^0 - t_0$, where ϑ^0 is the minimal root of the equation $\epsilon(\vartheta) = 0$.*

Let l^0 be the optimizer of problem (1.56) with $\vartheta = \vartheta^0$.

Theorem 1.4.2. *The time-optimal control $u^0(t)$, which steers the system from x^0 to point x^F satisfies the maximum condition of type (1.50) with l, u_l, t, ϑ substituted by $l^0, u^0(t), t_0, \vartheta^0$ of Problem 1.4.3.*

With $l^0 \neq 0$ this maximum condition is known as *the maximum principle*, discussed in more detail in the next section.

Remark 1.4.3. The optimal time $\vartheta^0 - t_0$ may be discontinuous in the boundary points x^0, x^F.

Exercise 1.4.6. Construct an example of Problem 1.4.3 for which $\vartheta^0 - t_0$ is discontinuous in both x^0, x^F.

We shall further deal with various versions of the time-optimal control problem in detail, when solving specific cases.

In the next section we first introduce the *maximum principle* for linear systems as the necessary condition of optimality, then indicate when it is a sufficient condition. We shall mostly emphasize the *nondegenerate case* which is the main situation in applications, the one for which the maximum principle was indeed introduced and for which the optimal control may be found precisely from this principle. But we shall also indicate degenerate *abnormal* or *singular* situations, which may occur and for which the maximum principle is noninformative.

1.5 Optimal Open-Loop Control for Linear-Convex Systems: The Maximum Principle

The necessary conditions for optimality of control problems were formulated by Pontryagin and his associates in the form of the "Maximum Principle," [226]. These conditions apply to a broad class nonlinear systems under nonsmooth constraints on controls. In the linear case they may be extended to necessary and sufficient conditions of optimality as indicated in the following text. (See also [30, 81, 120, 195].)

1.5 Optimal Open-Loop Control for Linear-Convex Systems: The Maximum...

Here we deal with control problems for *linear-convex systems*, namely those of type (1.2), with convex constraints on the control. In this section the problems are solved in the class of open-loop controls $\mathcal{U}_O$, with no disturbance term, $v(t) \equiv 0$.

Problem 1.5.1. Consider system (1.2), with $T = [t_0, \vartheta]$. Given starting position $\{t_0, x^0\}$, find

$$\mathcal{J}(t_0, x^0) = \min_u \{\varphi(\vartheta, x[\vartheta]) \mid x(t_0) = x^0\} \tag{1.57}$$

under constraints (1.2), (1.18). Here $\varphi(t, x)$ is continuous in $\{t, x\}$ and convex in x, bounded from below, with bounded level sets.

The level sets of $\varphi(t, x)$ are bounded iff $0 \in \text{int}\left(\text{Dom}\varphi^*(t, \cdot)\right)$, $\forall t \in T$. Recall that

$$\varphi(t, x) = \max_l \{\langle x, l \rangle - \varphi^*(t, l) \mid l \in \mathbb{R}^n\}, \tag{1.58}$$

and $\varphi^*(t, l)$, the (Fenchel) *conjugate* of $\varphi(t, x)$ in the second variable, is defined by

$$\varphi^*(t, l) = \max_x \{\langle x, l \rangle - \varphi(t, x) \mid x \in \mathbb{R}^n\}.$$

The attainability of the maximum is ensured by the properties of $\varphi(t, x)$.

1.5.1 The Necessary Conditions of Optimality

Substituting ϑ for t and $x[\vartheta]$ for x in (1.58) and using the integral representation (1.3) for $x[\vartheta]$, we come to

$$\mathcal{J}(t_0, x^0) = \min_u \max_l \Phi(u(\cdot), l \mid t_0, x^0), \tag{1.59}$$

in which $u(t) \in \mathcal{P}(t)$, $l \in \mathbb{R}^n$ and

$$\Phi(u(\cdot), l \mid t_0, x^0) = \langle l, G(\vartheta, t_0) x^0 \rangle + \int_{t_0}^{\vartheta} \langle l, G(\vartheta, s) B(s) u(s) \rangle ds - \varphi^*(\vartheta, l).$$

In (1.59), the minimization is with respect to $u(t) \in \mathcal{P}(t)$. As before, $u(\cdot) = u(t), t \in T$, denotes the function $u(t)$ regarded as an element of an appropriate function space, which here is $L_2(T)$ and $\mathcal{P}(\cdot) = \{p(\cdot)\}$ is the class of all measurable functions $p(t)$ that satisfy $p(t) \in \mathcal{P}(t)$ a.e. in T.

The functional $\Phi(u(\cdot), l \mid t_0, x^0)$ is concave in l (since $\varphi^*(\vartheta, l)$ is convex in l) and linear in $u(\cdot)$. Since $u(\cdot)$ belongs to a *weakly compact set* $\mathcal{P}(\cdot)$ in $L_2(T)$, one may apply the minmax theorem of Ky Fan [70] and interchange the order of min and max in (1.59):

$$\mathcal{I}(t_0, x^0) = \min_u \max_l \Phi(u(\cdot), l \mid t_0, x^0) = \max_l \min_u \Phi(u(\cdot), l \mid t_0, x^0) = \max_l \Phi_0(l \mid t_0, x^0), \tag{1.60}$$

in which

$$\Phi_0(l \mid t_0, x^0) = \langle \psi[t_0], x^0 \rangle - \int_{t_0}^{\vartheta} \rho(-\psi[t] \mid B(t)\mathcal{P}(t))dt - \varphi^*(\vartheta, l). \tag{1.61}$$

Here $\psi[t] = \psi(t, \vartheta, l)$ is the solution to the *adjoint equation*

$$\dot{\psi}(t) = -A'(t)\psi(t), \quad \psi(\vartheta) = l, \tag{1.62}$$

for system (1.2). The second term in the right-hand side of (1.61) comes from the relations

$$\min_u \left\{ \int_{t_0}^{\vartheta} \langle l, G(\vartheta, s) B(s) u(s) \rangle ds \mid u(\cdot) \in \mathcal{P}(\cdot) \right\}$$

$$= \int_{t_0}^{\vartheta} \min_u \{ \langle l, G(\vartheta, s) B(s) u \rangle \mid u \in \mathcal{P}(s) \} ds$$

$$= -\int_{t_0}^{\vartheta} \max_u \{ \langle -l, G(\vartheta, t) B(t) u \rangle \mid u \in \mathcal{P}(s) \} ds$$

$$= -\int_{t_0}^{\vartheta} \rho(-\psi[t] \mid B(t)\mathcal{P}(t))dt. \tag{1.63}$$

In the above lines the order of operations of min and integration may be interchanged (see [238, p. 675]).

Let l^0 be a maximizer of $\Phi_0(l \mid t_0, x^0)$ and let $u^0(\cdot)$ be an *optimal control*, i.e., a minimizer of $\varphi(\vartheta, x[\vartheta]) = \max_l \Phi(u(\cdot), l \mid t_0, x^0)$.

Recall the definition of *saddle point* for $\Phi(u(\cdot), l \mid t_0, x^0) = \Phi(u, l)$.

Definition 1.5.1. A pair $\{u^s, l^s\}$ is said to be a saddle point for $\Phi(u, l)$ if

$$\Phi(u^s, l) \leq \Phi(u^s, l^s) \leq \Phi(u, l^s), \quad \forall u, l.$$

Lemma 1.5.1. *The pair* $(u^0(\cdot), l^0)$ *is a* **saddle point** *of* $\Phi(u(\cdot), l \mid t_0, x^0) = \Phi(u, l)$ *and* $\Phi(u^0(\cdot), l^0 \mid t_0, x^0) = \Phi(u^0, l^0) = \mathcal{I}(t_0, x^0).$

1.5 Optimal Open-Loop Control for Linear-Convex Systems: The Maximum...

Proof. From (1.60) we have

$$\Phi(u^0, l) \leq \max_l \Phi(u^0, l) = \min_u \max_l \Phi(u, l) = \mathcal{J}(t_0, x^0) =$$
$$= \max_l \min_u \Phi(u, l) = \min_u \Phi(u, l^0) \leq \Phi(u, l^0),$$

substituting $u = u^0$ and $l = l^0$ we obtain $\mathcal{J}(t_0, x^0) = \Phi(u^0, l^0)$, from which the lemma follows. $\square$

Thus

$$\Phi(u^0, l \mid t_0, x^0) \leq \Phi(u^0, l^0 \mid t_0, x^0) = \mathcal{J}(t_0, x^0) \leq \Phi(u, l^0 \mid t_0, x^0), \quad \forall u, l. \tag{1.64}$$

Concentrating on the *necessary conditions of optimality* for the control we first introduce the next assumption.

Let $v^0 = \min\{\varphi(\vartheta, x) \mid x \in \mathbb{R}^n\}$.

Assumption 1.5.1. *The "regularity condition"* $v^0 < \mathcal{J}(t_0, x^0)$ *holds.*

Lemma 1.5.2. *Under Assumption 1.5.1 the maximizer* $l^0 = l^0(t_0, x^0) \neq 0$.

Proof. Indeed, suppose $l^0 = 0$. From (1.60), $\mathcal{J}(t_0, x^0) = -\varphi^*(\vartheta, 0)$, so that

$$\mathcal{J}(t_0, x^0) = -\varphi^*(\vartheta, 0) = -\max\{-\varphi(\vartheta, x) \mid x \in \mathbb{R}^n\} = \min\{\varphi(\vartheta, x) \mid x \in \mathbb{R}^n\} = v^0.$$

contrary to the assumption. $\square$

This assumption means that $x^{abs} = \arg\min\{\varphi(\vartheta, x) \mid x \in \mathbb{R}^n\}$, the absolute minimizer of function $\varphi(\vartheta, x)$, does not lie in $X[\vartheta] = X(\vartheta, t_0, x^0)$—the reachability set of system (1.2).

The second inequality in (1.64) is

$$\Phi(u^0(\cdot), l^0 \mid t_0, x^0) \leq \Phi(u(\cdot), l^0 \mid t_0, x^0), \quad \forall u(t) \in \mathcal{P}(t). \tag{1.65}$$

Let $\psi^0[t] = \psi(t, \vartheta, l^0)$. Assumption 1.5.1, which we now assume, yields $\psi^0[t] \neq 0$ so that the solution of the adjoint equation is *nontrivial*.

Then, rewriting the inequality of (1.65) in detail, using (1.60)–(1.62) and omitting similar terms in the left and right-hand sides of this inequality, we arrive at the relation

$$\int_{t_0}^{\vartheta} \rho(-\psi^0[t] \mid B(t)\mathcal{P}(t))dt \geq \int_{t_0}^{\vartheta} -\psi^0[t]' B(t) u(t) dt, \quad \forall u(t) \in \mathcal{P}(t).$$

Here equality is attained at $u^0(\cdot)$ if and only if for almost all $t \in T$ the following pointwise *maximum condition* holds:

$$-\psi^0[t]' B(t) u^0(t) = \max\{-\psi^0[t]' B(t) u \mid u \in \mathcal{P}(t)\} = \rho(-\psi^0[t] \mid B(t)\mathcal{P}(t)). \tag{1.66}$$

The last relation may also be rewritten in the form of a *minimum condition*:

$$\psi^0[t]' B(t) u^0(t) = \min\{\psi^0[t]' B(t) u \mid u \in \mathcal{P}(t)\}. \tag{1.67}$$

Definition 1.5.2. An open-loop control $u(t) \in \mathcal{P}(t)$, $t \in T$, is said to satisfy **the maximum principle** if there exists a **nontrivial** solution $(\psi^0[t] \neq 0)$ of the adjoint equation (1.62), such that the following conditions hold:

(i) $u(t)$ satisfies the maximum condition (1.66), where
(ii) vector $\psi^0[\vartheta] = l^0$ is a maximizer of $\Phi_0(l)$ (1.61).

We have thus proved the next result.

Theorem 1.5.1. *Under Assumption 1.5.1 every minimizing control $u^0(t)$ of Problem 1.5.1 satisfies the maximum principle.*

Let $x_{u^0} = x^0[\vartheta]$ denote the end point at time ϑ of the trajectory $x^0[t] = x^0(t, t_0, x^0)$, emanating from position $\{t_0, x^0\}$ under an optimizing control $u^0(t)$ of the previous theorem. Then, under Assumption 1.5.1, we have

$$\max_l \{\langle l, x_{u^0} \rangle - \varphi^*(\vartheta, l)\} = \varphi(\vartheta, x_{u^0}) = \langle l^0, x_{u^0} \rangle - \varphi^*(\vartheta, l^0),$$

which leads to the next property.

Lemma 1.5.3. *An optimal control of Theorem 1.5.1 satisfies the equality*

$$\langle l^0, x_{u^0} \rangle = \varphi(\vartheta, x_{u^0}) + \varphi^*(\vartheta, l^0). \tag{1.68}$$

This is the *transversality condition* which allows nonsmooth functions φ.

Corollary 1.5.1. *Relation (1.68) is equivalent to condition (ii) in Definition 1.5.2 of the maximum principle.*

The proof follows from the definition of the conjugate function.

Lemma 1.5.4. *The pair $\{l^0, x_{u^0}\}$ of Lemma 1.5.3 satisfies the following relations:*

$$l^0 \in \partial \varphi(\vartheta, x_{u^0}), \text{ and } x_{u^0} \in \partial \varphi^*(\vartheta, l^0), \tag{1.69}$$

where $\partial \varphi(\vartheta, x_{u^0}), \partial \varphi^*(\vartheta, l^0)$ *are the* **subdifferentials** *in the second variable of $\varphi(\vartheta, x_{u^0})$ and $\varphi^*(\vartheta, l^0)$ at x_{u^0} and l^0.*

1.5 Optimal Open-Loop Control for Linear-Convex Systems: The Maximum...

If $\varphi(\vartheta, x)$ is differentiable in x, we have

$$l^0 = \varphi_x(\vartheta, x_{u^0}) = \frac{\partial \varphi(\vartheta, x_{u^0})}{\partial x}.$$

Exercise 1.5.1. Prove Lemma 1.5.4. (See Lemma 1.5.7 below.)

Relations (1.69) are simpler under some additional conditions.

Lemma 1.5.5 ([238], Sect. 11.C). *Suppose function $\varphi(x)$ is finite, coercive (this means $\liminf\{\varphi(x)/\|x\|\} \to \infty$ as $\|x\| \to \infty$), and also convex, of class C^2 (twice continuously differentiable) on $\mathbb{R}^n$ and its Hessian matrix $\nabla^2 \varphi(x)$ is positive definite for all x. Then its conjugate $\varphi^*(l)$ satisfies the same properties, namely, it is also a finite, coercive, convex function of class C^2 on $\mathbb{R}^n$ with its Hessian matrix $\nabla^2 \varphi^*(l)$ positive definite for every l.*

Corollary 1.5.2. *Suppose function $\varphi(\vartheta, x)$ satisfies conditions of Lemma 1.5.5 in the second variable. Then relations (1.69) of Lemma 1.5.4 have the form*

$$l^0 = \varphi_x(\vartheta, x_{u^0}) = \frac{\partial \varphi(\vartheta, x_{u^0})}{\partial x}, \quad x_{u^0} = \varphi_l^*(\vartheta, l^0) = \frac{\partial \varphi^*(\vartheta, l^0)}{\partial l}.$$

The last corollary indicates when the Fenchel transformation used in this chapter to calculate the conjugate φ^* coincides with the classical *Legendre transformation*.

1.5.2 Degenerate Cases

The first degenerate case. As we observe, the maximum principle requires $\psi^0[\vartheta] \neq 0$. Suppose, however, that $\psi^0[\vartheta] = l^0 = 0$. Then the maximum condition, which looks like

$$0 \times B(t)u^0 = \max\{0 \times B(t)u \mid u \in \mathcal{P}(t)\},$$

is degenerate. It is trivially fulfilled, but it is not the maximum principle. The optimal control then has to be found from considerations other than the maximum principle. As we shall now observe, this case occurs when Assumption 1.5.1 is not fulfilled.

Example 5.1. For the one-dimensional system

$$\dot{x}(t) = u(t), \quad x(0) > 0, \quad |u(t)| \leq \mu, \quad t \in [0, \vartheta],$$

consider the problem of minimizing $\varphi(\vartheta, x[\vartheta]) = \frac{1}{2}x^2[\vartheta]$.

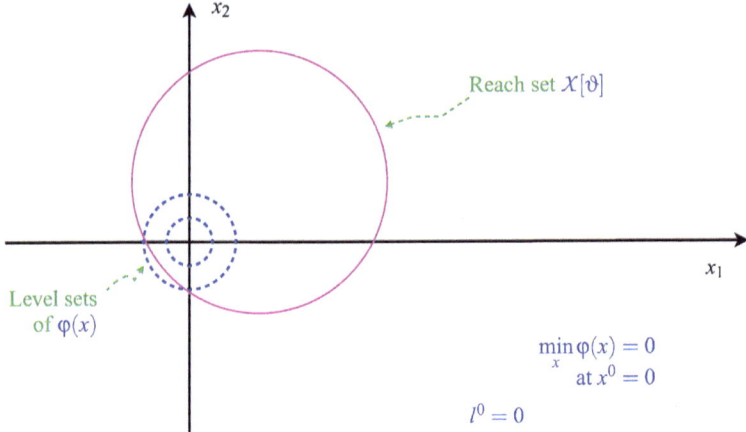

Fig. 1.2 Example 5.1

Here

$$\max_{l} \min_{u} \Phi(u, l) = \max_{l}\{l'x(0) - \mu\vartheta|l| - \frac{1}{2}\langle l, l\rangle^2\} = \max_{l} \Phi_0(l).$$

Assume $\mu\vartheta - x(0) > 0$. Then $\Phi_0(l) < 0$, $\forall l \neq 0$, and $\max_l \Phi_0(l) = \Phi_0(l^0) = 0$.
Hence

$$l^0 = 0, \text{ and } \max_{l} \min_{u} \Phi(u, l) = \Phi^0 = \mathcal{J}(0, x(0)) = 0.$$

The pair $l^0 = 0$ with **any** $u = u^0$ satisfies the *maximum condition* but not the maximum principle. Among these there exists an optimal control u^0 which must satisfy condition

$$\int_0^\vartheta u(t)dt + x(0) = 0$$

This is a *singular control* which has to be found from conditions *other than* the maximum principle (Fig. 1.2).
 We shall now see that there is a second degenerate case.
 The second degenerate case
 This is when $\psi^0[t] \neq 0$, but $B'(t)\psi^0[t] \equiv 0$.

Example 5.2. Consider system

$$\dot{x}_1 = u, \quad \dot{x}_2 = u, \quad |u| \leq \mu, \quad t \in [0, \vartheta],$$

so that $x \in \mathbb{R}^2$, $u \in \mathbb{R} = \mathbb{R}^1$. Here $A = 0$, $B'(t) \equiv b' = (1, 1)$ and the system **is not controllable**.

1.5 Optimal Open-Loop Control for Linear-Convex Systems: The Maximum...

Take $\varphi(\vartheta, x[\vartheta]) = \frac{1}{2}\langle x[\vartheta], x[\vartheta]\rangle$ and $x_1(0) > 0$, $x_2(0) < 0$. We have

$$\Phi^0 = \max_l \min_u \Phi(u, l) = \max_l \min_u \{(l_1 x_1(0) + l_2 x_2(0)) + \int_0^\vartheta (l_1 + l_2)u(t)dt - \frac{1}{2}\langle l, l\rangle\} =$$

$$= \max_l \{(l_1 x_1(0) + l_2 x_2(0)) - \mu\vartheta|l_1 + l_2| - \frac{1}{2}\langle l, l\rangle\}.$$

Calculating the maximum in l one may observe that

with $l_1 + l_2 > 0$ we have $\Phi^0 = \Phi_0(l^0 \mid 0, x(0)) = \frac{1}{2}((x_1(0) - \mu\vartheta)^2 + (|x_2(0)| + \mu\vartheta)^2)$,
with $l_1 + l_2 < 0$ we have $\Phi^0 = \frac{1}{2}((x_1(0) + \mu\vartheta)^2 + (-|x_2(0)| + \mu\vartheta)^2)$,
with $l_1 + l_2 = 0$ we have $\Phi^0 = \frac{1}{2}(x_1^2(0) + x_2^2(0))$.
With $x_1(0) + x_2(0) = 0$, the maximum is attained at $l_1^0 = x_1(0)$, $l_2^0 = x_2(0)$ and $\Phi^0 = \frac{1}{2}(x_1^2(0) + x_2^2(0))$.

The optimizing pair is $\{l^0, u^0\}$ and the maximum condition

$$l^0 b \times u^0 = \max_u \{l^0 b \times u \mid |u| \leq \mu\} = \max_u \{0 \times u\} = 0 \times u^0$$

is degenerate. However, $\psi^0[t] = l^0 \neq 0$ and condition (ii) of Definition 1.5.2, as well as (1.69) are satisfied. Here the maximum principle *is satisfied*, with any $|u(t)| \leq \mu$, which demonstrates that it is a *necessary* but not a sufficient condition for optimality. In this example the maximum principle is noninformative. The optimal control $u^0 \equiv 0$ is a *singular control* which means it has to be found from conditions other than the maximum principle (see Remark 1.5.3 below).

A similar situation arises with $x_1(0) = 0$, $x_2(0) < 0$. Then with $\vartheta\mu \geq |x_2(0)|/2$

$$\Phi^0 = \frac{1}{2}\langle x(\vartheta), x(\vartheta)\rangle, \quad x(\vartheta) = x(0) + bc(u(\cdot)), \quad c(u(\cdot))$$

$$= \int_0^\vartheta u(t)dt, \quad x_1(\vartheta) = -x_2(\vartheta) = x_* = -x_2(0)/2 > 0.$$

The optimal control u^0 must satisfy $c(u^0(\cdot)) = x_*$, and with $\vartheta\mu \geq |x_2(0)|/2$ it is singular, while $u^0(t) \neq 0$. There are many such controls (see Fig. 1.3).

We have thus indicated two types of degenerate problems. A protection from the first type was ensured by Assumption 1.5.1. A protection from the second type is ensured by the following assumption.

Let $b^{(i)}(t)$ be the i-th column of $B(t)$ and u_i the i-th coordinate of u, $i = 1, \ldots, p$.

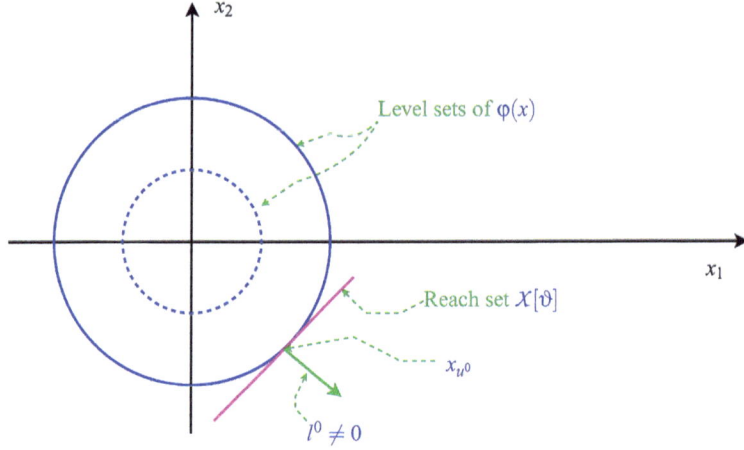

Fig. 1.3 Example 5.2

Assumption 1.5.2. *The system (1.44) is* **strongly completely controllable.** *This means that the condition of complete controllability of Corollary 1.3.1 is applicable to each column* $b^{(i)}(t)$ *of* $B(t)$.

This assumption implies that functions $\psi(s, \vartheta, l)' b^{(i)}(s) \neq 0$, $\forall l \neq 0$, $\forall i = 1, \ldots, n$, for almost all $t \in T$.

1.5.3 Sufficiency Conditions for Optimality

Some preliminary results

Lemma 1.5.6. *Suppose* $0 \in int\mathcal{P}(t) \subset \mathbb{R}^p$. *Then, under Assumption 1.5.2, the reachability set* $X[\vartheta] = X(\vartheta, t_0, x^0) \subset \mathbb{R}^n$ *has an interior point* x^0.

Proof. Since $\mathcal{P}(t)$ is continuous, there exists a number $\varepsilon > 0$, such that the ball

$$\mathcal{B}_\varepsilon(0) = \{x : \langle x, x \rangle \leq \varepsilon^2\} \subseteq \mathcal{P}(t), \quad \forall t \in [t_0, \vartheta].$$

Then $\rho(l \mid \mathcal{P}(t)) \geq \varepsilon \|l\|$, $\forall t$ and

$$\rho(l \mid X(\vartheta, t_0, 0)) = \int_{t_0}^{\vartheta} \rho(l \mid G(\vartheta, s) B(s) \mathcal{P}(s)) ds \geq \varepsilon \int_{t_0}^{\vartheta} \|l' G(\vartheta, s) B(s)\| ds.$$

Under the strong controllability condition (see Lemma 1.3.1), we have

$$\int_{t_1}^{t_2} \|l' G(\vartheta, s) B(s)\| ds = \int_{t_1}^{t_2} (l' D(s) D'(s) l)^{1/2} ds > 0, \quad \forall \|l\| \neq 0,$$

1.5 Optimal Open-Loop Control for Linear-Convex Systems: The Maximum...

whatever be $[t_1, t_2] \subseteq [t_0, \vartheta]$, $t_1 < t_2$, and since

$$\min_{l} \left\{ \int_{t_1}^{t_2} (l'D(s)D'(s)l)^{1/2} ds \mid \|l\| = 1 \right\} > 0,$$

this proves the lemma. □

Exercise 1.5.2. Prove that Lemma 1.5.6 is true when $\text{int} \mathcal{P}(t) \neq \emptyset$.

For any $l \neq 0$ we also have

$$\int_{t_0}^{\vartheta} \rho(l \mid G(\vartheta, s) B(s) \mathcal{P}(s)) ds = \int_{t_0}^{\vartheta} \max\{l' G(\vartheta, s) B(s) u(s) \mid u(s) \in \mathcal{P}(s)\} ds =$$

$$= \int_{t_0}^{\vartheta} l' G(\vartheta, s) B(s) u_l^0(s) ds > 0, \qquad (1.70)$$

where $u_l^0(s) \neq 0$, a.e. Thus we came to the next proposition.

Corollary 1.5.3. *Under the assumptions of Lemma 1.5.6 the maximum condition (1.70), taken for any $l \neq 0$, is achieved only by function $u_l^0(s) \neq 0$, a.e.*

Remark 1.5.1. Note that in Example 5.2, in the absence of controllability, we have $x \in \mathbb{R}^2$, but the reach set $\mathcal{X}[\vartheta]$ is one-dimensional and it has no interior points, but of course has a nonempty **relative interior** (see Fig. 1.3).

The Sufficiency Conditions

We shall now indicate, under Assumption 1.5.1, some conditions when the maximum principle turns out to be *sufficient* for optimality of the solution to Problem 1.5.1.

Theorem 1.5.2. *Suppose Assumption 1.5.1 is satisfied and the following conditions for a function $u^*(t)$ are true:*

(a) *$u^*(t)$ satisfies the maximum principle of Definition 1.5.1 for $l^0 \neq 0$—the maximizer of $\Phi_0(l)$;*

(b) *$u^*(t)$ satisfies the condition*

$$\max_{l} \Phi(u^*, l \mid t_0, x^0) = \max_{l} \Phi(u^*, l) = \Phi(u^*, l^0) = \Phi(u^*(\cdot), l^0 \mid t_0, x^0).$$

Then $u^*(t)$ is an optimal control of Problem 1.5.1 and $\mathcal{J}(t_0, x^0) = \Phi_0(l^0)$ is the optimal cost.

Proof. Denote $x^0[\vartheta], x^*[\vartheta]$ to be the vectors generated from $x[t_0] = x^0$ under an optimal control $u^0(t)$ and under control $u^*(t)$, respectively, with $x[\vartheta]$ being the one generated from the same point under any admissible control $u(t)$.

Then, on the one hand, since $u^*(t)$ satisfies condition (a) (the maximum principle for l^0), we have

$$\Phi(u^*(\cdot), l^0) = \langle \psi^0[t_0], x^0 \rangle + \int_{t_0}^{\vartheta} \psi^0[t]' B(t) u^*(t) dt - \varphi^*(\vartheta, l^0) =$$

$$= \langle \psi^0[t_0], x^0 \rangle + \int_{t_0}^{\vartheta} \min\{\psi^0[t]' B(t) u \mid u \in \mathcal{P}(t)\} dt - \varphi^*(\vartheta, l^0) = \Phi(u^0(\cdot), l^0) =$$

$$= \langle l^0, x^0[\vartheta] \rangle - \varphi^*(\vartheta, l^0) = \varphi(\vartheta, x^0[\vartheta]),$$

and on the other, due to condition (b),

$$\Phi(u^*(\cdot), l^0) = \max_l \Phi(u^*, l) = \max_l \{l' G(\vartheta, t_0) x^0 + \int_{t_0}^{\vartheta} l' G(\vartheta, t) B(t) u^*(t) dt - \varphi^*(\vartheta, l^0)\}$$

$$= \max_l \{\langle l, x^*[\vartheta] \rangle - \varphi^*(\vartheta, l) \mid l \in \mathbb{R}^n\} = \varphi(\vartheta, x^*[\vartheta]),$$

so that $u^*(t)$ is the optimal control. □

Remark 1.5.2. Condition (b) of Theorem 1.5.2 may be substituted by the following equivalent condition.

(b') $u^*(t)$ generates a trajectory $x^*[t]$ such that $l^0 \in \partial\varphi(\vartheta, x^*[\vartheta])$.

This statement is proved below, in Lemma 1.5.9.

However, the last theorem does not ensure that the maximum principle is nondegenerate and may serve *to determine* the control.

Remark 1.5.3. Indeed, to single out the optimal control $u^0 \equiv 0$ of Example 5.2 from all the others (recall that there the redundant maximum condition is satisfied by all the possible controls), one has to apply something else than the maximum principle. This could be condition (b) of the sufficiency Theorem 1.5.2, which would serve as a verification test.

Thus, taking control $u^*(t) \equiv k \neq 0, x_1(0) = -x_2(0)$, and calculating

$$\max_l \{\langle l, x(0) \rangle + k(l_1 + l_2)\vartheta - \frac{1}{2} \langle l, l \rangle\} = \varphi(x^*(\vartheta)),$$

we find

$$\varphi(x^*(\vartheta)) = (x_1(0) + k)^2 + (x_2(0) + k)^2 > \varphi(x^0(\vartheta)) = x_1^2(0) + x_2^2(0),$$

which indicates that the controls $u^*(t) \equiv k$ with $k \neq 0$ are not optimal, though they satisfy the maximum condition.

1.5 Optimal Open-Loop Control for Linear-Convex Systems: The Maximum...

Theorem 1.5.3. *Suppose control $u^*(t)$ satisfies conditions of Theorem 1.5.2 and the assumptions of Lemma 1.5.6 are satisfied. Then $u^*(t) \neq 0$ a.e. and it may be found from the maximum principle which is nondegenerate.*

Here is an example where the optimal control in not unique.

Example 5.3. For the two-dimensional system

$$\dot{x}(t) = u(t), \; x(0) = (1, -2)', \; |u_i(t)| \leq \mu, i = 1, 2, \; t \in [0, \vartheta],$$

find the optimal control that minimizes $\varphi(x[\vartheta]) = \max_i |x_i(\vartheta)|, \; i = 1, 2$.

Recall that here $\varphi^*(l) = I(l \mid \mathcal{B}^*(0))$, where $\mathcal{B}^*(0) = \{l : |l_1| + |l_2| \leq 1\}$ and $I(l \mid \mathcal{B})$ is the *indicator function* for set $\mathcal{B}$ (as defined in convex analysis), so that $I(l \mid \mathcal{B}) = 1$ if $l \in \mathcal{B}$ and $I(l \mid \mathcal{B}) = +\infty$ otherwise.

We have

$$\Phi^0 = \Phi(u^0, l^0) =$$

$$= \max\left\{\langle l, x(0)\rangle + \int_0^\vartheta \min\{\langle l, u(s)\rangle \mid |u(s)| \leq \mu\}ds - I(l \mid \mathcal{B}(0))\right\} = \max_l \Phi_0(l) =$$

$$= \max_l\{\langle l, x(0)\rangle - \mu(|l_1| + |l_2|)\vartheta \mid |l_1| + |l_2| \leq 1\} =$$

$$= \max\{\langle l, x(0)\rangle \mid |l_1| + |l_2| \leq 1\} - \mu\vartheta = \max_i |x_i(0)| - \mu\vartheta.$$

Taking further $\vartheta = 1, \mu = 1$, we have $l^0 = (0, -1) \neq 0$. (Note that here Assumptions 1.5.1, 1.5.2 are both satisfied, so that the maximum principle is nondegenerate.)

The minimum condition gives

$$\min\{\langle l^0, u(t)\rangle \mid |u(t)| \leq 1\} = -(|l_1^0| + |l_2^0|), \; u_1^*(t) \in [-1, 1], \; u_2^*(t) \equiv 1,$$

where l^0 is the maximizer of $\Phi_0(l)$. The maximum principle is thus satisfied with $u_2^*(t) \equiv 1$ and any control $u_1^*(t)$.

Let us now check condition (b) of Theorem 1.5.2, calculating

$$\max_l\{\langle l, x(0)\rangle + \int_0^\vartheta \langle l, u^*(s)\rangle ds \mid |l_1| + |l_2| \leq 1\} = \max_l \Phi(u^*, l).$$

This gives, for $u_2^*(t) \equiv 1$, and any $u_1^* \in [-1, 0]$, $\max_l \Phi(u^*, l) = \max\{|x_1(1)|, |x_2(1)|\} = 1$, which means *all such controls are optimal*.

At the same time, for $u_2^*(t) \equiv 1$, and any $u_1^* \in [k, 1], \; k \in (0, 1]$ we have $\max_l \Phi(u^*, l) = \max\{|x_1(1)|, |x_2(1)|\} \geq 1 + k > \Phi(u^*, l^0)$ so such controls are not optimal (Fig. 1.4).

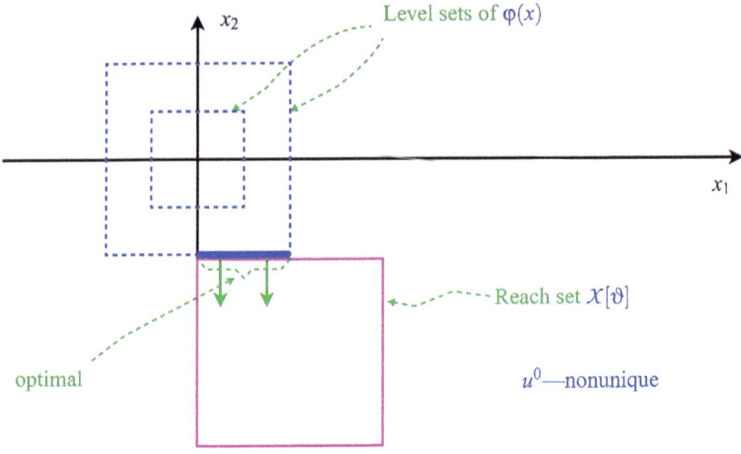

Fig. 1.4 Example 5.3

The next example illustrates the case when the optimal control is unique.

Exercise 1.5.3. For the two-dimensional system

$$\dot{x}(t) = u(t), \ x(0) = (1,0)', \ |u_i(t)| \le 1, i = 1,2, \ t \ge 0,$$

consider the problem of minimizing $\varphi(\vartheta, x[\vartheta]) = \frac{1}{2}\|x[\vartheta]\|^2$. For $\vartheta < 1$, show that the unique maximizer of $\Phi_0(l \mid t_0, x^0)$ is $l_0 = (1,0)'$. Hence $u(t), \ t \in [0, \vartheta]$, satisfies the maximum principle iff $u_1(t) \equiv -1$. The optimal control in addition satisfies $u_2^0(t) \equiv 0$. Prove that the optimal control $u_1^0(t) \equiv -1$, $u_2^0(t) \equiv 0$ is unique.

It is interesting to observe that the notion of saddle point allows us to give the following interpretation of sufficiency.

Theorem 1.5.4. *Suppose $\{u^s(\cdot), l^s\}$ is a saddle point of $\Phi(u(\cdot), l \mid t_0, x^0)$. Then $u^s(\cdot)$ is an optimal control of Problem 1.5.1 which satisfies the maximum condition of Definition 1.5.2 with $\psi^s[\vartheta] = l^s$.*

Note that Theorem 1.5.4 covers both degenerate cases. If Assumption 1.5.1 is satisfied, we moreover have $l^s \ne 0$.

Exercise 1.5.4. Prove Theorem 1.5.4.

1.5.4 A Geometrical Interpretation

A geometrical interpretation of the maximum principle and saddle point condition (1.64) is given in Lemma 1.5.7 and Exercise 1.5.4, and illustrated in Fig. 1.5.

1.5 Optimal Open-Loop Control for Linear-Convex Systems: The Maximum...

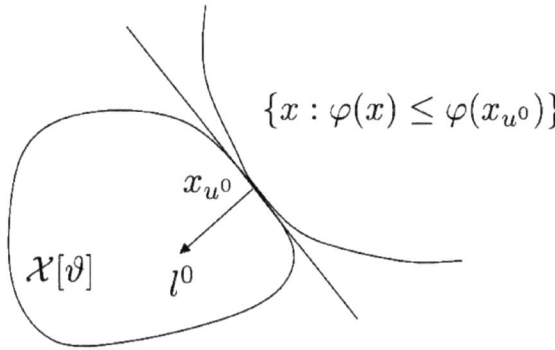

Fig. 1.5 x_{u^0} is the point of support for $X[\vartheta]$ in direction $-l^0$ and the set $\{\varphi(x) \leq \varphi(x_{u^0})\}$ in direction l^0

Lemma 1.5.7. *Let $x_{u^0} = x(\vartheta, t_0, x^0)$ be the state at time ϑ corresponding to a control u^0. Then l^0 satisfies the first inequality in (1.64),*

$$\Phi(u^0, l \mid t_0, x^0) \leq \Phi(u^0, l^0 \mid t_0, x^0), \quad \forall l,$$

if and only if

$$\varphi(\vartheta, x) \geq \varphi(\vartheta, x_{u^0}) + \langle l^0, x - x_{u^0}\rangle, \quad \forall x \in \mathbb{R}^n. \tag{1.71}$$

If $l^0 \neq 0$, then x_{u^0} is the point of support for the set $\{x : \varphi(\vartheta, x) \leq \varphi(\vartheta, x_{u^0})\}$ in direction l^0, i.e. $l^0 \in \partial \varphi(\vartheta, x_{u^0})$.

Proof. Evidently

$$\langle l^0, x_{u^0}\rangle - \varphi^*(\vartheta, l^0) = \Phi(u^0, l^0) = \max_l \Phi(u^0, l) = \max_l \{\langle l, x_{u^0}\rangle - \varphi^*(\vartheta, l)\} = \varphi(\vartheta, x_{u^0}),$$

is equivalent to

$$\varphi^*(\vartheta, l^0) = \langle l^0, x_{u^0}\rangle - \varphi(\vartheta, x_{u^0}),$$

which, since $\varphi^*(\vartheta, l^0) = \max_x \{\langle l^0, x\rangle - \varphi(\vartheta, x)\}$, is equivalent to (1.71).

Further on, if $\varphi(\vartheta, x) \leq \varphi(\vartheta, x_{u^0})$, by (1.71) one must have $\langle l^0, x - x_{u^0}\rangle \leq 0$, which proves the last assertion. □

Lemma 1.5.8. *Suppose $l^0 \neq 0$. Then u^0 satisfies the second inequality in (1.64),*

$$\Phi(u^0, l^0 \mid t_0, x^0) \leq \Phi(u, l^0 \mid t_0, x^0), \quad \forall u,$$

if and only if x_{u^0} is the point of support for the reach set $X[\vartheta]$ in direction $-l^0$, i.e.,

$$\langle l^0, x_{u^0}\rangle \leq \langle l^0, x\rangle, \quad \forall x \in X[\vartheta].$$

Exercise 1.5.5. Prove Lemma 1.5.8.

The next lemma can be used to verify whether a proposed control is indeed optimal.

Lemma 1.5.9. *Let $x^0(\vartheta)$ be the state at time ϑ corresponding to control u^0 and $l^0 \in \partial \varphi(\vartheta, x_{u^0})$. Then u^0 is optimal if it satisfies the maximum principle of Definition 1.5.2.*

Proof. By (1.71), for any $x(\vartheta) \in \mathbb{R}^n$, we have

$$\langle l^0, x(\vartheta) - x^0(\vartheta) \rangle \geq 0 \Rightarrow \varphi(\vartheta, x(\vartheta)) \geq \varphi(\vartheta, x^0(\vartheta));$$

and by Lemma 1.5.8 (the maximum principle in the form of a minimum rule),

$$x(\vartheta) \in X[\vartheta] \Rightarrow \langle l^0, x(\vartheta) - x^0(\vartheta) \rangle \geq 0;$$

whence

$$x(\vartheta) \in X[\vartheta] \Rightarrow \varphi(\vartheta, x(\vartheta)) \geq \varphi(\vartheta, x^0(\vartheta)),$$

so u^0 is optimal. □

Remark 1.5.4. Be careful, since, as indicated in Examples 5.1 and 5.2, the control u^0 of Lemma 1.5.9 may turn out to be singular and the maximum principle for u^0 may be degenerate.

1.5.5 Strictly Convex Constraints on Control

Recall that a compact set $\mathcal{P}$ is said to be *strictly convex*, if its boundary $\partial \mathcal{P}$ does not contain any line segment. That is, if $p', p'' \in \partial \mathcal{P}$, then $\lambda p' + (1-\lambda) p'' \notin \partial \mathcal{P}$, $\forall \lambda \in (0, 1)$.

Strictly convex sets may be presented as level sets of strictly convex functions. A function $\varphi(x)$ is *strictly convex* in the convex domain $\mathcal{D}$ if for points $p', p'' \in \mathcal{D}$ one has

$$\varphi(\lambda p' + (1 - \lambda) p'') < \lambda \varphi(p') + (1 - \lambda) \varphi(p''), \quad \forall \lambda \in (0, 1).$$

Note that the inequality here is strict.[4]

Lemma 1.5.10. *Suppose the function $\varphi(\vartheta, x)$ is strictly convex in x. Then the maximizer l^0 of $\Phi_0(l \mid t_0, x^0)$ is* **unique**.

---

[4] Recall that here, as mentioned in Sect. 1.2, set $\mathcal{P}(t)$ is assumed to have an interior point for all t.

1.5 Optimal Open-Loop Control for Linear-Convex Systems: The Maximum... 39

Exercise 1.5.6. Prove Lemma 1.5.2.

An example of a strictly convex function is the *strictly positive quadratic* function $\varphi(x) = \langle p - a, P(p - a)\rangle$ with $P = P' > 0$.

Denote a *degenerate ellipsoid* as

$$\mathcal{E}(p^*, P) = \{p : \langle p - p^*, P^{-1}(p - p^*)\rangle \leq 1\}, P = P' > 0.$$

Here p^* is the *center* and P is the *shape matrix*.

Another example is as follows. Take the function $\varphi(x) = d^2(x, \mathcal{E}(0, P))$, $P = P' > 0$, and a ray $\{x = \alpha h, h \notin \mathcal{E}(0, P) \ \alpha > 0\}$, with scalar α. Then function $f(\alpha) = \varphi(\alpha h)$ defined for $\alpha > 1$ will be strictly convex. (Prove this.)

The constraint on u may also be strictly convex.

Lemma 1.5.11. *Suppose $d \neq 0$ and $\mathcal{P}$ is strictly convex. Then the maximizer u^0 of the problem*

$$\max\{\langle d, u\rangle \mid u \in \mathcal{P}\} = \langle d, u^0\rangle$$

is unique.

Lemma 1.5.12. *Suppose the continuous function $d(t) \neq 0$ almost everywhere in T and function $\mathcal{P}(t)$ is Hausdorff-continuous with values in strictly convex compact sets. Then the optimizer $u^0(t)$ to*

$$\langle d(t), u^0(t)\rangle = \max\{\langle d(t), u\rangle \mid u \in \mathcal{P}(t)\}$$

is unique and integrable.

Exercise 1.5.7. Prove Lemmas 1.5.11, 1.5.12.

An example of a strictly convex set $\mathcal{P}$ is the nondegenerate ellipsoid, $\mathcal{P} = \mathcal{E}(p^*, P)$.

Another example is the intersection of m nondegenerate ellipsoids: $\cap_{i=1}^{m} \mathcal{E}(0, P_i)$, $P_i > 0$, $P_i = P_i'$. (Prove this.)

Lemmas 1.5.10–1.5.12 imply the following assertion.

Theorem 1.5.5. *Suppose Assumptions 1.5.1 and 1.5.2 hold, the sets $\mathcal{P}(t)$, $t \in T$, are strictly convex and the function $\varphi(\vartheta, x)$ is strictly convex in x. Then the optimal control $u^0(t)$ of Problem 1.5.1 is **unique**.*

Before completing this section we emphasize that finding the optimal control under hard bounds requires one to solve optimization problems contained in the maximum principle of type (1.50) or (1.66). These problems involve calculating a support function in the following form.

Problem 1.5.2. Given vector d and convex compact set $\mathcal{P}$ in $\mathbb{R}^p$, find

$$\rho(d \mid \mathcal{P}) = \max\{\langle d, u\rangle \mid u \in \mathcal{P}\} = \langle d, u^0\rangle. \quad (1.72)$$

This optimization problem is easily solved if $\mathcal{P}$ is defined by a standard norm in the space $l_q: \mathcal{P} = \{u: \|u\| \leq \mu\}$. For example,

- if $q = 1$, $\|u\| = \sum_{i=1}^{k} |u_i|$ and $u^0 = (0, \ldots, \mu \text{ sgn} d_j, \ldots, 0)'$, $j = \arg\max\{|d_i|, i = 1, \ldots, k\}$,
- if $q = 2$, $\|u\| = \left(\sum_{i=1}^{k} |u_i|^2\right)^{1/2}$ and $u^0 = \mu \, d(\langle d, d \rangle)^{-1/2}$,
- if $q = \infty$, $\|u\| = \max\{|u_i| \mid i = 1, \ldots, k\}$ and $u_i^0 = \mu \text{ sgn} d_i$, $i = 1, \ldots, k$.

Exercise 1.5.8. Solve Problem 1.5.2 for

- (i) $\mathcal{P} = \{u: \langle u-a, P(u-a)\rangle^{1/2} \leq \mu, \, P = P' > 0\}$.
- (ii) $\mathcal{P} = \{u: \left(\sum_{i=1}^{k}(u_i - a_i)^q\right)^{1/q} \leq \mu\}$, with $q \in (2, \infty)$.

for given vector a.

Summarizing this section we observe that the solution to the *primal* Problem 1.5.1 of optimal control was reduced to the *dual* optimization problem of maximizing the dual cost functional $\Phi_0(l \mid t_0, x^0)$. The dual problem leads us to a crucial element of the solution—the vector $\psi(\vartheta) = l^0$ that uniquely defines the trajectory $\psi[t] = \psi(t, \vartheta, l^0)$ of the adjoint equation (1.61) which, in its turn, determines the optimal control itself through the maximum principle.

In general, the dual problem for optimization *under constraints* is simpler than the primal problem. This justifies its usefulness. In most problems in the forthcoming sections the primal problem is to minimize a cost over a class of *functions* (the controls), while the dual problem is one of *finite-dimensional* optimization with either no constraints at all, or with simpler ones.

In this section the properties of system *linearity* and *convexity* of the constraints and of the "cost function" (the performance index) allowed us to indicate assumptions which allow the maximum principle to be not only a necessary condition of optimality (as in the general case), but also a sufficient condition.

A natural move is to apply the techniques of convex analysis and its *duality* methods used above to other types of optimal control problems in "linear-convex" systems. This is done in the next section.

1.6 Duality Methods of Convex Analysis in Problems of Optimal Control

In the investigation of optimization problems the notion of convexity turned out to be extremely important. This resulted in the development of *convex analysis*, with pioneering work by Fenchel, [73,74] Moreau, [217], and Rockafellar, [237,238], see also [48, 99, 227]. One of the key elements here is in introducing dual optimization problems solving which sometimes simplifies the optimization procedure.

1.6.1 The Primal and Dual Optimization Problems

In this section we elaborate further on the reduction of linear-convex problems of optimal control to respective dual problems of optimization.

Problem 1.6.1. Consider system (1.2) with $T = [t, \vartheta]$. Given starting position $\{t, x\}$, find

$$\mathcal{J}(t, x) = \min_u \left\{ \int_t^\vartheta (\varphi_1(s, H(s)x[s]) + \varphi_2(s, u(s)))ds + \varphi(\vartheta, x[\vartheta]) \right\}, \quad (1.73)$$

along the trajectories $x[s] = x(s, t, x)$ of system (1.2).

The $m \times n$ matrix $H(s)$ is continuous; $\varphi_1(t, z), \varphi(\vartheta, x), \varphi_2(t, u)$ are continuous in t and convex in $z \in \mathbb{R}^m$, $x \in \mathbb{R}^n$, and $u \in \mathbb{R}^p$, respectively, and assumed to have bounded level sets.

Note that there are no bounds on u, x. In convex analysis $\varphi_1(t, z), \varphi(t, x), \varphi_2(t, u)$ are called *normal integrands*, [238].

However, boundedness of the level sets is equivalent to Assumption 1.6.1 which we further presume to hold without additional reference.

Assumption 1.6.1. *The following inclusions are true*

$$0 \in intDom\varphi_1^*(t, \cdot), \ 0 \in intDom\varphi^*(\vartheta, \cdot), \ 0 \in intDom\varphi_2^*(t, \cdot).$$

Here and in the sequel all conjugates φ^*, φ_i^* and all sets Dom$(\cdot)$ are taken with respect to the second variable unless otherwise indicated. Next, we reduce Problem 1.6.1 to a dual problem of optimization. We first rewrite functions $\varphi_1(t, z), \varphi(t, x), \varphi_2(t, u)$ in terms of their conjugates:

$$\varphi_1(s, H(s)x[s]) = \max_\lambda \{\langle \lambda(s), H(s)x[s]\rangle - \varphi_1^*(s, \lambda(s)) \mid \lambda(s) \in \mathbb{R}^m\}, \quad (1.74)$$

$$\varphi(\vartheta, x[\vartheta]) = \max_l \{\langle l, x[\vartheta]\rangle - \varphi^*(\vartheta, l) \mid l \in \mathbb{R}^n\}, \quad (1.75)$$

$$\varphi_2(s, u) = \max_p \{\langle p(s), u\rangle - \varphi_2^*(s, p(s)) \mid p(s) \in \mathbb{R}^p\}. \quad (1.76)$$

Using formula (1.3), substitute for $x[s], x[\vartheta]$ first in (1.74) and (1.75), and then substitute these results into (1.73). This leads to the following problem of minimaximization:

$$\mathcal{J}(t, x) = \min_u \max_{l, \lambda} \Phi(u(\cdot), l, \lambda(\cdot) \mid t, x), \quad (1.77)$$

in which the variables range over $u(s) \in \mathbb{R}^p$, $\lambda(s) \in \mathbb{R}^m$, $s \in [t, \vartheta]$, and $l \in \mathbb{R}^n$, and

$$\Phi(u(\cdot), l, \lambda(\cdot) \mid t, x) = \langle l, G(\vartheta, t)x \rangle + \int_t^\vartheta \langle l, G(\vartheta, \xi) B(\xi) u(\xi) \rangle d\xi - \varphi^*(\vartheta, l) +$$

$$+ \int_t^\vartheta \left(\left\langle \lambda(s), H(s) G(s, t) x + H(s) \int_t^s G(s, \xi) B(\xi) u(\xi) d\xi \right\rangle - \varphi_1^*(s, \lambda(s)) + \varphi_2(s, u(s)) \right) ds. \tag{1.78}$$

We now introduce a new adjoint system for Problem 1.6.1,

$$\dot{\psi}(s) = -A'(s) \psi(s) - H'(s) \lambda(s), \quad \psi[\vartheta] = l. \tag{1.79}$$

Its trajectory $\psi[t] = \psi(t, \vartheta, l)$ is described by an integral formula analogous to (1.3),

$$\psi[t] = G'(\vartheta, t) l + \int_t^\vartheta G'(s, t) H'(s) \lambda(s) ds, \tag{1.80}$$

with $G(s, t)$ defined by (1.4), (1.5).

Interchanging the order of integration in the second integral transforms (1.78) into

$$\Phi(u(\cdot), l, \lambda(\cdot) \mid t, x)$$
$$= \langle \psi[t], x \rangle + \int_t^\vartheta \left(\psi[\xi]' B(\xi) u(\xi) + \varphi_2(\xi, u(\xi)) - \varphi_1^*(\xi, \lambda(\xi)) \right) d\xi - \varphi^*(\vartheta, l).$$

The functional $\Phi(u(\cdot), l, \lambda(\cdot) \mid t, x)$ is convex in u and concave in $\{l, \lambda(\cdot)\}$.

Since we assumed $0 \in \text{int}\text{Dom}\varphi_u^*(s, \cdot)$, we observe that

$$\varphi_2^*(s, p) = \max_u \{ \langle p, u \rangle - \varphi_2(s, u) \} = -\min_u \{ -\langle p, u \rangle + \varphi_2(s, u) \},$$

which yields

$$\min_u \varphi_2(s, u) = -\varphi_2^*(s, 0).$$

The last equality indicates that the minimizing set

$$\mathcal{U}^0(s) = \{ u : u \in \arg\min \varphi_2(s, u) \}$$

is bounded for each $s \in T = [t, \vartheta]$. Together with the continuity of $\varphi_2(s, u)$ in s this yields the Hausdorff-continuity of $\mathcal{U}^0(s)$, which means that in the minimaximization (1.77) the functions $u(s)$ may be selected in a weakly compact

1.6 Duality Methods of Convex Analysis in Problems of Optimal Control

subset set of $\mathcal{L}_2[T]$. The last fact allows us to apply Ky Fan's minmax theorem to (1.77), which yields

$$\mathcal{J}(t,x) = \min_u \max_{l,\lambda} \Phi(u(\cdot), l, \lambda(\cdot) \mid t, x) = \max_{l,\lambda} \min_u \Phi(u(\cdot), l, \lambda(\cdot) \mid t, x)$$
$$= \max_{l,\lambda} \Phi_0(l, \lambda(\cdot) \mid t, x), \tag{1.81}$$

in which

$$\Phi_0(l, \lambda(\cdot) \mid t, x) = \langle \psi[t], x \rangle - \int_t^\vartheta \Big(\varphi_2^*(s, -B'(s)\psi[s]) + \varphi_1^*(s, \lambda(s)) \Big) ds - \varphi^*(\vartheta, l). \tag{1.82}$$

Here we have used the relations

$$\min_{u(\cdot)} \int_t^\vartheta (\psi[s]'B(s)u(s) + \varphi_2(s, u(s))) ds = \int_t^\vartheta \min_u (\psi[s]'B(s)u(s) + \varphi_2(s, u(s))) ds =$$
$$= \int_t^\vartheta (-\varphi_2^*(s, -B'(s)\psi[s])) ds. \tag{1.83}$$

In summary we have two optimization problems: the *Primal* and the *Dual*.

Primal Problem: Find the optimal control $u^0(s)$, $s \in [t, \vartheta]$ that minimizes the primal functional

$$\Pi(t, x, u) = \int_t^\vartheta (\varphi_1(s, H(s)x[s]) + \varphi_2(s, u(s))) ds + \varphi(\vartheta, x[\vartheta]),$$

along trajectories of the primal system

$$\dot{x} = A(s)x + B(s)u + v(s), \quad s \in [t, \vartheta], \tag{1.84}$$

with given starting position $\{t, x\}$, $x(t) = x$, so that

$$\mathcal{J}(t, x) = \min_u \{\Pi(t, x, u) \mid u \in \mathbb{R}^n, (1.84)\}.$$

Dual Problem: Find the optimal pair $\{l, \lambda(s), s \in [t, \vartheta]\}$ that maximizes the dual functional

$$\Phi_0(l, \lambda(\cdot) \mid t, x) = \langle \psi[t], x \rangle - \int_t^\vartheta \Big(\varphi_2^*(s, -B'(s)\psi[s]) + \varphi_1^*(s, \lambda(s)) \Big) ds - \varphi^*(\vartheta, l),$$

along the trajectories of the dual (adjoint) system

$$\dot{\psi}(s) = -A'(s)\psi(s) - H'(s)\lambda(s),$$

with boundary value $\psi(\vartheta) = l$, (ψ is a column vector).

The attainability of the maximum in $\{l, \lambda(\cdot)\}$ in (1.81), (1.82) and of the minimum in $u(\cdot)$ in (1.83) is ensured by Assumption 1.6.1. This pointwise minimaximization in $u(s), \lambda(s)$, $s \in [t, \vartheta]$ with a further maximization in l yields functions $u^0(s), \lambda^0(s)$ and also a vector l^0. The indicated *optimizers* need not be unique. However, due to the properties of systems (1.2), (1.79) and functions $\varphi_1(t, x), \varphi(t, x), \varphi_2(t, u)$, there always exists a pair of *realizations* $u^0(s), \lambda^0(s)$ that are integrable and, in fact, may even turn out to be continuous.

Exercise 1.6.1. Prove the existence of integrable realizations $u^0(s), \lambda^0(s), s \in [t, \vartheta]$, for the optimizers of (1.81)–(1.83).

Exercise 1.6.2. Prove that whatever be $l^0 \neq 0$, one has $\psi(t) \not\equiv 0$, $t \in [t_0, \vartheta]$, whatever be $\lambda(t)$.

The relations above indicate the following *necessary conditions of optimality* of the control realization $u^0(s), s \in [t, \vartheta]$.

Theorem 1.6.1. *Suppose $u^0(t)$ is the optimal control for Problem 1.6.1 and $\{l^0, \lambda^0(\cdot)\}$ are the optimizers of the dual functional $\Phi_0(l, \lambda(\cdot) \mid t, x)$ that generate the solution $\psi^0[s] = \psi(s, \vartheta)$ under $l = l^0$, $\lambda(\cdot) = \lambda^0(\cdot)$. Then $u^0(t)$ satisfies the following pointwise* **minimum condition**

$$\psi^0[s]' B(s) u^0(s) + \varphi_2(s, u^0(s)) = \min_u \{\psi^0[s]' B(s) u + \varphi_2(s, u) \mid u \in \mathbb{R}^p\}, \ s \in [t, \vartheta], \quad (1.85)$$

which is equivalent to the pointwise maximum condition

$$-\psi^0[s]' B(s) u^0(s) - \varphi_2(s, u^0(s)) = \max_u \{-\psi^0[s]' B(s) u - \varphi_2(s, u) \mid u \in \mathbb{R}^p\}$$

$$= \varphi_2^*(s, -B'(s)\psi^0[s]), \ s \in [t, \vartheta], \quad (1.86)$$

producing with $l^0 \neq 0$ **the maximum principle**.

In order to ensure conditions $l^0, \lambda^0 \neq 0$ we will need some additional assumptions. Let

$$\mu^0(t, x) = \Phi_0(0, 0 \mid t, x) = -\int_t^\vartheta \Big(\varphi_2^*(s, 0) + \varphi_1^*(s, 0)\Big) ds - \varphi^*(\vartheta, 0) = \mathcal{I}_0(t, x).$$

Assumption 1.6.2. *The inequality $\mu^0(t, x) = \mathcal{I}_0(t, x) < \mathcal{I}(t, x)$ is true.*

Lemma 1.6.1. *Under Assumption 1.6.2 the maximizer $\{l^0, \lambda^0(\cdot)\} \neq 0$.*

Denote

$$\mu_1(t, x) = \max\{\Phi_0(l, 0 \mid t, x) \mid l \in \mathbb{R}^n\},$$
$$\mu_2(t, x) = \max\{\Phi_0(0, \lambda(\cdot) \mid t, x) \mid \lambda(\cdot) \in L_2(T)\}.$$

1.6 Duality Methods of Convex Analysis in Problems of Optimal Control

Assumption 1.6.3. *The following inequalities hold:*

$$(i) \ \mu_1(t,x) < \mathcal{J}(t,x), \quad (ii) \ \mu_2(t,x) < \mathcal{J}(t,x).$$

Lemma 1.6.2. *(a) Suppose Assumption 1.6.3(i) holds. Then $l^0 \neq 0$.*
(b) Suppose Assumption 1.6.3(ii) holds. Then $\lambda^0(\cdot) \neq 0$.

Exercise 1.6.3. Prove Lemma 1.6.2.

Assumption 1.6.4. *(i) Functions $\varphi_1(t,z), \varphi(t,x)$ are strictly convex in z, x and continuous in t.*
(ii) Function $\varphi_2(t,u)$ is strictly convex in u and continuous in t.

Lemma 1.6.3. *Suppose Assumption 1.6.4(i) is true. Then the optimizers $\{l^0, \lambda^0(\cdot)\}$ is unique.*

Exercise 1.6.4. Prove Lemma 1.6.3.

Lemma 1.6.4. *Under the boundedness of level sets for $\varphi_1(t, H(t)x), \varphi(t,x)$, $\varphi_2(t,u)$ and with a given nonzero optimizer $\{l^0, \lambda^0(\cdot)\}$ one has the inclusion*

$$- B'(s)\psi^0[s] \in \text{int } Dom\varphi_2^*(s, \cdot), \quad s \in T. \tag{1.87}$$

Proof. Indeed, if inclusion (1.87) does not hold, it would contradict the assumption that $\{l^0, \lambda^0(\cdot)\} \neq 0$. This could be checked by direct substitution. □

Note that inclusion (1.87) is also the necessary and sufficient condition for the level sets of the function

$$\psi^0[s]' B(s)u + \varphi(s,u)$$

to be bounded and hence for the attainability of the minimum in u in (1.85) (which is also the maximum in (1.86)) for $s \in T$.

Lemma 1.6.5. *Under the assumptions of Lemma 1.6.4 the minimum in (1.85) is attained and the minimizer $u^0(s), s \in T$, may be selected as a piecewise continuous realization.*

This brings us to the *sufficient* conditions for optimality of the control realization $u^0(s), \ s \in [t, \vartheta]$.

Theorem 1.6.2. *Suppose Assumptions 1.6.2, 1.6.3(i) are true, the requirements of Lemma 1.6.4 hold and*

(i) a control $u^*(s), \ s \in T$, satisfies the maximum principle (1.86), with $\{l^0 \neq 0, \lambda^0(t) \neq 0\}$ as the maximizer for the Dual Problem,

(ii) the equality

$$\max_{l,\lambda} \Phi(u^*(\cdot), l, \lambda(\cdot) \mid t, x) = \Phi(u^*(\cdot), l^0, \lambda^0(\cdot) \mid t, x)$$

is true.

*Then $u^*(s)$, $s \in T$, is the **optimal control** for the Primal Problem.*

Proof. The proof is similar to that of the previous section (see Theorem 1.5.2). □

An additional requirement for functions $\varphi_1(t, z)$, $\varphi(t, x)$, $\varphi_2(t, u)$ to be strictly convex in z, x and u results in the uniqueness of the optimal control.

Theorem 1.6.3. *Suppose under the conditions of Theorem 1.6.2 Assumption 1.6.4 is also true. Then solutions $u^0(s)$, and $\{l^0, \lambda^0(s)\}$ to both Primal and Dual Problems are unique.*

Exercise 1.6.5. Suppose that functions $\varphi_1(t, z)$, $\varphi(t, x)$, $\varphi_2(t, u)$ are *quadratic forms* in z and u:

$$\varphi_1(t, z) = \frac{1}{2} \langle z, M(t) z \rangle, \quad \varphi(t, x) = \frac{1}{2} \langle x - m, L(x - m) \rangle, \quad \varphi_2(t, u) = \frac{1}{2} \langle u, N(t) u \rangle,$$

with $z = Hx$, $M(t) = M'(t) > 0$, $N(t) = N'(t) > 0$, $L = L' > 0$, so that

$$\mathcal{J}(t, x) = \min_u \left\{ \int_t^\vartheta \frac{1}{2} (\langle H(s)x[s], M(s)H(s)x[s] \rangle + \langle u(s), N(s)u(s) \rangle) ds + \frac{1}{2} \langle x[\vartheta] - m, L(x[\vartheta] - m) \rangle \mid u(\cdot) \in \mathcal{L}_2[T] \right\}. \tag{1.88}$$

(i) Solve the **linear-quadratic problem** (1.88) by duality theory of this subsection.
(ii) Solve the same problem through methods of classical variational calculus.

1.6.2 General Remark: Feedforward Controls

Problem 1.5.1 of open-loop optimal control was solved in Sect. 1.5 for a fixed starting position $\{t_0 = t, x^0 = x\}$ with solution given as an optimal cost $\mathcal{J}(t_0, x^0)$ under control $u^0(\xi) = u^0(\xi \mid t_0, x^0)$, $\xi \in [t_0, \vartheta]$, in the class $u^0(\cdot) \in \mathcal{U}_{OO}$. Now, solving this problem of *feedforward control* for any interval $[t, \vartheta]$ and any starting position $\{t, x\}$, such solution may be used to create a *model-predictive control* $u_f(t, x) = u^0(t \mid t, x)$ which will be a function of $\{t, x\}$. However, such moves require a more general approach to variational problems of control that would indicate solutions of the same problem in the class $\mathcal{U}_{CC}$ of *closed-loop control strategies*. Such are the Dynamic Programming techniques discussed in Chap. 2.

Chapter 2
The Dynamic Programming Approach

Abstract This chapter describes general schemes of the Dynamic Programming approach. It introduces the notion of value function and its role in these schemes. They are dealt with under either classical conditions or directional differentiability of related functions, leaving more complicated cases to later chapters. Here the emphasis is on indicating solutions to forward and backward reachability problems for "linear-convex" systems and the design of closed-loop control strategies for optimal target and time-optimal feedback problems.

Keywords Dynamic programming • Value function • HJB equation • Reachability • Linear-convex systems • Colliding tubes • Terminal control • Time-optimal control

Dynamic Programming techniques provide a powerful conceptual and computational approach to a broad class of closed-loop optimal control problems. They are an application of the Hamiltonian formalism developed in calculus of variations and analytical dynamics [29, 33, 87, 269] to problems of feedback control using Hamilton–Jacobi–Bellman (HJB) equations also known as the Dynamic programming equations [16, 22, 24]. We start with the case when these partial Differential equations can be solved under either classical conditions that presume differentiability of value functions, or those where the value function is convex in the state variables and hence allows their directional differentiability.[1] The emphasis here is on linear-convex systems. For these systems it is not necessary to integrate the HJB equations since, as indicated here, the value function may be calculated through methods of convex analysis. These methods also apply to the calculation of support functions for the forward and backward reachability sets. The latter case is also useful for calculating feedback controls.

Note that Sects. 2.1, 2.3, 2.7, 2.8 of this chapter are applicable not only to linear systems but to nonlinear systems as well.

---

[1] The general nondifferentiable case for the value function is discussed later in Sect. 5.1.

2.1 The Dynamic Programming Equation

Consider system (1.1) expressed as

$$\dot{x} = f(t, x, u), \quad u \in \mathbf{U}(t), \quad t \in [\tau, \vartheta] = T_\tau, \; x \in \mathbb{R}^n, \; u \in \mathbb{R}^p \qquad (2.1)$$

with starting position $\{\tau, x\}$, $x(\tau) = x$. Here we have either $\mathbf{U}(t) = \mathbb{R}^p$ or $\mathbf{U}(t) = \mathcal{P}(t)$, where $\mathcal{P}(t)$ is compact-valued and Hausdorff-continuous.

The control will be sought for as either open-loop $u(t)$, $u(\cdot) \in \mathcal{U}_O$, or closed-loop $u(t, x)$, $u(\cdot, \cdot) \in \mathcal{U}_C$.

We wish to find the optimal control u that minimizes the Mayer–Bolza functional of the general form

$$\mathcal{J}(\tau, x \mid u(\cdot)) = \int_\tau^\vartheta L(t, x[t], u(t))\,dt + \varphi(\vartheta, x[\vartheta]). \qquad (2.2)$$

In (2.2) ϑ is a fixed terminal time, and $x[t] = x(t, \tau, x)$ is the trajectory of (2.1) starting from *initial* position $\{\tau, x\}$. It is customary to call $L(t, x, u)$ the *running cost* and $\varphi(\vartheta, x)$ the *terminal cost*. To begin, we suppose that the functions L, φ are differentiable and additional conditions are satisfied that justify the mathematical operations considered in this section. We will explicitly specify these conditions later.

Definition 2.1.1. The value function for the problem of minimizing (2.2) is defined as

$$V(\tau, x_\tau) = \inf_{u(\cdot)} \left(\int_\tau^\vartheta L(t, x[t], u(t))\,dt + \varphi(\vartheta, x[\vartheta]) \;\middle|\; u(\cdot) \in \mathbf{U}(\cdot), \; x[\tau] = x_\tau \right), \qquad (2.3)$$

where $\mathbf{U}(\cdot) = \{u(\cdot) : u(t) \in \mathbf{U}(t), \; t \in T_\tau\}$.

Thus $V(\tau, x_\tau)$ is the infimum of the cost (2.2) incurred by any feasible control $u(\cdot) \in \mathbf{U}(\cdot)$ and starting in state $x[\tau] = x_\tau$. For any $u(t) \in \mathbf{U}(t)$ and resulting trajectory $x[t] = x(t, \tau, x_\tau)$ we must have

$$V(\tau, x_\tau) \leq \int_\tau^\sigma L(t, x[t], u(t))\,dt + \left(\int_\sigma^\vartheta L(t, x[t], u(t))\,dt + \varphi(\vartheta, x[\vartheta]) \right),$$

2.1 The Dynamic Programming Equation

for $\tau < \sigma < \vartheta$. Minimizing the second term with respect to u gives

$$V(\tau, x_\tau)$$

$$\leq \int_\tau^\sigma L(t, x[t], u(t))dt$$

$$+ \inf \left\{ \int_\sigma^\vartheta L(t, x[t], u(t))dt + \varphi(\vartheta, x[\vartheta]) \, \bigg| \, u(\cdot) \in \mathbf{U}(\cdot), \; x[\sigma] = x(\sigma, \tau, x_\tau) \right\}$$

$$= \int_\tau^\sigma L(t, x[t], u(t))dt + V(\sigma, x(\sigma)). \tag{2.4}$$

On the other hand, since $V(\tau, x_\tau)$ is the infimum, for each $\varepsilon > 0$ there exists $\tilde{u}(\cdot)$ such that

$$V(\tau, x_\tau) + \varepsilon \geq \mathcal{J}(\tau, x \mid \tilde{u}(\cdot))$$

$$\geq \int_\tau^\sigma L(t, \tilde{x}[t], \tilde{u}(t))dt$$

$$+ \inf \left(\int_\sigma^\vartheta L(t, x[t], u(t))dt + \varphi(\vartheta, x(\vartheta)) \, \bigg| \, u(\cdot) \in \mathbf{U}(\cdot), \; \tilde{x}[\sigma] = \tilde{x}(\sigma, \tau, x_\tau) \right)$$

$$= \int_\tau^\sigma L(t, \tilde{x}[t], \tilde{u}(t))dt + V(\sigma, \tilde{x}[\sigma]), \tag{2.5}$$

with $\tilde{x}[t] = \tilde{x}(t, \tau, x_\tau)$ being the trajectory resulting from control $\tilde{u}(t)$. From (2.4) and (2.5)

$$V(\tau, x_\tau) \leq \int_\tau^\sigma L(t, \tilde{x}[t], \tilde{u}(t))dt + V(\sigma, \tilde{x}[\sigma]) \leq V(\tau, x_\tau) + \varepsilon.$$

Letting $\varepsilon \to 0$ leads to the *Dynamic Programming Equation*

$$V(\tau, x_\tau) = \inf \left(\int_\tau^\sigma L(t, x[t], u(t))dt + V(\sigma, x[\sigma]) \, \bigg| \, u(\cdot) \in \mathbf{U}(\cdot) \right). \tag{2.6}$$

Denote by $V(t, x \mid \vartheta, \varphi(\vartheta, \cdot))$ the value function for the problem to minimize (2.2) with terminal cost $\varphi(\vartheta, \cdot)$ (and the same running cost). With this notation, the value function (2.3) is $V(t, x \mid \vartheta, \varphi(\vartheta, \cdot))$ and the terminal cost serves as the *boundary condition*

$$V(\vartheta, x \mid \vartheta, \varphi(\vartheta, \cdot)) = \varphi(\vartheta, \cdot). \qquad (2.7)$$

Lastly, recognizing the right-hand side of (2.6) as the value function for the problem with terminal cost $V(\sigma, \cdot)$ allows us to express (2.6) as the *Principle of Optimality*:

$$V(\tau, x \mid \vartheta, V(\vartheta, \cdot)) = V(\tau, x \mid \sigma, V(\sigma, \cdot \mid \vartheta, V(\vartheta, \cdot))), \quad \tau \le \sigma \le \vartheta. \qquad (2.8)$$

Observe that (2.8) is the *semigroup* property for the mapping $V(\tau, \cdot) \mapsto V(\vartheta, \cdot)$. Substituting t for τ, $t + \sigma$ for σ and s for t, rewrite (2.6) as

$$V(t, x) = \inf \left(\int_t^{t+\sigma} L(s, x[s], u(s)) ds + V(t + \sigma, x[t + \sigma]) \,\Big|\, u(\cdot) \in \mathbf{U}(\cdot) \right). \qquad (2.9)$$

Here $\sigma \ge 0$ and $x[s] = x(s, t, x)$ is the trajectory of (2.1) starting from $\{t, x\}$. For $\sigma \ge 0$, (2.9) can be written as

$$\inf \left\{ \int_t^{t+\sigma} L(s, x[s], u(s)) ds + V(t + \sigma, x[\tau + \sigma]) - V(t, x) \,\Big|\, u(\cdot) \in \mathbf{U}(\cdot) \right\} = 0,$$

and so

$$\lim_{\sigma \to +0} \inf \left(\frac{1}{\sigma} \int_t^{t+\sigma} L(s, x[s], u(s)) ds + \frac{V(t + \sigma, x[t + \sigma]) - V(t, x)}{\sigma} \,\Big|\, u(\cdot) \in \mathbf{U}(\cdot) \right) = 0. \qquad (2.10)$$

Suppose now that $V(t, x)$ is continuously differentiable at (t, x). Then reversing the order of the operations lim and inf and passing to the limit with $\sigma \to +0$, we get

$$\inf_u \left(L(t, x[t], u) + \frac{d V(t, x[t])}{dt} \,\Big|\, u \in \mathbf{U}(t) \right) = 0. \qquad (2.11)$$

Expanding the total derivative $dV(t, x[t])/dt$ in terms of partial derivatives and noting that $\partial V(t, x[t])/\partial t$ does not depend on u, (2.11) may be rewritten as

$$\frac{\partial V(t, x)}{\partial t} + \inf \left\{ \left\langle \frac{\partial V(t, x)}{\partial x}, f(t, x, u) \right\rangle + L(t, x, u) \,\Big|\, u \in \mathbf{U}(t) \right\} = 0, \qquad (2.12)$$

2.1 The Dynamic Programming Equation

with boundary condition (2.7). This partial differential equation is known as the *(backward) Hamilton–Jacobi–Bellman equation* or simply the HJB equation. It is "backward" because the boundary condition (2.7) is at the terminal time; the "forward" HJB equation has boundary condition at the initial time.

Lemma 2.1.1. *If the value function $V(t,x)$ given by (2.3) is differentiable, it satisfies the HJB equation (2.12) with boundary condition (2.7).*

Exercise 2.1.1. Justify the reversal of lim and inf in (2.10).

Theorem 2.1.1. *(i) Suppose the value function $V(t,x)$ for problem (2.3) is differentiable at point $\{t,x\}$. Then it satisfies the inequality*

$$\frac{\partial V(t,x)}{\partial t} + \inf_w \left\{ \left\langle \frac{\partial V(t,x)}{\partial x}, f(t,x,w) \right\rangle + L(t,x,w) \mid w \in \mathbf{U}[t] \right\} \geq 0. \tag{2.13}$$

(ii) If in addition to (i) there exists an optimal control $u^0(s) \in \mathbf{U}(s)$, such that $u^0(s) \to w^0 \in U(t)$ as $s \to t+0$, then

$$\frac{\partial V}{\partial t} + \left\langle \frac{\partial V(t,x)}{\partial x}, f(t,x,w^0) \right\rangle + L(t,x,w^0) = 0. \tag{2.14}$$

Theorem 2.1.2 (The First Verification Theorem). *Suppose a differentiable function $\omega(t,x)$ satisfies the HJB equation (2.12) in domain $\mathcal{D} = T_\tau \times \mathbb{R}^n$ together with boundary condition (2.7). Then in this domain $\omega(t,x) \leq V(t,x)$. Moreover, if control $u^0(\cdot)$, with trajectory $x[s] = x^0(s,t,x)$, $s \geq t$, $x^0[t] = x$, satisfies*

$$L(s,x^0[s],u^0(s)) + \left\langle \frac{\partial \omega(s,x^0[s])}{\partial x}, f(s,x^0[s],u^0(s)) \right\rangle =$$

$$= \min \left\{ L(s,x^0[s],u) + \left\langle \frac{\partial \omega(s,x^0[s],u)}{\partial x}, f(s,x^0[s],u) \right\rangle \mid u \in \mathbf{U}(s) \right\}, \tag{2.15}$$

then $u^0(\cdot)$ is optimal and $\omega(t,x) = V(t,x)$.

Proof. Let $u(\cdot) \in \mathbf{U}(\cdot)$ be any control and $x[s] = x(s,t,x)$ the resulting trajectory starting at $x[t] = x$. From (2.12),

$$\omega(\vartheta, x[\vartheta]) - \omega(t,x) = \int_t^\vartheta \left(\frac{\partial \omega(s,x(s))}{\partial s} + \left\langle f(s,x(s),u(s)), \frac{\partial \omega(s,x(s))}{\partial x} \right\rangle \right) ds \geq$$

$$\geq -\int_t^\vartheta L(s,x(s),u(s))ds.$$

Since $\omega(\vartheta, x(\vartheta)) = \varphi(\vartheta, x[\vartheta])$, moving the integral to the left side of the last relation gives

$$\omega(t, x) \leq \int_t^\vartheta L(s, x(s), u(s)) ds + \varphi(\vartheta, x[\vartheta]) = \mathcal{J}(t, x \mid u(\cdot)).$$

Hence $\omega(t, x) \leq V(t, x)$.

Repeating this calculation for $u = u^0(\cdot)$, but using (2.15) instead of (2.12), yields $\omega(t, x) = \mathcal{J}(t, x \mid u^0(\cdot))$, which implies that $\omega(t, x) = V(t, x)$ and $u^0(\cdot)$ is optimal. □

Definition 2.1.2. A control in the form $u(t, x)$ is called a **closed-loop** or **feedback** control. It is sometimes known as **positional** control. A control $u = u(t)$ that is a function only of time is an **open-loop** control.

Corollary 2.1.1. *The optimal feedback (closed-loop) control for problem (2.3) is given by*

$$u^0(t, x) \in \arg\min \left\{ \left\langle \frac{\partial V}{\partial x}, f(t, x, u) \right\rangle + L(t, x, u) \,\bigg|\, u \in \mathbf{U}(t) \right\}. \tag{2.16}$$

Remark 2.1.1. Formally, given a closed-loop control $u^0(t, x)$, one may find the corresponding open-loop control $u^0 = u^0[s]$ by first obtaining the solution $x[s] = x(s, t, x)$ of the differential equation

$$\dot{x} = f(s, x, u^0(s, x)), \quad x[t] = x, \tag{2.17}$$

and then substituting to get $u^0[s] = u^0(s, x[s])$. However, one **must ensure that (2.17) has a solution** in some reasonable sense, particularly, when **the feedback control $u^0(s, x)$ is nonlinear** (even discontinuous) in x, **but the original system is linear in x**.

Consider now the problem of finding u that minimizes the *inverse* Mayer–Bolza functional

$$\mathcal{J}_0(t, x \mid u(\cdot)) = \varphi(t_0, x(t_0)) + \int_{t_0}^t L(s, x(s), u(s)) ds, \tag{2.18}$$

with given *final* position $\{t, x\}$, $x[t] = x$, and initial time $t_0 \leq t$. The value function for this problem is

$$V_0(t, x) = \inf_{u(\cdot)} \{ \mathcal{J}_0(t, x \mid u(\cdot)) \mid u(\cdot) \in \mathbf{U}(\cdot), \, x(t) = x \}. \tag{2.19}$$

2.1 The Dynamic Programming Equation

Mimicking the previous argument now leads to the Dynamic Programming Equation

$$V_0(t,x) = \inf\{V_0(t-\sigma, x[t-\sigma]) + \int_{t-\sigma}^{t} L(s, x[s], u(s))ds \mid u(\cdot) \in \mathbf{U}(\cdot), \ x(t) = x\},$$

from which, supposing $V_0(t,x)$ is differentiable, one obtains the *forward* HJB equation

$$\frac{\partial V_0}{\partial t}(t,x) + \sup\left\{\left\langle \frac{\partial V_0(t,x)}{\partial x}, f(t,x,u)\right\rangle - L(t,x,u) \,\Big|\, u \in \mathbf{U}(t)\right\} = 0, \quad (2.20)$$

with boundary condition

$$V_0(t_0, x) = \varphi(t_0, x). \quad (2.21)$$

The *Principle of Optimality* in forward time now has the form

$$V_0(t, x \mid t_0, V_0(t_0, \cdot)) = V_0(t, x \mid \tau, V_0(\tau, \cdot \mid t_0, V_0(t_0, \cdot))), \quad t_0 \leq \tau \leq t. \quad (2.22)$$

Exercise 2.1.2. Derive the Dynamic Programming Equation in forward time and then the forward HJB equation (2.20).

Theorem 2.1.3 (The Second Verification Theorem). *Suppose a differentiable function $\omega_0(t,x)$ satisfies the HJB equation (2.20) in domain $\mathcal{D} = T_{t_0} \times \mathbb{R}^n$, together with boundary condition (2.21). Then in this domain $\omega_0(t,x) \leq V_0(t,x)$. Moreover, if control $u^0(\cdot)$, with trajectory $x^0[s] = x^0(s,t,x)$, $x^0[t] = x$, $s \leq t$, satisfies*

$$-L(s, x^0[s], u^0(s)) + \left\langle \frac{\partial \omega_0}{\partial x}(s, x^0[s]), f(s, x^0[s], u^0(s))\right\rangle =$$

$$= \max\left\{-L(s, x^0[s], u) + \left\langle \frac{\partial \omega_0}{\partial x}(s, x^0[s], u), f(s, x^0[s], u)\right\rangle \,\Big|\, u \in \mathbf{U}(s)\right\}, \quad (2.23)$$

then $u^0(\cdot)$ is optimal and $\omega_0(t,x) = V_0(t,x)$.

Exercise 2.1.3. Prove Theorem 2.1.3.

Corollary 2.1.2. *The optimal feedback (closed-loop) control for problem (2.19) is*

$$u^0(t,x) \in \operatorname{Arg\,max}\left\{\left\langle \frac{\partial V_0}{\partial x}, f(t,x,u)\right\rangle - L(t,x,u) \,\Big|\, u \in \mathbf{U}(t)\right\}. \quad (2.24)$$

Remark 2.1.2. The optimal closed-loop control $u^0(t,x)$ found here is feasible only when accompanied by an existence theorem for the equation

$$\dot{x} = f(t, x, u^0(t, x)).$$

We shall now use the Dynamic Programming Verification Theorem to minimize a quadratic cost for a linear system.

2.2 The Linear-Quadratic Problem

Minimizing a quadratic integral cost over the trajectories of a linear control system is the basic problem in all courses on linear control. It has a solution in explicit form, fairly easily obtained and smooth enough to satisfy all requirements for a classical solution to the HJB equation of the previous section. This problem has a vast literature, but it is not the subject of this book, and is presented here as a necessary transitionary passage to main topics.

Consider the linear system (1.2) or (1.44), namely

$$\dot{x} = A(t)x + B(t)u, \quad t_0 \leq t \leq t_1, \quad x(t_0) = x_0, \qquad (2.25)$$

with control $u(t) \in \mathbb{R}^p$ and the quadratic integral cost functional

$$J(t_0, x_0, u(\cdot)) = \int_{t_0}^{t_1} \Big(\langle x, M(t)x \rangle + \langle u, N(t)u \rangle \Big) dt + \langle x(t_1), Tx(t_1) \rangle, \qquad (2.26)$$

in which $M(t) = M'(t) \geq 0$, $N(t) = N'(t) > 0$, are continuous in $t \in [t_0, t_1]$ and $T = T' > 0$.

Problem 2.2.1 (Linear-Quadratic Control Problem). Find a control $u(\cdot)$ that minimizes the cost (2.26).

Introduce the value function

$$V(t, x) = \min_{u(\cdot)} \{ J(t, x, u(\cdot)) \mid x(t) = x \}. \qquad (2.27)$$

Its backward HJB equation is (see (2.12))

$$\frac{\partial V}{\partial t} + \min_u \left\{ \left\langle \frac{\partial V}{\partial x}, A(t)x + B(t)u \right\rangle + \langle x, M(t)x \rangle + \langle u, N(t)u \rangle \right\} = 0, \qquad (2.28)$$

with boundary condition

$$V(t_1, x) = \langle x, Tx \rangle. \qquad (2.29)$$

2.2 The Linear-Quadratic Problem

The internal minimization problem in (2.28) is to find

$$u(t,x) = \arg\min\left\{\left\langle\frac{\partial V}{\partial x}, A(t)x + B(t)u\right\rangle + \langle u, N(t)u\rangle \,\Big|\, u \in \mathbb{R}^p\right\}.$$

Solving this minimization by differentiating with respect to u and equating the derivative to zero gives

$$u(t,x) = -\frac{1}{2}N^{-1}(t)B'(t)\frac{\partial V}{\partial x}. \qquad (2.30)$$

Substituting (2.30) into (2.28), we come to

$$\frac{\partial V}{\partial t} + \left\langle\frac{\partial V}{\partial x}, Ax\right\rangle - \frac{1}{4}\left\langle\frac{\partial V}{\partial x}, BN^{-1}B'\frac{\partial V}{\partial x}\right\rangle + \langle x, Mx\rangle = 0. \qquad (2.31)$$

We follow Theorem 2.1.2 and look for $V(t,x)$ as a quadratic form $V(t,x) = \langle x, P(t)x\rangle$, with $P(t) = P'(t)$. Then

$$\frac{\partial V}{\partial t} = \langle x, \dot{P}(t)x\rangle, \qquad \frac{\partial V}{\partial x} = 2P(t)x.$$

Substituting these into (2.31) gives

$$\langle x, \dot{P}(t)x\rangle + 2\langle P(t)x, A(t)x\rangle - \langle P(t)x, BN^{-1}B'P(t)x\rangle + \langle x, Mx\rangle = 0,$$

and so we arrive at the (backward) matrix differential equation of the *Riccati type*

$$\dot{P} + PA + A'P - P'BN^{-1}B'P + M = 0, \qquad (2.32)$$

with terminal matrix boundary condition

$$P(t_1) = T. \qquad (2.33)$$

Consider the special case of Eq. (2.32) with $M(t) \equiv 0$. Differentiating the identity $I = P(t)P^{-1}(t)$ gives $\dot{P}P^{-1} + P\dot{P}^{-1} = 0$ or

$$\dot{P}^{-1} = -P^{-1}\dot{P}P^{-1}.$$

Multiplying both sides of Eq. (2.32) by P^{-1}, and using the last relation, we have

$$\dot{P}^{-1} = AP^{-1} + P^{-1}A' - BN^{-1}B', \qquad P^{-1}(t_1) = T^{-1}. \qquad (2.34)$$

The solution of this *linear* matrix equation is given by the integral formula

$$P^{-1}(t) = \mathbf{G}(t,t_1)T^{-1}\mathbf{G}'(t,t_1) + \int_t^{t_1} \mathbf{G}(t,s)B(s)N^{-1}(s)B'(s)\mathbf{G}'(t,s)ds, \quad (2.35)$$

in which $\mathbf{G}(t,s)$ is defined in (1.11), (1.12).

Theorem 2.2.1. *The value function for Problem 2.2.1 has the quadratic form $V(t,x) = \langle x, P(t)x \rangle$, in which $P(t)$ satisfies the matrix Riccati equation (2.32), with boundary condition (2.33). The optimal control, given by the linear feedback function (2.30), is*

$$u(t,x) = -N^{-1}(t)B'(t)P(t)x.$$

If, furthermore, $M(t) \equiv 0$, then $P^{-1}(t)$ is given by formula (2.35).

The next problem is similar to Problem 2.2.1 and involves the quadratic cost functional

$$J_0(t_1, x_1, u(\cdot)) = \langle x(t_0), Lx(t_0) \rangle + \int_{t_0}^{t_1} \left(\langle x, M(t)x \rangle + \langle u, N(t)u \rangle \right) dt, \quad (2.36)$$

with final condition $x(t_1) = x^{(1)}$. Here $L = L' > 0$.

Problem 2.2.2 (Inverse Linear-Quadratic Control Problem). Find a control $u(\cdot)$ that minimizes the cost (2.36).

The value function for Problem 2.2.2 is

$$V_0(t,x) = \min_{u(\cdot)} \{ J_0(t, x, u(\cdot)) \mid x(t) = x \}. \quad (2.37)$$

The corresponding forward HJB equation is

$$\frac{\partial V_0}{\partial t} + \max_u \left\{ \left\langle \frac{\partial V_0}{\partial x}, A(t)x + B(t)u \right\rangle - \langle x, M(t)x \rangle - \langle u, N(t)u \rangle \right\} = 0, \quad (2.38)$$

with boundary condition

$$V_0(t_0, x) = \langle x, Lx \rangle. \quad (2.39)$$

The internal maximization in (2.38) gives

$$u(t,x) = \frac{1}{2} N^{-1}(t) B'(t) \frac{\partial V_0}{\partial x}, \quad (2.40)$$

2.2 The Linear-Quadratic Problem

so that $V_0(t, x)$ satisfies the relation

$$\frac{\partial V_0}{\partial t} + \left\langle \frac{\partial V_0}{\partial x}, A(t)x \right\rangle + \frac{1}{4}\left\langle \frac{\partial V_0}{\partial x}, B(t)(N(t))^{-1}B'(t)\frac{\partial V_0}{\partial x} \right\rangle - \langle x, M(t)x \rangle = 0, \tag{2.41}$$

with $V_0(t_0, x) = \langle x, Lx \rangle$. As before, assuming $V_0(t, x) = \langle x, P_0(t)x \rangle$ and repeating the type of substitutions given above, we arrive at the (forward) *Riccati equation*

$$\dot{P}_0 + P_0 A + A' P_0 + P_0 B N^{-1} B' P_0 - M = 0, \quad P_0(t_0) = L. \tag{2.42}$$

The solution to Problem 2.2.2 may be used to determine the reach set $X[t]$ of system (2.25) under the integral constraint

$$J_0(t, x(t), u(\cdot)) \leq \mu.$$

In fact,

$$X[t] = \{x : V_0(t, x) \leq \mu\} = \{x : \langle x, P_0(t)x \rangle \leq \mu\}. \tag{2.43}$$

Exercise 2.2.1. Prove (2.43).

Exercise 2.2.2. Solve Problem 2.2.1 for the cost

$$J(t_0, x_0, u(\cdot)) =$$

$$= \int_{t_0}^{t_1} \left(\langle x - x^*(t), M(t)(x - x^*(t)) \rangle + \langle u, N(t)u \rangle \right) dt + \langle x(t_1) - a, T(x(t_1) - a) \rangle, \, x(t_0) = x^0, \tag{2.44}$$

in which $T = T' > 0$, $M(t) = M'(t) \geq 0$, $N(t) = N'(t) > 0$ are continuous in $t \in [t_0, t_1]$, and function $x^*(\cdot)$ and vector a are given.

Exercise 2.2.3. Solve Problem 2.2.2 for the cost

$$J_0(t_1, x_1, u(\cdot)) = \max\{\langle x(t_0), Lx(t_0)\rangle, \int_{t_0}^{t_1} \langle u, N(t)u \rangle dt\}, \quad x(t_1) = x^{(1)}. \tag{2.45}$$

We shall now use the Dynamic Programming approach to the calculation of reach sets under hard bounds on the controls.

2.3 Reachability Through the HJB Equation: Hard Bounds

For problems of reachability and closed-loop control we derive the main Dynamic Programming equation which is a first-order partial differential equation with appropriate boundary conditions.

2.3.1 Forward Reachability

We extend Definition 1.4.1 of reach sets to the nonlinear system (2.1), under the hard bound $u \in \mathcal{P}(t)$.

Definition 2.3.1. The **reach set** $X[\vartheta] = X(\vartheta, t_0, x^0)$ of system (2.1) **at given time** ϑ, **from position** $\{t_0, x^0\}$, is the set of **all points** x, for each of which there exists a trajectory $x[s] = x(s, t_0, x^0)$, generated by a control subject to the given constraint, that transfers the system from position $\{t_0, x^0\}$ to position $\{\vartheta, x\}$, $x = x[\vartheta]$:

$$X(\vartheta, t_0, x^0) = \{x : \exists u(\cdot) \in \mathcal{P}(\cdot) \text{ such that } x(\vartheta, t_0, x^0) = x\}.$$

The reach set $X[\vartheta] = X(\vartheta, t_0, \mathcal{X}^0)$ **from set-valued position** $\{t_0, \mathcal{X}^0\}$ **at given time** ϑ is the union

$$X(\vartheta, t_0, \mathcal{X}^0) = \bigcup \{X(\vartheta, t_0, x^0) \mid x^0 \in \mathcal{X}^0\}.$$

The set-valued function $X[t] = X(t, t_0, \mathcal{X}^0), t_0 \leq t \leq \vartheta$, is the **reach tube** from $\{t_0, \mathcal{X}^0\}$.

This definition of reach sets does not involve an optimization problem. However, we shall relate it to a problem of optimization that can be solved by Hamiltonian methods in their Dynamic Programming version. We characterize the (forward) reach set by the *value function*

$$V_0(\tau, x) = \min_{u(\cdot), x(t_0)} \{d^2(x(t_0), \mathcal{X}^0) \mid x(\tau) = x\} = \min_{x(t_0)} \{d^2(x(t_0), \mathcal{X}^0) \mid x \in X(\tau, t_0, x(t_0))\} \quad (2.46)$$

over all measurable functions $u(t)$ that satisfy (1.18). See Fig. 2.1. Above,

$$d^2(x, X) = \min\{\langle x - z, x - z\rangle \mid z \in X\}$$

is the square of the distance $d(x, X)$ from point x to set X. Observe that (2.46) is a special case of (2.19) with zero running cost in (2.18).

2.3 Reachability Through the HJB Equation: Hard Bounds

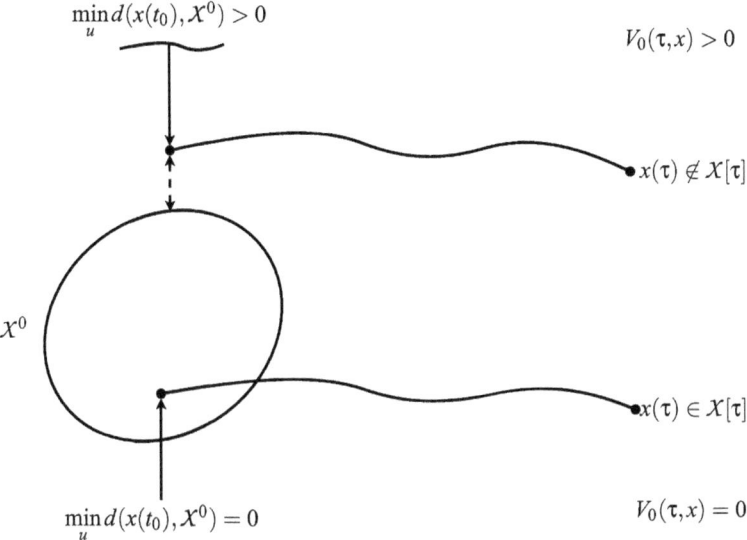

Fig. 2.1 The value function $V_0(\tau, x)$ in (2.46) with $x(t_0) \notin \mathcal{X}^0$

Lemma 2.3.1. *The following relation holds:*

$$\mathcal{X}(\tau, t_0, \mathcal{X}^0) = \{x : V_0(\tau, x) \leq 0\}. \tag{2.47}$$

As seen from Fig. 2.1, (2.47) follows from the definition of the reach set $\mathcal{X}(\tau, t_0, \mathcal{X}^0)$, which thus turns out to be the *level set* (at level zero) of the function $V_0(\tau, \cdot)$.

For the value function $V_0(t, x)$ we again use the notation

$$V_0(t, x) = V_0(t, x \mid t_0, V_0(t_0, \cdot)),$$

to emphasize its dependence on the boundary condition $V_0(t_0, x) = d^2(x, \mathcal{X}^0)$. The next assertion merely restates (2.22).

Theorem 2.3.1. *The value function $V_0(t, x)$ satisfies the Principle of Optimality in forward time and has the semigroup form*

$$V_0(t, x \mid t_0, V_0(t_0, \cdot)) = V_0(t, x \mid \tau, V_0(\tau, \cdot \mid t_0, V_0(t_0, \cdot))), \quad t_0 \leq \tau \leq t. \tag{2.48}$$

This theorem together with Lemma 2.3.1 also implies the following assertion.

Lemma 2.3.2. *The reachability set $\mathcal{X}[t] = \mathcal{X}(t, t_0, \mathcal{X}^0)$ satisfies the next relation*

$$\mathcal{X}[t] = \mathcal{X}(t, \tau, \mathcal{X}(\tau, t_0, \mathcal{X}^0)). \tag{2.49}$$

The solution of the reachability problem, namely the one of calculating the reach set, now depends on the properties of either the "classical" or the generalized "viscosity" solutions of the forward HJB equation

$$\frac{\partial V_0(t,x)}{\partial t} + \max_u \left\{ \left\langle \frac{\partial V_0(t,x)}{\partial x}, f(t,x,u) \right\rangle \mid u \in \mathcal{P}(t) \right\} = 0, \qquad (2.50)$$

with boundary condition

$$V_0(t_0, x) = d^2(x, \mathcal{X}^0). \qquad (2.51)$$

Equation (2.50) follows from (2.48) or it may be seen as a special case of (2.20). This equation is solvable in the classical sense (that is, (2.50) holds everywhere), if the partials of $V(t,x)$ in t,x exist and are continuous. Otherwise (2.50) is a symbolic relation for the generalized HJB equation which has to be described in terms of subdifferentials, Dini derivatives or their equivalents. However, the typical situation is that V *is not differentiable*. The treatment of Eq. (2.50) then involves the notion of generalized "viscosity" or "minmax" solution for this equation (see [16, 50, 247]). The last notion will be later treated in a separate Sect. 5.1 which deals with nondifferentiable value functions and generalized solutions. In the case of linear systems with convex constraints the value functions are convex in the state space variables and hence directionally differentiable these solutions belong the generalized, which property is checked directly.

The calculation of reach sets and tubes can thus be reduced to the calculation of the value function $V_0(t,x)$ or its level sets $\mathcal{X}[t] = \{x : V_0(t,x) \leq 0\}$ and their evolution in time t. On the other hand, we note that $\mathcal{X}[t]$ may also be treated as the cross-section ("cut") $\mathcal{X}[t] = \mathcal{X}(t, t_0, \mathcal{X}^0)$ of the solution tube $\mathcal{X}(\cdot, t_0, \mathcal{X}^0)$ to the differential inclusion

$$\dot{x} \in F(t,x), \quad t \geq t_0, \quad x^0 \in \mathcal{X}^0, \qquad (2.52)$$

in which $F(t,x) = \cup \{f(t,x,u) \mid u \in \mathcal{P}(t)\}$.

A set-valued function $\mathcal{X}[t]$ is a solution to Eq. (2.52) on the interval $T = [t_0, \tau]$ if it satisfies this equation for almost all $t \in T$. Under the previous assumptions on f and $\mathcal{P}$ there is a unique solution $\mathcal{X}[\cdot] = \mathcal{X}(\cdot, t_0, \mathcal{X}^0)$ issuing from $\mathcal{X}^0$ for any time interval $[t_0, \tau]$. The solution $\mathcal{X}[\cdot]$ may also be based on using the "funnel equation" described in the next lines (see also Sect. 8.1 and [158]).

Assumption 2.3.1. $F(t,x)$ *is convex compact-valued* ($F : [t_0, t_1] \times \mathbb{R}^n \to \text{conv} \mathbb{R}^n$), *continuous in* t, x *in the Hausdorff metric and Lipschitz-continuous in* x:

$$h(F(t,x), F(t,y)) \leq L\langle x-y, x-y \rangle^{1/2}, \quad L < \infty. \qquad (2.53)$$

2.3 Reachability Through the HJB Equation: Hard Bounds

Here h is the *Hausdorff distance*

$$h(Q, M) = \max\{h_+(Q, M), h_+(M, Q)\},$$

and the Hausdorff *semidistance* between sets Q, M is

$$h_+(Q, M) = \max_x \min_z \{\langle x - z, x - z \rangle^{1/2} \mid x \in Q, z \in M\}.$$

The Hausdorff distance is *a metric* in the space of compact sets.
The "ordinary distance" $d(Q, M)$ between sets Q, M, is

$$d(Q, M) = \min_x \min_z \{\langle x - z, x - z \rangle^{1/2} \mid x \in Q, z \in M\}.$$

Theorem 2.3.2. *Under Assumption 2.3.1 the forward reach set $X[t]$ is compact and the set-valued tube $X[\cdot]$ is the unique solution to the following "funnel" equation for the differential inclusion (2.52):*

$$\lim_{\sigma \to +0} \sigma^{-1} h(X[t + \sigma], \cup\{(x + \sigma F(t, x)) \mid x \in X[t]\}) = 0, \qquad (2.54)$$

with boundary condition $X[t_0] = X^0$.

Theorem 2.3.2 was introduced in [75, 224], see also [26, 158]. Note that the set-valued function $X[t]$ of the last theorem is continuous in the Hausdorff metric. Thus it is possible to single out two alternative approaches to the treatment of reachability: namely, through the solution of the HJB equation (2.50) (the calculation of the value function $V_0(t, x)$ and its level sets) or through the calculation of solution tubes $X[t]$ to the differential inclusion (2.52) by dealing, for example, with the funnel equation (2.54). In either approach, the calculation of reach sets is solved in *forward* time. A reciprocal approach considers reach sets in *backward (reverse)* time.

2.3.2 Backward Reachability or the Solvability Problem

Whereas the reach set is the set of all states that can be reached from a given initial set, the backward reach set comprises all those states, taken at instant τ, *from which* it is possible to reach a given "target" set M taken to be a convex compact in $\mathbb{R}^n$.

Definition 2.3.2. Given a closed target set $M \subseteq \mathbb{R}^n$, the **backward reach (solvability) set** $W[\tau] = W(\tau, t_1, M)$ **at given time** τ, **from set-valued position** $\{t_1, M\}$, is the set of states $x \in \mathbb{R}^n$ for each of which there exists a control $u(t) \in \mathcal{P}(t)$ that steers system (2.1) from state $x(\tau) = x$ to $x(t_1) \in M$. The set-valued function $W[t] = W(t, t_1, M), \tau \leq t \leq t_1$, is **the solvability (or backward reach) tube** from $\{t_1, M\}$.

Backward reach sets $W[\tau]$ are also known as **weakly invariant sets** relative to position $\{t_1, M\}$.

Definition 2.3.3. A set $W[\tau]$ is said to be **weakly invariant** relative to system (2.1) [(1.2), (1.7)] and position $\{t_1, M\}$, if it consists of all those points $x = x(\tau)$ (positions $\{\tau, x\}$), from which set M is reachable at time t_1 **by some** of the possible controls.

Hence, from each position $\{\tau, x\} \in W[\tau]$ there exists at least one trajectory that stays in the backward reach tube and reaches M at $t = t_1$. The backward reach set $W(\tau, t_1, M)$ is the largest or, in other words, *the inclusion maximal* weakly invariant set relative to M.

The calculation of backward reach sets may be achieved through the value function

$$V(\tau, x) = \min_u \{d^2(x(t_1), M) \mid x(\tau) = x\},$$

$$V(\tau, x) = V(\tau, x \mid t_1, V(t_1, \cdot)), \quad V(t_1, x) = d^2(x, M), \tag{2.55}$$

in which the minimization is over all $u(\cdot) \in \mathcal{P}(\cdot)$. Similarly to Lemma 2.3.1, the backward reach set is a level set.

Lemma 2.3.3. *The following relation is true:*

$$W(\tau, t_1, M) = \{x : V(\tau, x) \leq 0\}, \tag{2.56}$$

The next result is the analog of Theorem 2.3.1.

Theorem 2.3.3. *The set-valued mapping $W(t, t_1, M)$ satisfies the semigroup property*

$$W(t, t_1, M) = W(t, \tau, W(\tau, t_1, M)), \quad t \leq \tau \leq t_1. \tag{2.57}$$

*(in backward time) and the value function $V(t, x | t_1, V(t_1, \cdot))$ satisfies the **Principle of Optimality** in the semigroup form*

$$V(t, x \mid t_1, V(t_1, \cdot)) = V(t, x \mid \tau, V(\tau, \cdot \mid t_1, V(t_1, \cdot))), \quad t \leq \tau \leq t_1. \tag{2.58}$$

Associated with the Principle of Optimality (2.58) is the backward HJB equation

$$\frac{\partial V(t, x)}{\partial t} + \min_u \left\{ \left\langle \frac{\partial V}{\partial x}, f(t, x, u) \right\rangle \right\} = 0, \tag{2.59}$$

with boundary condition

$$d^2(x, M) = V(t_1, x). \tag{2.60}$$

2.3 Reachability Through the HJB Equation: Hard Bounds

Associated with the semigroup (2.57) is the "backward" funnel equation for the differential inclusion (2.52):

$$\lim_{\sigma \to +0} \sigma^{-1} h(W[t - \sigma], \cup\{(x - \sigma F(t, x)) | x \in W[t]\}) = 0, \quad (2.61)$$

with boundary condition $W[t_1] = \mathcal{M}$.

Lemma 2.3.4. *The set-valued function $W[t] = W(t, t_1, \mathcal{M})$, $W[t_1] = \mathcal{M}$, is the unique solution to Eq. (2.61).*

Remark 2.3.1. The above assertions emphasize *two ways* of describing reachability sets and tubes, namely *through set-valued functions* like $W[t]$ or *through "ordinary" functions* like $V(t, x)$. These are connected through (2.56).

Finally we formulate an important property of reach sets for our systems in the absence unknown disturbances ($v = 0$).[2] This is when the forward and backward **reach sets** $X(\tau, t_0, X^0)$, $W(\tau, t_1, \mathcal{M})$ have the next connection.

Theorem 2.3.4. *Suppose $X(t_1, t_0, X^0) = X[t_1] = \mathcal{M}$. Then $W[t_0] = W(t_0, t_1, \mathcal{M}) = X^0$.*

Each of these sets $X(\tau, t_0, X^0)$, $W(\tau, t_1, \mathcal{M})$ does not depend on whether it is calculated in the class of open-loop controls $\mathcal{U}_O$ or closed-loop controls $\mathcal{U}_C$.

Exercise 2.3.1. Prove Theorem 2.3.4.

Strongly Invariant Sets. In contrast with the weakly invariant backward reach sets $W[\tau]$ from which the target $\mathcal{M}$ is reachable at time t_1 with *some* control we now introduce the notion of strongly invariant sets.

Definition 2.3.4. *A set $W_s[\tau]$ is **strongly invariant** relative to system (2.10) and position $\{t_1, \mathcal{M}\}$, if from each of its points $x = x(\tau)$ set $\mathcal{M}$ is reachable at time t_1 **by all** possible controls.*

*A set $W_s[\tau]$ that consists of all such points $x = x(\tau)$ is said to be the **inclusion-maximal** strongly invariant set relative to position $\{t_1, \mathcal{M}\}$.*

The strongly invariant set relative to $\mathcal{M}$ is given by the value function

$$V_s(\tau, x) = \max_u \{d^2(x(t_1), \mathcal{M}) \mid x(\tau) = x\}, \quad (2.62)$$

in which the maximization is over all $u(\cdot) \in \mathcal{P}(\cdot)$, $u(\cdot) \in \mathcal{U}_O(\cdot)$.

Consider the backward HJB equation

$$\frac{\partial V_s(t, x)}{\partial t} + \max_u \left\{ \left\langle \frac{\partial V_s}{\partial x}, f(t, x, u) \right\rangle \right\} = 0, \quad (2.63)$$

---

[2] Problems of reachability under unknown but bounded disturbances are beyond the scope of this book, along the approaches of which they are treated in papers [133, 176, 183].

with boundary condition

$$V_s(t_1, x) = d^2(x, \mathcal{M}). \tag{2.64}$$

Theorem 2.3.5. *If Eq. (2.64) has a unique (viscosity or classical) solution $V_s(t, x)$, the strongly invariant set $\mathcal{W}_s[\tau] = \mathcal{W}_s(\tau, t_1, \mathcal{M})$ is the level set*

$$\mathcal{W}_s[\tau] = \{x : V_s(\tau, x) \leq 0\}.$$

The next result is obvious.

Lemma 2.3.5. *The set $\mathcal{W}_s[\tau] \neq \emptyset$ iff there exists a point x such that the reach set $X(t_1, \tau, x) \subseteq \mathcal{M}$.*

Exercise 2.3.2. Does the mapping $\mathcal{W}_s(\tau, t_1, \mathcal{M})$ satisfy a semigroup property similar to (2.57)?

Solutions of the HJB equations above require rather subtle analytical or numerical techniques. Some solutions may however be obtained through methods of convex analysis. In particular, this is the case for *linear-convex systems,* which are when the system is linear in x, u and the constraints on u, x^0 are convex. These will be studied next.

2.4 Reachability for Linear-Convex Systems

In this section we calculate the value functions through methods of convex analysis and also find the support functions for the forward and backward reachability sets.

2.4.1 Forward Reachability: Calculating the Value Functions

Consider system (2.25), namely,

$$\dot{x} = A(t)x + B(t)u, \tag{2.65}$$

with continuous matrix coefficients $A(t), B(t)$ and hard bound $u(t) \in \mathcal{P}(t)$, with $\mathcal{P}(t)$ convex, compact, and Hausdorff-continuous.

The Value Function (Forward)

Problem 2.4.1. Given starting set-valued position $\{t_0, \mathcal{X}^0\}$, terminal time ϑ and condition $x(\vartheta) = x$, calculate the value function $V_0(\vartheta, x)$ of (2.46) at any position $\{\vartheta, x\}$.

2.4 Reachability for Linear-Convex Systems

After solving this problem we may use (2.47) to find the forward reach set $X[t] = X(t, t_0, X^0)$ as the level set $X[t] = \{x : V_0(\vartheta, x) \le 0\}$.

The HJB equation (2.50) now has the form

$$\frac{\partial V_0(t,x)}{\partial t} + \max\left\{\left\langle \frac{\partial V_0(t,x)}{\partial x}, A(t)x + B(t)u \right\rangle \middle| u \in \mathcal{P}(t)\right\} = 0,$$

or, since

$$\max\{\langle l, u \rangle \mid u \in \mathcal{P}(t)\} = \rho(l \mid \mathcal{P})$$

is the support function of set $\mathcal{P}(t)$, we may write

$$\frac{\partial V_0(t,x)}{\partial t} + \left\langle \frac{\partial V_0(t,x)}{\partial x}, A(t)x \right\rangle + \rho\left(\frac{\partial V_0(t,x)}{\partial x} \,\middle|\, B(t)\mathcal{P}(t)\right) = 0. \tag{2.66}$$

The boundary condition for (2.66) is

$$V_0(t_0, x) = d^2(x, X^0) = \min_{q}\{\langle x - q, x - q \rangle \mid q \in X^0\}. \tag{2.67}$$

Instead of solving the HJB equation, $V_0(\tau, x)$ may be calculated through duality techniques of convex analysis along the scheme given, for example, in [174, Sect. 1.5].

Indeed, observing that the Fenchel conjugate for function $\varphi(\tau, x) = d^2(x, X^0)$ in the second variable is $\varphi^*(\tau, l) = \rho(l \mid X^0) + \frac{1}{4}\langle l, l \rangle$, we may write, making use of the minmax theorem of [72] to interchange min and max below,

$$V_0(\tau, x) = \min_{u}\{d^2(x(t_0), X^0) \mid x(\tau) = x\}$$

$$= \min_{u}\max_{l}\{\langle l, x(t_0)\rangle - \frac{1}{4}\langle l, l\rangle - \rho(l \mid X^0) \mid x(\tau) = x\}$$

$$= \max_{l}\min_{u} \Psi(\tau, x, l, u(\cdot)), \tag{2.68}$$

where

$$\Psi(\tau, x, l, u(\cdot)) = \langle s(\tau, t_0, l), x\rangle - \int_{t_0}^{\tau} \langle s(t, t_0, l), B(t)u(t)\rangle dt - \frac{1}{4}\langle l, l\rangle - \rho(l \mid X^0).$$

Here $s[t] = s(t, t_0, l)$ is the solution of the adjoint equation

$$\dot{s} = -A'(t)s, \quad s(t_0) = l, \tag{2.69}$$

and s, l are column vectors.

The function $\Psi(\tau, x, l, u(\cdot))$ is concave in l and $\Psi(\tau, x, l, u(\cdot)) \to -\infty$ as $\langle l, l \rangle \to \infty$. It is also convex in u with $u(t)$ bounded by $\mathcal{P}(t)$. This allows us again, in view of ([72]), to interchange min and max in (2.68). Thus

$$V_0(\tau, x) = \max_l \Phi_0[\tau, x, l], \qquad (2.70)$$

with

$$\Phi_0[\tau, x, l] = \min\{\Psi(\tau, x, l, u(\cdot)) \mid u(t) \in \mathcal{P}(t), \ t_0 \le t \le \tau\} =$$

$$= \langle s(\tau, t_0, l), x \rangle - \int_{t_0}^{\tau} \rho(B'(t)s(t, t_0, l) \mid \mathcal{P}(t))dt - \frac{1}{4}\langle l, l \rangle - \rho(l \mid \mathcal{X}^0).$$

We now show by a direct substitution of $V_0(t, x)$ into (2.66) that it satisfies this equation with boundary condition (2.67). Indeed, let $l^0 = l^0(t, x)$ be the maximizer of function $\Phi_0[t, x, l]$ in l (see (2.70)). The structure of $\Phi_0[t, x, l]$ indicates that l^0 is *unique*. (Prove this assertion.)

Next we calculate the partial derivatives $\partial V_0(t, x)/\partial t$, $\partial V_0(t, x)/\partial x$. According to the rules of differentiating the maximum of a function like in (2.70), (see [61]), we have

$$\frac{\partial V(t, x)}{\partial t} = \frac{\partial \Phi_0[t, x, l^0]}{\partial t} = \langle \dot{s}(t, t_0, l^0), x \rangle - \rho(s(t, t_0, l^0) \mid B(t)\mathcal{P}(t)),$$

$$\frac{\partial V(t, x)}{\partial x} = \frac{\partial \Phi_0[t, x, l^0]}{\partial x} = s(t, t_0, l^0).$$

Substituting these partials into (2.66) and keeping (2.69) in view, one observes that $V_0(t, x)$ is indeed a solution to Eq. (2.66).

The boundary condition (2.67), namely,

$$V(t_0, x) = \max\{\langle l, x \rangle - \frac{1}{4}\langle l, l \rangle - \rho(l \mid \mathcal{X}^0) \mid l \in \mathbb{R}^n\} =$$

$$= \max\{\langle l, x \rangle - \varphi^*(t_0, l) \mid l \in \mathbb{R}^n\} = d^2(x, \mathcal{X}^0),$$

is also fulfilled. We summarize these results in the next assertion.

Theorem 2.4.1. *For the linear system (1.44) the value function $V_0(\tau, x)$ of the forward HJB equation (2.66) has the following properties:*

(i) $V_0(\tau, x)$ is given by (2.70).
(ii) The maximizer $l^0(\tau, x) = \arg\max_l \Phi_0[\tau, x, l]$ is unique and continuous in τ, x.
(iii) $V_0(\tau, x)$ is a proper convex function in x and is therefore directionally differentiable along any direction $\{1, l\}$.
(iv) $V_0(t, x)$ satisfies the HJB equation (2.66) for all t, x and also the boundary condition (2.67).

2.4 Reachability for Linear-Convex Systems

Proof. The second assertion follows from the strict convexity of $\Phi_0(\tau, x, l)$ in l due to the quadratic term and from the continuity of $\Phi(\tau, x, l)$ in τ, x. Property (iii) may be checked directly from the explicit formulas for the partials of $V_0(t, x)$, while (iv) is verified through direct substitution (Check whether $V_0(t, x)$ is unique). □

Remark 2.4.1. Note that for a time-invariant system one has

$$V_0(\tau, x) = V_0(\tau, x \mid t_0, \mathcal{X}^0) = V_0(\tau - t_0, x \mid 0, \mathcal{X}^0).$$

2.4.2 The Conjugate of the Value Function V_0

Having calculated $V_0(t, x)$ according to (2.70), let us now find its conjugate in the second variable, denoted by $V_0^*(t, l)$. We have

$$V_0^*(t, l) = \max_x \{\langle l, x\rangle - V_0(t, x) \mid x \in \mathbb{R}^n\} =$$

$$\max_x \{\langle l, x\rangle - \max_\lambda \{\Phi_0[t, x, \lambda] \mid \lambda \in \mathbb{R}^n\} \mid x \in \mathbb{R}^n\}$$

$$= \max_x \min_\lambda \{\langle l, x\rangle - \Phi_0[t, x, \lambda] \mid x, \lambda \in \mathbb{R}^n\}$$

$$= \min_\lambda \max_x \{\langle l, x\rangle - \langle s(t, t_0, \lambda), x\rangle + \int_{t_0}^t \rho(B'(\xi)s(\xi, t_0, \lambda) \mid \mathcal{P}(\xi)) d\xi$$

$$+ \frac{1}{4}\langle \lambda, \lambda\rangle + \rho(\lambda \mid \mathcal{X}^0) \mid x, \lambda \in \mathbb{R}^n\}.$$

The minmax is here attained at $s(t, t_0, \lambda) = l$ (for values of λ where this equality is not true the internal maximum in x is $+\infty$). Hence, keeping in mind that $s(t, t_0, \lambda) = G'(t_0, t)\lambda = l$, we come to the formula

$$V_0^*(t, l) = \int_{t_0}^t \rho(l \mid G(t, \xi)B(\xi)\mathcal{P}(\xi))d\xi + \rho(l \mid G(t, t_0)\mathcal{X}^0) + \frac{1}{4}\langle l, G(t, t_0)G'(t, t_0)l\rangle, \tag{2.71}$$

Theorem 2.4.2. (i) *The conjugate* $V_0^*(t, l)$ *to the value function* $V_0(t, x)$ *is given by (2.71).*

(ii) *The forward reach set* $\mathcal{X}[t] = \mathcal{X}(t, t_0, \mathcal{X}^0)$ *is given by*

$$\mathcal{X}[t] = \mathcal{X}(t, t_0, \mathcal{X}^0) = \{x : \langle l, x\rangle - V_0^*(t, l) \leq 0 \mid \forall l \in \mathbb{R}^n\} \tag{2.72}$$

Proof. Assertion (ii) follows from

$$V_0(t,x) = \max_l \{\langle l, x \rangle - V_0^*(t,l) \mid l \in \mathbb{R}^n\},$$

since V_0 is conjugate to V_0^*. □

2.4.3 The Value Function (Backward)

A scheme similar to the previous one works for the solution $V(t,x)$ of the backward equation (2.59) written for linear system (2.65) as

$$\frac{\partial V(t,x)}{\partial t} + \left\langle \frac{\partial V(t,x)}{\partial x}, A(t)x \right\rangle - \rho\left(-\frac{\partial V(t,x)}{\partial x} \,\bigg|\, B(t)\mathcal{P}(t)\right) = 0. \qquad (2.73)$$

The boundary condition for (2.55) or (2.73) is

$$V(t_1, x) = d^2(x, \mathcal{M}) = \min_q \{\langle x - q, x - q \rangle \mid q \in \mathcal{M}\}. \qquad (2.74)$$

Omitting the calculation, which is similar to the above, we come to

$$V(\tau, x) = \max_l \Phi[\tau, x, l], \qquad (2.75)$$

with

$$\Phi[\tau, x, l] = \langle s(\tau, t_1, l), x \rangle - \int_\tau^{t_1} \rho(-s(t, t_1, l) \mid B(t)\mathcal{P}(t)) dt - \frac{1}{4}\langle l, l \rangle - \rho(l \mid \mathcal{M}).$$

Theorem 2.4.3. *The value function $V(\tau, x)$ for the backward HJB equation (2.73), (2.74) has the following properties:*

(i) *$V(\tau, x)$ is given by (2.75).*
(ii) *The maximizer $l^0(\tau, x) = \arg\max_l \Phi[\tau, x, l]$ is unique and continuous in τ, x.*
(iii) *$V(\tau, x)$ is a proper convex function in x and is directionally differentiable for any τ, x.*
(iv) *$V(t, x)$ satisfies the HJB equation (2.73) for all t, x and also the boundary condition (2.74).*

Exercise 2.4.1. Find the conjugate function $V^*(t,l)$.

Formulas of type (2.70), (2.75) may be also calculated with criteria $d(x(t_0), \mathcal{X}^0)$, $d(x(t_1), \mathcal{M})$ instead of $d^2(x(t_0), \mathcal{X}^0), d^2(x(t_1), \mathcal{M})$ in the definitions of $V_0(t,x), V(\tau, x)$. In this case the term $\frac{1}{4}\langle l, l \rangle$ has to be omitted and the maxima taken over a unit sphere $\langle l, l \rangle \leq 1$. These calculations lead to the next result.

2.4 Reachability for Linear-Convex Systems

Theorem 2.4.4. Let $V_f(t, x), V_b(t, x)$ be the solutions of HJB equations (2.66), (2.73) with boundary conditions

$$V_f(t_0, x) = d(x, \mathcal{X}^0), \quad V_b(t_1, x) = d(x, \mathcal{M}), \tag{2.76}$$

respectively. Then the following relations hold:

$$V_f(\tau, x) = \min_u \{d(x(t_0), \mathcal{X}^0) \mid x(\tau) = x\} = \max_l \{\Phi_f[\tau, x, l] \mid \langle l, l \rangle \leq 1\} =$$

$$= \max_l \{\Phi_f[\tau, x, l] - I(l \mid \mathcal{B}(0)) \mid l \in \mathbb{R}^n\}, \tag{2.77}$$

with $I(l \mid \mathcal{B}(0))$ being the indicator function for the unit ball $\mathcal{B}(0)$ and

$$\Phi_f[\tau, x, l] = \langle s(\tau, t_0, l), x \rangle - \int_{t_0}^{\tau} \rho(s(t, t_0, l) \mid B(t)\mathcal{P}(t))dt - \rho(l \mid \mathcal{X}^0).$$

Similarly,

$$V_b(\tau, x) = \min_u \{d(x(t_1), \mathcal{M}) \mid x(\tau) = x\} = \max_l \{\Phi_b[\tau, x, l] \mid \langle l, l \rangle \leq 1\} =$$

$$= \max_l \{\Phi_b[\tau, x, l] - I(l \mid \mathcal{B}(0)) \mid l \in \mathbb{R}^n\}, \tag{2.78}$$

with

$$\Phi_b[\tau, x, l] = \langle s(\tau, t_1, l), x \rangle - \int_{\tau}^{t_1} \rho(-s(t, t_1, l) \mid B(t)\mathcal{P}(t))dt - \rho(l \mid \mathcal{M}).$$

The conjugate value functions are

$$V_f^*(\tau, l) = \int_{t_0}^{\tau} \rho(l \mid G(\tau, \xi)B(\xi)\mathcal{P}(\xi))d\xi + \rho(l \mid G(\tau, t_0)\mathcal{X}^0) + I(l \mid G(\tau, t_0)\mathcal{B}(0)), \tag{2.79}$$

$$V_b^*(\tau, l) = \int_{\tau}^{t_1} \rho(-l \mid G(\tau, \xi)B(\xi)\mathcal{P}(\xi))d\xi + \rho(l \mid G(\tau, t_1)\mathcal{M}) + I(l \mid G(\tau, t_1)\mathcal{B}(0)). \tag{2.80}$$

Here, as in the above, $G(\tau, \xi)$ is defined by (1.11), (1.12). The forward and backward reach sets may now be described as follows.

Theorem 2.4.5. *The support functions for the forward and backward reach sets are*

$$\rho(l \mid X[\tau]) = \rho(l \mid X(\tau, t_0, \mathcal{X}^0)) = \int_{t_0}^{\tau} \rho(l \mid G(\tau, \xi)B(\xi)\mathcal{P}(\xi))d\xi + \rho(l \mid G(\tau, t_0)\mathcal{X}^0), \tag{2.81}$$

$$\rho(l \mid W[\tau]) = \rho(l \mid W(\tau, t_1, \mathcal{M})) = \int_{\tau}^{t_1} \rho(-l \mid G(\tau, \xi)B(\xi)\mathcal{P}(\xi))d\xi + \rho(l \mid G(\tau, t_1)\mathcal{M}). \tag{2.82}$$

Exercise 2.4.2. Prove Theorems 2.4.4, 2.4.5.

Figure 2.2 illustrates the four-dimensional value function of type $V_0(t,x)$, (see (2.46)), for system

$$\dot{x}_1 = x_3, \quad \dot{x}_3 = -(k_1+k_2)x_1 + k_2 x_3,$$

$$\dot{x}_2 = x_4, \quad \dot{x}_4 = k_1 x_1 - (k_1+k_2)x_3 + u,$$

$k_1, k_2 > 0$, under bounded control $|u| \leq \mu$ through its two-dimensional projections on subspaces $L\{x_1, x_2\}$ and $L\{x_3, x_4\}$. Shaded are the projections of related level sets.

Having described the forward and backward reachable tubes, we shall now analyze the notion of *colliding tubes*, then use it to calculate all the points of the reachable sets.

2.5 Colliding Tubes: Calculating All Reachable Points

In this section we introduce the *Principle of Colliding Tubes* investigating the interaction between the forward reach tube emanating from a starting point with the backward reach tube emanating from a terminal point. Using this principle, one may calculate optimal trajectories from starting point to terminal point without calculating the controls.

Colliding Tubes

Consider two set-valued positions of system (2.25): the starting position $\{t_0, \mathcal{X}^0\}$ and the terminal position $\{t_1, \mathcal{M}\}$ with $t_0 \leq t_1$. For each of these one may specify a reach tube—in forward and backward time, respectively:

$$\mathcal{X}[t] = \mathcal{X}(t, t_0, \mathcal{X}^0), \text{ and } \mathcal{W}[t] = \mathcal{W}(t, t_1, \mathcal{M}), \quad t \in [t_0, t_1].$$

Lemma 2.5.1. *Suppose at any time* $t = t^*$ *the intersection* $\mathcal{X}[t^*] \cap \mathcal{W}[t^*] \neq \emptyset$. *Then*

$$\mathcal{X}[t] \cap \mathcal{W}[t] \neq \emptyset, \quad \forall t \in [t_0, t_1].$$

The proof of this statement follows from the definitions of forward and backward reach sets. This yields the next conclusion.

2.5 Colliding Tubes: Calculating All Reachable Points

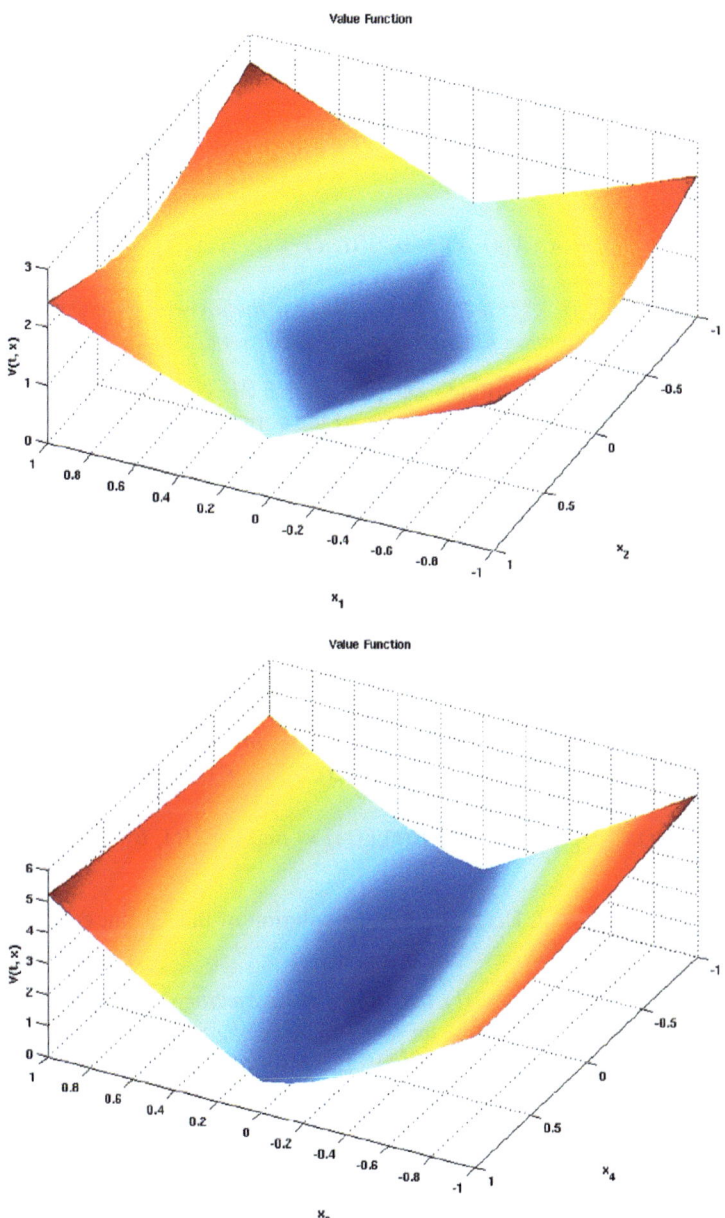

Fig. 2.2 Two-dimensional projections of the four-dimensional value function with their level sets

Corollary 2.5.1. *The tube*

$$Z[t] = X[t] \cap W[t], \ t \in [t_0, t_1],$$

consists of all trajectories of system (2.1) that connect the set X^0 at $t = t_0$ to the set $\mathcal{M}$ at $t = t_1$. Here

$$Z[t] = \{x : V_0(t, x) \leq 0, \ V(t, x) \leq 0\} = X[t] \cap W[t],$$

for all $t \in [t_0, t_1]$; $V_0(t, x)$ is the value function for the forward reach set (2.46) and $V(t, x)$ is the value function for the backward reach set (2.55).

In particular $Z[t_0] = X^0 \cap W[t_0]$, $Z[t_1] = X[t_1] \cap \mathcal{M}$. Note that the knowledge of tube $Z(\cdot)$ does not require any calculation of the corresponding controls.

The problem of calculating all reachable points may be separated into two classes, namely the one of calculating the *boundary points* of the reach set $X[t_1] = X(t_1, t_0, x^0)$ and that of calculating its *interior points*.

The first class of problems is easily solved in the following way. Suppose x^1 is a point on the boundary of $X[t_1]$. The corresponding support vector (or vectors) l^* are given by

$$l^* \in \{l : \rho(l \mid X[t_1]) = \langle l, x^1 \rangle\} = \mathbf{L}^* \qquad (2.83)$$

Taking any $l^* \in \mathbf{L}^*$ and applying the *maximum principle* in its simplest form (see Sect. 1.4, (1.46)), we observe that the desired control $u^*(t)$ satisfies the relation

$$\langle \psi^*(t), B(t)u^*(t) \rangle = \rho(\psi^*(t) \mid B(t)\mathcal{P}(t)) = \max\{\langle \psi^*(t), B(t)u \rangle \mid u \in \mathcal{P}(t)\} \qquad (2.84)$$

for almost all $t \in [t_0, t_1]$. Here $\psi^*(t)$ is the solution of the adjoint equation (1.47) (1.61), with $\psi^*(t_1) = l^*$.[3]

Lemma 2.5.2. *The control $u^*(t)$ that steers the trajectory $x(t)$ from point $x^0 = x(t_0)$ to given point $x^1 = x(t_1)$ on the boundary of reach set $X[t_1]$ satisfies the maximum principle (2.84), in which $\psi^*[t]$ is the solution to Eq. (1.47) ((1.62)) with boundary condition $\psi^*(t_1) = l^*$, and vector l^* given by (2.83).*

Note that $u^*(t)$ will be defined for almost all $t \in [t_0, \vartheta]$ provided that each coordinate function $h_i(t)$ of the row $h(t) = (\psi^*(t))'B(t)$ is nonzero almost everywhere. This property is ensured by *strong controllability* of the system under consideration (see Sect. 1.3.3).

---

[3] With additional information on $\mathcal{P}(t)$ (see Remark 1.5.3), in degenerate cases the control $u^*(t)$ may be written down in more detail.

2.5 Colliding Tubes: Calculating All Reachable Points

Suppose now that x^1 is an *interior point* of $X[t_1]$ ($x^1 \in \text{int} X[t_1] = \text{int} X(t_1, t_0, x^0)$) and we are to find the control that steers $x(t)$ from $x(t_0) = x^0$ to $x(t_1) = x^1$. We may then proceed as follows.

First find the forward reach sets $X[\tau] = X(\tau, t_0, x^0)$ from *point* x^0 and the backward reach sets $W[\tau] = W(\tau, t_1, x^1)$ from *point* x^1. This may be done, for example, by applying formulas (2.81) and (2.82). The latter yield two relations:

$$\rho(l \mid X[\tau]) = \langle l, G(\tau, t_0) x^0 \rangle + \int_{t_0}^{\tau} \rho(l \mid G(\tau, s) B(s) \mathcal{P}(s)) ds, \qquad (2.85)$$

and

$$\rho(l \mid W[\tau]) = \langle l, G(\tau, t_1) x^1 \rangle + \int_{\tau}^{t_1} \rho(-l \mid G(\tau, s) B(s) \mathcal{P}(s)) ds, \qquad (2.86)$$

where tube $X[\tau]$ develops from $X[t_0] = x^0$ with τ increasing and tube $W[\tau]$ develops from $W[t_1] = x^1$, with τ decreasing.

We thus have two intersecting tubes ($X[\tau] \cap W[\tau] \neq \emptyset$, $\forall \tau \in [t_0, \vartheta]$), heading towards each other in opposite directions. If x^0 lies on the boundary of $W[t_0]$, then we arrive at the situation described in the previous case, Lemma 2.5.2, with x^0 and x^1 interchanged and problem solved in backward time. Hence we further assume $x^0 \in \text{int} W[t_0]$.

Then we have $x^1 \in \text{int} X[t_1]$ and $x^0 \in \text{int} W[t_0]$ and there exists an instant $\tau^* \in [t_0, t_1]$ when the boundaries of the two tubes collide. Namely, developing $W[t]$ from t_1 backwards towards t_0, we first have $W[t] \subset X[t]$ till we reach instant $t = \tau^*$, when $W[t]$ bumps into the boundary of $X[t]$, touching it from inside at point x^*, so that

$$W[\tau^*] \subseteq X[\tau^*], \text{ and } \exists l^* : \langle l^*, x^* \rangle = \rho(l^* \mid X[\tau^*]) = \rho(l^* \mid W[\tau^*]). \qquad (2.87)$$

Let us find this instant τ^*. With $W[\tau] \subset X[\tau]$, we have

$$k(\tau, l, x^0, x^1) = \rho(l \mid X[\tau]) - \rho(l \mid W[\tau]) > 0,$$

with

$$k(\tau, l, x^0, x^1)$$
$$= \langle l, G(\tau, t_0) x^0 - G(\tau, t_1) x^1 \rangle + \int_{t_0}^{\tau} \rho(l \mid G(\tau, s) B(s) \mathcal{P}(s)) ds$$
$$- \int_{\tau}^{t_1} \rho(-l \mid G(\tau, s) B(s) \mathcal{P}(s)) ds,$$

and

$$\gamma(\tau, x^0, x^1) = \min_l \{k(\tau, l, x^0, x^1) \mid \langle l, l \rangle = 1\} > 0. \qquad (2.88)$$

With x^0, x^1 fixed, instant τ will then be the largest root of equation $\gamma(\tau, x^0, x^1) = 0$—the first time when $\mathcal{W}[\tau]$ touches $\mathcal{X}[\tau]$ from inside.

Function $\gamma(\tau, x^0, x^1)$ is defined for $\tau \in [t_0, t_1]$, with $\gamma(t_0, x^0, x^1) < 0$, $\gamma(t_1, x^0, x^1) > 0$, and it is continuous in τ. Therefore, point τ^*, where $\gamma(\tau^*, x^0, x^1) = 0$ exists. The maximizer l^* of (2.88) is the support vector to $\mathcal{X}[\tau^*]$ at touching point, which is precisely the vector $x^* = x(\tau^*)$, that ensures (2.87).

Lemma 2.5.3. *The first instant τ^* of collision for $\mathcal{W}[\tau] \subseteq \mathcal{X}[\tau]$ with boundary of tube $\mathcal{X}[\tau]$ from inside is the largest root of equation $\gamma(\tau^*, x^0, x^1) = 0$. The point x^* of such collision satisfies the maximum rule*

$$\max\{\langle l^*, x \rangle | x \in \mathcal{X}[\tau^*]\} = \langle l^*, x^* \rangle = \max\{\langle l^*, x \rangle | x \in \mathcal{W}[\tau^*]\}.$$

The calculation of the control $u^*(t)$ which steers $x(t)$ from $x(t_0) = x^0$ to $x(t_1) = x^1$ is now reduced to the class of points considered in Lemma 2.5.2, namely the points that lie on the boundaries of the respective reach sets.

Here we first specify the control $u^*(t)$ for interval $t \in [t_0, \tau^*]$, working with point x^*, which lies on the boundary of set $\mathcal{X}[\tau^*]$. Then $u^*(t)$ satisfies the maximum principle

$$\langle l^*, G(\tau^*, t) B(t) u^*(t) \rangle = \max\{\langle l^*, G(\tau^*, t) B(t) u \rangle | u \in \mathcal{P}(t)\}$$
$$= \rho(l^* \mid G(\tau^*, t) B(t) \mathcal{P}(t)), \ t \leq \tau^*. \qquad (2.89)$$

Applying similar reasoning to the interval $[\tau^*, t_1]$, with "starting point" $x^1 = x(t_1)$ and $x^*(\tau^*)$ being on the boundary of reach set $\mathcal{W}[\tau^*]$ with same support vector l^*, we may calculate $u^*(t)$ for $t \in [\tau^*, t_1]$. This leads again to relation (maximum principle) (2.89), but now taken for $t > \tau^*$.

Theorem 2.5.1. *The control $u^*(t)$ that steers the trajectory $x(t)$ from $x(t_0) = x^0$ to $x(t_1) = x^1$ satisfies the maximum principle relation (2.89), where l^* is the minimizer of problem (2.88), with $\tau = \tau^*$ being the largest root of equation $\gamma(\tau, x^0, x^1) = 0$.*

Here Fig. 2.3 illustrates the idea of colliding tubes.[4] Namely, given in green is the backward reach tube from the terminal point x^m at target set $\mathcal{M}$ (marked in red). Given in blue is the forward reach tube emanating from the starting point x^0 located in set $\mathcal{X}^0$. The intersection of these two tubes is the union of all system trajectories that connect the starting point with the terminal point.

---

[4]This example is animated in the toolbox [132].

2.6 The Closed-Loop Target Control

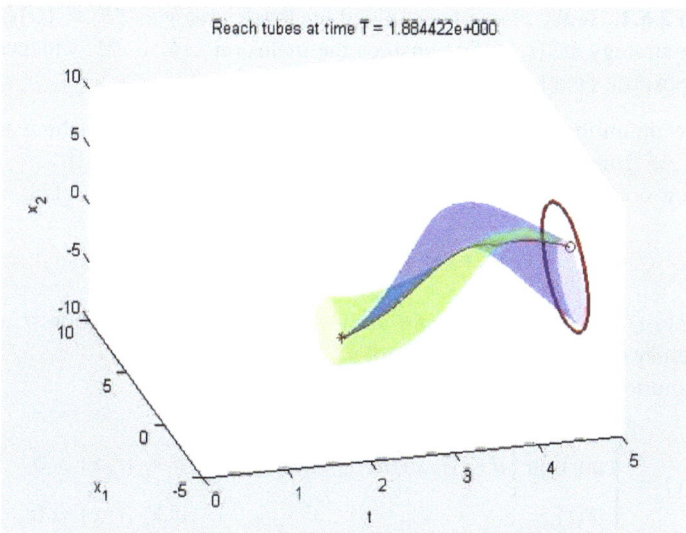

Fig. 2.3 Colliding tubes for control "from point to point"

The original equations for this picture are ($x \in \mathbb{R}^2$):

$$\dot{x} = Ax + Bu, \quad A = \begin{pmatrix} 0 & 1 \\ -2 & 0 \end{pmatrix}, \quad B = \begin{pmatrix} 0 \\ 1 \end{pmatrix}, \quad x^0 = \begin{pmatrix} 3 \\ 1 \end{pmatrix}, \quad x^m = B = \begin{pmatrix} 2 \\ 0 \end{pmatrix}.$$

with $-2 \leq u \leq 2$, $t \in [0, 5]$. The time of switching from one tube to the other is $\tau = 0.886$.

We shall now indicate how to construct feedback strategies for target control by using backward reach sets.

2.6 The Closed-Loop Target Control

Here we introduce a scheme for finding the closed-loop control for minimizing at given time the distance from a given target set. This would require to know the backward reach set. Once this set is known, there would be no need of integrating the HJB equation which produces it as a level set. The problem will be therefore shifted to computation of these level sets without solving the HJB equation. Such computation could be done through ellipsoidal methods, as explained in the next chapter.

Consider system (1.2) with target set $\mathcal{M}$. Let $\mathcal{W}[\tau] = \mathcal{W}(\tau, \vartheta, \mathcal{M})$ be its backward reach set (the solvability set) from position $\{\vartheta, \mathcal{M}\}$.

Problem 2.6.1. In the class of set-valued feedback strategies $\mathcal{U}_C = \{U(t,x)\}$ find feedback strategy $U^*(t,x)$ that ensures the inclusion $x[\vartheta] \in \mathcal{M}$, whatever be the starting position $\{\tau, x\}$, $x \in \mathcal{W}[\tau]$.

From the definition of $\mathcal{W}[\tau]$ it follows that the problem of reaching target set $\mathcal{W}[\vartheta] = \mathcal{M}$ from position $\{\tau, x\}$ is solvable if and only if $x \in \mathcal{W}[\tau]$.

We now construct a solution strategy $U^*(t,x)$ for Problem 2.6.1. Let

$$V_C(t,x) = d(G(\vartheta,t)x, G(\vartheta,t)\mathcal{W}[t])$$

and let $dV_C(t,x)/dt \mid_u$ be the *total derivative* of $V_C(t,x)$ at position $\{t,x\}$, along the trajectory of Eq. (1.2) under control u.

The solution strategy is defined as

$$U^*(t,x) = \begin{cases} \arg\min\left\{dV_C(t,x)/dt \mid_u \middle| u \in \mathcal{P}(t)\right\}, & \text{if } V_C(t,x) > 0, \\ \mathcal{P}(t), & \text{if } V_C(t,x) = 0. \end{cases} \quad (2.90)$$

Observe that the set-valued function $U^*(t,x)$ is upper semicontinuous in $\{t,x\}$ with respect to inclusion. This ensures the existence and extendability of solution to the differential inclusion

$$\dot{x} \in A(t)x + B(t)U^*(t,x), \quad (2.91)$$

from $x(\tau) = x \in \mathcal{W}[\tau]$. The next fact turns out to be true.

Lemma 2.6.1. *With $U^*(t,x)$ selected according to (2.90) we have*

$$\left.\frac{dV_C(t,x)}{dt}\right|_u \leq 0,$$

for all $u \in U^*(t,x)$.

It now remains to verify that u may indeed be selected so as to ensure the derivative $dV_C(t, x^*[t])/dt \mid_u$ is non-positive within $t \in [\tau, \vartheta]$. Calculate

$$dV_C(t,x)/dt \mid_u = \frac{\partial V_C(t,x)}{\partial t} + \left\langle \frac{\partial V_C(t,x)}{\partial x}, A(t)x + B(t)u \right\rangle \quad (2.92)$$

in which

$$V_C(t,x) = \max\{0, \max\{\langle G'(\vartheta,t)l, x\rangle - \rho(G'(\vartheta,t)l \mid \mathcal{W}[t]) \mid \langle l,l\rangle \leq 1\}\}.$$

Note that we need to ensure $dV_C(t,x)/dt \mid_u \leq 0$ only where $V_C(t,x) > 0$. Now suppose $V_C(t,x) > 0$.

2.6 The Closed-Loop Target Control

Then let

$$l^0(t, x) = \arg\max\{\langle l, G(\vartheta, t)x\rangle - \rho(l \mid G(\vartheta, t)\mathcal{W}[t]) \mid \langle l, l\rangle \leq 1\}. \qquad (2.93)$$

Under the constraint $\langle l, l\rangle \leq 1$ the vector $l^0 = l^0(t, x)$ is unique. Applying the rules for differentiating a "maximum function," [61], we further have

$$\frac{\partial V_C(t, x)}{\partial t} = \langle l^0, -G(\vartheta, t)A(t)x\rangle - \frac{\partial \rho(l^0 \mid G(\vartheta, t)\mathcal{W}[t])}{\partial t},$$

$$\frac{\partial V_C(t, x)}{\partial x} = G'(\vartheta, t)l^0.$$

Since

$$\dot{x} = A(t)x + B(t)u, \quad \mathcal{W}[t] = G(t, \vartheta)\mathcal{M} - \int_t^\vartheta G(t, s)B(s)\mathcal{P}(s)ds,$$

then from (2.82), (2.80) we have

$$\rho(l^0 \mid G(\vartheta, t)\mathcal{W}[t]) = \rho(l \mid \mathcal{M}) + \int_t^\vartheta \rho(-l^0 \mid G(\vartheta, s)B(s)\mathcal{P}(s))ds,$$

so that

$$\frac{\partial \rho(l^0 \mid G(\vartheta, t)\mathcal{W}[t])}{\partial t} = -\rho(-l^0 \mid G(\vartheta, t)B(t)\mathcal{P}(t)).$$

Substituting the results into Eq. (2.92) we finally have

$$dV_C(t, x)/dt \mid_u = \rho(-l^0 \mid G(\vartheta, t)B(t)\mathcal{P}(t)) + \langle l^0, G(\vartheta, t)B(t)u\rangle.$$

Taking

$$U^*(t, x) = \arg\max\{\langle l^0, G(\vartheta, t)B(t)u\rangle \mid u \in \mathcal{P}(t)\}, \qquad (2.94)$$

we get

$$dV_C(t, x)/dt \mid_u = 0, \; u \in U^*(t, x).$$

To prove that $U^*(t, x)$ is a solution to Problem 2.6.1, first integrate $dV_C(t, x)/dt$ from τ to ϑ along the trajectories $x^*[t]$ of (2.91) emanating from $x^*(\tau) = x$. Then, with $U^*(t, x)$ selected according to (2.90), one gets

$$\int_\tau^\vartheta dV_C(t, x^*[t]) = V_C(\vartheta, x^*[\vartheta]) - V_C(\tau, x) \leq 0,$$

so $V_C(\vartheta, x^*[\vartheta]) \leq V_C(\tau, x)$, which implies that once $V_C(\tau, x) \leq 0$ (which means $x^*[\tau] = x \in \mathcal{W}[\tau]$), one will have $V_C(\vartheta, x^*[\vartheta]) \leq 0$ (which means $x^*[\vartheta] \in \mathcal{W}[\vartheta] = \mathcal{M}$).

Note that under $V_C(t, x(t)) \leq 0$, with control selected due to (2.90), the trajectory $x(t)$ never leaves the tube $\mathcal{W}[t]$. So if, say $V_C(t, x(t)) = 0$, but at some point $t' = t + \sigma$ we have

$$V_C(t, x(t')) = d(G(\vartheta, t')x(t'), G(\vartheta, t')\mathcal{W}[t']) > 0,$$

then, due to the directional differentiability of function $V_C(t, x(t))$, we would have at some point $t'' \in (t, t']$ that the derivative $dV_C(t'')/dt > 0$, which is not possible due to (2.90).

Theorem 2.6.1. *The closed-loop strategy $U^*(t, x)$ which solves Problem 2.6.1 is defined by (2.90), (2.94), in which $l^0 = l^0(t, x)$ is the optimizer of (2.93) and $\mathcal{W}[t]$ is the backward reach tube from position $\{\vartheta, \mathcal{M}\}$.*

Remark 2.6.1. A strategy $U^*(t, x)$ given by (2.94), (2.93), is said to be produced by the **extremal aiming rule** (see [121]).

Exercise 2.6.1. Find the closed-loop strategy that steers point $x \in \mathcal{W}[\tau]$ to a given point $x^* \in \mathcal{M}$ (provided x, x^* are such that the problem is solvable).

2.7 Reachability Within an Interval

The problem of reachability may be formulated as a request to know what states may be reached *within a time interval* rather than at fixed time.

Thus we consider a generalization of Problem 2.6.1.

Problem 2.7.1. Find the set $\mathbf{W}[\tau, \Theta] = \mathbf{W}(\tau, \Theta, \mathcal{M})$ of all points for which there exists a control $u^*(t)$, which ensures the inclusion $x[\vartheta] = x(\vartheta, \tau, x) \in \mathcal{M}$, for **some** $\vartheta \in \Theta = [\alpha, \beta]$, $\beta \geq \alpha \geq \tau$.

This is the *backward reach set relative to* $\{\Theta, \mathcal{M}\}$. Unlike Problem 2.6.1 where the terminal time is fixed at ϑ, in Problem 2.7.1 the terminal time can be any instant within the interval $\Theta = [\alpha, \beta]$.

The set $\mathbf{W}[\tau, \Theta]$ is *weakly invariant* relative to $\{\Theta, \mathcal{M}\}$, namely, such that for each point $x \in \mathbf{W}[\tau, \Theta]$ there exists at least one trajectory that reaches set $\mathcal{M}$ at some time $\vartheta \in \Theta$.

Remark 2.7.1. Note that the maximal backward reach sets $\mathbf{W}[\tau, \Theta]$ **are the same** whether calculated in the class of open-loop or closed-loop controls.

Problem 2.7.2. In the class of set-valued feedback strategies $\mathcal{U}_C = \{U(t, x)\}$ find feedback strategy $U^*(t, x)$ which ensures the inclusion $x[\vartheta] \in \mathcal{M}$ for **some** $\vartheta \in \Theta$, whatever be the starting position $\{\tau, x\}$, $x \in \mathbf{W}[\tau, \Theta]$.

2.7 Reachability Within an Interval

Problem 2.7.1 may be solved by calculating the set $\mathbf{W}[\tau, \Theta]$. Clearly,

$$\mathbf{W}[\tau, \Theta] = \bigcup \{\mathcal{W}(\tau, \vartheta, \mathcal{M}) \mid \vartheta \in \Theta\}. \tag{2.95}$$

Let

$$\mathbf{V}(\tau, x) = \min_{\vartheta} \{V(\tau, x \mid \vartheta, V(\vartheta, \cdot)) \mid \vartheta \in \Theta\}, \tag{2.96}$$

with $V(\tau, x \mid \vartheta, V(\vartheta, \cdot))$ taken from (2.55). That is,

$$\mathbf{V}(\tau, x) =$$

$$= \min_{u} \min_{\vartheta} \{d^2(x[\vartheta], \mathcal{M}) \mid x[\tau] = x, \vartheta \in \Theta\} = \min_{\vartheta} \min_{u} \{d^2(x[\vartheta], \mathcal{M}) \mid x[\tau] = x, \vartheta \in \Theta\}. \tag{2.97}$$

Lemma 2.7.1.

$$\mathbf{W}[\tau, \Theta] = \{x : \mathbf{V}(\tau, x) \le 0\}. \tag{2.98}$$

Returning to formula (2.55), introduce the notation

$$V(\tau, \vartheta, x) = \min_{u} \{d^2(x[\vartheta], \mathcal{M}) \mid x[\tau] = x\}.$$

Hence

$$\mathbf{V}(\tau, x) = \min_{\vartheta} \{V(\tau, \vartheta, x) \mid \vartheta \in \Theta\} = V(\tau, \vartheta^0, x),$$

where $\vartheta^0 = \vartheta^0(\tau, x)$. The solution to Problem 2.7.2 is now reduced to Problem 2.6.1 with $\vartheta = \vartheta^0$.

Exercise 2.7.1. Solve Problem 2.7.2 in detail, by applying the scheme of Sect. 2.6 to $\mathbf{W}[\tau, \Theta]$.

The notion of *strong invariance* is applicable also to reachability within an interval.

Definition 2.7.1. A set $\mathbf{W}_s[\tau, \Theta]$ is **strongly invariant** relative to set $\mathcal{M}$ within interval Θ, if for each of its points x the entire reach set $X(\vartheta, \tau, x) = X[\vartheta] \subseteq \mathcal{M}$ for some $\vartheta \in \Theta$.

If $\mathbf{W}_s[\tau, \Theta]$ contains **all** such points, then it is said to be the **maximal strongly invariant** set relative to $\mathcal{M}$, within interval Θ.

Define value function

$$\mathbf{V}_s(\tau, x) = \min_{\vartheta}\{V_s(\tau, \vartheta, x) \mid \vartheta \in \Theta\}, \qquad (2.99)$$

$$V_s(\tau, \vartheta, x) = \max_u\{d(x(\vartheta), \mathcal{M}) \mid u(\cdot) \in \mathcal{P}(\cdot), x(\tau) = x\}.$$

Lemma 2.7.2. $\mathbf{W}_s[\tau, \Theta]$ *is the level set*

$$\mathbf{W}_s[\tau, \Theta] = \{x : \mathbf{V}_s(\tau, x) \le 0\}. \qquad (2.100)$$

Denoting $\mathcal{W}_s(\tau, \vartheta, \mathcal{M}) = \{x : V_s(\tau, \vartheta, x) \le 0\}$, we have

$$\mathbf{W}_s[\tau, \Theta] = \bigcup\{\mathcal{W}_s(\tau, \vartheta, \mathcal{M}) \mid \vartheta \in \Theta\}.$$

Following (2.63), we also have (for ϑ fixed)

$$\frac{\partial V_s(t, \vartheta, x)}{\partial t} + \max_u\left\{\left\langle\frac{\partial V_s(t, \vartheta, x)}{\partial x}, A(t)x + B(t)u\right\rangle\right\} = 0, \qquad (2.101)$$

with boundary condition

$$d(x, \mathcal{M}) = V_s(\vartheta, \vartheta, x). \qquad (2.102)$$

Let us now find the support function $\rho(l \mid \mathcal{W}_s(\tau, \vartheta, \mathcal{M}))$. Following the scheme of calculating (2.78), (2.80), but with $\min_u$ substituted for $\max_u$, we get

$$V_s(\tau, \vartheta, x) = \max_l\{\Phi_s[\tau, \vartheta, x, l] \mid \langle l, l \rangle \le 1\}, \qquad (2.103)$$

with

$$\Phi_s[\tau, \vartheta, x, l] = \langle s(\tau, \vartheta, l), x\rangle + \int_\tau^\vartheta \rho(s(t, \vartheta, l) \mid B(t)\mathcal{P}(t))dt - \rho(l \mid \mathcal{M}).$$

Hence

$$\mathcal{W}_s(\tau, \vartheta, \mathcal{M}) = \{x : \Phi_s[\tau, \vartheta, x, l] \le 0\}$$

for all $\langle l, l \rangle \le 1$ and therefore, since the function $\Phi_s[\tau, \vartheta, x, l]$ is homogeneous in l, also for all $l \in \mathbb{R}^n$.

After a substitution similar to (2.71) we observe that $\forall x \in \mathcal{W}_s(\tau, \vartheta, \mathcal{M})$,

$$\langle l, x \rangle \le \rho(l \mid G(t, \vartheta)\mathcal{M}) - \int_t^\vartheta \rho(l \mid G(t, \xi)B(\xi)\mathcal{P}(\xi))d\xi = k(t, \vartheta, l). \qquad (2.104)$$

This yields the next result.

2.7 Reachability Within an Interval

Lemma 2.7.3. *If $k(t, \vartheta, l)$ is convex in l, one has*

$$\rho(l \mid \mathcal{W}_s'(\tau, \vartheta, \mathcal{M})) = k(t, \vartheta, l),$$

while in the absence of convexity in l one has

$$\rho(l \mid \mathcal{W}_s'(\tau, \vartheta, \mathcal{M})) = k^{**}(t, \vartheta, l) \quad (2.105)$$

where the second conjugate is taken in the variable l.

This lemma follows from a well-known result in convex analysis on lower envelopes of convex functions (see [237, 238]). Namely, the second Fenchel conjugate $k^{**}(t, \vartheta, l)$ is the *lower convex envelope* of $k(t, \vartheta, l)$ in the variable l. This means epigraph epi k_l^{**} in the variable l is *the convex hull* of the set epi k_l. Recall that the epigraph of a function $k(l)$ is the set of points lying "above" the graph of k:

$$\text{epi } k := \{\{l, \alpha\} \in \mathbb{R}^n \times \mathbb{R} : \alpha \geq k(l)\}.$$

Theorem 2.7.1. *The value function $\mathbf{V}_s(\tau, x)$, (2.99), that determines $\mathbf{W}_s[\tau, \Theta]$ through (2.100) is given by (2.101)–(2.103).*

The next item is reachability within an interval (see Remark 1.4.2).

Definition 2.7.2. The **reach set** $\mathbf{X}[\Theta] = \mathbf{X}(\Theta, t_0, \mathcal{X}^0)$ of system (1.44) under constraint $u(s) \in \mathcal{P}(s)$, $s \geq t_0$, **within given time-interval** $\Theta = [\alpha, \beta]$, $\beta \geq \alpha \geq t_0$, **from set-valued position** $\{t_0, \mathcal{X}^0\}$, is the set of **all points** x, for each of which there exists a trajectory $x[s] = x(s, t_0, x^0)$, for some $x^0 \in \mathcal{X}^0$, generated by some control subjected to the given constraint, that transfers the system from position $\{t_0, x^0\}$ to position $\{\vartheta, x\}$, $x = x[\vartheta]$, **for some** $\vartheta \in \Theta$:

$$\mathbf{X}[\Theta] = \mathbf{X}(\Theta, t_0, \mathcal{X}^0) = \{x : \exists\, x^0 \in \mathcal{X}^0, \, u(\cdot) \in \mathcal{P}(\cdot), \, \vartheta \in \Theta, \, x = x(\vartheta; t_0, x^0)\}.$$

Problem 2.7.3. *Find the reach set $\mathbf{X}[\Theta]$.*

Denote

$$\mathbf{V}^0(\Theta, x) = \min\{V_0(\vartheta, x \mid t_0, V_0(t_0, \cdot)) \mid \vartheta \in \Theta\}, \quad (2.106)$$

where $V_0(\vartheta, x \mid t_0, V_0(t_0, \cdot))$ is taken from (2.48).
That is,

$$\mathbf{V}^0(\Theta, x) = \min_u \{\min_\vartheta \{d^2(x[\tau], \mathcal{X}^0) \mid x = x[\vartheta], \vartheta \in \Theta\}\} = \quad (2.107)$$

$$= \min_\vartheta \{\min_u \{d^2(x[\tau], \mathcal{X}^0) \mid x = x[\vartheta]\} \mid \vartheta \in \Theta\}.$$

Lemma 2.7.4. *The reach set* $\mathbf{X}[\Theta]$ *within a given time interval* Θ *is the level set*

$$\mathbf{X}[\Theta] = \{x : \mathbf{V}^0(\Theta, x) \leq 0\}. \tag{2.108}$$

The exact calculation of sets $\mathbf{X}[\Theta], \mathbf{W}[\tau, \Theta]$ may be achieved through formulas (2.106), (2.96) applied to functions $V_0(\tau, x)$, $V(\tau, x)$ given in (2.68), (2.77).

Reachability from a Tube

Consider a tube $\{\Theta_0, \mathcal{X}^0(\cdot)\} = \{t, x : t \in \Theta_0 = [t', t''], x \in \mathcal{X}^0(t)\}$ where compact set-valued function $\mathcal{X}^0(t)$ is Hausdorff-continuous on Θ_0.

Definition 2.7.3. *The reachability set* $\mathbf{X}[\Theta, \Theta_0] = \mathcal{X}(\Theta, \Theta_0, \mathcal{X}^0(\cdot))$, $\Theta \cap \Theta_0 \neq \phi$, *from tube* $\{\Theta_0, \mathcal{X}^0(\cdot)\}$ *within given time-interval* Θ *is the union*

$$\mathbf{X}[\Theta, \Theta_0] = \bigcup \{x(\vartheta, t_0, x^0) \mid \vartheta \in \Theta, \; t_0 \in \Theta_0, \; x^0 \in \mathcal{X}^0(t_0)\}.$$

Problem 2.7.4. *Find the reach set* $\mathbf{X}[\Theta, \Theta_0]$.

Taking

$$\mathbf{V}^0(\Theta, \Theta_0, x) = \min_{u}\{\min_{\vartheta, t_0}\{d^2(x[t_0], \mathcal{X}^0(t_0)) \mid x[\vartheta] = x, \vartheta \in \Theta, \; t_0 \in \Theta_0\}\} = $$
$$= \min_{\vartheta, t_0}\{\min_{u}\{d^2(x[t_0], \mathcal{X}^0[t_0]) \mid x[\vartheta] = x\} \mid \vartheta \in \Theta, t_0 \in \Theta_0\}\} \tag{2.109}$$

we come to the next relation.

Lemma 2.7.5. *The reach set from given tube* Θ_0 *within given interval* Θ *is the level set*

$$\mathbf{X}[\Theta, \Theta_0] = \{x : \mathbf{V}^0(\Theta, \Theta_0, x) \leq 0\}. \tag{2.110}$$

Sets $\mathbf{W}[\tau, \Theta], \mathbf{X}[\Theta]$ of course need not be convex.

The Total Reachability Sets

Definition 2.7.4. *The* **total backward reachability set** *from position* $\{\vartheta, \mathcal{M}\}$ *is the union*

$$\mathbf{W}_-[\vartheta] = \mathbf{W}_-(\vartheta, \mathcal{M}) = \bigcup\{\mathcal{W}(\tau, \vartheta, \mathcal{M}) \mid \tau \in (-\infty, \vartheta]\}.$$

2.8 Dynamic Programming: Time-Optimal Control

This is the set of all points from which set $\mathcal{M}$ may be reached at time ϑ from **some** finite $\tau \in (-\infty, \vartheta]$. For an autonomous system the set $\mathbf{W}_-[\vartheta]$ does not depend on ϑ.

Denote $\mathbf{V}_-(\vartheta, x) = \inf_\tau \{V(\tau, x|\vartheta, V(\vartheta, \cdot)) \mid \tau \in (-\infty, \vartheta]\}$, $V(\vartheta, x) = d^2(x, \mathcal{M})$.

Lemma 2.7.6. *The total backward reachability set $\mathbf{W}_-[\vartheta]$ is the level set*

$$\mathbf{W}_-[\vartheta] = \{x : \mathbf{V}_-(\vartheta, x) \le 0\}.$$

Definition 2.7.5. The **total forward reachability set** $\mathbf{X}_+[t_0]$ from position $\{t_0, \mathcal{X}^0\}$ is the union

$$\mathbf{X}_+[t_0] = \mathbf{X}_+(t_0, \mathcal{X}^0) = \bigcup \{\mathcal{X}(\vartheta, t_0, \mathcal{X}^0) \mid \vartheta \in [t_0, \infty)\}.$$

This is the set of all points that can be reached at some time $\vartheta \ge t_0$. For an autonomous system the set $\mathbf{X}_+[t_0]$ does not depend on t_0.

Denote $\mathbf{V}_+(t_0, x) = \inf\{V_0(\vartheta, x \mid t_0, V_0(t_0, \cdot)) \mid \vartheta \in [t_0, \infty)\}$, where $V_0(t_0, x) = d^2(x, \mathcal{X}^0)$.

Lemma 2.7.7. *The total reachability set $\mathbf{X}_+[t_0]$ is the level set*

$$\mathbf{X}_+[t_0] = \{x : \mathbf{V}_+(t_0, x) \le 0\}.$$

Exercise 2.7.2. For a linear autonomous system (1.44) with $u(t) \in \mathcal{P} = const$, $0 \in \text{int}\mathcal{P}$ and $\mathcal{X}^0 \ne \{0\}$ given, describe conditions when $\mathbf{X}_+ = \mathbf{X}_+[0] = \mathbb{R}^n$ and when $\mathbf{X}_+ \ne \mathbb{R}^n$.

Exercise 2.7.3. Suppose in Definition 2.7.2 we have $\Theta = [t_0, \sigma]$. Find a direct HJB-type relation for $\mathbf{V}^0(\sigma, x) = \mathbf{V}^0(\Theta, x)$ without reducing it to (2.106).

2.8 Dynamic Programming: Time-Optimal Control

We shall now use the notion of reachability to solve the problem of closed-loop time-optimal control.

Problem 2.8.1. Given system (2.25), hard bound $u \in \mathcal{P}$, starting position $\{t, x\}$ and target set $\mathcal{M}$, $x \notin \mathcal{M}$, find

$$\tau^0(t, x) = \min\{\tau : x(\tau) \in \mathcal{M}\}. \tag{2.111}$$

Here target set $\mathcal{M}$ is taken similar to the above, being convex and compact. In particular, one may have $\mathcal{M} = \{m\}$ as an isolated point.

The solution scheme is as follows: for the given starting position $\{t, x\}$, we construct the reachability tube $X[\tau] = X(\tau, t, x)$, $\tau \geq t$, then find the first instant of time $\tau^*(t, x)$ when $X[\tau]$ touches M. The value $\tau^*(t, x) - t$ will be the minimal time for reaching set M from position $\{t, x\}$, namely, $\tau^*(t, x) = \tau^0(t, x)$.

To calculate $X[\tau]$, recall problem (2.46), (2.47). Then

$$X[\tau] = \{z : V_0(\tau, z) \leq 0\},$$

where function

$$V_0(\tau, z) = V_0(\tau, z | t, x) = \min_u \{d^2(x(t), x) \mid x(\tau) = z\}$$

satisfies the HJB equation (2.50) with boundary condition

$$V_0(t, z) = d^2(z, x).$$

Let us now look how to find the first instant of time, when $X[t]$ touches M.

From Sect. 2.3 it follows that for inclusion $m \in X[\tau]$ to be true it is necessary and sufficient that $V_0(\tau, m) = 0$. Now, having noted that $V_0(t, m) > 0$, solve equation $V_0(\tau, m) = 0$ and find the smallest root $\tau^0 > t$ of this equation. The number τ^0 exists if point m is reachable in finite time.

Suppose set $M = \{m\}$, then $\tau^0 = \tau^0(t, x)$ is already the solution to Problem 2.8.1. If not, consider equation

$$\min\{V_0(\tau, m) \mid m \in M\} = f(\tau) = 0, \tag{2.112}$$

Here $f(\tau) = f(\tau \mid t, x) > 0$ and $f(\tau) = 0$ for some $\tau > 0$, provided set M is reachable in finite time.[5]

Taking the smallest root $\tau^0(t, x)$ of equation

$$f(\tau \mid t, x) = 0, \tag{2.113}$$

we come to the solution of Problem 2.8.1.

Theorem 2.8.1. *The solution of Problem 2.8.1 is the smallest root $\tau^0(t, x)$ of Eq. (2.113).*

In order to better understand Eq. (2.113) we look at a simple example.

---

[5] A closed set Q is said to be reachable in finite time if the intersection $Q \cap \mathbf{X}_+(t, x) \neq \emptyset$ for some $\tau > t$. Here $\mathbf{X}_+(t, x)$ is the total (forward) reachability set from position $\{t, x\}$.

Example 8.1

Consider the one-dimensional system

$$\dot{x} = u + f(t), \quad |u| \leq \mu,$$

with target set $\mathcal{M} = m = \{0\}$.

The value function $V_0(t,z)$ for this case, calculated according to (2.77), is as follows:

$$V_0(t,z) = \min_u \{d(x(0),x) \mid x(t) = z\} =$$

$$= \max_l \{l(z-x) - \mu \int_0^t |l|\,ds - \int_0^t lf(s)\,ds \mid |l| \leq 1\}.$$

This gives:

$$V_0(t,z) = \begin{cases} z - x - \mu t - \int_0^t f(s)\,ds & \text{if } z > x + \int_0^t f(s)\,ds, \\ -z + x - \mu t + \int_0^t f(s)\,ds & \text{if } z < x + \int_0^t f(s)\,ds. \end{cases}$$

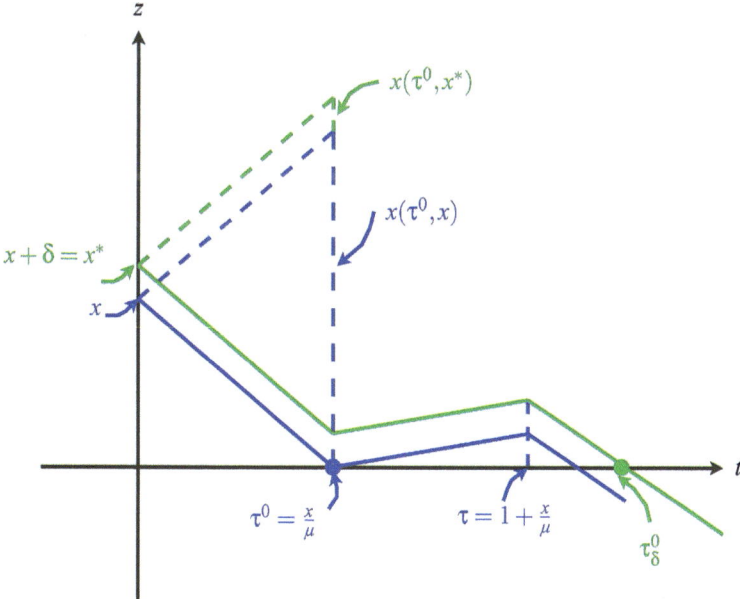

Fig. 2.4 Discontinuity of optimal time τ^0 in x

Now, take for $x > 0$

$$f(t) = \begin{cases} 0 & \text{if } 0 \le t \le x/\mu, \\ (\mu + \epsilon) & \text{if } 1 + x/\mu > t > x/\mu, \ \epsilon > 0, \\ -r & \text{if } t \ge 1 + x/\mu, \ r > 0. \end{cases}$$

Then one may notice that the minimal time for reaching $m = 0$ is $\tau^0(t, x) = x/\mu$. However, with starting position $\{t, x^*\}$ instead of $\{t, x\}$, where $x^* = x + \delta > x$, $\delta > 0$, the minimal time will be $\tau^0(t, x^*) = \tau^0(t, x) + 1 + (\epsilon + \delta)/(r + \mu)$. Hence, function $\tau^0(t, x)$ *is discontinuous in x*. This example yields the next remark.

Remark 2.8.1. In general the **minimal time** $\tau^0(t, x)$ **is not robust** relative to the starting point x.

Exercise 2.8.1. Indicate additional conditions for Problem 2.8.1 that ensure continuity of function $\tau^0(t, x)$ in $\{t, x\}$ (Fig. 2.4).

Chapter 3
Ellipsoidal Techniques: Reachability and Control Synthesis

Abstract This chapter describes the ellipsoidal techniques for control problems introduced in earlier chapters. We derive formulas for reachability sets using the properties of ellipsoids and relations from convex analysis. The formulas are derived through *inductive procedures*. They allow calculation of both *external* and *internal* ellipsoidal approximations of forward and backward reachability sets with any desired level of accuracy. The approximations are illustrated on examples explained in detail, then followed by ellipsoid-based formulas for problems of reachability and control synthesis.

Keywords Ellipsoidal approximations • Parametrization • Parallel calculation • Ellipsoidal maximum principle • Reachability • Control synthesis

In this chapter we begin to apply ellipsoidal techniques to problems of control. We consider systems that operate in the absence of uncertainty. We propose methods to represent trajectory tubes for controlled systems through parameterized varieties of ellipsoidal-valued tubes which achieve external and internal approximation of set-valued solution functions with any desired accuracy. This is done by using intersections and unions of a finite number of ellipsoidal tubes. Then, by appropriately tending the number of approximating ellipsoids to infinity, one may approach the exact solutions. We describe the construction of forward and backward set-valued trajectories of reach sets (with these being taken either at given instants of time or calculated for preassigned time-intervals). The routes for constructing the external and internal ellipsoidal approximations employ algorithms that are based on recurrent procedures. They do not require to be calculated "afresh" for any new instant of time. The structure of the suggested solutions also allows their natural parallelization. The results of Sects. 3.1–3.11 of this chapter are based on those given in [177, 181, 182]. On the other hand, for the problem of internal ellipsoidal approximations there exists a "second approach" which introduces other formulas than in Sects. 3.7–3.10. A concise review of this approach is given in Sect. 3.12. The "second approach" may also be recommended as a solution tool for problems of reachability and control synthesis. This approach was discussed in [45, 174].

We thus start to introduce an array of methods for treating trajectory tubes rather than isolated trajectories. Of special importance are methods that permit *systems of high dimensions* in limited time. Such types of examples are presented in Chap. 4, being solved with the aid of the *ellipsoidal toolbox*, [132].

3.1 Linear Systems Under Ellipsoidal Constraints

Consider the linear *time-variant* system model (1.1) described as

$$\dot{x}(t) = A(t)x(t) + B(t)u + f(t). \tag{3.1}$$

The values u of the controls are restricted for almost all t by *hard bounds*

$$u \in Q(t), \tag{3.2}$$

where $Q(t)$ is a set-valued function $Q : T \to \text{conv}\mathbb{R}^q$, Hausdorff-continuous in t, $f(t)$ is a given integrable function.

Considering all the possible functions of time $u(t)$, $t \in T$,—the *open-loop controls*—restricted by constraint (3.2), we came to a *linear differential inclusion*

$$\dot{x}(t) \in A(t)x(t) + B(t)Q(t) + f(t) \tag{3.3}$$

under condition

$$x^{(0)} = x(t_0) \in \mathcal{X}^0,$$

where $\mathcal{X}^0 \in \text{conv}\mathbb{R}^n$.

In this chapter we further restrict the constraints on u, x^0 to be *ellipsoidal-valued*, namely:

$$< u - q(t), Q^{-1}(t)(u - q(t)) > \le 1, \tag{3.4}$$

and

$$< x^{(0)} - x^0, (X^0)^{-1}(x^{(0)} - x^0) > \le 1, \tag{3.5}$$

where continuous functions $q(t)$ and vector x^0 are given together with continuous, positive, symmetric matrix-valued function $Q'(t) = Q(t) > 0$ and matrix $(X^0)' = X^0 > 0$. Here the respective ellipsoids are *nondegenerate*.

In terms of inclusions we have

$$u \in \mathcal{E}(q(t), Q(t)), \tag{3.6}$$

$$x(t_0) = x^{(0)} \in \mathcal{E}(x^0, X^0), \tag{3.7}$$

3.1 Linear Systems Under Ellipsoidal Constraints

or, in terms of support functions, the inequalities

$$<l,u> \leq <l,q(t)> + <l,Q(t)l>^{\frac{1}{2}}, \tag{3.8}$$

$$<l,x^{(0)}> \leq <l,x^0> + <l,X^0 l>^{\frac{1}{2}}, \tag{3.9}$$

$$\forall l \in \mathbb{R}^n .$$

Remark 3.1.1. Note that constraints in the form (3.8), (3.9) also allow *degenerate* matrices $Q(t) \geq 0$, $X^0 \geq 0$. In this case ellipsoids $\mathcal{E}(q(t), Q(t))$ or $\mathcal{E}(x^0, X^0)$ turn out to be *elliptical cylinders*. This case will be discussed separately in Chap. 4.

Let us now start with the *reachability* issue described in Chap. 1, Sect. 1.4. Note that we further take $q(t) \neq 0$ and presume $f(t) \equiv 0$, since keeping it here does not add much to the procedure. The case $f(t) \neq 0$ will be important in Chap. 10, while considering systems under uncertainty.

The reach set $\mathcal{X}[\tau]$ may be treated as the cut $\mathcal{X}[\tau] = \mathcal{X}(\tau, t_0, \mathcal{E}(x^0, X^0))$ of the solution tube $\mathcal{X}(\cdot) = \{\mathcal{X}[t] : t \geq t_0\}$ to the differential inclusion ($t \geq t_0$),

$$\dot{x} \in A(t)x + \mathcal{E}(B(t)q(t), B(t)Q(t)B'(t)), \quad x^{(0)} \in \mathcal{E}(x^0, X^0). \tag{3.10}$$

The reach set $\mathcal{X}(t, t_0, \mathcal{E}(x^0, X^0))$ may also be presented through a set-valued ("Aumann") integral (see [9]).

Lemma 3.1.1. *The following relation is true*

$$\mathcal{X}(t, t_0, \mathcal{E}(x^0, X^0)) = x^*(t) + G(t, t_0)\mathcal{E}(0, X^0) + \int_{t_0}^t G(t, s)\mathcal{E}(0, B(s)Q(s)B'(s))ds, \tag{3.11}$$

where

$$x^*(t) = G(t, t_0)x^0 + \int_{t_0}^t G(t, s)B(s)q(s)ds. \tag{3.12}$$

A standard direct calculation indicates

Lemma 3.1.2. *The support function*

$$\rho(l|\mathcal{X}(t, t_0, \mathcal{E}(x^0, X^0))) = \tag{3.13}$$

$$<l, x^*(t)> + <l, G(t, t_0)X^0 G'(t, t_0)l>^{1/2} +$$

$$+ \int_{t_0}^t <l, G(t, s)B(s)Q(s)B'(s)G'(t, s)l>^{1/2} ds.$$

A direct consequence of the last representation leads to the next fact.

Lemma 3.1.3. *The reach set* $X[t] = X(t, t_0, \mathcal{E}(x^0, X^0))$ *is a convex compact set in* $\mathbb{R}^n$ *that evolves continuously in* t.

The continuity in time of a convex compact set $X[t]$ is understood here as the continuity in time of the support function $\rho(l|X[t])$ uniformly in $l :< l, l >\leq 1$. For an ellipsoidal-valued function $\mathcal{E}(q(t), Q(t))$ continuity means that the center $q(t)$ and "shape matrix" $Q(t)$ are continuous.

An obvious consequence of Corollary 1.3.1 is the next property.

Lemma 3.1.4. *System (3.1) with* $f(t) = 0$, *is completely controllable iff the quadratic form*

$$W_Q(\tau, \sigma) = \int_\sigma^\tau <l, G(\tau, s) B(s) Q(s) B'(s) G'(\tau, s) l> ds, \; l \in \mathbb{R}^n,$$

is positive definite for any $\tau > \sigma$, *provided* $Q(s) = Q'(s) > 0$.

The boundary $\partial X[\tau]$ of set $X[\tau]$ may be here defined as $\partial X[\tau] = X[\tau] \setminus int X[\tau]$. Under the controllability condition of the above lemma set $X[\tau]$ has a non-void interior $int X[\tau] \neq \emptyset$ for $\tau > t_0$.

We further assume the following property.

Assumption 3.1.1. *The symmetric matrix* $W_Q(\tau, \sigma)$ *is positive definite whenever* $\tau > \sigma$.

Points on the boundary of the reach set $X[t]$ satisfy some important properties. Namely, consider point $x^* \in \partial X[\tau]$. Then there exists a related *support vector* l^* such that

$$<l^*, x^*> = \rho(l^*|X[\tau]) = \max\{<l^*, x> \,|x \in X[\tau]\}. \quad (3.14)$$

Denote $u = u^*(t)$ to be the control which transfers system (3.10) from state $x(t_0) = x^{(0)}$ to $x(\tau) = x^*$. Then the Maximum Principle of Sect. 2.5 appears as follows [120, 195, 226].

Theorem 3.1.1. *Suppose the state* $x(\tau) = x^*$ *is given and* $x^* \in \partial X[\tau]$. *Then the control* $u = u^*(t)$ *and the initial state* $x(t_0) = x^{(0)}$, *which yield the unique trajectory* $x^*(t)$ *that reaches point* $x^* = x^*(\tau)$ *from point* $x^{(0)} = x(t_0)$, *and ensures relations (3.14), satisfy the following "maximum principle" for the control:*

$$<l^*, G(\tau, t) B(t) u^*(t)> = \quad (3.15)$$

$$= \max\{<l^*, G(\tau, t) B(t) u> \,|u \in \mathcal{E}(q(t), Q(t))\} =$$

$$=<l^*, G(\tau, t) B(t) q(t)> + <l^*, G(\tau, t) B(t) Q(t) B'(t) G'(\tau, t) l^*>^{1/2}$$

for all $t \in [t_0, \tau]$, and the "maximum" (transversality) condition for the initial state:

$$< l^*, G(\tau, t_0)x^{(0)} > = \max\{< l^*, x > | x \in G(\tau, t_0)\mathcal{E}(x^0, X^0)\} = \qquad (3.16)$$

$$= < l^*, G(\tau, t_0)x^0 > + < l^*, G(\tau, t_0)X^0 G'(\tau, t_0)l^* >^{1/2}.$$

Here l^* is the support vector for set $X[t]$ at point x^* that satisfies relation (3.14).

The product $l'G(\tau, t) = s[t] = s(t, \tau, l)$ may be presented as the backward solution to the "adjoint system"

$$\dot{s} = -sA(t), \quad s[\tau] = l',$$

where s is a row-vector.

Remark 3.1.2. The ellipsoidal nature of the constraints on $u(t), x^{(0)}$ yield uniqueness of the optimal control and trajectory. (Prove this fact.)

3.2 Ellipsoidal Approximation of Reach Sets

Observe that although the initial set $\mathcal{E}(x^0, X^0)$ and the control set $\mathcal{E}(q(t), Q(t))$ are ellipsoids, the reach set $X[t] = X(t, t_0, \mathcal{E}(x^0, X^0))$ will not generally be an ellipsoid. (Since already the sum of two ellipsoids, $\mathcal{E}_1 + \mathcal{E}_2$, is generally not an ellipsoid.)

As indicated in [174], the reach set $X[t]$ may be approximated both externally and internally by ellipsoids $\mathcal{E}_-$ and $\mathcal{E}_+$, with $\mathcal{E}_- \subseteq X[t] \subseteq \mathcal{E}_+$.

Definition 3.2.1. An external approximation $\mathcal{E}_+$ of a reach set $X[t]$ is **tight** if there exists a vector $l \in \mathbb{R}^n$ such that

$$\rho(\pm l | \mathcal{E}_+) = \rho(\pm l | X[t]).$$

The last definition is relevant for reach sets which are compact convex bodies, symmetrical around the center $x^*(t)$ (see formula (3.14) of the above). However it does not produce a unique ellipsoid. A more precise notion is given through the next definitions, where tightness is defined within a certain subclass $\mathbf{E}_+$ of ellipsoids that does not coincide with the variety of all possible ellipsoids.

Definition 3.2.2. An external approximation $\mathcal{E}_+$ is **tight** in the class $\mathbf{E}_+$, if for any ellipsoid $\mathcal{E} \in \mathbf{E}_+$, the inclusions $X[t] \subseteq \mathcal{E} \subseteq \mathcal{E}_+$ imply $\mathcal{E} = \mathcal{E}_+$.

This paper is concerned with external approximations, where class $\mathbf{E}_+ = \{\mathcal{E}_+\}$ is described through the following definition.

Definition 3.2.3. The class $\mathbf{E}_+ = \{\mathcal{E}_+\}$ consists of ellipsoids that are of the form $\mathcal{E}_+[t] = \mathcal{E}(x^\star, X_+[t])$, where $x^\star(t)$ satisfies the equation[1]

$$\dot{x}^\star = A(t)x^\star + B(t)q(t), \ x^\star(t_0) = x^0, \ t \geq t_0, \qquad (3.17)$$

and

$$X_+[t] = X_+(t|p[\cdot]) = \qquad (3.18)$$

$$= \left(\int_{t_0}^t p(s)ds + p_0(t)\right)\left(\int_{t_0}^t p^{-1}(s)G(t,s)B(s)Q(s)B'(s)G'(t,s)ds\right.$$

$$\left. + p_0^{-1}(t)G(t,t_0)X^0G'(t,t_0)\right).$$

Here $X^0, Q(s), s \in [t_0, t]$ are any positive definite matrices with function $Q(s)$ continuous, $q(t)$ is any continuous function, $p(s) > 0$, $s \in [t_0, t]$, is any integrable function and $p_0(t) > 0$.

In particular, this means that if an ellipsoid $\mathcal{E}(x^\star, X_+^*) \supseteq X[t]$ is tight in $\mathbf{E}_+$, then there exists no other ellipsoid of type $\mathcal{E}(x^\star, kX_+^*), k \leq 1$ that satisfies the inclusions $X[t] \subset \mathcal{E}(x^\star, kX_+^*) \subset \mathcal{E}(x^\star, X_+^*)$ (ellipsoid $\mathcal{E}(x^\star, X_+^*)$ touches set $X[t]$).

Definition 3.2.4. We further say that external ellipsoids are **tight** if they are tight in $\mathbf{E}_+$.

Functions $p(t)$ are parameterizing parameters which generate the class $\mathbf{E}_+$ of approximating ellipsoids and we actually further deal only with ellipsoids $\mathcal{E}_+ \in \mathbf{E}_+$. For problems of this book Definition 3.2.1 is motivated by 3.2.4, so we usually apply Definition 3.2.1, only within the class $\mathbf{E}_+$. Then, however, with direction l given, the related tight ellipsoid will be unique.

The class $\mathbf{E}_+$ is rich enough to arrange effective approximation schemes, though it does not include all possible ellipsoids. In fact, all the schemes of books [44, 174] on external ellipsoids actually do not go beyond the class $\mathbf{E}_+$. A justification for using class $\mathbf{E}_+$ is due to the next proposition.

Theorem 3.2.1. *The following inclusions are true:*

$$X[t] \subseteq \mathcal{E}(x^\star(t), X_+(t \ |p[\cdot])), \ \forall p_0, p(s) > 0, \ t_0 \leq s \leq t. \qquad (3.19)$$

Moreover,

$$X[t] = \cap\{\mathcal{E}(x^\star(t), X_+(t \mid p[\cdot])|p_0, p(s) > 0, \ s \in [t_0, t]\}. \qquad (3.20)$$

---

[1] In the following formulas symbol $p[\cdot]$ stands for the pair $\{p_0, \ p(s), s \in [t_0, t]\}$.

3.2 Ellipsoidal Approximation of Reach Sets

Relations (3.19), (3.20) are true for all $p_0 \geq 0$ and $p(s)$ ranging over all continuous positive functions. These results follow from *ellipsoidal calculus* (see [174], Sects. 2.1, 2.7). What also holds is the following characterization of tight external ellipsoids.

Theorem 3.2.2. *For a given time τ the tight external ellipsoids $\mathcal{E}_+[\tau] = \mathcal{E}(x^\star(\tau), X_+[\tau])$, with $\mathcal{E}_+[\tau] \in \mathbf{E}_+$, are those for which the functions p and p_0 are selected as*

$$p(s) = < l, G(\tau, s) B(s) Q(s) B'(s) G'(\tau, s) l >^{1/2}, \quad t_0 \leq s \leq \tau, \tag{3.21}$$

$$p_0(\tau) = < l, G(\tau, t_0) X^0 G'(\tau, t_0) l >^{1/2}, \tag{3.22}$$

with vector l given. Each ellipsoid $\mathcal{E}_+[\tau]$ touches $X[\tau]$ at points

$$\{x^* :< l, x^* >= \rho(l | X[\tau])\},$$

so that

$$\rho(l | X[\tau]) = < l, x^*(\tau) > + < l, X_+[\tau] l >^{1/2} = \rho(l | \mathcal{E}(x^\star(\tau), X_+[\tau])).$$

Remark 3.2.1. Note that the result above requires the evaluation of the integrals in (3.18) for each time τ and vector l. If the computation burden for each evaluation of (3.18) is C, and we estimate the reach tube via (3.19) for N values of time τ and L values of l, the total computation burden would be $C \times N \times L$. A solution to the next problem would reduce this burden to $C \times L$.

Problem 3.2.1. Given a unit-vector function $l^*(t)$, $< l^*, l^* >= 1$, continuously differentiable in t, find an external ellipsoid $\mathcal{E}_+^*[t] \supseteq X[t]$ that would ensure **for all** $t \geq t_0$, the equality

$$\rho(l^*(t) | X[t]) = \rho(l^*(t) | \mathcal{E}_+[t]) = < l^*(t), x^*(t) >, \tag{3.23}$$

so that the supporting hyperplane for $X[t]$ generated by $l^*(t)$, namely, the plane $\langle x - x^*(t), l^*(t) \rangle = 0$ that touches $X[t]$ at point $x^*(t)$, would also be a supporting hyperplane for $\mathcal{E}_+^*[t]$ and touch it at the same point.

This problem is solvable in the class $\mathbf{E}_+$. In order to solve it, we first make use of Theorem 3.2.1. However, the function $p(s)$ and vector p_0 (3.21), (3.22), used for the parametrization in (3.18), will now depend on an additional variable t, so that $p(s) = p_t(s)$ will be a function of two variables, s, t, and $p_0 = p_0(t)$—since the requirement is that relation (3.23) should now hold *for all* $t \geq t_0$.

Theorem 3.2.3. *With $l(t) = l^*(t)$ given, the solution to Problem 3.2.1 is an ellipsoid $\mathcal{E}_+[t] = \mathcal{E}(x^*(t), X_+^*[t])$, where*

$$X_+^*[t] = \qquad (3.24)$$

$$\left(\int_{t_0}^t p_t^*(s)ds + p_0^*(t)\right)\left(\int_{t_0}^t (p_t^*(s))^{-1} G(t,s)B(s)Q(s)B'(s)G'(t,s)ds\right.$$

$$\left. + p_0^{*-1}(t)G(t,t_0)X^0 G'(t,t_0)\right),$$

and

$$p_t^*(s) = <l^*(t), G(t,s)B(s)Q(s)B'(s)G'(t,s)l^*(t)>^{1/2}, \qquad (3.25)$$

$$p_0^*(t) = <l^*(t_0), G(t,t_0)X^0 G'(t,t_0)l^*(t_0)>^{1/2}.$$

The proof is obtained by direct substitution, namely, by substituting (3.25) into (3.24) and further comparing the result with (3.13), (3.23). Relations (3.24), (3.25) need to be solved "afresh" for each t. It may be more convenient for computational purposes to have them given in the form of recurrent relations generated through differential equations.

Remark 3.2.2. In all the ellipsoidal approximations considered in this book the center of the approximating ellipsoid is always the same, being given by $x^*(t)$ of (3.17).

The discussions in the next section will therefore actually concern only the relations for the matrices $X_+[t], X_+^*[t]$ of these ellipsoids ($x^*(t) \equiv 0$).

3.3 Recurrent Relations: External Approximations

We start with a particular case.

Assumption 3.3.1. *The function $l^*(t)$ is of the following form $l^*(t) = G'(t_0, t)l$, with $l \in \mathbb{R}^n$ given. For the time-invariant case $l^*(t) = e^{-A'(t-t_0)}l$.*

Such curves $l(t)$ will be further referred to as "good ones." Under Assumption 3.3.1 the vector $l^*(t)$ may be expressed as the solution to equation

$$\dot{l}^* = -A'(t)l^*, \quad l^*(t_0) = l,$$

which is the adjoint to the homogeneous part of Eq. (1.1).

3.3 Recurrent Relations: External Approximations

Then $p_t^*(s)$, $p_0^*(t)$, $X_+^*[t]$ of (3.24), (3.25) transform into

$$p_t^*(s) = \qquad (3.26)$$

$$=<l,G(t_0,s)B(s)Q(s)B'(s)G'(t_0,s)l>^{1/2}= p^*(s); \quad p_0^*(t) =<l,X^0l>^{1/2}= p_0^*,$$

and

$$X_+^*[t] = G(t,t_0)X_+(t)G'(t,t_0), \qquad (3.27)$$

$$X_+[t] = \left(\int_{t_0}^t p^*(s)ds + p_0^*\right)\Psi(t), \qquad (3.28)$$

where

$$\Psi(t) =$$

$$= \int_{t_0}^t <l,G(t_0,s)B(s)Q(s)B'(s)G'(t_0,s)l>^{-1/2} G(t_0,s)B(s)Q(s)B'(s)G'(t_0,s)ds +$$

$$+ <l,X^0l>^{-1/2} X^0.$$

In this particular case $p_t^*(s)$ does not depend on t ($p_{t'}^*(s) = p_{t''}^*(s)$ for $t' \neq t''$) and the lower index t may be dropped.

Direct differentiation of $X_+[t]$ yields

$$\dot{X}_+[t] = \qquad (3.29)$$

$$= \pi^*(t)X_+[t] + \pi^{*-1}(t)G(t_0,t)B(t)Q(t)B'(t)G'(t_0,t), \quad X_+[t_0] = X^0,$$

where

$$\pi^*(t) = p^*(t)\left(\int_{t_0}^t p^*(s)ds + p_0^*\right)^{-1}.$$

Calculating

$$<l,X_+[t]l> = \left(\int_{t_0}^t p^*(s)ds + p_0^*\right) <l,\Psi(t)l> = \left(\int_{t_0}^t p^*(s)ds + p_0^*\right)^2,$$

one may observe that

$$\pi^*(t) =<l,G(t_0,t)B(t)Q(t)B'(t)G'(t_0,t)l>^{1/2}<l,X_+[t]l>^{-1/2}. \qquad (3.30)$$

In order to pass to the matrix function $X_+^*[t]$ we note that

$$\dot{X}_+^*[t] = A(t)G(t,t_0)X_+[t]G'(t,t_0) + G(t,t_0)X_+[t]G'(t,t_0)A'(t) + G(t,t_0)\dot{X}_+[t]G'(t,t_0).$$

After a substitution from (3.29) this gives

$$\dot{X}_+^* = A(t)X_+^* + X_+^* A'(t) + \quad (3.31)$$

$$+\pi^*(t)X_+^* + \pi^{*-1}(t)B(t)Q(t)B'(t), \quad X_+^*(t_0) = X^0.$$

We shall denote the ellipsoid constructed with matrix $X_+^*(t)$ described by Eqs. (3.31), (3.30), as $E(x^*(t), X_+^*(t))$. We now summarize the last results.

Theorem 3.3.1. *Under Assumption 3.3.1 the solution to Problem 3.2.1 is given by the ellipsoid $E_+^*[t] = E(x^*(t), X_+^*[t])$, where $x^*(t)$ satisfies Eq. (3.17) and $X_+^*[t]$ is a solution to Eqs. (3.31), (3.30).*

Since set $X_+^*[t]$ depends on vector $l \in \mathbb{R}^n$, we further denote $X_+^*[t] = X_+^*[t]_l$, using it in the following text whenever it will be necessary.

Theorem 3.3.2. *For any $t \geq t_0$ the reach set $X[t]$ may be described as*

$$X[t] = \cap\{E(x^*, X_+^*[t]) \mid l :< l, l >= 1\}. \quad (3.32)$$

This is a direct consequence of Theorem 3.2.2 and of the selection of good curves for representing the solution. Differentiating $\pi^*(t)$, according to (3.30), we arrive at the next result.

Corollary 3.3.1. *If $B(t) \equiv B \equiv const$ and function $\pi^*(t)$ is differentiable, then it satisfies the differential equation*

$$\dot{\pi}^* = f(t)\pi^* - \pi^{*2}, \quad \pi(t_0) = 1, \quad (3.33)$$

where

$$f(t) = \frac{<l, G(t_0,t)(-A(t)B(t)Q(t)B'(t) - B(t)Q(t)B'(t)A'(t) + B(t)\dot{Q}(t)B'(t))G'(t_0,t)l>}{2<l, G(t_0,t)B(t)Q(t)B'(t)G'(t_0,t)l>}.$$

Corollary 3.3.2. *With $A, B = const$, $Q(t) = Q = const$, and l being an eigenvector of A' ($A'l = \lambda l$), with real eigenvalue λ, the function $f(t)$ will be*

$$f(t) = -\lambda.$$

Thus, if $l^*(t)$ satisfies Assumption 3.3.1, the complexity of computing a tight, external ellipsoidal approximation to the reach set for all t, is the same as computing the solution to the differential equation (3.31).

3.3 Recurrent Relations: External Approximations

In the general case, differentiating relation (3.24) for $X_+^*[t]$, we have

$$\dot{X}_+^*[t] =$$

$$= \left(p_t^*(t) + \int_{t_0}^t (\partial p_t^*(s)/\partial t)\,ds + \dot{p}_0^*\right)\left(\int_{t_0}^t p_t^{*-1}(s)G(t,s)B(s)Q(s)B'(s)G'(t,s)\,ds\right.$$

$$\left. + p_0^{*-1}(t)G(t,t_0)X^0 G'(t,t_0)\right)$$

$$+ \left(\int_{t_0}^t p_t^*(s)\,ds + p_0^*(t)\right)\left((p_t^*(t))^{-1}B(t)Q(t)B'(t) + \right.$$

$$\int_{t_0}^t (\partial((p_t^*(s))^{-1}G(t,s)B(s)Q(s)B'(s)G'(t,s))/\partial t)\,ds +$$

$$\left. + d(p_0^{*-1}(t)G(t,t_0)X^0 G'(t,t_0))/dt\right),$$

which, using the notations

$$X_+^*[t] =$$

$$= \left(\int_{t_0}^t p_t^*(s)\,ds + p_0^*(t)\right)\left(\int_{t_0}^t (p_t^*(s))^{-1}\right.$$

$$\left. G(t,s)B(s)Q(s)B'(s)G'(t,s)\,ds + p_0^{*-1}(t)G(t,t_0)X^0 G'(t,t_0)\right),$$

$$\pi^*(t) = p_t^*(t)\left(\int_{t_0}^t p_t^*(s)\,ds + p_0^*(t)\right)^{-1}, \qquad (3.34)$$

gives

$$\dot{X}_+^* = \qquad (3.35)$$

$$= A(t)X_+^* + X_+^* A'(t) + \pi^*(t)X_+^* + \pi^{*-1}(t)B(t)Q(t)B'(t)$$

$$+ G(t,t_0)\phi(t,l(t),B(\cdot)Q(\cdot)B'(\cdot))G'(t,t_0),$$

$$X_+^*(t_0) = X^0.$$

Here

$$\phi(t, l(t), B(t)Q(\cdot)B'(t)) = \quad (3.36)$$

$$= \left(\int_{t_0}^{t} (\partial p_t^*(s)/\partial t)ds + \dot{p}_0^*(t)\right)\left(\int_{t_0}^{t} (p_t^*(s))^{-1} P(t_0, s)ds + (p_0^*(t))^{-1} P_0\right) +$$

$$+ \left(\int_{t_0}^{t} p_t^*(s)ds + p_0^*(t)\right)\left(\int_{t_0}^{t} (\partial (p_t^*(s))^{-1}/\partial t) P(t_0, s)ds + (d(p_0^*(t))^{-1}/dt) P_0\right)$$

and

$$P(t_0, s) = G(t_0, s)B(s)Q(s)B'(s)G'(t_0, s), \quad P_0 = X^0.$$

It is important to note that under Assumption 3.3.1 ($l^*(t) = G'(t_0, t)l$) the terms $\partial p_t^*(s)/\partial t = 0$, $\dot{p}_0^*(t) = 0$, so that $\phi(t, l(t), B(\cdot)Q(\cdot)B'(\cdot)) = 0$.

Recall that an external ellipsoid constructed in the general case, with $X_+^*(t)$ taken from Eq. (3.35), (3.34), is denoted as $\mathcal{E}^*(x^*(t), X_+^*(t))$, in contrast with an ellipsoid constructed along good curves, with $X_+^*(t)$ taken from Eqs. (3.31), (3.30) and denoted earlier as $E_+^*[t]$.

Theorem 3.3.3. *The solution $\mathcal{E}_+^*[t] = \mathcal{E}(x^*(t), X_+^*(t))$ to Problem 3.2.1 is given by vector function $x^*(t)$ of (3.17) and matrix function $X_+^*(t)$ that satisfies Eq. (3.35), with $\pi^*(t)$ defined in (3.34) and $p_t^*(s)$, $p_0^*(t)$ in (3.25). Under Assumption 3.3.1 Eq. (3.35) appears with term $\phi \equiv 0$.*

Throughout the previous discussion we have observed that under Assumption 3.3.1 the tight external ellipsoidal approximation $\mathcal{E}(x^*(t), X_+^*(t))$ is governed by the simple ordinary differential equations (3.31). Moreover, in this case the points $x^l(t)$ of support for the hyperplanes generated by vector $l(t)$ run along *a system trajectory* of (3.1) which is generated by a control that satisfies the Maximum Principle (3.15) and the relation (3.16) of Theorem 3.1.1.

In connection with this fact the following question arises:

Problem 3.3.1. Given is a curve $l(t)$ such that the supporting hyperplane generated by vectors $l(t)$ touches the reach set $X[t]$ at the point of support $x^l(t)$. What should be the curve $l(t)$, so that $x^l(t)$ would be a system trajectory for (3.1)?

Let us investigate this question when $x^0 = 0$, $q(t) \equiv 0$. Suppose a curve $l(t)$ is such that ($x_t^{l0} = x^l(t_0)$)

$$x^l(t) = G(t, t_0)x_t^{l0} + \int_{t_0}^{t} G(t, \tau) B(\tau) u_t(\tau) d\tau \quad (3.37)$$

is a vector generated by the pair $\{x_t^{l0}, u_t(\cdot)\}$. In order that $x^l(t)$ be a support vector to the supporting hyperplane generated by $l(t)$ it is necessary and sufficient that this

3.3 Recurrent Relations: External Approximations

pair satisfy at each time t the maximum principle (3.15) and the condition (3.16) under constraints

$$x_t^{l0} \in \mathcal{E}(0, X^0), \quad u_t(\tau) \in \mathcal{E}(0, Q(\tau)),$$

which means

$$u_t(\tau) = \frac{Q(\tau)B'(\tau)G'(t,\tau)l(t)}{\langle l(t), G(t,\tau)B(\tau)Q(\tau)B'(\tau)G'(t,\tau)l(t)\rangle^{1/2}}, \qquad (3.38)$$

with $t_0 \le \tau \le t$, and

$$x_t^{l0} = \frac{X^0 G'(t,t_0)l(t)}{<l(t), G(t,t_0)X^0 G'(t,t_0)l(t)>^{1/2}}. \qquad (3.39)$$

Note that since $x^l(t)$ varies in t, the variables $x_t^{l0} = x^l(t_0), u_t(\tau)$ should in general also be taken as being dependent on t.

Now the question is: what should be the function $l(t)$ so that $x^l(t)$ of (3.37)–(3.39) is a trajectory of system (3.1)?

Differentiating (3.37), we come to

$$\dot{x}^l(t) = A(t)\left(G(t,t_0)x_t^{l0} + \qquad (3.40) \right.$$

$$\left. + \int_{t_0}^t G(t,\tau)B(\tau)u_t(\tau)d\tau \right) + B(t)u_t(t) + Y(t),$$

where

$$Y(t) = G(t,t_0)(\partial x_t^{l0}/\partial t) + \int_{t_0}^t G(t,\tau)B(\tau)(\partial u_t(\tau)/\partial t)d\tau. \qquad (3.41)$$

In order that $x^l(t)$ would be a system trajectory for (3.1) it is necessary and sufficient that $Y(t) \equiv 0$. Here the control at time t is $u_t(t)$ and the initial position is x_t^{l0}.

Let us look for the class of functions $l(t)$ that yield $Y(t) = Y_l(t) \equiv 0$. (Since we are about to vary the problem parameters depending on choice of $l(\cdot)$, we further denote $Y(t) = Y_l(t)$ and $p_t(\tau) = p_t(\tau, l)$, $p_0(t) = p_0(t, l)$, see below, at (3.43).)

For any t we have

$$Y_l(t) = G(t,t_0)(\partial (X^0 G'(t,t_0)l(t)(p_0(t,l(t)))^{-1})/\partial t) + \qquad (3.42)$$

$$+ \int_{t_0}^t G(t,\tau)B(\tau)(\partial(Q(\tau)B'(\tau)G'(t,\tau)l(t)(p_t(\tau,l(t)))^{-1})/\partial t)d\tau,$$

where, as in (3.25),

$$p_t(s, l(t)) = <l(t), G(t,s)B(s)Q(s)B'(s)G'(t,s)l(t)>^{1/2}, \qquad (3.43)$$

$$p_0(t, l(t)) = <l(t), G(t,t_0)X^0 G'(t,t_0)l(t)>^{1/2}.$$

Continuing the calculation of $Y_l(t)$, after some transformations we come to

$$Y_l(t) = \Phi_0(t, t_0, l(t))(p_0(t, l(t)))^{-3} \qquad (3.44)$$

$$+ \int_{t_0}^{t} \Phi(t, \tau, l(t))(p_t(\tau, l(t)))^{-3} d\tau,$$

where

$$\Phi_0(t, t_0, l) = P_0(t, t_0)(A'(t)l + dl/dt)\langle l, G(t, t_0)X^0 G'(t, t_0)l\rangle -$$

$$- P_0(t, t_0)l\Big((l'A(t) + dl'/dt)P_0(t, t_0)l\Big),$$

and

$$\Phi(t, \tau, l) = P(t, \tau)(l'A(t) + dl'/dt)'\langle l, P(t, \tau)l\rangle -$$

$$- P(t, \tau)l\Big((l'A(t) + dl'/dt)P(t, \tau)l\Big).$$

Here, in accordance with previous notation,

$$P_0(t, t_0) = G(t, t_0)X^0 G'(t, t_0), \quad P(t, \tau) = G(t, \tau)B(\tau)Q(\tau)B'(\tau)G'(t, \tau).$$

Let us now suppose that $l(t)$ satisfies the equation

$$dl/dt + A'(t)l = k(t)l, \qquad (3.45)$$

where $k(t)$ is a continuous scalar function. Then, by direct substitution, we observe that $Y_l(t) \equiv 0$.

On the other hand, suppose $Y_l(t) \equiv 0$. Let us prove that then $l(t)$ satisfies Eq. (3.45). By contradiction, suppose a function $l_0(t)$ yields $Y_{l_0}(t) \equiv 0$, but does not satisfy (3.45) for any $k(t)$ (including $k(t) \equiv 0$). Then we should have

$$(l'_0(t)A(t) + dl'_0(t)/dt)Y_{l_0}(t) \equiv 0,$$

3.3 Recurrent Relations: External Approximations

so that the expression

$$(l_0'(t)A(t) + dl_0'(t)/dt) \times$$

$$\times \left(\Phi_0(t, t_0, l_0(t))(p_0(t, l_0(t)))^{-3} + \int_{t_0}^{t} \Phi(t, \tau, l_0(t))(p_t(\tau, l_0(t)))^{-3} d\tau \right) =$$

$$= \left((l_0'(t)A(t) + dl_0'(t)/dt) P_0(t, t_0)(A'(t)l_0(t) + dl_0(t)/dt) < l_0(t), P_0(t, t_0)l_0(t) > - \right.$$

$$\left. -(l_0'(t) P_0(t, t_0)(A'(t)l_0(t) + dl_0(t)/dt))^2 \right) p_0(t, l_0(t))^{-3} +$$

$$+ \int_{t_0}^{t} \left((l_0'(t)A(t) + dl_0'(t)/dt) P(t, \tau)(A'(t)l_0(t) + dl_0(t)/dt) < l_0(t), P(t, \tau)l_0(t) > - \right.$$

$$\left. -((l_0'(t)A(t) + dl_0'(t)/dt) P(t, \tau)l)^2 \right) (p_t(\tau, l_0(t)))^{-3} d\tau,$$

should be equal to zero. But due to the Hölder inequality (applied in finite-dimensional version for the first term and in infinite-dimensional version for the integral term) the above expression is equal to zero if and only if $l_0(t)$ satisfies relation (3.45) for some $k(t)$: $(A'(t)l + dl(t)/dt)$ is collinear with $l(t)$.[2] This contradicts the assumption $Y_l(t) \equiv 0$ and proves the following issues.

Theorem 3.3.4. *Given a curve $l(t)$, the function $x_l(t)$ of (3.37)–(3.39), formed of support vectors to the hyperplanes generated by $l(t)$, is a system trajectory for (3.1) if and only if $l(t)$ is a solution to the differential equation (3.45).*

Through similar calculations the following assertion may be proved

Theorem 3.3.5. *In order that $G(t, t_0)\phi(t, l(t), B(\cdot)Q(\cdot)B'(\cdot))G'(t, t_0) \equiv 0$, it is necessary and sufficient that $l(t)$ satisfies Eq. (3.45).*

Exercise 4.1. Work out the proofs of Theorems 3.3.4, 3.3.5 in detail.

We have thus come to the condition that $l(t)$ should satisfy (3.45). This equation may be interpreted as follows. Consider the transformation

$$l_k(t) = \exp\left(-\int_{t_0}^{t} k(s)ds\right) l(t). \tag{3.46}$$

---

[2] Here we also take into account that $(l'(t)A(t) + dl(t)/dt)Y_l(t)$ is bounded with $p_0(t) > 0$ and $p_t(\tau) > 0$ almost everywhere due to the controllability assumption.

Then

$$dl_k(t)/dt = \exp(-\int_{t_0}^t k(s)ds)dl(t)/dt - k(t)\exp(-\int_{t_0}^t k(s)ds)l(t) =$$

$$= (-A'(t) + k(t)I)\exp(-\int_{t_0}^t k(s)ds)l(t) - k(t)\exp(-\int_{t_0}^t k(s)ds)l(t) = -A'(t)l_k(t)$$

which means that $l_k(t)$ is again a solution to the adjoint equation $dl/dt = -A'(t)l$, but in a new scale. However, the nature of transformation (3.46) is such that $l_k(t) = \gamma(t)l(t)$, $\gamma(t) > 0$.

Lemma 3.3.1. *The solutions $l_k(t)$ to Eq. (3.45) generate the same support hyperplanes to the reach set $X[t]$ and the same points of support $x^{lk}(t)$, whatever be the functions $k(t)$ (provided all these solutions have the same initial condition $l_k(t_0) = l_0$).*

In particular, with $k(t) \equiv 0$, we come to functions $l(t)$ of Assumption 3.3.1 and with $Al = kl$ (k is an eigenvalue of constant matrix A) to equation $dl/dt = 0$ and to condition $l(t) = l = const$.

In general the function $l(t)$ of Assumption 3.3.1 is not normalized so that $< l(t), l(t) > \neq 1$. In order to generate a function $l_u(t)$ with vectors $l_u(t)$ of unit length, take the substitution $l_u(t) = l(t) < l(t), l(t) >^{-1/2}$, where $dl/dt + A'(t)l = 0$. Direct calculations indicate the following statement.

Lemma 3.3.2. *Suppose function $l_u(t)$ satisfies Eq. (3.45) with*

$$k(t) = < l(t), A'(t)l(t) >< l(t), l(t) >^{-1},$$

where $l(t)$ satisfies Assumption 3.3.1. Then $< l_u(t), l_u(t) > \equiv 1$.

Remark 3.3.1. Lemma 3.3.1 indicates that the necessary and sufficient conditions for the solution of Problem 3.3.1 do not lead us beyond the class of functions given by Assumption 3.3.1.

Relations (3.35), (3.17), above describe the evolution of the basic parameters $x^\star(t)$, $X_+^\star[t]$ of the ellipsoids $\mathcal{E}_+^\star[t]$. This allows to proceed with the next topic.

3.4 The Evolution of Approximating Ellipsoids

It is known that the dynamics of the reach set $X[t]$, $X[t_0] = \mathcal{E}^0 = \mathcal{E}(x^0, X^0)$ may be described by an evolution equation for the "integral funnel" of the differential inclusion

$$\dot{x} \in A(t)x + B(t)\mathcal{Q}(t), \qquad (3.47)$$

3.4 The Evolution of Approximating Ellipsoids

where $Q(t) = \mathcal{E}(q(t), Q(t))$, so that the set-valued function $X[t]$ satisfies for all t the relation

$$\lim_{\varepsilon \to 0} \varepsilon^{-1} h(X[t+\varepsilon], (I + \varepsilon A(t))X[t] + \varepsilon \mathcal{E}(B(t)q(t), B(t)Q(t)B'(t))) = 0, \quad (3.48)$$

with initial condition $X[t_0] = \mathcal{E}^0$ [158].

Here, as before, $h(Q, \mathcal{M})$ stands for the *Hausdorff distance* between sets $Q, \mathcal{M}$. Recall that

$$h(Q, \mathcal{M}) = \max\{h_+(Q, \mathcal{M}), h_-(Q, \mathcal{M})\}, \quad h_+(Q, \mathcal{M}) = h_-(\mathcal{M}, Q),$$

while the *Hausdorff semidistance* is defined as

$$h_-(\mathcal{M}, Q) = \max_x \min_z \{<x-z, x-z>^{1/2} \mid x \in Q, z \in \mathcal{M}\}.$$

Equation (3.48) has a unique solution.

The idea of constructing *external* ellipsoidal approximations for $X[t]$ is also reflected in the following evolution equation, where we will be interested only in *ellipsoidal-valued solutions*:

$$\lim_{\varepsilon \to 0} \varepsilon^{-1} h_-(\mathcal{E}[t+\varepsilon], (I + \varepsilon A(t))\mathcal{E}[t] + \varepsilon \mathcal{E}(B(t)q(t), B(t)Q(t)B'(t))) = 0 \quad (3.49)$$

with initial condition $\mathcal{E}[t_0] = \mathcal{E}(x^0, X^0) = \mathcal{E}^0$.

Definition 3.4.1. A set-valued function $\mathcal{E}_+[t]$ is said to be the solution to Eq. (3.49) if

(i) $\mathcal{E}_+[t]$ satisfies (3.49) for all t,
(ii) $\mathcal{E}_+[t]$ is ellipsoidal-valued and $\mathcal{E}_+[t] \in \mathbf{E}_+$.
  A solution to Eq. (3.49) is said to be **minimal** in $\mathbf{E}_+$ if together with conditions (i), (ii) it satisfies condition
(iii) $\mathcal{E}_+[t]$ is a minimal in $\mathbf{E}_+$ solution to (3.49) with respect to inclusion.

Condition *(iii)* means that there is no other ellipsoid $\mathcal{E}[t] \neq \mathcal{E}_+[t]$, $\mathcal{E}[t] \in \mathbf{E}_+$ that satisfies both Eq. (3.49) and the inclusions $\mathcal{E}_+[t] \supseteq \mathcal{E}[t] \supseteq X[t]$. This also means that among the minimal solutions $\mathcal{E}_+[t]$ to (3.49) there are non-dominated (tight) ellipsoidal tubes $\mathcal{E}_+[t]$ that contain $X[t]$. For a given initial set $\mathcal{E}^0 = \mathcal{E}[t_0]$ the solutions to (3.49) as well as its minimal solutions *may not be unique.*

Remark 3.4.1. Note that Eq. (3.49) is written in terms of Hausdorff *semidistance* h_- rather than Hausdorff distance h as in (3.48).

Denote an ellipsoidal-valued tube $E[t] = \mathcal{E}(x^\star(t), X[t])$, that starts at $\{t_0, \mathcal{E}^0\}$, $\mathcal{E}^0 = \mathcal{E}(x^0, X^0)$ and is generated by given functions $x^\star(t), X[t]$ (matrix $X[t] = X'[t] > 0$) as $E[t] = E(t|t_0, \mathcal{E}^0) = \mathcal{E}(x^\star(t), X[t])$.

The evolution Eq. (3.49) defines a generalized dynamic system in the following sense.

Lemma 3.4.1. *Each solution $E(t|t_0, \mathcal{E}^0) = E[t]$ to Eq. (3.49), in the sense of Definition 3.4.1, defines a map with* **the semigroup property**

$$E(t|t_0, \mathcal{E}^0) = E(t|E(\tau, |t_0, \mathcal{E}^0)), \quad t_0 \leq \tau \leq t. \tag{3.50}$$

This follows from Definition 3.4.1. Equation (3.49) implies that a solution $E[t]$ to this equation should satisfy the inclusion

$$E[t + \epsilon] + o(\epsilon)\mathcal{B}(0) \supseteq (I + \epsilon A(t))E[t] + \epsilon \mathcal{E}(B(t)q(t), B(t)Q(t)B'(t)),$$

where $\epsilon^{-1}o(\epsilon) \to 0$ with $\epsilon \to 0$ and $\mathcal{B}(0) = \{x : \langle x, x \rangle \leq 1\}$ is the unit ball in $\mathbb{R}^n$. Further denote $q_B(t) = B(t)q(t)$, $Q_B(t) = B(t)Q(t)B'(t)$.

Given a continuously differentiable function $l(t), t \geq t_0$, and assuming that $E[t] = \mathcal{E}(x^*(t), X[t])$ is defined up to time t, let us select $E[t + \epsilon]$ as an external ellipsoidal approximation of the sum

$$Z(t + \epsilon) = (I + \epsilon A(t))\mathcal{E}(x^*(t), X[t]) + \epsilon \mathcal{E}(q_B(t), Q_B(t)),$$

requiring that it touches this sum $Z(t + \epsilon)$ (which, by the way, is not an ellipsoid), according to relation

$$\rho(l(t + \epsilon)|E[t + \epsilon]) = \rho(l(t + \epsilon)|Z(t + \epsilon)), \tag{3.51}$$

that is at those points $z(t + \epsilon)$ where $Z(t + \epsilon)$ is touched by its supporting hyperplane generated by vector $l(t + \epsilon)$. Namely,

$$\{z(t, \epsilon) :< l(t + \epsilon), z(t, \epsilon) >=< l(t + \epsilon), x^*(t + \epsilon) > + < l(t + \epsilon), X[t + \epsilon]l(t + \epsilon) >^{1/2}=$$

$$=< l(t + \epsilon), x^*(t + \epsilon) > + < l(t + \epsilon), (I + \epsilon A(t))X[t](I + \epsilon A(t))'l(t + \epsilon) >^{1/2} +$$

$$+\epsilon < l(t + \epsilon), Q_B(t)l(t + \epsilon) >^{1/2} + o_1(\epsilon)\}.$$

In the latter case, according to the definitions of tightness, the ellipsoid $E[t + \epsilon]$ is among the tight external approximations for $Z(t + \epsilon)$ (relative to terms of order higher than ϵ).[3]

The requirement above is ensured if $E[t + \epsilon]$ externally approximates the sum

$$Z(t + \epsilon) = G(t + \epsilon, t)\mathcal{E}(x^*(t), X[t]) + \int_t^{t+\epsilon} G(t + \epsilon, s)\mathcal{E}(q_B(s), Q_B(s))ds$$

---

[3] Here and in the sequel the terms of type $o_i(\epsilon)$ are assumed to be such that $o_i(\epsilon)\epsilon^{-1} \to 0$ with $\epsilon \to 0$.

3.4 The Evolution of Approximating Ellipsoids

with the same criterion (3.51) as in the previous case. Select

$$X[t+\epsilon] = (1 + p^{-1}(t+\epsilon))G(t+\epsilon,t)X[t]G'(t+\epsilon,t) + \quad (3.52)$$
$$\epsilon^2(1 + p(t+\epsilon))G(t+\epsilon,t)Q_B(t)G'(t+\epsilon,t),$$

where

$$p(t+\epsilon) = \frac{<l(t+\epsilon), G(t+\epsilon,t)X[t]G'(t+\epsilon,t)l(t+\epsilon)>^{1/2}}{\epsilon <l(t+\epsilon), G(t+\epsilon,t)Q_B(t)G'(t+\epsilon,t)l(t+\epsilon)>^{1/2}}. \quad (3.53)$$

Indeed, with $p(t+\epsilon)$ chosen as in (3.53), a direct substitution of $p(t+\epsilon)$ into (3.52) gives

$$<l(t+\epsilon), X[t+\epsilon]l(t+\epsilon)>^{1/2} = <l(t+\epsilon), G(t+\epsilon,t)X[t]G'(t+\epsilon,t)l(t+\epsilon)>^{1/2} +$$
$$+\epsilon <l(t+\epsilon), G(t+\epsilon,t)Q_B(t)G'(t+\epsilon,t)l(t+\epsilon)>^{1/2}. \quad (3.54)$$

Estimating the Hausdorff distance

$$R(t,\epsilon) = h\left(\int_t^{t+\epsilon} G(t+\epsilon,s)\mathcal{E}(0, Q_B(s))ds, \epsilon^2 G(t+\epsilon,t)Q_B(t)G'(t+\epsilon,t)\right),$$

by direct calculation, we have the next result.

Lemma 3.4.2. *The following estimate holds,*

$$R(t,\epsilon) \leq K\epsilon^2, \quad K > 0. \quad (3.55)$$

In view of (3.53), (3.55), and the fact that $G(t+\epsilon,t) = I + \epsilon A(t) + o_2(\epsilon)$, we arrive at the desired result below.

Lemma 3.4.3. *With $p(t+\epsilon)$ selected as in (3.53), relations (3.52)–(3.55) reflect equalities*

$$\rho(l(t+\epsilon)|\mathcal{E}(x^*(t+\epsilon), X[t+\epsilon])) = \quad (3.56)$$
$$= \rho(l(t+\epsilon)|Z[t+\epsilon]) + o_3(\epsilon) = \rho(l(t+\epsilon)|Z(t+\epsilon)) + o_4(\epsilon).$$

Lemma 3.4.3 implies

$$X[t+\epsilon] = (1+p^{-1}(t+\epsilon))(I+\epsilon A(t))X[t](I+\epsilon A(t))' + \quad (3.57)$$
$$+\epsilon^2(1+p(t+\epsilon))(I+\epsilon A(t))'Q_B(t)(I+\epsilon A(t)) + o_5(\epsilon).$$

Substituting $p(t + \epsilon)$ of (3.53) into (3.57), denoting $\epsilon p(t + \epsilon) = \pi^{-1}(t + \epsilon)$, and expanding $X[t + \epsilon], G(t + \epsilon, s), \pi(t + \epsilon)$ in ϵ, one arrives at

$$X[t + \epsilon] = X[t] + \epsilon A(t)X[t] + \epsilon X[t]A'(t) + \pi(t)X[t] + \pi^{-1}(t)Q_B(t) + o_6(\epsilon). \tag{3.58}$$

This further gives, by rewriting the last expression, dividing it by ϵ, and passing to the limit with $\epsilon \to 0$,

$$\dot{X} = A(t)X + XA'(t) + \pi(t)X + \pi^{-1}(t)Q_B(t), \tag{3.59}$$

where

$$\pi(t) = \frac{<l(t), Q_B(t)l(t)>^{1/2}}{<l(t), X[t]l(t)>^{1/2}}. \tag{3.60}$$

Starting the construction of $E[t]$ with $E[t_0] = \mathcal{E}^0$, we have thus constructed the ellipsoid $E[t]$ as a solution to (3.49) with its evolution selected by the requirement (3.51), (3.54), (3.56), which resulted in Eqs. (3.59), (3.60), $X[t_0] = X^0$.

Equation (3.59) obviously coincides with (3.35) if $\phi = 0$. However, as indicated in Sects. 3.3, 3.4, the ellipsoid $E[t]$ constructed from Eq. (3.35), $\phi = 0$, which then turns into (3.31), touches the reach set $X[t]$ at points generated by support vector $l(t) = l^*(t)$ if and only if the latter is of the type $l^*(t) = G'(t_0, t)l$ (for any preassigned $l \in \mathbb{R}^n$). Then it even satisfies *for all* $t \geq t_0$ the requirement (iii) of Definition 3.4.1. On the other hand, as we have just shown, an ellipsoidal tube $E[t] = \mathcal{E}(x^*(t), X[t])$ evolves due to funnel equation (3.49) under Definition 3.4.1, only if it satisfies Eqs. (3.59), (3.60). Therefore, an ellipsoid $E[t]$ which satisfies Definition 3.4.1 may touch the exact reach tube *only if it follows a good curve* $l(t)$.

The above observations lead to the next result.

Theorem 3.4.1. *An ellipsoid* $E[t] = \mathcal{E}(x^*(t), X[t])$ *is a **minimal solution** to the ellipsoidal funnel equation (3.49) only if it is constructed following Eqs. (3.31), (3.59), (3.60), with* $X[t_0] = X^0$, *where curve* $l(t) = G'(t_0, t)l$ *for some* $l \in \mathbb{R}^n$ *(it is a good curve).*

Note that the opposite of the last Theorem is also true.

Theorem 3.4.2. *In order that* $\mathcal{E}(x^*[t], X_+^*[t])$ *would be described by Eqs. (3.31), (3.59), (3.60) it is necessary and sufficient that* $l(t)$ *satisfy Eq. (3.45) with some* $k(t)$. *The ellipsoid* $\mathcal{E}(x^*[t], X_+^*[t])$ *will be tight in* $\mathbf{E}_+$.

The solutions given in this theorem are thus *tight* in the sense that there exists no other ellipsoid in $\mathbf{E}_+$ which could be squeezed in between $\mathcal{E}(x^*(t), X_+^*[t])$ and $X[t]$.

From Theorems 3.4.1, 3.4.2 and Lemma 3.4.1 it follows that the mapping $E[t] = E(t|t_0, X^0)$ constructed as in Theorem 3.4.1, with $X[t] = X_+^*[t]$, satisfies the semigroup property (3.50). This may be also checked by direct calculation, using (3.26)–(3.28).

Remark 3.4.2. The ellipsoid $E[t]$ of 3.4.1 solves Problem 3.2.1 for any (non-normalized) curve $l(t)$. In order to have a solution for a normalized curve with $<l(t), l(t)> = 1$, one has to follow Lemmas 3.3.1, 3.3.2.

In the general case an ellipsoidal tube $\mathcal{E}_+^*[t] = \mathcal{E}(x^*, X_+^*[t])$ may be such that it touches $X[t]$ along some curve according to the requirement of Problem 3.2.1. But this curve may not be a good one. Then equations for $X_+^*[t]$ will differ from (3.31), (3.59), but will coincide with (3.35), where $\phi \neq 0$. And $\mathcal{E}_+^*[t]$ need not satisfy the semigroup property. An example of such case is when the touching curve is "drawn" on the surface of exact reach tube by an external *minimum-volume ellipsoid*. This touching curve is not a good one. The failure to distinguish good curves from any curves led to confusion and controversial statements in some publications in the 1980s on minimum-volume external ellipsoidal approximations.

But what if Eqs. (3.31), (3.59) are still used to approximate the exact reach set from above, disregarding the effect of being tight? Then the ellipsoid $\mathcal{E}_+[t]$ under consideration will be a conservative upper estimate for the exact reach set so that $X[t] \subseteq \mathcal{E}_+[t]$, with no promise of being tight.

3.5 The Ellipsoidal Maximum Principle and the Reachability Tube

We may now summarize the earlier results emphasizing that they were directed at the calculation of reachability *tubes* rather than fixed-time reach sets. Theorems 3.4.1, 3.4.2, lead to the next proposition.

Theorem 3.5.1. *The solution of Problem 3.2.1 allows the following conclusions:*

(i) *the ellipsoid $\mathcal{E}_+^*[t] = \mathcal{E}^*(x^*(t), X_+^*(t))$ constructed from Eqs. (3.24), (3.25) for any given $l^*(t) = l(t), X_+^*(t_0) = X^0$, is a solution to Problem 3.2.1 and $X_+^*(t)$ satisfies the "general" equation (3.35) with parameters $\pi^*(t)$ defined in (3.34).*

(ii) *The ellipsoid $E[t] = E(x^*(t), X_+^*(t))$ constructed from Eqs. (3.31), (3.30) (or (3.59), (3.60), which are the same) is an upper bound for the solution $\mathcal{E}_+^*[t] = \mathcal{E}^*(x^*(t), X_+^*(t))$ of (3.24), (3.25), so that for any given function $l^*(t)$ we have*

$$\rho(l(t)|E(x^*(t), X_+^*(t))) \geq \rho(l^*(t)|\mathcal{E}^*(x^*(t), X_+^*(t))), \qquad (3.61)$$

with equality reached for all $t \geq t_0$ provided $l^*(t)$ is chosen according to Assumption 3.3.1. In the last case the external ellipsoids are **tight**.

(iii) *In order that $\mathcal{E}(x^*(t), X_+^*(t))$ be described by Eqs. (3.31), (3.30), it is necessary and sufficient that $l(t)$ satisfy Eq. (3.45) with some function $k(t)$.*

108   3 Ellipsoidal Techniques: Reachability and Control Synthesis

Remark 3.5.1. If in the previous theorem $l^*(t)$ is chosen with $k(t) \equiv 0$, and $X_+^*(t)$ is its related matrix with $\pi^*(t)$ being its related parameterizing function, then we further drop the asterisks, assuming $l^*(t) = l(t)$, $X_+^*(t) = X_+(t)$, $\pi^*(t) = \pi(t)$ and are using Eqs. (3.31), (3.30) (or (3.59), (3.60)). For vector $x^\star(t)$ of the center of ellipsoid $\mathcal{E}(x^\star(t), X_+^*(t))$ the asterisk remains since it does not depend on l.

The solutions to the problem of the last theorem are tight in the sense of Definitions 3.2.1, 3.2.4. It is also useful to note the following.

Lemma 3.5.1. *Each ellipsoidal tube of type* $E(t|t_0, \mathcal{E}^0) = E[t]$, *generated as a solution to Eqs. (3.31), (3.30), defines a map with* **the semigroup property**

$$E(t|t_0, \mathcal{E}^0) = E(t|E(\tau, |t_0, \mathcal{E}^0)), \quad t_0 \leq \tau \leq t. \tag{3.62}$$

We now indicate how to calculate the reach tube with the results above.

An application of Theorem 3.1.1 under Assumption 3.3.1 yields the following conclusion.

Theorem 3.5.2. *Suppose Assumption 3.3.1 is fulfilled. Then the points* $x^{l^*}(t)$ *of support for vector* $l = l^*(t)$, *namely, those for which the equalities*

$$< l^*(t), x^{l^*}(t) > = \rho(l^*(t)|X[t]) = \rho(l^*(t)|E(x^\star(t), X_+[t])) \tag{3.63}$$

are true **for all** $t \geq t_0$, *are reached from initial state* $x^l(t_0) = x^{l0}$ *using* **a system trajectory** *with control* $u = u^*(t)$ *which satisfies* **the maximum principle** *(3.15) and condition (3.16) which now have the form* $l^*(t) = G'(t_0, t)l$

$$< l, G(t_0, s) B(s) u^*(s) > = \max\{< l, G(t_0, s) B(s) u > | u \in \mathcal{E}(q(t), Q(t))\}, \tag{3.64}$$

and

$$< l, x^{l0} > = \max\{< l, x > | x \in \mathcal{E}(x^0, X^0)\}, \tag{3.65}$$

for any $t \geq t_0$. *For all* $s \geq t_0$ *the control* $u^*(s), s \in [t_0, t]$ *may be taken to be the same, whatever be the value* t.

From the *maximum principle* (Theorem 3.1.1) one may directly calculate the optimal control $u^*(s)$ which is

$$u^*(s) = \frac{Q(s) B'(s) G'(t_0, s) l}{< l, G(t_0, s) B(s) Q(s) B'(s) G'(t_0, s) l >^{1/2}}, \quad t_0 \leq s \leq t, \tag{3.66}$$

and the optimal trajectory

$$x^l(t) = x^\star(t) + \int_{t_0}^t \frac{G(t, s) B(s) Q(s) B'(s) G'(t_0, s) l}{< l, G(t_0, s) B(s) Q(s) B'(s) G'(t_0, s) l >^{1/2}} ds, \quad t \geq t_0, \tag{3.67}$$

3.5 The Ellipsoidal Maximum Principle and the Reachability Tube

where

$$x^l(t_0) = x^{l0} = \frac{X^0 l}{<l, X^0 l>^{1/2}} + x^0 = x^*(t_0). \tag{3.68}$$

These relations are not very convenient for calculation, being expressed in a *non-recurrent* form.

However, one should note that due to (3.38), (3.39), the *trajectory* $x^l(t)$ of Theorem 3.5.2 also satisfies the following **"ellipsoidal" maximum principle**.

Theorem 3.5.3. *The next condition holds*

$$<l^*(t), x^l(t)> = \max\{<l^*(t), x> \,|\, x \in E(x^*(t), X_+[t])\}, \tag{3.69}$$

and is attained at

$$x^l(t) = x^*(t) + X_+[t]l^*(t) <l^*(t), X_+[t]l^*(t)>^{-1/2}, \tag{3.70}$$

where $l^*(t) = G'(t_0, t)l$,

$$X_+[t_0] = X^0, \quad x^l(t_0) = x^0 + X^0 l <l, X^0 l>^{-1/2}, \tag{3.71}$$

and $X_+[t]$ *may be calculated through Eq. (3.31) with explicitly known parameter (3.30).*

Thus the "trick" is to substitute at each point of the boundary of the reach set $X[t]$ the "original" maximum principle (3.15) of Theorem 3.1.1 by the "ellipsoidal" maximum principle (3.69).

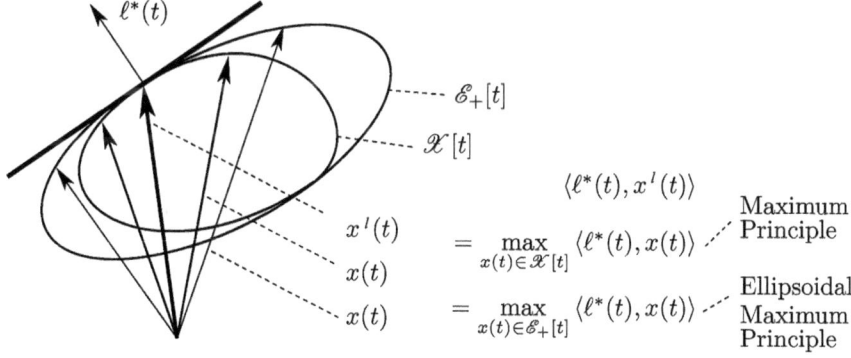

MP: $\langle \ell^*(t), x^l(t) \rangle = \max\{\langle \ell^*(t), x(t) \rangle \,|\, x(t) \in \mathscr{X}[t]\}$

EMP: $\langle \ell^*(t), x^l(t) \rangle = \max\{\langle \ell^*(t), x(t) \rangle \,|\, x(t) \in \mathscr{E}_+[t]\}$

The maximum principles

This results in the next proposition.

Theorem 3.5.4. *The trajectory $x^l(t)$ that runs along the boundary $\partial X[t]$ of the reach set $X[t]$ and touches $X[t]$ at the points of support for the vector $l^*(t) = G'(t,t_0)l$ is given by equalities (3.71), (3.70), where $X_+[t]$ is the solution to Eqs. (3.31), (3.30) (or, what is the same, to Eqs. (3.59), (3.60)).*

Due to Lemma 3.3.1 the last theorem may be complemented by

Corollary 3.5.1. *Given vector $l^0 = l^*(t)$, the related ellipsoid $E(x^*(t), X_+[t])$ and vector $x^l(t) \in \partial X[t]$, generated by function $l^*(t)$ of Eq. (3.45), do not depend on the choice of function $k(t)$ in this equation.*

Denoting $x^l(t) = x[t, l]$, we thus come to a two-parametric surface $x[t, l]$ that defines the boundary $\partial X[\cdot]$ of the *reachability tube* $X[\cdot] = \cup\{\partial X[t], \; t \geq t_0\}$. With $t = t'$ fixed and $l \in \mathbf{S} = \{l :< l, l >= 1\}$ varying, the vector $x[t', l]$ runs along the boundary $\partial X[t']$. On the other hand, with $l = l'$ fixed and with t varying, the vector $x[t, l']$ moves along one of the system trajectories $x^l(t)$ that touch the reachability set $X[t]$ of system (3.1) with control $u^*(t)$ of (3.66) and x^{l0} of (3.68). Then

$$\cup\{x[t,l]|l \in \mathbf{S}\} = \partial X[t], \; \cup\{x[t,l]|l \in \mathbf{S}, \; t \geq t_0\} = \partial X.$$

Remark 3.5.2. Relation (3.70) is given **in a recurrent form** and throughout the calculation of curves $x^l[t]$ and surface $x[t, l]$ one need not compute the respective controls $u^*(t)$.

Remark 3.5.3. This remark concerns the question of *optimality* of external ellipsoids. Suppose for each time t a globally volume-optimal ellipsoid $\mathcal{E}(x^*(t), X_v(t))$ is constructed. It will obviously touch the reach set $X[t]$ at each instant t and thus produce a certain curve $x_v(t)$ on the surface $\partial X[t]$ of the reach tube $X[\cdot]$. However, an example given in [174, Sect. 2.7, Example 2.7.1] indicates that the volume-optimal ellipsoid may not exist in the class $\mathbf{E}_+$. The conclusion which follows is that in general the volume-optimal curve $x_v(t)$ *is not a "good one"* in the sense of Assumption 3.3.1. Therefore, one should not expect that matrix $X_v(t)$ of the volume-optimal ellipsoid to be described by Eq. (3.31) ((3.59)), as claimed by some authors.

On the other hand, the tight ellipsoids of Sect. 3.2 do satisfy Definition 3.2.1. This implies

$$d[l] = \rho(l|X[t]) + \rho(-l|X[t]) = \rho(l|E(x^*(t), X_{+l}(t))) + \rho(-l|E(x^*(t), X_{+l}(t))) =$$

$$= 2 < l, X_{+l}(t)l >^{1/2}$$

where $d[l]$ is the length of the *exact projection of set $X[t]$ on the direction l*. Here matrix $X_{+l}(t)$ is calculated due to (3.31), (3.30), for the given vector $l \in \mathbb{R}^n$.

3.6 Example 3.6

For an illustration of the results consider system

$$\dot{x}_1 = x_2, \quad \dot{x}_2 = u, \tag{3.72}$$

$$x_1(0) = x_1^0, \quad x_2(0) = x_2^0; \quad |u| \leq \mu, \ \mu > 0.$$

Here we have:

$$x_1(t) = x_1^0 + x_2^0 t + \int_0^t (t-\tau) u(\tau) d\tau,$$

$$x_2(t) = x_2^0 + \int_0^t u(\tau) d\tau.$$

The support function

$$\rho(l|X[t]) = \max\{< l, x(t) > \ | \ |u| \leq \mu\}$$

may be calculated directly and is given by formula

$$\rho(l|X[t]) = l_1 x_1^0 + (l_1 t + l_2) x_2^0 + \mu \int_0^t |l_1(t-\tau) + l_2| d\tau.$$

The boundary of the reach set $X[t]$ may be calculated from the formula ([174])

$$\min_l \{\rho(l|X[t]) - l_1 x_1 - l_2 x_2| < l, l >= 1\} = 0. \tag{3.73}$$

Then the direct calculation of the minimum gives a parametric representation for the boundary $\partial X[t]$ by introducing parameter $\sigma = l_2^0/l_1^0$, where l_1^0, l_2^0 are the minimizers in (3.73). This gives two curves

$$x_1(t) = x_1^0 + x_2^0 t + \mu(t^2/2 - \sigma^2), \tag{3.74}$$

$$x_2(t) = x_2^0 + 2\mu\sigma + \mu t,$$

and

$$x_1(t) = x_1^0 + x_2^0 t - \mu(t^2/2 - \sigma^2), \tag{3.75}$$

$$x_2(t) = x_2^0 - 2\mu\sigma - \mu t.$$

for $\sigma \in [-t, 0]$.[4]

---

[4] Note that for $l_2^0/l_1^0 > 0$ or $l_2^0/l_1^0 < -t$ we have $\sigma \notin [-t, 0]$. For such vectors the point of support $x^l[t]$ will be at either of the vertices of set $X[t]$.

Fig. 3.1 The reach set at given time

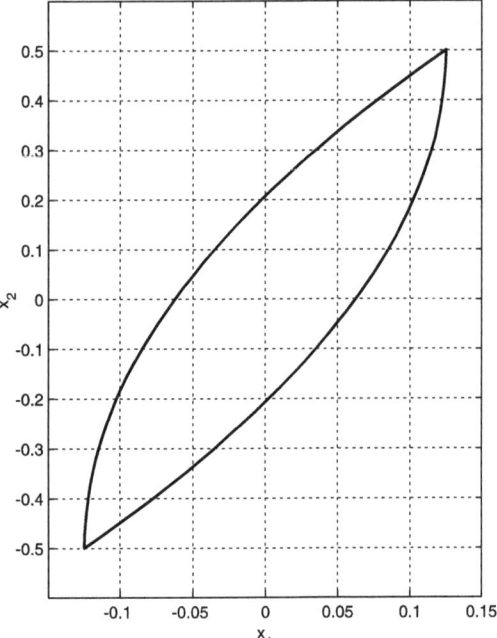

These curves (parameterized by σ) form the upper and lower boundaries of $X[t]$ with parameter t fixed (see Fig. 3.1). With t increasing this allows to draw the reachability tube $X[t]$ as a set-valued function of time t, see Fig. 3.2.

The formula $l(t) = e^{-A't}l$ of Assumption 3.3.1 here transforms into $(l_1(t) = l_1; l_2(t) = l_2 - tl_1)$. This allows to write down the relations for the points of support to the hyperplanes generated by vector $l(t)$. A substitution into (3.74), (3.75), gives for $l_2^0/l_1^0 \in [0, t]$

$$x_1^l(t) = x_1^0 + x_2^0 t + \mu(t^2/2 - (tl_1^0 - l_2^0)/l_1^0)^2), \tag{3.76}$$

$$x_2^l(t) = x_2^0 + 2\mu(l_2^0 - tl_1^0)/l_1^0 + \mu t,$$

and

$$x_1^l(t) = x_1^0 + x_2^0 t - \mu(t^2/2 - ((tl_1^0 - l_2^0)/l_1^0)^2), \tag{3.77}$$

$$x_2^l(t) = x_2^0 - 2\mu(l_2^0 - tl_1^0)/l_1^0 - \mu t.$$

With vector $l \in \mathbb{R}^2$ fixed, this gives a parametric family of curves $x^l(t)$ that cover the surface of the reach tube $X[\cdot]$ and are the points of support for the hyperplanes generated by vectors $l \in \mathbb{R}^2$ through formula $l(t) = (l_1, -tl_1 + l_2)$. These curves are shown in thick lines in Fig. 3.2.

3.6 Example 3.6

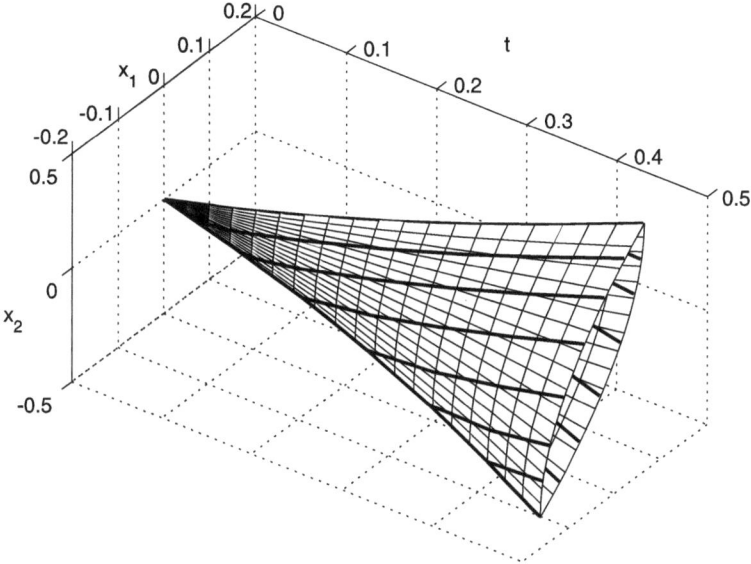

Fig. 3.2 The reach tube

Let us now choose two pairs of vectors $l \in \mathbb{R}^2$, for example

$$l^* = (-10, -4), \quad l^* = (10, 4); \quad l^{**} = (-4, 0), \quad l^{**} = (4, 0).$$

Each of these pairs generates a tight ellipsoid $\mathcal{E}^*[t]$ and $\mathcal{E}^{**}[t]$ due to Eq. (3.71), where the elements $x_{i,j}^*$ of X_+^* satisfy the equations

$$\dot{x}_{11}^* = x_{12}^* + x_{21}^* + \pi(t)x_{11}^*, \quad \dot{x}_{12}^* = x_{22}^* + \pi x_{12}^*,$$

$$\dot{x}_{21}^* = x_{22}^* + \pi(t)x_{21}^*, \quad \dot{x}_{22}^* = \pi(t)x_{22}^* + (\pi(t))^{-1}\mu^2,$$

with $X_+^*(0) = 0$ and

$$\pi(t) = f_1(t)/f_2(t),$$

with

$$f_1(t) = \mu|l_2 - tl_1|; \quad f_2(t) = (x_{11}^* l_1^2 + 2x_{12}^* l_1(l_2 - tl_1) + x_{22}^*(l_2 - tl_1)^2)^{1/2}.$$

Ellipsoid $\mathcal{E}^{**}[t]$ is expressed similarly.

Here $\mathcal{E}^*[t]$ touches the reach set $X[t]$ at points $x_l^*(t)$ generated by vector l^* through formulas (3.75), (3.77) and $\mathcal{E}^{**}[t]$ touches it at points $x_l^{**}(t)$ generated by vector l^{**} through formulas (3.74), (3.76) (see cross-section in Fig. 3.3 for $t = 0.5$).

Fig. 3.3 Two tight ellipsoids

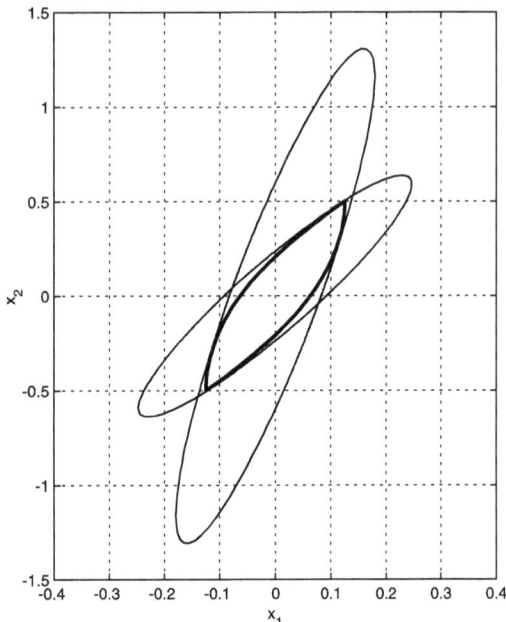

For a given time t the intersection

$$\mathcal{E}^*[t] \cap \mathcal{E}^{**}[t] \supset \mathcal{X}[t]$$

gives an approximation of $\mathcal{X}[t]$ better than the four supporting hyperplanes that could be placed at the same points of support as for the ellipsoids (see Fig. 3.3). At the same time, by taking more ellipsoids for other values of l we achieve, as a consequence of Theorem 3.2.1, a more accurate approximation of set $\mathcal{X}[t]$ (see Fig. 3.4 for seven ellipsoids) with exact representation achievable if number of appropriately selected ellipsoids tends to infinity.

Having introduced a family of external ellipsoidal approximations for reach sets and specified its basic properties, we now indicate that similar properties are also true for *internal approximations* which are often required whenever one has to deal with *guaranteed performance*. This is a more difficult problem, though, as compared with external approximations.

3.7 Reachability Sets: Internal Approximations

Existing approaches to the calculation of internal ellipsoidal approximations for the sum of a pair of ellipsoids are given in [174, Sect. 2.3, pp. 121–127]. However, the important question for dealing with dynamics is how to effectively compute

3.7 Reachability Sets: Internal Approximations

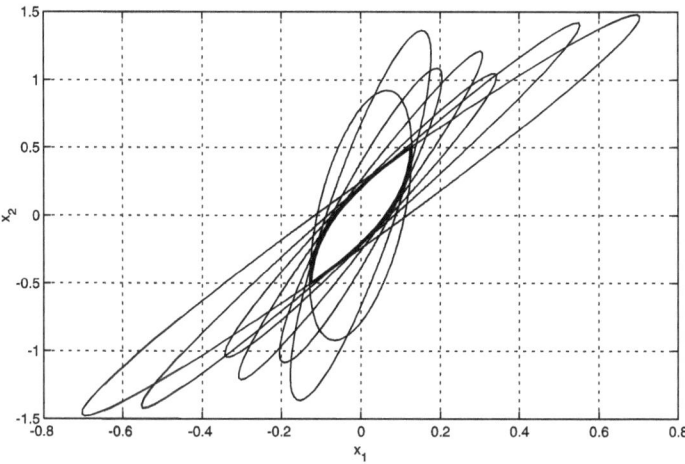

Fig. 3.4 Many tight ellipsoids

an *internal approximation* of the sum of *any* finite number of ellipsoids and going further of an *integral of ellipsoidal-valued functions*. It is also important, as in the external case, to single out families of tight internal approximations of reach tubes or their neighborhoods through such ellipsoidal-valued functions that would touch their boundary from inside at *any point* on their surface. Both of the questions are answered in the forthcoming sections.

Thus, given as before is system (3.1) or differential inclusion (3.3), with $A(t) \equiv 0$. The problem is to approximate the respective reach sets and tubes internally, through appropriate tight ellipsoidal sets and ellipsoidal-valued tubes.

The text of this section relies on the next constructions introduced earlier in [177, 182]. Though described there for nondegenerate ellipsoids, they are true for degenerate ellipsoids as well.

Consider the internal approximation of sum

$$\mathcal{E}(x^0, X_0) + \int_{t_0}^{t} \mathcal{E}(B(s)q(s), B(s)Q(s)B'(s))ds,$$

of an n-dimensional ellipsoid $\mathcal{E}(x^0, X_0)$ and a set-valued integral of an ellipsoidal-valued function $\mathcal{E}(B(s)q(s), B(s)Q(s)B'(s)) = \mathcal{E}(q_B(s), Q_B(s))$ with continuous $n \times p$-matrix function $B(s) > 0$. (Recall notations $q_B(t) = B(t)q(t)$, $Q_B(t) = B(t)Q(t)B'(t)$.)

Theorem 3.7.1. (i) *The following inclusion is true*

$$\mathcal{E}(x^0, X^-(t)) \subseteq x(t) + \mathcal{E}(0, X_0) + \int_{t_0}^{t} \mathcal{E}(0, Q_B(\tau))d\tau,$$

$$x(t) = x^0 + \int_{t_0}^{t} q_B(s)ds,$$

whatever be the matrix

$$X^-(t) = \tag{3.78}$$

$$= \left(S_0 X_0^{1/2} + \int_{t_0}^t S(\tau) Q_B^{1/2}(\tau) d\tau\right)' \left(S_0 X_0^{1/2} + \int_{t_0}^t S(\tau) Q_B^{1/2}(\tau) d\tau\right).$$

Here $X_0 = (X_0)' > 0$ is an $n \times n$ matrix, $Q(\tau) = Q'(\tau)$, $\tau \in [t_0, t]$ are any positive definite, $p \times p$ matrices with continuous function $Q(\tau)$, matrices $S_0 S_0' = I$, $S'(\tau) S(\tau) \equiv I$ are orthogonal $n \times n$ matrices and $S(\tau)$ are continuous in time.

(ii) *For a given vector $l \in \mathbb{R}^n$ relation*

$$<l, x(t)> + \rho(l|\mathcal{E}(0, X^-(t))) \leq$$

$$\leq \rho(l|\mathcal{E}(0, X_0)) + \int_{t_0}^t \rho(l|\mathcal{E}(0, Q_B(\tau)) d\tau, \quad l \in \mathbb{R}^n, \tag{3.79}$$

turns into an equality iff matrices $S_0, S(\tau)$ may be chosen such that for some scalar function $\lambda(\tau) > 0$ the equality

$$S(\tau) Q_B^{1/2}(\tau) l = \lambda(\tau) S_0 X_0^{1/2} l \tag{3.80}$$

would be fulfilled for all $\tau \in [t_0, t]$.

Remark 3.7.1. Note that with $m < n$ matrices $Q_B(t) = B(t) Q(t) B'(t)$ turn out to be degenerate, of rank $\leq m$. Then the square root $Q_B^{1/2}$ is understood to be an $n \times n$ matrix which may be calculated through the "square root" version of the singular value matrix decomposition of Q_B which gives $Q_B^{1/2} Q_B^{1/2} = Q_B$ (see [245, pp. 80–83]).

We now pass to the definition of internally tight ellipsoids.

Definition 3.7.1. *An internal approximation $\mathcal{E}_-$ of a reach set $X[t]$ is **tight** if there exists a vector $l \in \mathbb{R}^n$ such that*

$$\rho(\pm l | \mathcal{E}_-) = \rho(\pm l | X[t]).$$

This definition may be appropriate for the reach sets of the present paper, but the respective ellipsoids may not be unique. A more general definition follows.

Definition 3.7.2. *An internal approximation $\mathcal{E}_-$ is **tight** in the class $\mathbf{E}_-$, if for any ellipsoid $\mathcal{E} \in \mathbf{E}_-$, $X[t] \supseteq \mathcal{E} \supseteq \mathcal{E}_-$ implies $\mathcal{E} = \mathcal{E}_-$.*

This section is concerned with internal approximations, where class $\mathbf{E}_- = \{\mathcal{E}_-\}$ is described within the following definition.

3.7 Reachability Sets: Internal Approximations

Definition 3.7.3. The class $\mathbf{E}_- = \{\mathcal{E}_-\}$ consists of ellipsoids that are of the form $\mathcal{E}_-[t] = \mathcal{E}(x^*(t), X^-(t))$, where $x^*(t)$ satisfies the equation

$$\dot{x}^* = B(t)q(t), \quad x^*(t_0) = x^0, \ t \geq t_0,$$

$X^-(t)$ is of the form (3.78), $q(t) \in \mathbb{R}^p$ is any Lebesgue-measurable function.

In particular, this means that if an ellipsoid $\mathcal{E}(x, X) \subseteq X[t]$ is tight in $\mathbf{E}_-$, then there exists no other ellipsoid of type $\mathcal{E}(x, kX), k > 1$, that satisfies the inclusions $X[t] \supseteq \mathcal{E}(x, kX) \supseteq \mathcal{E}(x, X)$ (ellipsoid $\mathcal{E}(x, X)$ touches set $X[t]$).

Definition 3.7.4. The *internal ellipsoids* are said to be **tight** *if they are tight in* $\mathbf{E}_-$.

We actually further deal only with ellipsoids $\mathcal{E}_- \in \mathbf{E}_-$. For the problems of this book Definition 3.7.1 follows from 3.7.4.

The class $\mathbf{E}_-$ is rich enough to arrange effective approximation schemes, though it does not include all possible ellipsoids. A justification for using this class is due to the propositions of Theorem 3.7.3 which also gives conditions for the internal ellipsoids $\mathcal{E}(0, X^-(\tau))$ to be *tight* in the previous sense.

Let us now return to equation

$$\dot{x} = A(t)x + B(t)u, \quad t_0 \leq t \leq t_1, \tag{3.81}$$

of Sect. 3.1.

Then the problem consists in finding the internal ellipsoid $\mathcal{E}(x^*(t), X^-(t))$ for the *reach set*

$$X[t] = G(t, t_0)\mathcal{E}(x^0, X_0) + \int_{t_0}^{t} G(t, \tau)B(\tau)\mathcal{E}(q(\tau), Q(\tau))d\tau.$$

Since for a matrix-valued map we have $B\mathcal{E}(q, Q) = \mathcal{E}(q_B, Q_B)$, the formula of Theorem 3.7.1 given there for $X^-(t)$ will now transform into

$$X^-(t) = G(t, t_0)\left(X_0^{1/2}S_0'(t_0) + \int_{t_0}^{t} G(t_0, \tau)Q_B^{1/2}(\tau)S'(\tau)d\tau\right) \times$$

$$\times \left(S_0(t_0)X_0^{1/2} + \int_{t_0}^{t} S(\tau)Q_B^{1/2}(\tau)G'(t_0, \tau)\right)G'(t, t_0), \tag{3.82}$$

and

$$x^*(t) = G(t, t_0)x^0 + \int_{t_0}^{t} G(t, \tau)q_B(\tau)d\tau. \tag{3.83}$$

Theorem 3.7.1 now transforms into the following statement.

Theorem 3.7.2. *The internal ellipsoids for the reach set $X[t]$ satisfy the inclusion*

$$\mathcal{E}(x^\star(t), X^-(t)) \subseteq \mathcal{E}(G(t,t_0)x^0, G(t,t_0)X_0G'(t,t_0)) + \quad (3.84)$$

$$+ \int_{t_0}^t \mathcal{E}(G(t,\tau)q_B(\tau), G(t,\tau)Q_B(\tau)G'(t,\tau))d\tau = X[t],$$

where $X^-(t), x^\star(t)$ are given by (3.82), (3.83), with $S_0, S(\tau)$ being any orthogonal matrices and $S(\tau)$ continuous in time.

The tightness conditions now transfer into the next proposition.

Theorem 3.7.3. *For a given instant t the internal ellipsoid $\mathcal{E}(x^\star(t), X^-(t))$ will be tight and will touch $X[t]$ at the point of support x^l of the tangent hyperplane generated by given vector $l^\star$, namely,*

$$\rho(l^\star | X[t]) = \quad (3.85)$$

$$= <l^\star, x^\star(t)> + <l^\star, X_0 l^\star>^{1/2} + \int_{t_0}^t <l^\star, G(t,\tau)Q_B(\tau)G'(t,\tau)l^\star>^{1/2} d\tau =$$

$$\rho(l^\star | \mathcal{E}(x^\star(t), X^-(t))) = <l^\star, x^\star(t)> + <l^\star, X^-(t)l^\star>^{1/2} = <l^\star, x^l>,$$

iff $S_0, S(\tau)$ satisfy the relation

$$S(\tau)Q_B^{1/2}(\tau)G'(t,\tau)l^\star = \lambda(\tau)S_0 X_0^{1/2} G'(t,t_0)l^\star, \quad t_0 \leq \tau \leq t, \quad (3.86)$$

for some function $\lambda(\tau) > 0$.

Direct calculation indicates the following.

Lemma 3.7.1. *The function $\lambda(\tau)$ of Theorem 3.7.3 is given by*

$$\lambda(\tau) = <l^\star, G(t,\tau)Q_B(\tau)G'(t,\tau)l^\star>^{1/2} <l^\star, G(t,t_0)X_0 G'(t,t_0)l^\star>^{(-1/2)}, \quad t_0 \leq \tau \leq t. \quad (3.87)$$

The previous Theorems 3.7.2, 3.7.3 were formulated *for a fixed instant of time t and a fixed support vector $l^\star$*. It is important to realize what would happen if $l^\star$ varies in time.

3.8 Example 3.8

Consider system

$$\dot{x}_1 = x_2, \quad \dot{x}_2 = u, \quad (3.88)$$

$$x_1(0) = x_1^0, \quad x_2(0) = x_2^0; \quad |u| \leq \mu, \quad \mu > 0.$$

3.8 Example 3.8

Here

$$x_1(t) = x_1^0 + x_2^0 t + \int_0^t (t-\tau)u(\tau)d\tau,$$

$$x_2(t) = x_2^0 + \int_0^t u(\tau)d\tau.$$

Assume $X_0 = \mathcal{B}_\epsilon(0) = \{x : <x, x> \le \epsilon^2\}$. The support function

$$\rho(l|X[t]) = \max\{<l, x(t)> \mid |u| \le \mu, \ x^0 \in X_0\}$$

of the reach set $X[t] = X(t, 0, X_0)$ may be calculated directly and is given by

$$\rho(l|X[t]) = \epsilon(l_1^2 + (l_1 t + l_2)^2)^{1/2} + \int_0^t |l_1(t-\tau) + l_2| d\tau.$$

The boundary $\partial X[t]$ of the reach set $X[t]$ is the set of vectors such that

$$\min_l \{\rho(l|X[t]) - l_1 x_1 - l_2 x_2 | \langle l, l \rangle = 1\} = 0. \quad (3.89)$$

This leads to the next parametric presentation of $\partial X[t]$ through two bounding curves (one with sign plus in $\pm$ and the other with minus)

$$x_1(t) = x_1^0 + x_2^0 t \pm \mu(t^2/2 - \sigma^2), \quad (3.90)$$

$$x_2(t) = x_2^0 \pm 2\mu\sigma \pm \mu t,$$

taken for values of parameter $\sigma \in (-t, 0)$. The values $\sigma \notin (-t, 0)$ correspond to two points—the two vertices of $X[t]$.
Here $\sigma = l_2^0/l_1^0$ where l_1^0, l_2^0 are the minimizers in (3.89).
Solving the problem for all $t > 0$, set $l^0 = l(t)$. Then

$$x_1(t) = \quad (3.91)$$

$$= \epsilon(l_1(t)(t^2+1) + l_2(t)t)/(l_1^2(t) + (l_1(t)t + l_2(t))^2)^{1/2} \pm \mu(t^2/2 - l_2^2(t)/l_1^2(t)),$$

$$x_2(t) =$$

$$= \epsilon(l_1(t)t + l_2(t))/(l_1(t)^2 + (l_1(t)t + l_2(t))^2)^{1/2} \pm 2\mu l_2(t)/l_1(t) \pm \mu t,$$

where $l \in \mathbb{R}^2$, $t \ge 0$.

Here, for each t, vector $l^0 = l(t)$ is the support vector to $X[t]$ at point $x(t) \in \partial X[t]$. Moreover, with t fixed, and $x = x^{(l)}(t)$ running along $\partial X[t]$ (which is a closed curve in $\mathbb{R}^n$ for a given t), its support vector $l^0 = l(t)$ will sweep out *all* directions in $\mathbb{R}^n$. Therefore, considering any function $l(t)$ of t with $t \geq t_0$, we may be sure that there exists a corresponding trajectory $x^{(l)}(t) \in \partial X[t]$ for $t \geq t_0$.

Proceeding further, we select $l(t)$ satisfying Assumption 3.3.1, namely, as $l(t) = e^{-A't}l^*$. This transforms into $l_1(t) = l_1^*$, $l_2(t) = l_2^* - tl_1^*$ and (3.91) simplifies to

$$x_1(t) = \tag{3.92}$$

$$= \epsilon(l_1^* + tl_2^*)/(l_1^{*2} + l_2^{*2})^{1/2} \pm \mu(t^2/2 - (tl_1^* - l_2^{*2}(t))^2/l_1^{*2}(t)),$$

$$x_2(t) = \epsilon l_2^*/(l_1^{*2} + l_2^{*2})^{1/2} \pm 2\mu(l_2^* - tl_1^*)/l_1^* \pm \mu t.$$

These relations depend only on the two-dimensional vector l^*. They produce a parametric family of curves $\{x_1(t), x_2(t)\}$ that cover all the surface of the reach tube $X[t]$ so that vectors $x(t)$, $x(t) = \{x_1(t), x_2(t)\}$ are the points of support for the hyperplanes generated by vectors $l(t) = (l_1^*, -tl_1^* + l_2^*)'$. The reach tube that starts at $X^0 \neq \{0\}$ with these curves on its surface is shown in Fig. 3.5.

We now construct the tight internal ellipsoidal approximations for $X[t]$ that touch the boundary $\partial X[t]$ *from inside* at points of support taken for a given vector $l = l^*$. The support function $\rho(l^* \mid X[t])$ for the exact reach set may be rewritten as

$$\rho(l^* \mid X[t]) = \epsilon \langle l^*, Il^* \rangle^{1/2} + \mu \int_0^t \langle l^*, Q_B(\tau)l^* \rangle^{1/2} d\tau, \tag{3.93}$$

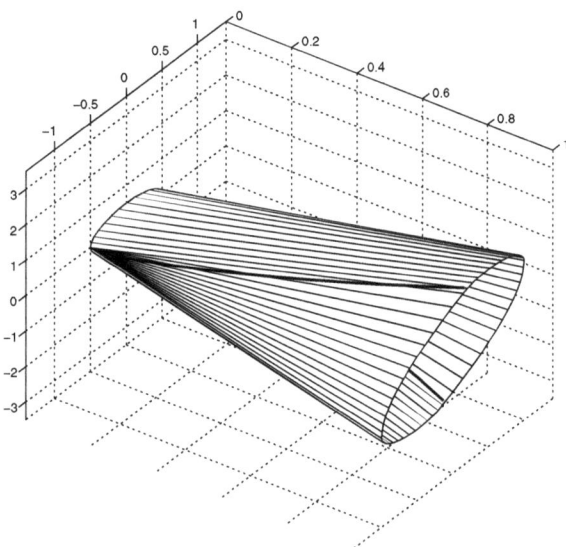

Fig. 3.5 Reach tube

3.8 Example 3.8

where $B' = (0, 1)$ and

$$Q_B(\tau) = \begin{pmatrix} \tau^2, & \tau \\ \tau, & 1 \end{pmatrix}$$

and $Q_B^{1/2}(\tau) = Q_B(\tau)(1 + \tau^2)^{-1/2}$.

According to (3.102) and in view of Assumption 3.3.1, we have (taking $S_0 = I$),

$$X^-(\tau) =$$

$$= \left(\epsilon I + \int_0^\tau Q_B^{1/2}(\tau) S'(\tau) d\tau \right) \left(\epsilon I + \int_0^\tau S(\tau) Q_B^{1/2}(\tau) d\tau \right), \quad (3.94)$$

where matrix $S(\tau)$ must satisfy the conditions

$$S'(\tau) S(\tau) = I, \quad S(\tau) Q_B^{1/2}(\tau) l^* = \epsilon \lambda(\tau) l^*, \quad \tau \geq 0 \quad (3.95)$$

for some $\lambda(\tau) > 0$. Calculations give

$$\epsilon^2 \lambda^2(\tau) = \langle l^*, Q_B(\tau) l^* \rangle \langle l^*, l^* \rangle^{-1} = (l_1^* \tau + l_2^*)^2 (l_1^{*2} + l_2^{*2})^{-1}. \quad (3.96)$$

Denote

$$p(\tau) = Q_B^{1/2}(\tau) l^* = \begin{pmatrix} \tau^2 l_1^* + \tau l_2^* \\ \tau l_1^* + l_2^* \end{pmatrix} (1 + \tau^2)^{-1/2} = r_p(\tau) \begin{pmatrix} \cos \phi_p(\tau) \\ \sin \phi_p(\tau) \end{pmatrix},$$

where

$$r_p(\tau) = |l_1^* \tau + l_2^*|, \quad \phi_p(\tau) = \pm \arccos(\pm \tau / (1 + \tau^2)^{1/2})$$

and also

$$l^* = \langle l^*, l^* \rangle^{1/2} \begin{pmatrix} \cos \phi_l(\tau) \\ \sin \phi_l(\tau) \end{pmatrix}, \quad \phi_l = \pm \arccos(l_1^* / (l_1^{*2} + l_2^{*2})^{1/2}).$$

Selecting further the orthogonal matrix-valued function $S(\tau)$ as

$$S(\tau) = \begin{pmatrix} \cos \alpha(\tau), & -\sin \alpha(\tau) \\ \sin \alpha(\tau), & \cos \alpha(\tau) \end{pmatrix},$$

we may rewrite the second relation (3.95) as

$$\begin{pmatrix} \cos(\phi_p(\tau) + \alpha(\tau)) \\ \sin(\phi_p(\tau) + \alpha(\tau)) \end{pmatrix} r_p(\tau) = \epsilon \lambda(\tau) (l_1^{*2} + l_2^{*2})^{1/2} \begin{pmatrix} \cos \phi_l(\tau) \\ \sin \phi_l(\tau) \end{pmatrix}, \quad (3.97)$$

where $\tau \in [t_0, t]$. Here $\alpha(\tau)$ has to be selected from equality

$$\phi_p(\tau) + \alpha(\tau) = \phi_l(\tau), \quad \tau \in [t_0, t], \tag{3.98}$$

and $\lambda(\tau)$ is given by (3.96).

Equations (3.97), (3.98) need no recalculation for new values of t.

Thus we have found an orthogonal matrix function $S(\tau)$

$$S(\tau) = \begin{pmatrix} \cos(\phi_l(\tau) - \phi_p(\tau)), & -\sin(\phi_l(\tau) - \phi_p(\tau)) \\ \sin(\phi_l(\tau) - \phi_p(\tau)), & \cos(\phi_l(\tau) - \phi_p(\tau)) \end{pmatrix}, \tag{3.99}$$

that depends on l^*, that is continuous in τ and satisfies (3.97).

Matrix $X_-(t)$ may now be calculated from equations

$$\dot{X}_- = (\dot{X}_-^*(t))' X_-^*(t) + X_-^{*'}(t) \dot{X}_-^*(t), \quad X_-(0) = \epsilon^2 I, \tag{3.100}$$

where

$$\dot{X}_-^* = S(t) Q_B^{1/2}(t), \quad X_-^*(0) = \epsilon I.$$

The internal ellipsoids for the reach set $X[t] = X(t, 0, \mathcal{X}^0)$ are shown in Figs. 3.6, 3.7, and 3.8 for $\mathcal{X}^0 = \mathcal{E}(0, \epsilon I)$, with epsilon increasing from $\epsilon = 0$ (Fig. 3.6) to $\epsilon = 0.175$ (Fig. 3.7), and $\epsilon = 1$ (Fig. 3.8). The tube in Fig. 3.5 corresponds to the epsilon of Fig. 3.7. One may also observe that the exact reach sets $X(t, 0, \{0\})$ taken for $\epsilon = 0$ are located within the sets $X(t, 0, \mathcal{X}^0) = X[t]$, calculated for $\epsilon \neq 0$ (see Figs. 3.7 and 3.8).

3.9 Reachability Tubes: Recurrent Relations—Internal Approximations

We start with a question similar to Problem 3.2.1, but now formulated for internal approximations.

Problem 3.9.1. Given a vector function $l^*(t)$, continuously differentiable in t, find an internal ellipsoid $\mathcal{E}(x_-(t), X_-(t)) \subseteq X[t]$ that would ensure **for all** $t \geq t_0$, the equality

$$\rho(l^*(t)|X[t]) = \rho(l^*(t)|\mathcal{E}(x_-(t), X_-(t))) = \langle l^*(t), x^l(t) \rangle, \tag{3.101}$$

so that the supporting hyperplane for $X[t]$ generated by $l^*(t)$, namely, the plane $\langle x - x^l(t), l^*(t) \rangle = 0$ that touches $X[t]$ at point $x^l(t)$, would also be a supporting hyperplane for $\mathcal{E}(x_-(t), X_-(t))$ and touch it at the same point.

3.9 Reachability Tubes: Recurrent Relations—Internal Approximations

Fig. 3.6 Internal ellipsoids for $\mathcal{X}^0 = \mathcal{E}(0,0)$

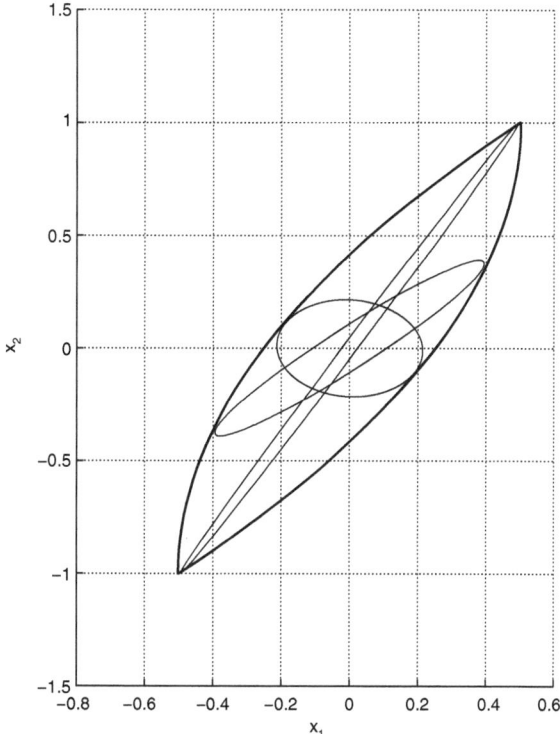

This problem is solvable in the class $\mathbf{E}_-$. In order to solve this problem, we shall refer to Theorems 3.7.2, 3.7.3. However, functions $S(\tau), \lambda(\tau)$ used for the parametrization in (3.82), (3.86) should now be functions of two variables, namely, of τ, t, since the requirement is that relation (3.101) should now hold for all $t \geq t_0$ (and therefore S_0 should also depend on t). We may therefore still apply Theorems 3.7.2, 3.7.3 but now with $S_0, S(\tau), \lambda(\tau)$ substituted by $S_{0t}, S_t(\tau), \lambda_t(\tau)$.

Theorem 3.9.1. *With $l = l^*(t)$ given, the solution to Problem 3.9.1 is an ellipsoid $\mathcal{E}(x_-(t), X_-(t))$, where $x_-(t) = x^*(t)$ and*

$$X_-(t) = \tag{3.102}$$

$$= G(t, t_0)\left(X_0^{1/2} S'_{0t}(t_0) + \int_{t_0}^t G(t_0, \tau) Q_B^{1/2}(\tau) S'_t(\tau) d\tau\right) \times$$

$$\times \left(S_{0t}(t_0) X_0^{1/2} + \int_{t_0}^t S_t(\tau) Q_B^{1/2}(\tau) G'(t_0, \tau)\right) G'(t, t_0),$$

Fig. 3.7 Internal ellipsoids for $\mathcal{X}^0 = \mathcal{E}(0,0)$, $\mathcal{E}(0, \ 0.175)$

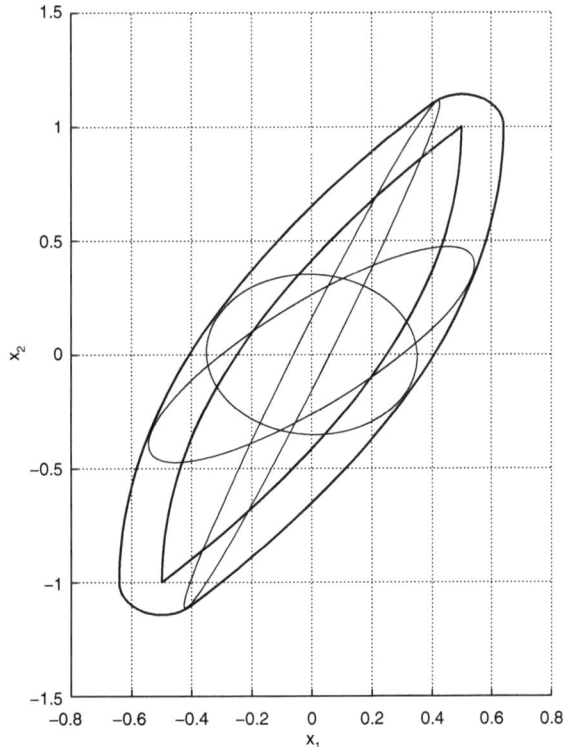

with S_{0t}, $S_t(\tau)$ satisfying relations

$$S_t(\tau) P_B^{1/2}(\tau) G'(t,\tau) l^*(t) = \lambda_t(\tau) S_{0t} X_0^{1/2} G'(t,t_0) l^*(t), \qquad (3.103)$$

and $S'_{0t} S_{0t} = I$, $S'_t(\tau) S_t(\tau) \equiv I$ for all $t \geq t_0$, $\tau \in [t_0, t]$, where

$$\lambda_t(\tau) = \qquad (3.104)$$

$$= <l^*(t), G(t,\tau) Q_B(\tau) G'(t,\tau) l^*(t) >^{1/2} < l^*(t), G(t,t_0) X_0 G'(t,t_0) l^*(t) >^{(-1/2)}.$$

The proof follows by direct substitution. The last relations are given in a "static" form and Theorem 3.9.1 indicates that the calculation of parameters S_{0t}, $S_t(\tau)$, $\lambda_t(\tau)$ has to be done "afresh" for every new instant of time t. We shall now investigate whether the calculations can be made in a recurrent form, without having to perform the additional recalculation.

In all the ellipsoidal approximations considered in this book the center of the approximating ellipsoid is always the same, being given by $x^\star(t)$ of (3.83). The discussions shall therefore concern only the relations for $X_-(t)$.

3.9 Reachability Tubes: Recurrent Relations—Internal Approximations

Fig. 3.8 Internal ellipsoids for $\mathcal{X}^0 = \mathcal{E}(0,0), \ \mathcal{E}(0,1)$

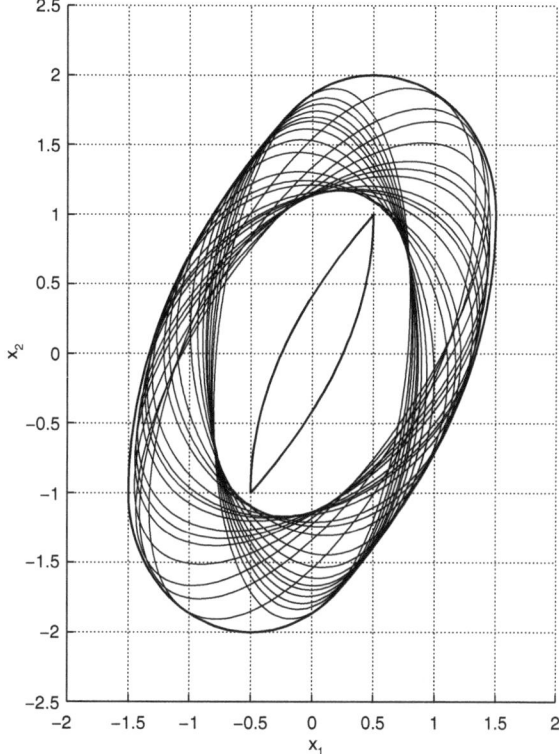

Remark 3.9.1. Without conditions (3.103), (3.104) equality (3.101) turns into an inequality

$$\rho(l^*(t)|X[t]) \geq \rho(l^*(t)|\mathcal{E}(x_-(t), X_-(t))) = \langle l^*(t), x^l(t) \rangle, \tag{3.105}$$

for any orthogonal matrix-valued function $S(t)$.

Remark 3.9.2. The results of Sects. 3.7 and 3.9 are also true for degenerate ellipsoids $\mathcal{E}(x^0, X_0)$, $\mathcal{E}(q(t), Q(t))$. This will further allow to treat systems with box-valued constraints in the form of parallelotopes as well as to treat zonotope-valued constraints (see Chap. 5, Sects. 5.4, 5.5).

We will now pass to the treatment of internal approximations of reachability tubes. We start with a particular function $l^*(t)$, namely, the one that satisfies the Assumption 3.3.1, where the requirement is that $l^*(t)$ is of the form $l^*(t) = G'(t_0, t)l$.

Substituting $l^*(t)$ in (3.103), (3.104), we observe that the relations for calculating $S_t(\tau), \lambda_t(\tau)$ transform into

$$S_t(\tau) Q_B^{1/2}(\tau) G'(t_0, \tau) l = \lambda_t(\tau) S_{0t} X_0^{1/2} l, \tag{3.106}$$

$$S'_{0t} S_{0t} = I, \ S'_t(\tau) S_t(\tau) \equiv I,$$

and
$$\lambda_t(\tau) = <l, G(t_0,\tau)Q_B(\tau)G'(t_0,\tau)l>^{1/2} <l, X_0 l>^{-1/2}. \qquad (3.107)$$

Here the known vectors and functions used for calculating $S_{0t}, S_t(\tau), \lambda_t(\tau)$ do not depend on t. But then the matrix S_{0t} and functions $S_t(\tau), \lambda_t(\tau)$ do not depend on t either, no matter what is the interval $[t_0, t]$. Therefore, the lower indices t in S_{0t}, S_t, λ_t may be dropped and the matrix $X_-(t) = X_-^{(l)}$ will depend only on l. Then we have
$$X_-^{(l)}(t) = G(t,t_0)\mathcal{K}(t)G'(t,t_0),$$
where
$$\mathcal{K}(t) = K'(t)K(t), \quad \dot{\mathcal{K}} = \dot{K}'K + K'\dot{K},$$
and
$$K(t) = S_0 X_0^{1/2} + \int_{t_0}^t S(\tau)Q_B^{1/2}(\tau)G'(t_0,\tau)d\tau. \qquad (3.108)$$

Differentiating $X_-^{(l)}(t)$ in view of the last notations, we come to
$$\dot{X}_-^{(l)} = A(t)X_-^{(l)} + X_-^{(l)}A'(t) + G(t,t_0)(\dot{K}'(t)K(t) + K(t)\dot{K}'(t))G'(t,t_0), \qquad (3.109)$$
where
$$\dot{K} = H(t)G'(t_0,t), \quad H(t) = S(t)Q_B^{1/2}(t), \quad K(t_0) = S_0 X_0^{1/2}. \qquad (3.110)$$

The differentiation of (3.83) also gives
$$\dot{x}^\star = A(t)x^\star + B(t)q(t), \quad x^\star(t_0) = x^0. \qquad (3.111)$$

This leads to the following theorem.

Theorem 3.9.2. *Under Assumption 3.3.1 the solution to Problem 3.9.1 is given by ellipsoid $\mathcal{E}(x^\star(t), X_-^{(l)}(t))$ where $X_-^{(l)}(t), x^\star(t)$ are given by Eqs. (3.109), (3.111), and functions $S(t), \lambda(t)$ involved in the calculation of $K(t)$ satisfy together with S_0 the relations (3.106), (3.107), where the lower indices t in S_{0t}, S_t, λ_t are to be dropped.*

We may also denote $X_-^*(t) = K(t)G'(t,t_0)$, so that $X_-^{(l)}(t) = (X_-^*(t))'X_-^*(t)$.

Lemma 3.9.1. *Function $X_-^*(t)$ may be expressed through equation*
$$\dot{X}_-^* = X_-^* A'(t) + H(t), \quad X_-^*(t_0) = S_0 X_0^{1/2}. \qquad (3.112)$$

This yields the next result.

3.9 Reachability Tubes: Recurrent Relations—Internal Approximations

Lemma 3.9.2. *The ellipsoid* $\mathcal{E}(x^\star(t), X_-^{(l)}(t))$ *of Theorem 3.9.2, given by Eqs. (3.109)–(3.112) depends on the selection of orthogonal matrix function* $S(t)$ *with orthogonal* S_0, *and for any such* $S_0, S(t)$ *the inclusion*

$$\mathcal{E}(x^\star(t), X_-^{(l)}(t)) \subseteq X[t], \quad t \geq t_0,$$

is true with equality (3.105) attained for $X_-(t) = X_-^{(l)}(t)$ *and* $x_-(t) = x^\star(t)$ *under conditions (3.106), (3.107).*

Let us now suppose that $l(t)$ of Problem 3.9.1 is the vector function that generates *any* continuous curve of related support vectors on the surface of $X[t]$. Namely, Assumption 3.3.1 *is not fulfilled*. Then one has to use formula (3.102), having in mind that $S_{0t}, S_t(\tau)$ depend on t. After a differentiation of (3.102) in t, one may observe that (3.110) transforms into

$$\dot{X}_- = A(t)X_- + X_- A'(t) + H'(t)X_-^*(t) + X_-^*(t)H(t) + \Psi(t,\cdot), \qquad (3.113)$$

$$X_-(t_0) = X_0,$$

where

$$\Psi(t,\cdot) = \Psi^*(t,\cdot) + \Psi^{*'}(t,\cdot)$$

and

$$\Psi^*(t,\cdot) = G(t,t_0)\left(X_0^{1/2}(\partial S'_{0t}(t_0)/\partial t) + \int_{t_0}^t G(t_0,\tau)Q_B^{1/2}(\tau)(\partial S'_t(\tau)/\partial t)d\tau\right) \times$$

$$\left(S_{0t}(t_0)X_0^{1/2} + \int_{t_0}^t S_t(\tau)Q_B^{1/2}(\tau)G'(t_0,\tau)d\tau\right)G'(t,t_0).$$

Lemma 3.9.3. *Under Assumption 3.3.1 the functional* $\Psi(t,\cdot) \equiv 0$.

Similarly to Sects. 3.3 and 3.4, we come to the proposition.

Theorem 3.9.3. *Let* $l(t)$ *generate a curve* $x^l(t)$ *of related support points located on the surface of set* $X[t]$, *forming a system trajectory for (3.81) due to some control* $u(t)$. *Then Assumption 3.3.1 is satisfied and the functional* $\Psi(t,\cdot) \equiv 0$ *iff* $l(t)$ *is a "good" curve.*

We would finally like to emphasize that the suggested approach appears to be appropriate for *parallel computations* (see Chap. 4, Sect. 4.5).

3.10 Backward Reachability: Ellipsoidal Approximations

The backward reachability or "solvability" sets $W[\tau] = W[\tau, t_1, \mathcal{M}]$ were introduced earlier in Sect. 2.3. As indicated in this section, their calculation is crucial for constructing strategies of feedback target control. We shall now indicate the ellipsoidal techniques that allow effective calculation of feedback control strategies for such problems. The suggested methods apply to systems of fairly high dimensions.

Ellipsoidal approximations of solvability sets are derived through procedures similar to forward reach sets. Here are the respective results. A detailed derivation of the coming relations may serve to be one of the good exercises on the topics of this book.

Keeping notation $Q_B(t) = B(t)Q(t)B'(t)$, we have the next assertion.

Theorem 3.10.1. *Given is system (3.1) with ellipsoidal bounds* $u(t) \in \mathcal{E}(q(t), Q(t))$ *and target set* $\mathcal{E}(m, M)$. *Then the following results are true.*

(i) *With* $\mathcal{E}(w^\star(\tau), W_+^{(l)}(\tau))$ *and* $\mathcal{E}(w^\star(\tau), W_-^{(l)}(\tau))$ *being the external and internal ellipsoidal approximations of set* $W[\tau]$ *the next inclusions are true*

$$\mathcal{E}(w^\star(\tau), W_-^{(l)}(\tau)) \subseteq W[\tau] \subseteq \mathcal{E}(w^\star(\tau), W_+^{(l)}(\tau)). \quad (3.114)$$

(ii) *Vector* $w^\star(t)$ *satisfies equation*

$$\dot{w}^\star = A(t)w^\star + B(t)q(t), \quad w^\star(t_1) = m, \quad (3.115)$$

while matrix $W_+^{(l)}(t)$ *satisfies equation*

$$\dot{W}_+^{(l)} = A(t)W_+^{(l)} + W_+^{(l)}A'(t) - \pi(t)W_+^{(l)} - \pi^{-1}(t)Q_B(t), \quad (3.116)$$

where

$$\pi(t) = <l(t), Q_B(t)l(t)>^{1/2} <l(t), W_+^{(l)}(t)l(t)>^{-1/2}, \quad W_+^{(l)}(t_1) = M, \quad (3.117)$$

with $l(t) = G'(t_1, t)l$, $l \in \mathbb{R}^n$.

(iii) *For each vector* $l \in \mathbb{R}^n$ *the next equality is true*

$$\rho(l \mid \mathcal{E}(w^\star(t), W_+^{(l)}(t))) = \rho(l \mid W[t]), \quad (3.118)$$

(ellipsoid $\mathcal{E}(w(t), W_+^{(l)}(t))$ *is tight).*

(iv) *For any vector* l *an internal ellipsoid* $\mathcal{E}(w^\star(\tau), W_-^{(l)}(\tau)) \subseteq W[\tau]$ *is defined through matrix*

3.10 Backward Reachability: Ellipsoidal Approximations

$$W_-^{(l)}(t) =$$

$$= G(t,t_1)\left(M^{1/2}S_m' - \int_t^{t_1} G(t_1,t)Q_B(\tau)^{1/2}S'(\tau)d\tau\right)$$

$$\left(S_m M^{1/2} - \int_t^{t_1} S(\tau)Q_B(\tau)^{1/2}G'(t_1,\tau)d\tau\right)G'(t,t_1),$$

which satisfies equation

$$\dot{W}_-^{(l)} = \qquad (3.119)$$

$$= A(t)W_-^{(l)} + W_-^{(l)}A'(t) + G(t,t_1)\dot{\mathcal{H}}^{(l)}(t)G'(t,t_1), \quad W_-^{(l)}(t_1) = M,$$

with

$$\mathcal{H}^{(l)}(t) = H'(t)H(t), \quad \dot{\mathcal{H}}^{(l)} = \dot{H}'H + H'\dot{H},$$

$$H(t) = S_m M^{1/2} - \int_t^{t_1} S(\tau)Q_B(\tau)^{1/2}G'(t_1,\tau)d\tau, \qquad (3.120)$$

and

$$\dot{H} = S(t)Q_B^{1/2}(t)G'(t_1,t), \quad H(t_1) = S_m M^{1/2},$$

where S_m is any orthogonal matrix and $S(t)$ is any orthogonal matrix-valued function.

(v) *Suppose orthogonal matrices* $S_m = S_m^{(l)}$, $S(t) = S^{(l)}(t)$ *satisfy conditions*

$$(S_m^{(l)})'S_m^{(l)} = I, \quad (S^{(l)}(t))'S^{(l)}(t) \equiv I,$$

$$S^{(l)}(t)Q_B(t)^{1/2}G'(t_1,t)l < l, G(t_1,t)Q_B(t)^{1/2}G'(t_1,t)l >^{-1/2} =$$

$$= S_m^{(l)}M^{1/2}l < l, Ml >^{-1/2},$$

for all $t \in [\tau, t_1]$ *and therefore depend on* l.

Then for each vector $l \in \mathbb{R}^n$ *the equality condition for support functions is true:*

$$\rho(l|\mathcal{E}(w^\star(t), W_-^{(l)}(t))) = \rho(l|\mathcal{W}[t]) \qquad (3.121)$$

(the ellipsoid $\mathcal{E}(w^\star(t), W_-^{(l)}(t))$ *is tight).*

(vi) *The next relations are true*

$$\cup \{\mathcal{E}(w^\star(t), W_-^{(l)}(\tau)) | <l,l> \le 1\} = \mathcal{W}[\tau] = \cap\{\mathcal{E}(w^\star(t), W_+^{(l)}(\tau)) | <l,l> \le 1\}. \tag{3.122}$$

Further, passing to notations

$$\mathbf{H}(t) = W_-^{-1/2}(t)S(t)Q_B^{1/2}, \quad W_-^{(l)}(t) = W_-(t)$$

we also come to equation

$$\dot{W}_- = A(t)W_- + W_-A'(t) - \mathbf{H}'(t)W_- - W_-\mathbf{H}(t), \quad W_-(t_1) = M, \tag{3.123}$$

which may be used as well as (3.119).

The given theorem is formulated for tight ellipsoids placed along good curves of type $l(t) = G'(t_1,t)l$. Equality (3.122) follows from conditions (3.121).

Exercise 3.10.1. Derive the formulas of Theorem 3.10.1 for $A(t) \ne 0$.

In Fig. 3.9 presented is a picture related to calculating the reachability tube for system[5]

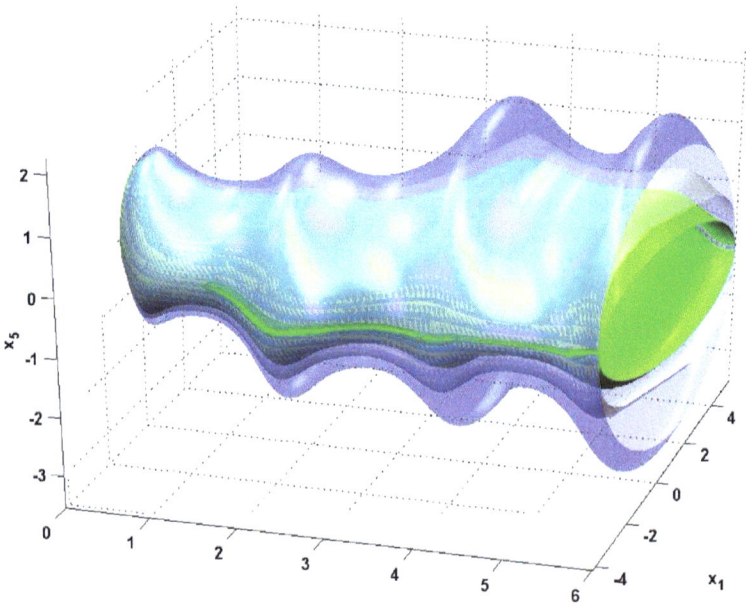

Fig. 3.9 Reachability tube with one external and one internal approximation

---

[5]Thus example was worked out by P. Gagarinov.

$$A(t) = \begin{pmatrix} 0 & 1 & 0 & 0 & 0 & 0 & 0 & 0 \\ -1 & 0 & 0 & 0 & 0 & 0 & 0 & 0 \\ 0 & 0 & 0 & 0 & 1 & 0 & 0 & 0 \\ -1 & 0 & -9 & 0 & 0 & 0 & 0 & 0 \\ 0 & 0 & 0 & 0 & 0 & 1 & 0 & 0 \\ 0 & 0 & -1 & 0 & -3 & 0 & 0 & 0 \\ 0 & 0 & 0 & 0 & 0 & 0 & 0 & 1 \\ 0 & 0 & 0 & 0 & -1 & 0 & -2 & 0 \end{pmatrix}, \quad B(t) = \begin{pmatrix} 1 \\ 0 \\ 0 \\ 0 \\ 0.3 \\ 0 \\ 0 \\ 0 \end{pmatrix},$$

$X_0 = I \in \mathbb{R}^{8\times 8}$, $x_0' = (0, \ldots, 0) \in DR^8$, $P(t) = 1$, $p(t) = cos(2*t); t \in [0, 6]$,

where given is the cross-section ("cut") of its dynamic projection at time $t = t^* = 6$ on the subspace $\mathcal{L}(t^*)$ orthogonal to vector $(1, 0, 0, 0, 1, 0, 0, 0)'$, with tight internal and external ellipsoids touching the exact reach set along "good" direction at $l'(t^*) = (1, 0, 0, 0, 1, 0, 0, 0)$.

Remark 3.10.1. In the last two sections we have studied the behavior of **tight internal ellipsoidal approximations** of reach sets and reach tubes. It shows that Eqs. (3.110)–(3.112) with appropriately chosen parameter $S(t)$ (an orthogonal matrix-valued function restricted by an equality) generate a family of internal ellipsoids that touch the reach tube or its neighborhood **from inside** along a special family of "good" curves that cover the whole tube. Such "good" curves are the same as for the external approximations. The suggested techniques allow an on-line calculation of the internal ellipsoids without additional computational burden present in other approaches. The calculation of similar ellipsoids along any other given smooth curve on the boundary of the reach tube requires additional burden as compared with the "good" ones. The internal approximations of this paper rely on relations different from those indicated in [35, 45, 174] and are relevant for solving various classes of control and design problems requiring guaranteed performance.

3.11 The Problem of Control Synthesis: Solution Through Internal Ellipsoids

We now pass to the construction of *synthesizing target control strategies* through ellipsoidal methods. Without loss of generality, following the transformations of Sect. 1.1, we may take $A(t) \equiv 0$. This is done by applying transformation $\bar{x} = G(t_1, t)x$ and getting for $\bar{x}$ the reduced equation, but keeping for it the original notation x. Such a move will also allow to avoid some cumbersome calculations and to present a clearer demonstration of the main relations, since now the "good" curves will turn out to be straight lines and we will have $W^{(l)}_-(t) \equiv \mathcal{H}^{(l)}(t)$, $G(t_1, t) \equiv I$.

Problem 3.11.1. Given is the system

$$\dot{x} = B(t)u, \quad u \in \mathcal{E}(q_B(t), Q_B(t)) = \mathcal{Q}(t) \qquad (3.124)$$

and target set $\mathcal{M} = \mathcal{E}(m, M)$. One is to find a synthesizing strategy $u = U^*(t, x)$, that transfers system (3.124) from any position $\{\tau, x(\tau)\}$, $x(\tau) \in \mathcal{W}[\tau]$, to the set $\mathcal{M}$ at time t_1, so that $x(t_1) \in \mathcal{M}$. The specified strategy $u = U^*(t, x)$ must ensure the existence of solutions to the differential inclusion

$$\dot{x} \in B(t)U^*(t, x).$$

The solution to the given problem is well known and indicated in Sect. 2.6. It may be found, once we know the solvability tube $\mathcal{W}[t]$, $t \in [\tau, t_1]$.

Namely, denoting

$$V_C(t, x) = d(x, \mathcal{W}[t])$$

and taking the total derivative $dV_C(t, x)/dt \mid_u$ at position $\{t, x\}$, due to Eq. (3.124), under control u, one gets the desired strategy as

$$U_e^*(t, x) = \begin{cases} \{u : dV_C(t, x)/dt \mid_u \leq 0 \mid u \in Q(t)\}, & \text{if } V_C(t, x) > 0 \ (x \notin \mathcal{W}[t]), \\ Q(t), & \text{if } V_C(t, x) = 0 \ (x \in \mathcal{W}[t]). \end{cases}$$

(3.125)

It will be further indicated that also

$$U_e^*(t, x) = \begin{cases} \arg\min \{dV_C(t, x)/dt \mid_u \mid u \in Q(t)\}, & \text{if } V_C(t, x) > 0 \ (x \notin \mathcal{W}[t]), \\ Q(t), & \text{if } V_C(t, x) = 0 \ (x \in \mathcal{W}[t]). \end{cases}$$

(3.126)

This is the so-called extremal aiming rule (due to Krasovski, see [121, 123]).

We shall now demonstrate that the solution to Problem 3.11.1 may be obtained by operating *only with ellipsoidal functions*.

Consider position $\{\tau, x(\tau)\}$, and a parameterized family of ellipsoidal tubes $\mathcal{E}(w^\star(t), W_-^{(l)}(t))$, $t \in [\tau, t_1]$, constructed according to Theorem 3.10.1. For a given position $\{\tau, x\}$, $x = x(\tau)$ the solution to Problem 3.11.1 exists if $x \in \mathcal{W}[\tau]$.

With $x \in \text{int } \mathcal{W}[\tau]$ one may choose any control $u \in Q(\tau)$, so the selection of control solutions will evolve around the boundary of $\mathcal{W}[\tau]$ and beyond. We shall start by supposing that given is position $\{\tau, x^*\}$ with $x = x^* \in \partial \mathcal{W}[\tau]$ on the boundary of $\mathcal{W}[\tau]$. We may then introduce an internal ellipsoid $\mathcal{E}(w^\star(\tau), W_-(\tau))$ that touches its boundary at this x^*. To do this we first note that with $A(t) \equiv 0$ the "good" curves that cover the tube $\mathcal{W}[t]$, $t \in [\tau, t_1]$ are such that $l(t) \equiv l^*$, is constant. Hence we may further take $l = l^*$ to be the support vector of $x^* \in \partial \mathcal{W}[\tau]$ so that the ellipsoid that touches would be with $W_-(\tau) = W_-^{(l^*)}(\tau)$, namely

$$\langle l^*, x(t) \rangle = \rho(l^* \mid W_-^{(l^*)}[\tau]).$$

(3.127)

And such property will hold within the whole interval $t \in [\tau, t_1]$:

$$\rho(l^* \mid \mathcal{E}(w^\star(t), W^{(l^*)}(t)) = \langle l^*, w^\star(t) \rangle + \langle l^*, W_-^{(l^*)}(t) l^* \rangle^{1/2},$$

along the selected "good curve" $l(t) \equiv l^*$.

3.11 The Problem of Control Synthesis: Solution Through Internal Ellipsoids

The selection of vector $l = l^*$ and the related ellipsoid $\mathcal{E}(w^\star(t), W_-^{(l^*)}(\tau))$ is a separate element of the solution process which is discussed further. But first recall the following. Suppose for $W_-(t) = W'_\star(t) W_\star(t)$ we have

$$W_\star(t) = S_m M^{1/2} - \int_t^{t_1} S(\xi) Q_B^{1/2}(\xi) d\xi, \tag{3.128}$$

Then *for any* orthogonal-valued matrix S_m and matrix function $S(t)$, $t \in [\tau, t_1]$ the ellipsoid $\mathcal{E}(w^\star(t), W_-^{(l^*)}(t)) \subseteq \mathcal{W}[t]$ will be an internal approximation of $\mathcal{W}[t]$. However, if for a given vector $l \equiv l^*$ we select S_m, $S(t)$ coordinated according to relations

$$S(t) Q_B^{1/2}(t) l = \lambda(\tau) S_m M^{1/2} l, \quad \lambda(t) = <l, Q_B(t)l>^{1/2} <l, Ml>^{-1/2}. \tag{3.129}$$

Then, due to these relations, vectors $S_m M^{1/2} l$ and $S(t) Q_B^{1/2}(t) l$, $\forall t \in [\tau, t_1]$ will be collinear. Hence, taking such $S_m = S_m^*$, $S(t) = S^*(t)$ together with $l = l^*$, and taking $W_-(t) = W_-^{(l^*)}(t)$, we will have

$$\langle l^*, W_-^{(l^*)}(t) l^* \rangle^{1/2} = \langle l^*, M l^* \rangle^{1/2} + \int_t^{t_1} \langle l^*, Q_B(s) l^* \rangle^{1/2} ds, \tag{3.130}$$

and since

$$\dot{W}_-(t) = \dot{W}'_\star(t) W_\star(t) + W'_\star(t) \dot{W}_\star(t)$$

we further get by direct calculation,

$$\langle l^*, \dot{W}_-^{(l^*)}(t) l^* \rangle^{1/2} = -\langle l^*, Q_B(t) l^* \rangle^{1/2}. \tag{3.131}$$

Also note that if items $l(t)$ and $S_m, S(t)$ are not coordinated, as in (3.129), then (3.130), (3.131) will turn into inequalities with $=$ substituted by $\leq$.

Also observe that for any $\{l^* : \langle l^*, l^* \rangle = 1\}$, coordinated with related $S_m^*, S^*(t)$ according to (3.129), we also have

$$\langle l^*, W_-^{(l^*)}(t)) l^* \rangle \geq \langle l, W_-^{(l^*)}(t)) l \rangle \tag{3.132}$$

where l need not be coordinated with $S_m^*, S^*(t)$.

We further work with the selected tight tube $\mathcal{E}_-^{(l^*)}[t]$ substituted in place of the solvability tube $\mathcal{W}[t]$ of (3.129), while following exactly the respective "aiming" scheme. To use the ellipsoidal tube is more convenient than the exact solvability tube, since in the ellipsoidal case all the auxiliary operations may be presented *in explicit form.*

To indicate the ways of finding the control solutions, we begin with *the nondegenerate case* specified as follows:

Assumption 3.11.1. *Given set $\mathcal{E}_-^{(l^*)}[\tau] = \mathcal{E}(w^*(\tau), W_-^{(l^*)}(\tau))$ that touches $\mathcal{W}[\tau]$ internally, there exists $\varepsilon > 0$ such that $W_-^{(l^*)}(t) \geq \varepsilon I$, $\forall t \in [\tau, t_1]$.*

Checking whether $x = x(\tau) \in \mathcal{W}[\tau]$.
We first find if the distance $d[\tau, x] = d(x(\tau), \mathcal{W}[\tau]) > 0$. Then

$$V_C(\tau, x) = d[\tau, x] = \max\{\langle l, x\rangle - \rho(l \mid \mathcal{W}[\tau]) \mid \langle l, l\rangle = 1\}$$
$$= \langle l^0, x\rangle - \rho(l^0 \mid \mathcal{W}[\tau]) = \varepsilon > 0, \quad (3.133)$$

and

$$\langle l^0, x[\tau]\rangle = \varepsilon + \langle l^0, x^0[\tau]\rangle, \quad \langle l^0, x^0[\tau]\rangle = \rho(l^0 \mid \mathcal{W}[\tau])$$

so that x^0 is the point in $\mathcal{W}[\tau]$ closest to $x = x(\tau)$ and $x^0 \in \partial \mathcal{W}[\tau]$. With l^0 calculated, the solution $U^*(t, x)$ to Problem 3.11.1 would be given by (3.125).

Now we will do the calculations of (3.133) through ellipsoids.

Synthesizing control strategies through ellipsoids.
We first have to find an internal ellipsoid $\mathcal{E}_-^{(l^*)}[\tau] = \mathcal{E}(w^*(t), W_-^{(l^*)}(t))$ that touches $\mathcal{W}[\tau]$ from inside at closest point $x^0[\tau]$. And this will happen once $l^0 = l^*$, with tube $\mathcal{E}_-^{(l^*)}[t]$ taken due to coordinated $l^*, S_m^*, S^*(t), t \in [\tau, t_1]$.

To ensure $l^0 = l^*$ there are various options. Thus, for example

(i) we may take the distances $d_{ell}[\tau] = d(x, \mathcal{E}_-^{(l)}[\tau])$ over all internal ellipsoids $\mathcal{E}_-^{(l)}[\tau]) = \mathcal{E}(w^*(\tau), W_-^{(l)}(\tau))$ whatever be $l, S_m, S(\cdot)$ and find its minimum over unit vectors l and orthogonal matrices $S_m, S(t)$, arriving at

$$V_C(\tau, x) = d[\tau, x] = \min\{d_{ell}[\tau] \mid \langle l, l\rangle = 1, S_m S_m' = I, S(t) S'(t) = I,$$
$$t \in [\tau, t_1]\} = d_{ell}^0[\tau].$$

or (ii) we may select among internal ellipsoids $\mathcal{E}_-^{(l)}[\tau]$ that internally touch the boundary $\partial \mathcal{W}[\tau]$ the one that does it for $l = l^* = l^0$. This may not need to calculate the whole reach set.[6] Then vector l^* with related $W_-^{(l^*)}[\tau]$ are those that satisfy

$$V_C(\tau, x) = \min\{d(x, x^l) \mid \langle l, l\rangle = 1\} = d(x, x^*), \quad x^l = x^*, \, l = l^*.$$

The corresponding equations for finding an ellipsoid $\mathcal{E}_-^{(l)}[t] = \mathcal{E}(w^*(t), W_-^{(l)}(t))$ which touches $\mathcal{W}[t]$ at point $x^l \in \partial \mathcal{W}[\tau]$ with support vector l are

$$\begin{cases} w^*(t_1) = m, \\ \dot{w}^* = q_B(t); \end{cases} \quad \begin{cases} W_*(t_1) = S_m M^{1/2}, \\ \dot{W}_*(t) = S(t) Q_B^{1/2}(t). \end{cases} \quad (3.134)$$

---

[6] If $\mathcal{W}[t]$ is already calculated, the procedure is simple.

3.11 The Problem of Control Synthesis: Solution Through Internal Ellipsoids

Here

$$W_\star(t) = W_\star(t_1) - \int_t^{t_1} S(\xi) Q_B^{1/2}(\xi) d\xi, \quad W_-(t) = W_\star'(t) W_\star(t),$$

$Q_B(t) = B(t) Q(t) B'(t)$ and vectors $S_m M^{1/2} l$, $S(\xi) Q_B^{1/2}(\xi) l$ should be collinear.

Now suppose Assumption 3.11.1 is true and the tight tube $\mathcal{E}_-^{(l^*)}[t]$, $l^* = l^0$ is already selected. Then the ellipsoidal control strategy $U_e^\star(t, x)$ is defined, starting from $t = \tau$, in the following way. Taking

$$V_e(t, x) = d(x, \mathcal{E}(w^\star(t), W_-^{(l^*)}(t))) = \min_z \{\langle x - z, x - z \rangle \mid z \in \mathcal{E}_-^{(l^*)}[t]\} =$$

$$= \max_l \{\langle l, x - w^\star(t) \rangle - \rho(l \mid \mathcal{E}_-(0, W_-^{(l^*)}(t))) \mid \langle l, l \rangle = 1\} = \langle l^0, x \rangle - \langle l^0, W_-^{(l^*)}(t) l^0 \rangle^{1/2},$$
(3.135)

we observe that the maximizer l^0 of this problem is unique and due to (3.132) we have

$$\langle l^0, x \rangle = \langle l^0, w^\star(t) \rangle + \langle l^0, W_-^{(l^*)}(t) l^0 \rangle^{1/2} = \rho(l^0 \mid \mathcal{E}_-(w^\star(t), W_-^{(l^*)}(t))) \leq$$

$$\leq \rho(l^0 \mid \mathcal{W}(t))$$

Then, following the reasoning of Sect. 2.6, we get for $d[x(t), \mathcal{E}(w^\star(t), W_-^{(l^*)}(t))] > 0$ the strategy

$$U_e^\star(t, x) = \begin{cases} \arg\min \left\{ dV_e(t, x)/dt \mid_u \mid u \in \mathcal{E}(q(t), Q(t)) \right\}, & \text{if } x \notin \mathcal{E}(w^\star(t), W_-^{(l^*)}(t)), \\ \mathcal{E}(q(t), Q(t)), & \text{if } x \in \mathcal{E}(w^\star(t), W_-^{(l^*)}(t)). \end{cases}$$
(3.136)

Since $l^* = l^0$ this gives

$$U_e^\star(t, x) = \begin{cases} q(t) - Q(t) B'(t) l^0 < l^0, Q_B(t) l^0 >^{-1/2}, & \text{if } x \notin \mathcal{E}(w^\star(t), W_-^{(l^0)}(t)), \\ \mathcal{E}(q(t), Q(t)), & \text{if } x \in \mathcal{E}(w^\star(t), W_-^{(l^0)}(t)). \end{cases}$$
(3.137)

We further show that the *ellipsoidal strategy* $U_e^\star(t, x)$ of (3.137) does solve the problem of control synthesis if the starting position $\{\tau, x\}$ is such that $x = x(\tau) \in \partial \mathcal{E}(w^\star(\tau), W_-^{(l^0)}(\tau))$ for related l^0 and also $x(\tau) \in \partial \mathcal{W}[\tau]$. Indeed, suppose $x(\tau) \in \mathcal{E}_-^{(l^0)}[\tau]$, and $x[t] = x(t, \tau, x(\tau))$ that satisfies Eq. (3.124) with $u \in U_e^\star(t, x)$, $\tau \leq t \leq t_1$, is the respective trajectory that emanates from $\{\tau, x(\tau)\}$. We will demonstrate that any such solution $x[t]$ of (3.124) will guarantee the inclusion $x[t_1] \in \mathcal{E}(m, M)$. To do that we will need, as in Sect. 2.6, to know the total derivative $dV_e(t, x)/dt$ for $V_e(t, x) > 0$.

Recalling

$$V_e(t, x[t]) = d[t] = d_0(t, x[t]) = <l^0, x[t]> - \rho(l^0 \mid \mathcal{E}_-^{(l^0)}[t]),$$

for $d[t] > 0$, and due to the uniqueness of the maximizer $l^* = l^0 \neq 0$, we have

$$\frac{d}{dt}d[t] = \frac{d}{dt}(<l^0, x[t]> - \rho(l^0 \mid \mathcal{E}_-^{(l^0)}[t])),$$

so that

$$\frac{d}{dt}d[t] = <l^0, \dot{x}[t]> - \frac{\partial}{\partial t}\rho(l^0 \mid \mathcal{E}_-(w^*(t), W^{(l^0)}(t))) = \quad (3.138)$$

$$= <l^0, B(t)(u - q(t))> - \frac{d}{dt} <l^0, W_-^{(l^0)}(t)l^0>^{1/2}.$$

Then from (3.131), with $u = u_e^0 \in U_e^*(t, x)$ we have:

$$<l^0, B(t)(u_e^0 - q(t))> - \frac{d}{dt}<l^0, W_-^{(l^0)}(t)l^0>^{1/2} \leq \langle l^0, B(t)(u_e^0 - q(t))\rangle + \langle l^0, Q_B(t)l^0\rangle. \quad (3.139)$$

Here $l^0 = l^*$ is coordinated with $S_m^*, S^*(t)$, so (3.131) gives an inequality.

From here it follows that with $u = u_e^0 \in U_e^*(t, x)$ of (3.137), and vector $l^0 = l^0(t, x)$ of (3.135) we ensure the inequality

$$\left.\frac{d}{dt}d[t]\right|_u \leq 0.$$

Summarizing, we finally have

$$U_e^*(t, x) = \arg\min\{<l^0, B(t)u> \mid u \in \mathcal{E}(q(t), Q(t))\}. \quad (3.140)$$

Integrating $dd[t]/dt = dd_0(t, x(t))/dt$ from τ to t_1, along any trajectory $x(t, \tau, x(\tau))$ generated by $u = u_e^0 \in U_e^*(t, x)$, we get

$$V_e(t_1, x(t_1)) = d_0(t_1, x(t_1)) = d(x(t_1), \mathcal{E}(m, M)) \leq d_0(\tau, x(\tau)) = V_e(\tau, x(\tau)) = 0$$

which means $x(t_1) \in \mathcal{E}(m, M)$, provided $x(\tau) \in \mathcal{E}_-^{(l^0)}[\tau] \subseteq \mathcal{W}[\tau]$, whatever be the solution to

$$\dot{x} \in B(t)U_e^*(t, x), \quad t \in [\tau, t_1],$$

generated from position $\{\tau, x\}$, $x \in \mathcal{W}[\tau]$.

3.11 The Problem of Control Synthesis: Solution Through Internal Ellipsoids   137

We thus came to

Theorem 3.11.1. *The solution to Problem 3.11.1 (with $A(t) \equiv 0$) is given by the "ellipsoidal strategy" $U_e^*(t, x)$ of (3.140).*

Exercise 3.11.1. Prove an analogy of Theorem 3.11.1 for system (3.1) with $A(t) \neq 0$.

We now indicate some techniques for finding l^0.

Calculating l^0

The scalar product $\langle x, z \rangle$ taken above, in solving Problem 3.11.1, may be substituted for one of more general kind, namely by $\langle x, Tz \rangle$, $T = T' > 0$, $\langle x, Tx \rangle = \|x\|_T$, so that in the above we took $T = I$. However we will also need to consider $T \neq I$, since the calculation of vector l^0 is especially simple if one takes $T^{-1} = W_-^{(l^0)}(t)$.

Therefore, before moving towards the proof that control (3.136) does indeed solve Problem 3.11.1, we first look at how to calculate l^0. Namely, we have to find the distance $d[t] > 0$ in the more general form as

$$d[t] = V_T(t, x) = d_T[t, x] = d_T(x, \mathcal{E}(w^*(t), W_-^{(l^*)}(t))) =$$

$$= \max\{< Tl, x - w^*(t) > -\langle Tl, W_-^{(l^*)}(t)Tl \rangle^{1/2}) \mid < l, Tl > \leq 1\}, \quad (3.141)$$

where $V_T(t, x) = V_e(t, x)$ if $T = I$. This problem is solved through simple optimization techniques. Here is its solution.

An auxiliary optimization problem. Consider a nondegenerate ellipsoid $\mathcal{E} = \mathcal{E}(y^*, Y)$, $Y = Y' > 0$. We need to find

$$\mathcal{V}(t, x) = \min_y \{\langle x - y, T(x - y) \rangle^{1/2} \mid y \in \mathcal{E}\}.$$

Clearly, with $V(t, x) > 0$,

$$\min_y \{\|x - y\|_T \mid y \in \mathcal{E}\} = \min_y \{\max_l \{\langle x - y, l \rangle_T \mid \|l\|_T \leq 1\} \mid y \in \mathcal{E}\} =$$

$$= \max_l \{\langle x, l \rangle_T + \min_y \{-\langle y, l \rangle_T) \mid y \in \mathcal{E}\} \mid \|l\|_T \leq 1\} =$$

$$= \max_l \{\langle x, l \rangle_T - \rho(Tl|\mathcal{E}) \mid \|l\|_T \leq 1\} = \langle x, l^0 \rangle_T - \rho(Tl^0| \mathcal{E}).$$

Here l^0 is the maximizer of function $\langle x, l \rangle_T - \rho(Tl|\mathcal{E})$.

In view of

$$\rho(Tl|\mathcal{E}) = \langle y^*, Tl \rangle + \langle Tl, YTl \rangle^{1/2} = \langle y^*, l \rangle_T + \|Tl\|_Y,$$

we find

$$\mathcal{V}(t,x) = \max_l \{ \langle x - y^*, l \rangle_T - \|Tl\|_Y \mid \|l\|_T \leq 1 \}. \qquad (3.142)$$

Lemma 3.11.1. *For $d_T[t] > 0$ the distance between given point $x \in \mathbb{R}^n$ and ellipsoid $E = E(y^*, Y) \subset \mathbb{R}^n, Y = Y' > 0$ is the solution to optimization problem (3.142), with $T = T' > 0$.*

Finding the solution to (3.142) may be pursued with various matrices T. We emphasize two cases–when $T = I$ and $T = Y^{-1}$.

Case (a) : $T = I$

Referring to Lemma 3.11.1, we use the standard Euclid metric, taking, $y^* = w^*$, $Y = T$. Then relations (3.137)–(3.140) remain without formal change.

Now $\mathcal{V}(t, x)$ of (3.142) transforms into

$$\mathcal{V}(t,x) = \max_l \left\{ \langle x - w^*, l \rangle - \|l\|_{W_-^{(l^*)}} \mid \|l\| \leq 1 \right\} = \qquad (3.143)$$

$$= d(x - w^*, \mathcal{E}(0, W_-^{(l^*)})) = \min_p \{ \|z - p\| \mid \langle p, (W_-^{(l^*)})^{-1} p \rangle \leq 1, \ z = x - w^* \}.$$

For $\|z\|_T > 1$, $T = (W_-^{(l^*)})^{-1}$, this may be reduced to finding $\min_p \{ \|z - p\|^2 \mid \langle p, (W_-^{(l^*)})^{-1} p \rangle = 1 \} > 0$ through conventional Lagrangian techniques. The related Lagrangian is

$$\mathcal{L}(p, \lambda) = \langle z - p, z - p \rangle + \lambda(\langle p, (W_-^{(l^*)})^{-1} p \rangle - 1)$$

with

$$\mathcal{L}'_p(p, \lambda) = 2((z - p) + \lambda (W_-^{(l^*)})^{-1} p) = 0, \text{ so that } p = (I - \lambda (W_-^{(l^*)})^{-1})^{-1} z.$$

Multiplier λ satisfies equation $h(\lambda) = 0$, where

$$h(\lambda) = \langle (I - \lambda (W_-^{(l^*)})^{-1})^{-1} z, (W^{(l^*)})^{-1} (I - \lambda (W_-^{(l^*)})^{-1})^{-1} z \rangle - 1 = \langle p, (W_-^{(l^*)})^{-1} p \rangle - 1.$$

Here $h(0) > 0$ and with $\lambda \to \infty$ we have $\lim h(\lambda) = -1$, so there exists a root $\lambda^0 > 0$ of equation $h(\lambda) = 0$. This root λ^0 is unique, since $h'(\lambda) < 0$. We have thus found

$$d(x - w^*, \mathcal{E}(0, W_-^{(l^*)})) = \|z - p^0\|, \quad p^0 = (I - \lambda^0 (W_-^{(l^*)})^{-1})^{-1} z$$

and the element of $\mathcal{E}(w^*, W_-^{(l^*)})$ closest to x is $x^* = w^* + p^0$.

3.11 The Problem of Control Synthesis: Solution Through Internal Ellipsoids

Summarizing we come to the conclusion

Lemma 3.11.2. *(i) The vector l^0 which solves (3.143) (or (3.142)) under $T = I$ is*

$$l^0 = \begin{cases} 0, & \text{if } \langle x - w^\star, (W_-^{(l^\star)})^{-1}(x - w^\star) \rangle \leq 1 \ (x \in \mathcal{E}(w^\star, W_-^{(l^\star)})), \\ (x - x^\star)\|x - x^\star\|^{-1}, & \text{otherwise,} \end{cases}$$
(3.144)

Here $x^\star \in \mathcal{E}(w^\star, W_-^{(l^\star)})$ is the closest to x in the Euclid metric and

$$x^\star = w^\star + p^0, \quad p^0 = (I - \lambda^0 (W_-^{(l^\star)})^{-1})^{-1}(x - w^\star).$$

(ii) The related controls are then specified by (3.137) with $w^\star = w^\star(t)$, $W_-^{(l^\star)} = W_-^{(l^\star)}(t)$, $p^0 = p^0(t)$.

Exercise 3.11.2. Work out the formulas for the "ellipsoidal strategy" $U^\star(t, x)$ of Problem 3.11.1 for $A(t) \neq 0$.

Case (b) : $T = Y^{-1}$
This is the simpler case. Indeed, substituting Y^{-1} for T, we transform (3.142) into

$$\mathcal{V}(t, x) = \max_l \{\langle x - y^\star, l \rangle_T - \|l\|_T \mid \|l\|_T \leq 1\}.$$
(3.145)

Due to the inequality of Cauchy–Bunyakovski–Schwartz we now have

$$\langle x - y^\star, l \rangle_T - \|l\|_T \leq \|x - y^\star\|_T \|l\|_T - \|l\|_T,$$

with equality attained at $l = \lambda(x - y^\star)$, $\lambda \geq 0$. Since $\langle l, Tl \rangle = 1$, this gives $\lambda = \|x - y^\star\|_T^{-1}$, so that under $T = Y^{-1}$ the maximizer of (3.142) is

$$l_T^0 = \begin{cases} 0, & \text{if } \|x - y^\star\|_T \leq 1, \\ (x - y^\star)\|x - y^\star\|_T^{-1}, & \text{otherwise.} \end{cases}$$
(3.146)

Hence, with $\mathcal{V}(t, x) > 0$ ($x \notin \mathcal{E}(y^\star, Y)$), we have

$$\mathcal{V}(t, x) = \max_l \{\|x - y^\star\|_T \|l\|_T - \|l\|_T \mid \langle l, Tl \rangle \leq 1\} = \|x - y^\star\|_T - 1.$$
(3.147)

Otherwise, with $x \in \mathcal{E}$, we have $\mathcal{V}(t, x) = 0$.
The element $y^0 = y^\star + q^0 \in \mathcal{E}(y^\star, Y)$ closest to x is found as follows: find

$$\min_y \{\|x - y\|_{Y^{-1}} \mid y \in \mathcal{E}(y^\star, Y)\} = \min_q \{\|x - (y^\star + q)\|_{Y^{-1}} \mid q \in \mathcal{E}(0, Y)\}$$

$$= \|x - (y^\star + q^0)\|_{Y^{-1}}.$$

Then $y^0 - y^* = q^0$, so that $\|q^0\|_{Y^{-1}} = \|y^0 - y^*\|_{Y^{-1}} = 1$ (y^0 lies on the boundary of $\mathcal{E}(y^*, Y)$).

Lemma 3.11.3. *(i) Under $T = Y^{-1}$ the distance in metric $\|x - y\|_T$ between $x \notin \mathcal{E}$ and ellipsoid $\mathcal{E} = \mathcal{E}(y^*, Y)$ is*

$$\mathcal{V}(t, x) = d(x, \mathcal{E})_T = \max\{0, \|x - y^0\|_T - 1\}.$$

(ii) The optimizer l_T^0 in problem (3.147) under metric $\|x - y\|_T$, $T = Y^{-1}$ is given by (3.146).

(iii) The element of $\mathcal{E}$ closest to x under metric $\|x - y\|_T$ is

$$y^0 - y^* = q^0, \quad \|q^0\|_T = 1.$$

Exercise 3.11.3. In the previous lemma show that vectors l_T^0 of (ii) and p^0 of (iii) are collinear.

We now return to the original Problem and apply the previous Lemma 3.11.3 to find the synthesizing control $U_e^*(t, x)$. In terms of Problem 3.11.1 we have: $x = x(t)$, $y^* = w^*(t)$, $Y = W_{-}^{(l^*)}(t)$.

Now, in order to rewrite these results in terms of Lemma 3.11.3, we need to substitute its l^0 by $Tl^0 = l_T^0$, $T = (W_{-}^{(l^*)})^{-1}$, with metric $\|x - y\|_T = \langle x - y, (W_{-}^{(l^*)})^{-1}(x - y)\rangle^{1/2}$. We thus get

$$l_T^0 = \begin{cases} 0, & \text{if } \|x(t) - w^*(t)\|_T \leq 1, \\ (x(t) - w^*(t))\|x(t) - w^*(t)\|_T^{-1}, & \text{otherwise.} \end{cases} \quad (3.148)$$

The related control $u_{Te}^*(t, x)$, for $x(t) \notin \mathcal{E}$, is then found due to (3.140), as

$$u_{Te}^*(t, x) = q(t) - Q(t)B'(t)l_T^0 \langle l_T^0, Q_B(t)l_T^0 \rangle^{-1/2}.$$

After following (3.140) and recalling $Q_B(t) = B(t)Q(t)B'(t)$, this yields the solution strategy

$$U_{Te}^*(t, x) = \begin{cases} q(t) & \text{if } x \in \mathcal{E}(w^*(t), W_{-}^{(l^*)}(t)), \\ q(t) - Q(t)B'(t)(x(t) - w^*(t))\|Q^{1/2}(t)B'(t)(x(t) - w^*(t))\|_T^{-1}, \\ \quad \text{if } x \notin \mathcal{E}(w^*(t), W_{-}^{(l^*)}(t)). \end{cases}$$

(3.149)

Exercise 3.11.4. By direct substitution of $u = U_{Te}^*(t, x)$ into (3.124) prove the following:

(i) the existence of solution to (3.124) under such substitution,
(ii) that $u = U_{Te}^*(t, x)$ does solve Problem 3.11.1.

Remark 3.11.1. A solution in the absence of Assumption 3.11.1 has to deal with degenerate internal ellipsoids. The calculation of related controls then has to undergo a procedure of regularization. This issue will be treated at the end of Chap. 4.

The given scheme thus describes the applicability of ellipsoidal techniques to the calculation of trajectory tubes for controlled systems with further design of synthesizing controls within a fairly simple scheme as compared to existing methods. This scheme is realized in the *ellipsoidal toolbox* used further to solve the presented examples (see [45, 132]).

3.12 Internal Approximations: The Second Scheme

Apart from the techniques for internal approximation of sets and tubes given in Sects. 3.7–3.10, there exists another set of formulas which may be used for the same purpose. The restrictive element in this "second approach" is that here the ellipsoidal approximations are derived through operations with only a *pair* of ellipsoids rather than *any number* of these as in the above. The treatment of internals for sums of ellipsoids may therefore be achieved only through an inductive sequence of pairwise operations rather than through a single operation. Nevertheless this scheme turns out to be useful and is also being recommended. The scheme was discussed in [174] (see also [45]).

We shall calculate the reachability set $X[t] = X(t, t_0, X_0)$ for system (3.1) with $v(t) \equiv 0$. Using the funnel equation (2.54), we have

$$(I + \sigma A(t))X[t] + \sigma \mathcal{E}(q_B(t), Q_B(t)) \subseteq X[t + \sigma] + o_1(\sigma)\mathcal{B}(0), \quad (3.150)$$

where $\mathcal{B}(0) = \{x : \langle x, x \rangle \leq 1\}$,

$$\sigma^{-1} o_1(\sigma) \to 0 \text{ if } \sigma \to 0.$$

With $X[t]$ being an ellipsoid of type

$$X[t] = \mathcal{E}(x^\star(t), X^-(t)), \quad X^-(t) > 0,$$

we may apply the following formula. For two given ellipsoids $\mathcal{E}(a^{(1)}, X_1)$, $\mathcal{E}(a^{(2)}, X_2)$, $X_1 = X_1' > 0$, $X_2 = X_2' > 0$, it indicates the matrix

$$X^-(H) = H^{-1}[(HX_1H')^{1/2} + (HX_2H')^{1/2}]^2 H'^{-1}, \quad (3.151)$$

where $H \in \Sigma$ and

$$\Sigma = \{H \in \mathbb{R}^{n \times n} : H' = H, |H| \neq 0\}.$$

(Matrices H are therefore symmetrical and nondegenerate.)

Matrix X^- ensures the relations

$$\mathcal{E}(a^{(1)} + a^{(2)}, X^-(H)) \subseteq \mathcal{E}(a^{(1)}, X_1) + \mathcal{E}(a^{(2)}, X_2), \quad \forall H \in \Sigma, \quad (3.152)$$

and

$$E_{1,2} = \bigcup \{\mathcal{E}(a^{(1)} + a^{(2)}, X^-(H)) \mid H \in \Sigma\} = \mathcal{E}(a^{(1)}, X_1) + \mathcal{E}(a^{(2)}, X_2). \quad (3.153)$$

If one of the two ellipsoids is degenerate, say $X_2 \geq 0$, then the union $E_{1,2}$ at the left-hand side of (3.153) should be substituted by its *closure* $\overline{E_{1,2}}$.

Applying the last relations to (3.150), with $X[t] = \mathcal{E}(x^\star(t), X^-(t))$ being an ellipsoid, we take

$$X_1 = (I + \sigma A(t))X^-[t](I + \sigma A(t))', \quad X_2 = \sigma^2 B(t) Q(t) B'(t).$$

Then in view of (3.152) we have

$$X[t + \sigma] + o_1(\sigma)\mathcal{B}(0) \supseteq \mathcal{E}(x^\star(t + \sigma), X^-(t + \sigma)),$$

where

$$\begin{aligned}X^-(t + \sigma) = {} & H^{-1}(t)[H(t)X^-(t)H'(t) + \\ & + \sigma(A(t)X^-(t) + X^-(t)A'(t)) + \sigma^2 A(t)X^-(t)A'(t) + \\ & + \sigma(H(t)(I + \sigma A(t))X^-(t)(I + \sigma A'(t))H'(t))^{\frac{1}{2}}(H(t)Q_B(t)H'(t))^{\frac{1}{2}} + \\ & + \sigma(H(t)Q_B(t)H'(t))^{\frac{1}{2}}(H(t)(I + \sigma A(t))X^-(t)(I + \sigma A'(t))H'(t))^{\frac{1}{2}} + \\ & + \sigma^2 H(t)QB(t)H'(t)]H'^{-1}(t),\end{aligned}$$

and

$$x^\star(t + \sigma) = (I + A(t)\sigma)x^\star(t) + \sigma B(t)q(t).$$

From here, after subtracting $X^-(t)$ from both sides, then dividing them by σ and applying a limit transition with $\sigma \to 0$, we come to ordinary differential equations

$$\begin{aligned}\dot{X}^-(t) = {} & A(t)X^- + X^- A'(t) + H^{-1}((H(t)X^-(t)H'(t))^{\frac{1}{2}}(H(t)Q_B(t)H'(t))^{\frac{1}{2}} + \\ & + (H(t)Q_B(t)H'(t))^{\frac{1}{2}}(H(t)X^-(t)H(t))^{\frac{1}{2}})(H^{-1})', \quad (3.154)\end{aligned}$$

$$\dot{x}^\star = A(t)x^\star + B(t)q(t), \quad (3.155)$$

3.12 Internal Approximations: The Second Scheme

with initial conditions

$$x^\star(t_0) = x^0, \quad X^-(t_0) = X^0.$$

What follows from here is the inclusion

$$X[t] \supseteq \mathcal{E}(x^\star(t), X^-(t)), \qquad (3.156)$$

where $x^\star(t), X^-(t)$ satisfy (3.154), (3.155) and $H(t)$ is a continuous function of t with values in Σ—the variety of symmetric nondegenerate matrices.

This leads us to the next conclusion.

Theorem 3.12.1. *The internal approximation of the reachability set $X[t] = X(t, t_0, \mathcal{E}(x^0, X^0))$ for system (3.1) is given by the inclusion (3.156), where $x^\star(t), X^-(t)$ satisfy Eqs. (3.155) and (3.154). Moreover, the following representation is true*

$$X[t] = \overline{\cup\{\mathcal{E}(x^\star(t), X^-(t)) | H(\cdot) \in \Sigma\}}, \qquad (3.157)$$

where the union is taken over all measurable matrix-valued functions with values in Σ.

Remark 3.12.1. The closure of the union of sets at the right-hand side of equality (3.157) appears due to the fact that matrix $Q_B = B(t)Q(t)B'(t)$ may turn out to be degenerate. Then the exact reachability set $X[t]$ may have kinks while being approximated by internal ellipsoids all of which are nondegenerate. This effect is absent in the "first approach" of Sects. 3.7–3.10 which allows degenerate elements among the internal approximating ellipsoids.

One should note that internal approximations of reachability sets by nondegenerate ellipsoids require that the "reach set" itself would be nondegenerate.

Definition 3.12.1. A reachability set $X[t]$ is said to be nondegenerate if there exists an n-dimensional ball $\mathcal{B}_\varepsilon(c) = \{x :< x-c, x-c >\leq \varepsilon\}$ of radius $\varepsilon > 0$, such that $\mathcal{B}_\varepsilon(c) \subseteq X[t]$.

In other words, set $X[t]$ a nonempty interior. The last property may arise either from the properties of the system itself and its constraints or from the degeneracy of the starting set $X[t_0]$, or from both factors. To check degeneracy due to the system itself it suffices to deal with reach sets of type $X[t] = X(t, t_0, \{0\})$, emanating from point $x(t_0) = 0$.

Consider system

$$\dot{x} \in A(t)x + B(t)\mathcal{E}(q(t), Q(t)), \qquad (3.158)$$

$$x(t_0) = 0,$$

where $B(t)$ is continuous and $q(t) \in \mathbb{R}^p, Q(t) \in \mathbb{R}^{p \times p}, p < n$ are also continuous.

The parameters of this system allow to generate the set-valued integral

$$X^*[t] = \int_{t_0}^{t} G(t,s)B(s)\mathcal{E}(0,Q(s))ds, \qquad (3.159)$$

where matrix $G(t,s)$ is defined in Sect. 1.1.

Assumption 3.12.1. *There exists a continuous scalar function $\beta(t) > 0$, $t > t_0$, such that the support function*

$$\rho(l|X^*[t]) \geq \beta(t) <l,l>^{1/2},$$

for all $l \in \mathbb{R}^n$, $t \geq t_0$.

With $Q(t) > 0$, $t > t_0$, this assumption implies that the reachability domain $X[t]$ of system (3.158) has a nonempty interior $(int X[t] \neq \emptyset)$ and is therefore nondegenerate. Assumption 3.12.1 is actually equivalent to the requirement that system (3.158) with unbounded control $u(t)$ would be *completely controllable* [109, 195], on every finite time interval $[t_0, t]$.

Exercise 3.12.1. Prove that under Assumption 3.12.1 and with $Q(t) > 0$, $t > 0$, $x(t_0) = 0$ the reachability set $X[t]$ for system (3.158) is nondegenerate. (Use Exercise 1.4.2.)

In a similar way the "second approach" may be applied to the approximation of *backward reachability sets* $\mathcal{W}[\tau]$. A reasoning similar to the above in this section gives the result.

Theorem 3.12.2. *The internal approximation of the backward reachability set $\mathcal{W}[t]$ is given by the inclusion*

$$\mathcal{E}(w^*(t), W_-[t]) \subseteq \mathcal{W}[t],$$

where $\{w^*(t), W_-[t]\}$ satisfy Eq. (3.114) and

$$\dot{W}_-(t) = A(t)W_-(t) + W_-(t)A'(t)- \qquad (3.160)$$

$$H^{-1}(t)[(H(t)W_-(t)H'(t))^{\frac{1}{2}}(H(t)Q_B(t)H'(t))^{\frac{1}{2}}$$

$$+ (H(t)Q_B(t)H'(t))^{\frac{1}{2}}(H(t)W_-(t)H(t))^{\frac{1}{2}}]H'^{-1}(t) ,$$

with boundary conditions $x(t_1) = m$, $W_-[t_1] = M$.
Moreover, the following representation is true

$$\mathcal{W}[t] = \overline{\cup\{\mathcal{E}(w^*(t), W_-(t))| H(\cdot) \in \Sigma\}}, \qquad (3.161)$$

where the union is taken over all measurable matrix-valued functions with values in Σ.

3.12 Internal Approximations: The Second Scheme

Problem 3.11.1 of ellipsoidal control synthesis is then solved according to the scheme of Sect. 3.10.

Exercise 3.12.2. (i) Prove formulas (3.160) and (3.161).
(ii) Design an ellipsoidal synthesis for Problem 3.11.1 using Eq. (3.161) and prove that it solves the problem.[7]

---

[7] Such a proof is indicated in [174, Sect. 3.6, p. 212].

Chapter 4
Solution Examples on Ellipsoidal Methods: Computation in High Dimensions

Abstract In this chapter we describe solution examples for controlled systems that illustrate the contents of Chaps. 1–3. These include the multiple integrator, planar Newtonian motions and calming down a chain of springs. Special sections are devoted to relevant computational formulas and high-dimensional systems. Also discussed are possible degeneracy effects in computation and the means of their avoidance.

Keywords Multiple integrator • Newtonian motion • Oscillating system • Graphic illustration • Regularization • High dimensions

In this chapter we first present examples of control problems discussed above. All these problems are treated in finite time. The solutions are reached through the theory given in Chaps. 1, 2 and the computation is done using methods of Chap. 3.[1]

The second part deals with specifics of calculation for systems of high dimensions.

We begin with the examples.

4.1 The Multiple Integrator

The calculation of reach sets for a two-dimensional system with one integrator was treated in detail in Chap. 3, Sect. 3.9. We continue now by dealing with *the multiple integrator with single input* given by system ($i = 1, \ldots, n - 1; n \geq 2$)

$$\begin{cases} \dot{x}_i = x_{i+1}, \\ \dot{x}_n = b(t)u \end{cases} \tag{4.1}$$

---

[1] The illustrations for the presented solutions were provided by N.Bairamov, A.Mesiats, D.Odinokov, and V.Stepanovich.

with $b(t)$ taken continuous. The fundamental transition matrix for this system is

$$G(t,s) = \begin{pmatrix} 1 & t-s & \cdots & (t-s)^{n-1}/(n-1)! \\ 0 & 1 & \cdots & (t-s)^{n-2}/(n-2)! \\ \vdots & \vdots & \ddots & \vdots \\ 0 & 0 & 0 & 1 \end{pmatrix}$$

Here, with initial conditions $x_1^0 = 0, \ldots, x_n^0 = 0$, we have

$$x_1(t) = \int_0^t \cdots \int_0^{\xi_{n-1}} b(\xi_n) u(\xi_n) d\xi_n \ldots d\xi_1 = \int_0^t \left((t-s)^{n-1}((n-1)!)^{-1} b(s) u(s) \right) ds$$

which is an iterated integration of multiplicity n.

Exercise 4.1.1. Is system (4.1) controllable?

Example 4.1. For system (4.1) assume $x = (x_1, \ldots, x_n)'$, $x \in \mathbb{R}^n$, and $n \times n$ - matrix $X^0 = X^{0'} > 0$. Find the forward reach set $X[t] = X(t, t_0, X^0)$ from starting position $\{t_0, X^0\}$, under constraint $|u(s)| \leq \mu$, with

$$X^0 = \{x : \langle x - x^0, (X^0)^{-1}(x - x^0) \rangle \leq 1\}.$$

Note that system (4.1) is written in the form

$$\dot{x} = Ax + B(t)u, \quad u \in Q(t), \quad t \geq t_0, \tag{4.2}$$

where $Q(t) = \{u : |u| \leq \mu\}$ and

$$A = \begin{pmatrix} 0 & 1 & 0 & \cdots & 0 & 0 \\ 0 & 0 & 1 & \cdots & 0 & 0 \\ 0 & 0 & 0 & \cdots & 0 & 0 \\ \vdots & \vdots & \vdots & \ddots & \vdots & \vdots \\ 0 & 0 & 0 & \cdots & 0 & 1 \\ 0 & 0 & 0 & \cdots & 0 & 0 \end{pmatrix}, \quad B(t) = \begin{pmatrix} 0 \\ 0 \\ 0 \\ \vdots \\ 0 \\ b(t) \end{pmatrix}.$$

In order to calculate the reach sets for this system we shall follow Chap. 1 and transform the coordinates so that the new system has $A \equiv 0$ and a new matrix B. Applying transformation $z(t) = G(t_0, t)x(t)$, $t \geq t_0$ to (4.2), (see Sect. 1.1, formula (1.6)), we get

$$\dot{z} = \mathbf{B}(t)u, \quad \mathbf{B}(t) = G(t_0, t)B(t). \tag{4.3}$$

4.1 The Multiple Integrator

We consider this system without loss of generality, under the same starting set X^0 and same constraint on control u as before. Now, with $t \geq t_0$, the *forward* reach set is

$$X_z[t] = X^0 + \int_{t_0}^t \mathbf{B}(s)Q(s)ds, \tag{4.4}$$

while in the original coordinates

$$X[t] = X(t, t_0, X^0) = G(t, t_0)X^0 + \int_{t_0}^t G(t, s)B(s)Q(s)ds. \tag{4.5}$$

Recall that here the transformation

$$z(t) = G(t_0, t)x(t) = \begin{pmatrix} 1 & t-s & \cdots & (t_0-t)^{n-1}/(n-1)! \\ 0 & 1 & \cdots & (t_0-t)^{n-2}/(n-2)! \\ \vdots & \vdots & \ddots & \vdots \\ 0 & 0 & \cdots & 1 \end{pmatrix} x(t)$$

and

$$\mathbf{B}(t) = \begin{pmatrix} (t_0-t)^{n-1}/(n-1)! \\ (t_0-t)^{n-2}/(n-2)! \\ \vdots \\ 1 \end{pmatrix} b(t).$$

We now return in (4.3)–(4.4) from boldface letter $\mathbf{B}$ to ordinary B, getting $\dot{x} = B(t)u$, where $B(t)$ is time-dependent even when the original system (4.2) is time-invariant.

We begin with some examples on calculating forward reach sets through ellipsoidal approximations.

Given $X^0 = \mathcal{E}(x^0, X^0)$, $Q(t) = \mathcal{E}(q(t), Q(t))$, $\mathcal{M} = \mathcal{E}(m, M)$, we now assume

$$Q(t) = \mathcal{E}(q_b(t), Q_B(t)), \quad Q_B(t) = B(t)Q(t)B'(t), \quad q_b(t) = B(t)q(t),$$

$X_z^0 = X^0$, so that we deal with system

$$\dot{z} = B(t)u, \quad z(t_0) \in \mathcal{E}(x^0, X^0), \quad u \in Q(t), \quad t \geq t_0. \tag{4.6}$$

The required reach set is now

$$X_z[t] = \mathcal{E}(x^0, X^0) + \int_{t_0}^t Q_B(s)ds$$

with its internal and external approximating ellipsoids being $\mathcal{E}(x_z^*(t), X_{z-}(t))$, $\mathcal{E}(x_z^*(t), X_{z+}(t))$. These are given by

$$x_z^*(t) = x^0 + \int_{t_0}^{t} q_b(s) ds, \qquad (4.7)$$

$$X_{z-}(t) = X_{z-}^{*'}(t) X_{z-}^{*}(t), \qquad X_{z-}^{*}(t) = (X^0)^{1/2} + \int_{t_0}^{t} S(s) Q_B^{1/2}(s) ds, \quad (S_0 = I), \qquad (4.8)$$

$$X_{z+}(t) = \left(\langle l, X^0 l \rangle^{1/2} + \int_{t_0}^{t} \langle l, Q_B(s) l \rangle^{1/2} ds \right) \left(\frac{X^0}{\langle l, X^0 l \rangle^{1/2}} + \int_{t_0}^{t} \frac{Q_B(s)}{\langle l, Q_B(s) l \rangle^{1/2}} ds \right), \qquad (4.9)$$

where $S(t) = S(t; l)$ is an orthogonal matrix, so that $S'(t)S(t) = I$.
The reach set $\mathcal{X}_z[t]$ is approximated as

$$\mathcal{E}(x_z^*(t), X_{z-}(t)) \subseteq \mathcal{X}_z[t] \subseteq \mathcal{E}(x_z^*(t), X_{z+}(t)), \qquad (4.10)$$

where tightness ensures that the support functions

$$\rho(l \mid \mathcal{E}(x_z^*(t), X_{z-}(t))) = \rho(l \mid \mathcal{X}_z[t]) = \rho(l \mid \mathcal{E}(x_z^*(t), X_{z+}(t)))$$

along good curves $l(t) \in \mathbb{R}^n$. (See Sect. 3.3.) Here the good curves $l_z(t) \equiv l_z(t_0) = l$ turn out to be constant. Therefore the previous relations turn into equalities for any l under proper selection of parameterizing coefficients $\pi(t)$, $S(t)$, in ODEs that describe functions $x_z^*(t)$, $X_{z-}(t)$, $X_{z+}(t)$ for the approximating ellipsoids.
These ODEs are

$$\begin{cases} x_z^*(t_0) = x^0, \\ \dot{x}_z^* = q_b(t); \end{cases} \quad \begin{cases} X_{z-}^{*}(t_0) = (X^0)^{1/2}, \\ \dot{X}_{z-}^{*} = S(t) Q_B^{1/2}(t); \end{cases} \quad \begin{cases} X_{z+}(t_0) = X^0, \\ \dot{X}_{z+} = \pi(t) X_{z+} + \pi^{-1}(t) Q_B(t) \end{cases} \qquad (4.11)$$

with

$$\pi(t) = \langle l, Q_B(t) l \rangle^{1/2} \left(\langle l, X^0 l \rangle^{1/2} + \int_{t_0}^{t} \langle l, Q_B(s) l \rangle^{1/2} ds \right)^{-1}$$

and with matrices $S'(t)S(t) = I$, selected such that vectors $(X^0)^{1/2} l$, $S(t) Q_B^{1/2}(t) l$ are collinear.

4.1 The Multiple Integrator

Returning to the original coordinates $\{x\}$, we have

$$\mathcal{E}(x^*(t), X_+[t]) \subseteq X[t] \subseteq \mathcal{E}(x^*(t), X_-[t]), \tag{4.12}$$

where $x^*[t] = G(t, t_0)x_z^*[t]$, and

$$X_+[t] = G(t, t_0)X_{z+}[t]G'(t, t_0), \quad X_-[t] = G(t, t_0)X_{z-}[t]G'(t, t_0).$$

Remark 4.1.1. In further relations below we omit the lower indices z for all the related variables in (4.12).

For the *backward reach sets* the passage to new coordinates $\{z\}$ is given by $z(t) = G(\vartheta, t)x(t)$, $t \leq \vartheta$. Then, with given target set $\mathcal{M} = \mathcal{E}(m, M)$, we again have Eq. (4.6) and in these new coordinates the backward reach set $\mathcal{W}[t] = \mathcal{W}(t, \vartheta, \mathcal{M})$, $\mathcal{M} = \mathcal{E}(m, M)$ also allows tight ellipsoidal approximations

$$\mathcal{E}(w^*(t), W_-(t)) \subseteq \mathcal{W}[t] \subseteq \mathcal{E}(w^*(t), W_+(t)).$$

The parameterized functions $w^*(t), W_+(t), W_-(t)$ may also be calculated directly, through equations of Chap. 3 (see (3.97), (3.59), (3.98), (3.99)). The corresponding ODEs are

$$\begin{cases} w^*(\vartheta) = m, \\ \dot{w}^* = q_b(t); \end{cases} \quad \begin{cases} W_-^*(\vartheta) = M^{1/2}, \\ \dot{W}_-^* = S(t)Q_B^{1/2}(t); \end{cases} \quad \begin{cases} W_+(\vartheta) = M, \\ \dot{W}_+ = -\pi(t)W_+ - \pi^{-1}(t)Q_B(t). \end{cases} \tag{4.13}$$

Here

$$\pi(t) = \langle l, Q_B(t)l \rangle^{1/2} \left(\langle l, Ml \rangle^{1/2} + \int_t^\vartheta \langle l, Q_B(s)l \rangle^{1/2} ds \right)^{-1}$$

with

$$W_- = W_-^{*'}W_-^*, \quad W_-^* = (W^0)^{1/2} - \int_\tau^\vartheta S(t)Q_B^{1/2}(t)dt,$$

and $S'(t)S(t) = I$, while vectors $M^{1/2}l$, $S(t)Q_B^{1/2}(t)l$ are collinear.

4.1.1 Computational Results

Demonstrated here are the projections of reachability sets for system (4.2) of dimension Dim = 4 and 8 on the plane $\{x_1, x_2\}$ at time ϑ. N stands for the number

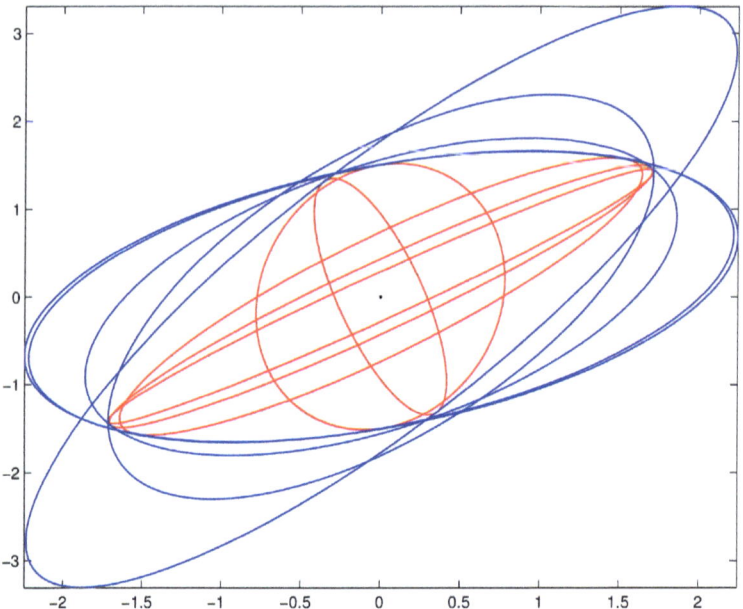

Fig. 4.1 $N = 5, Dim = 4$. Internal and external estimates

of approximating ellipsoids. The related parameters are $x^0 = 0$, $\tau = 0$, $\vartheta = 2$ for dimension 4 and $\vartheta = 3$ for dimension 8; $q = 0$, $Q = 1$,

$$X^0 = \begin{pmatrix} 1 & 0 & \cdots & 0 \\ 0 & 0.001 & \cdots & 0 \\ \vdots & \vdots & \ddots & \vdots \\ 0 & 0 & \cdots & 0.001 \end{pmatrix}.$$

The directions l that ensure tight approximations are selected for time ϑ to be a uniform partition on the unit circle in the plane $\{x_1, x_2\}$ (Figs. 4.1, 4.2, 4.3, 4.4, 4.5, 4.6, 4.7, and 4.8).

4.2 A Planar Motion Under Newton's Law

This is the motion of a particle of mass $m > 0$ given by system

$$\begin{cases} \dot{x}_1 = x_2, \\ m\dot{x}_2 = b_1 u_1, \\ \dot{x}_3 = x_4, \\ m\dot{x}_4 = b_2 u_2 \end{cases} \quad (4.14)$$

with $b_1, b_2 > 0$, $t \in [t_0, \vartheta]$.

4.2 A Planar Motion Under Newton's Law 153

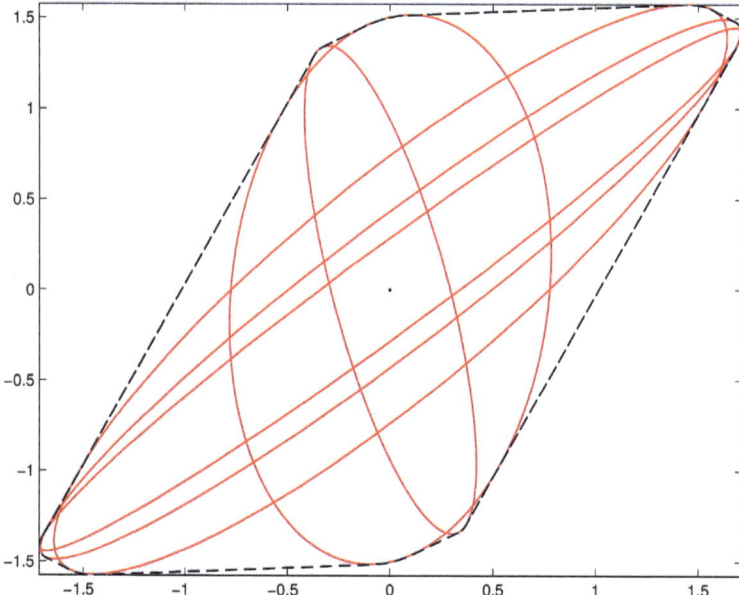

Fig. 4.2 $N = 5, Dim = 4$. Internal estimates and their convex hull

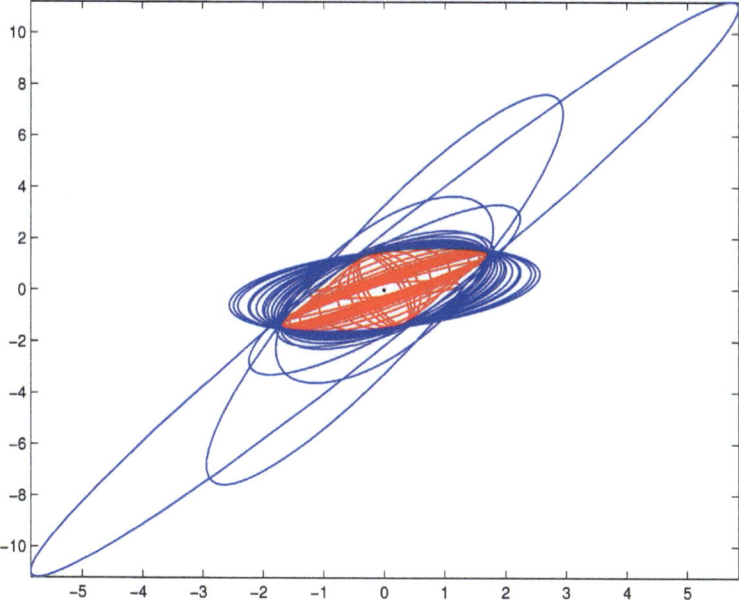

Fig. 4.3 $N = 20, Dim = 4$. Internal and external estimates

154 4 Solution Examples on Ellipsoidal Methods: Computation in High Dimensions

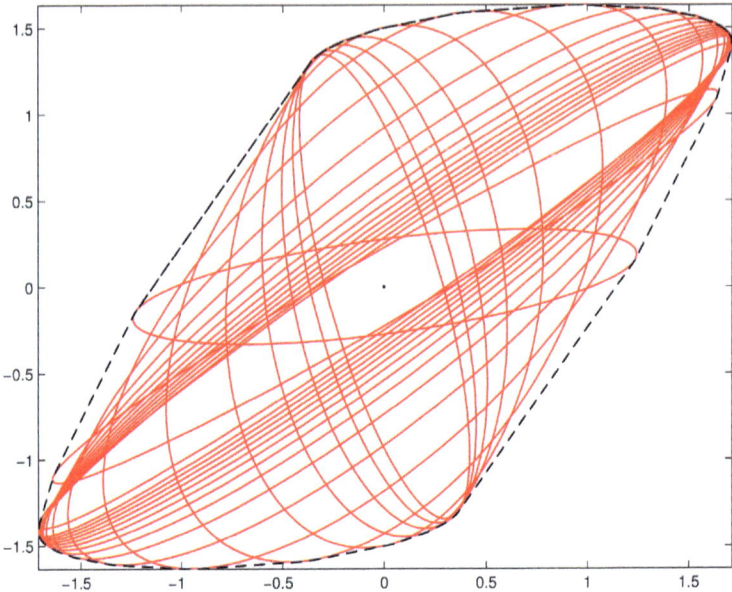

Fig. 4.4 $N = 20$, $Dim = 4$. Internal estimates and their convex hull

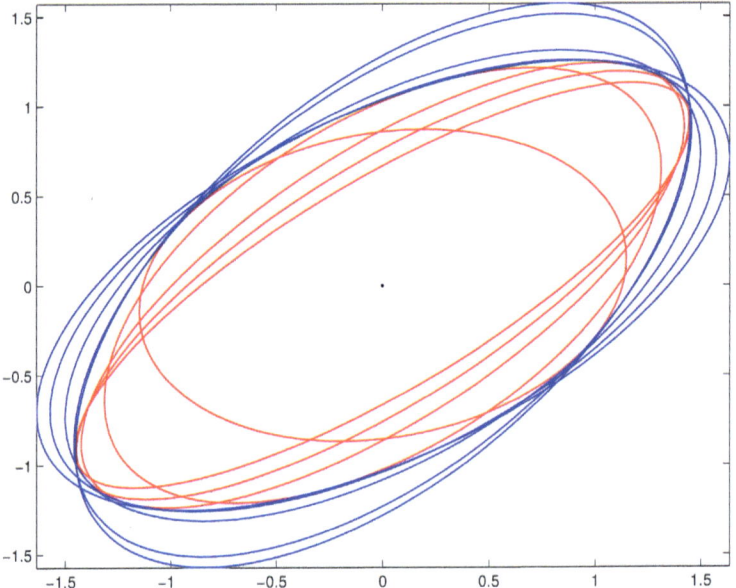

Fig. 4.5 $N = 5$, $Dim = 8$. Internal and external estimates

4.2 A Planar Motion Under Newton's Law

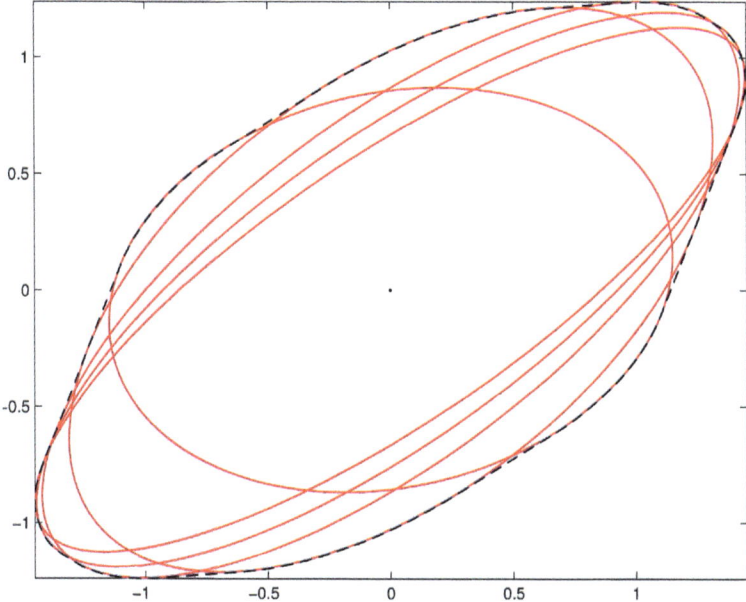

Fig. 4.6 $N = 5$, $Dim = 8$. Internal estimates and their convex hull

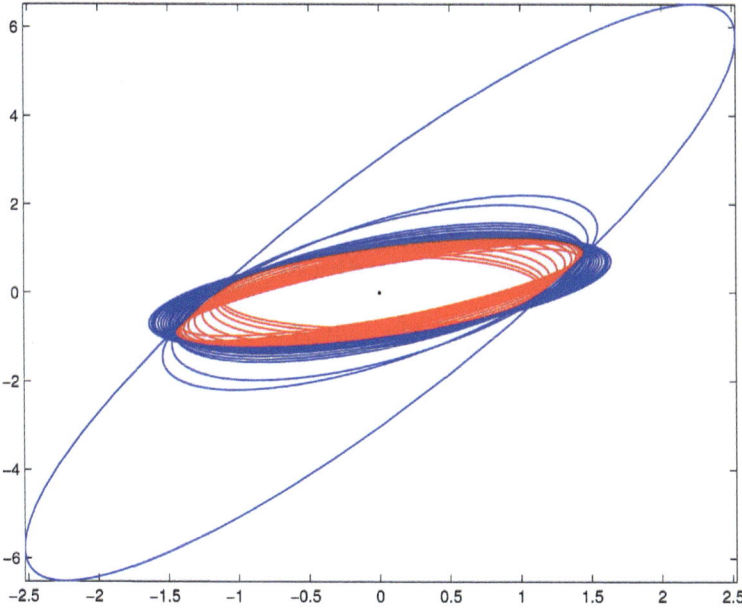

Fig. 4.7 $N = 20$, $Dim = 8$. Internal and external estimates

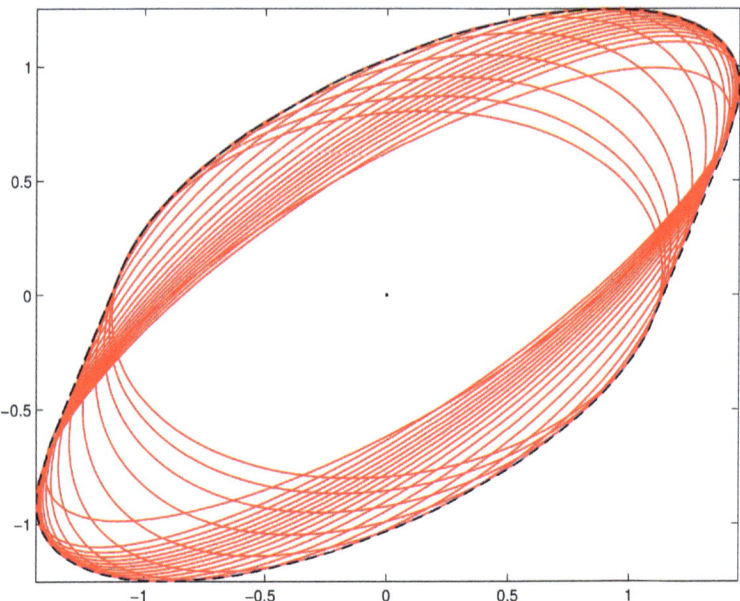

Fig. 4.8 $N = 20, Dim = 8$. Internal estimates and their convex hull

Taking further $m = 1$, we may transform this system to canonical first-order form, which is

$$\begin{cases} \dot{x}_1 = x_2, \\ \dot{x}_2 = b_1 u_1, \\ \dot{x}_3 = x_4, \\ \dot{x}_4 = b_2 u_2. \end{cases} \quad (4.15)$$

Our first task will be to find forward and backward reach sets

$$\mathcal{X}[\vartheta] = \mathcal{X}(\vartheta, t_0, \mathcal{X}^0), \quad \mathcal{X}^0 = \mathcal{E}(x^0, X^0), \quad \mathcal{W}[t] = \mathcal{W}(t, \vartheta, \mathcal{M}), \quad \mathcal{M} = \mathcal{E}(m, M)$$

under constraints

$$u(t) = u = \begin{pmatrix} u_1 \\ u_2 \end{pmatrix} \in \mathcal{Q}(t) = \mathcal{E}(q(t), Q(t)), \quad x(t_0) = x^{(0)} \in \mathcal{X}^0.$$

Denote

$$x = \begin{pmatrix} x_1 \\ x_2 \\ x_3 \\ x_4 \end{pmatrix}, \quad A = \begin{pmatrix} 0 & 1 & 0 & 0 \\ 0 & 0 & 0 & 0 \\ 0 & 0 & 0 & 1 \\ 0 & 0 & 0 & 0 \end{pmatrix}, \quad u = \begin{pmatrix} u_1 \\ u_2 \end{pmatrix}, \quad B = \begin{pmatrix} 0 & 0 \\ b_1 & 0 \\ 0 & 0 \\ 0 & b_2 \end{pmatrix}.$$

4.2 A Planar Motion Under Newton's Law

Then system (4.14) has the form (4.2), here with $B = const$. The fundamental matrix of its homogeneous part has the form $G(t, \tau) = \exp A(t - \tau)$, where $A^k = 0$, $k \geq 2$. Hence

$$G(t, \tau) = e^{A(t-\tau)} = I + A(t - \tau) = \begin{pmatrix} 1 & t-\tau & 0 & 0 \\ 0 & 1 & 0 & 0 \\ 0 & 0 & 1 & t-\tau \\ 0 & 0 & 0 & 1 \end{pmatrix}.$$

Using transformation $z = G(t_0, t)x$, $t \geq t_0$, taking $\mathbf{B}(t) = G(t_0, t)B$ the system (4.14) may be transformed, as in the previous section (see (4.6)), but without changing original notations, into

$$\dot{x} = \mathbf{B}(t)u(t), \; u(t) \in Q(t) = \mathcal{E}(q_B(t), Q_B(t)), \; t \geq t_0, \qquad (4.16)$$

with

$$\mathbf{B}(t) = G(t_0, t)B = \begin{pmatrix} (t_0 - t)b_1 & 0 \\ b_1 & 0 \\ 0 & (t_0 - t)b_2 \\ 0 & b_2 \end{pmatrix}, \; q_B(t) = \mathbf{B}(t)q,$$

$$\mathbf{B}(t)Q(t)\mathbf{B}'(t) = Q_B(t).$$

Returning now from boldface $\mathbf{B}$ to B, the forward reach set will be

$$\mathcal{X}_z[t] = \mathcal{X}_z(t, t_0, \mathcal{X}^0) = \mathcal{X}^0 + \int_{t_0}^{t} B(s)Q(s)ds.$$

With $\mathcal{X}^0 = \mathcal{E}(x^0, X^0)$ and support function

$$\rho(l \mid \mathcal{X}_z[t]) = \qquad (4.17)$$

$$\langle l, x^0 \rangle + \langle l, X^0 l \rangle^{1/2} + \int_{t_0}^{t} (\langle l, q_B(t) \rangle + \langle l, Q_B(t)l \rangle^{1/2})dt.$$

Similarly, for the backward reach set we use transformation $z = G(\vartheta, t)x$, $t \leq \vartheta$, and take $\mathbf{B}(t) = G(\vartheta, t)B$. Then system (4.14), without changing notations, may be again transformed into (4.17), but now with $t \leq \vartheta$. Here we also return from boldface $\mathbf{B}$ to B.

For target set $\mathcal{M} = \mathcal{E}(m, M)$ the backward reach set will be

$$\mathcal{W}_z[t] = \mathcal{W}_z(t, \vartheta, \mathcal{M}) = \mathcal{M} - \int_{t}^{\vartheta} B(s)Q(s)ds$$

with support function

$$\rho(l \mid \mathcal{W}_z[t]) = \tag{4.18}$$

$$\langle l, m \rangle + \langle l, Ml \rangle^{1/2} - \int_t^\vartheta (\langle l, q_B(t) \rangle - \langle l, Q_B(t)l \rangle^{1/2}) dt.$$

The ellipsoidal approximations of $X_z[t]$, $\mathcal{W}_z[t]$ are given by relations (4.7)–(4.13).

4.2.1 Computational Results

Forward Reach Sets

Presented are calculations of the forward reach set $X[1] = X(1, 0, \mathcal{E}(x^0, X^0))$ for the four-dimensional system

$$\dot{x} = B(t)u, \quad b_1 = 2, \ b_2 = 1, \ t \in [0, 1] \tag{4.19}$$

and

$$x^0 = \begin{pmatrix} 0 \\ 0 \\ 0 \\ 0 \end{pmatrix}, \quad X^0 = \begin{pmatrix} 3 & 0 & 0 & 0 \\ 0 & 1 & 0 & 0 \\ 0 & 0 & 1 & 0 \\ 0 & 0 & 0 & 3 \end{pmatrix}, \quad q = \begin{pmatrix} 1 \\ 1 \end{pmatrix}, \quad Q = \begin{pmatrix} 1 & 0 \\ 0 & 4 \end{pmatrix}.$$

Indicated are two-dimensional projections of external and internal ellipsoidal approximations for $X[t]$ on the system of coordinates $(x_1, \ldots, x_4)'$. Also shown are the convex hulls of the intersection of external estimates (as a red line, Figs. 4.9 and 4.11) and the union of internal estimates (as a blue line in Fig. 4.10 and dashed line in Fig. 4.11). Here internal ellipsoids are in green.

Indicated here are two two-dimensional projections of the reach tube for the system of the above (external and internal, Figs. 4.12 and coinciding 4.13).

Backward Reachability Sets and Control Synthesis: Aiming Methods

We are now passing to the calculation of control strategies. Our task in the simplified coordinates (from z to $x = G(t, \vartheta)z$) will be to transfer system (4.19) from given position $\{t_0, x\}$ to the final one $m \in \mathcal{M}$ within prescribed time $\vartheta - t_0$. As we have seen in Sects. 2.6 and 3.11, the task will be solvable iff $x \in \mathcal{W}[t_0]$, where $\mathcal{W}[t_0] = \mathcal{W}(t_0, \vartheta, \mathcal{M})$ is the backward reachability set from target set $\mathcal{M}$.

4.2 A Planar Motion Under Newton's Law

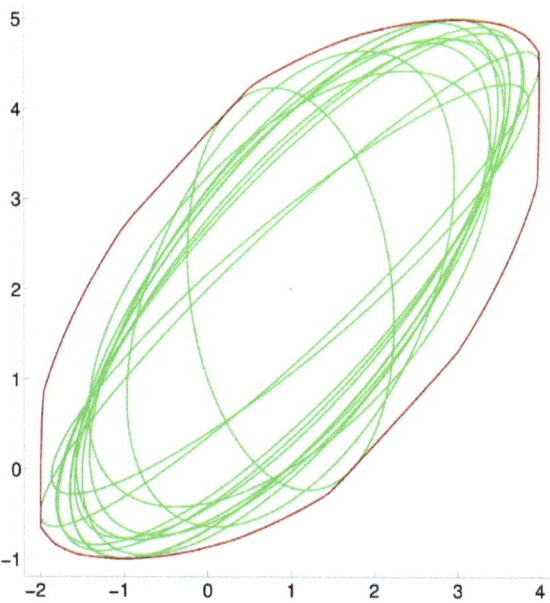

Fig. 4.9 $N = 15, Dim = 4, \{x_1, x_2\}$. External convex hull and internals

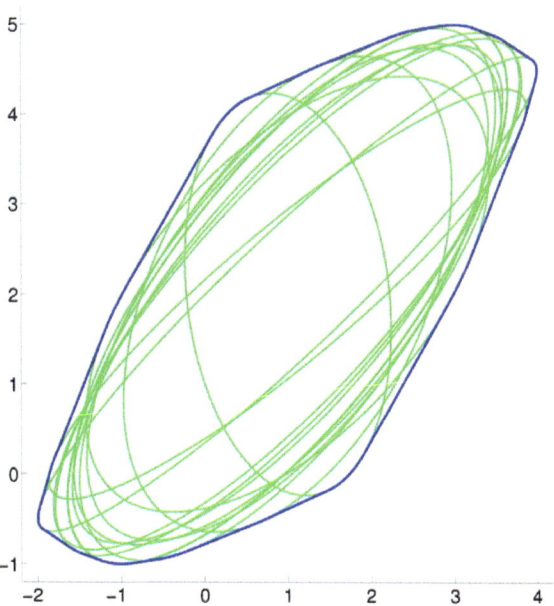

Fig. 4.10 $N = 15, DIM = 4, \{x_1, x_2\}$. Internals with their convex hull

Fig. 4.11 $N = 15$, $Dim = 4$, $\{x_1, x_2\}$. External and internal convex hulls

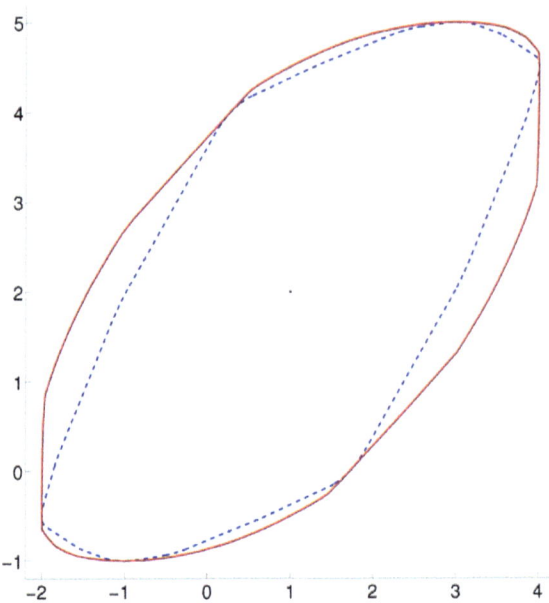

Fig. 4.12 $N = 15$, $Dim = 4$, $\{x_1, x_3\}$. Projection of reach tube—coordinates of motion

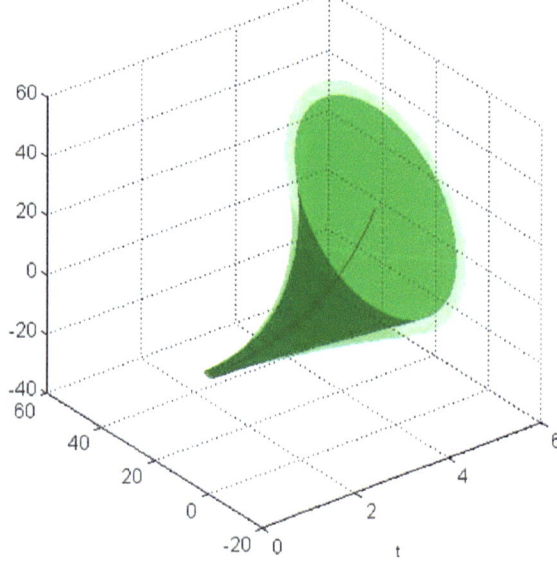

4.2 A Planar Motion Under Newton's Law

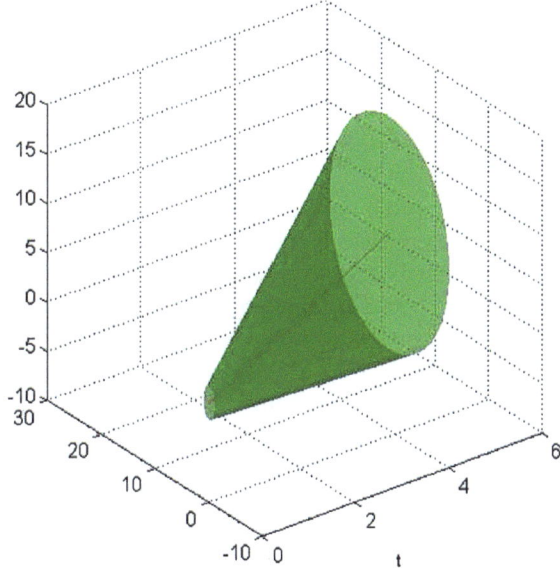

Fig. 4.13 $N = 15, Dim = 4, \{x_2, x_4\}$. Projection of reach tube—velocities of motion

We therefore have a problem in the next form: find feedback control strategy $u = U(t, x)$ which ensures the following conditions

$$\begin{cases} \dot{x}(t) = B(t)u(t), & t \in [t_0, \vartheta], \\ x(t) \in \mathbb{R}^n, \ x(\vartheta) \in \mathcal{E}(m, M), \ u(t) \in \mathcal{E}(q(t), Q(t)), \end{cases} \quad (4.20)$$

whatever be the starting position $\{t_0, x\}$, $x \in \mathcal{W}[t_0]$.

To find the feedback solution strategies we will apply the aiming methods described in Sects. 2.6 and 3.11. For calculating the solution we will substitute $\mathcal{W}[t]$ by its tight internal ellipsoidal approximation $\mathcal{E}_-[t] = \mathcal{E}(w^*(t), W_-[t])$. Then, as indicated in Sect. 3.11, the synthesizing feedback control $u = U(t, x)$ has the next form $(q(t) \in \mathcal{E}(q(t), Q(t)))$

$$U(t, x) = \begin{cases} q(t), & \text{if } x \in \mathcal{E}_-[t], \\ q(t) - Q(t)B'(t)l^0 \langle B'(t)l^0, Q(t)B'(t)l^0 \rangle^{-\frac{1}{2}}, & \text{if } x \notin \mathcal{E}_-[t]. \end{cases} \quad (4.21)$$

Here the unit vector $l^0 = (x(t) - s^0)/\|x(t) - s^0\|$, where

$$s^0 = \operatorname{argmin}\{\|x - s\| : s \in \mathcal{E}_-[t], x = x(t)\}$$

so that l^0 is the gradient of the distance function $d(x, \mathcal{E}_-[t])$ with t fixed and s^0 is the nearest point of $\mathcal{E}_-[t]$ from $x \notin \mathcal{E}_-[t]$.

Aiming method (a). As indicated in Sect. 3.11 (T=I) vector $s^0 = (Q + \lambda E)^{-1} Q(z - q) + q$ for $\mathcal{E}_-[t] = \mathcal{E}(q(t), Q(t))$ may be found from equation $h(\lambda) = 0$, where

$$h(\lambda) = \langle (Q + \lambda E)^{-1} Q(z - q), (Q + \lambda E)^{-1}(z - q) \rangle - 1.$$

and $\lambda > 0$ is its only positive root (prove that).

Here the bottleneck is to find a proper $\mathcal{E}_-[t]$ among such internal ellipsoids which would actually be the closest to x.

A second option—aiming method (b)—is to calculate the distance function in the metric generated by $\mathcal{E}(0, Q)$, namely,

$$s^0 = \operatorname{argmin}\{\|x - s\|_{Q^{-1}} : s \in \mathcal{E}_-[t], x = x(t)\}.$$

Then s^0 will lie on the line that connects point x and $w^\star$, hence

$$l^0 = (x - w^\star)/\|x - w^\star\|.$$

The final formula will then be as follows

$$U(t, x) = \begin{cases} q(t), & \text{if } x \in \mathcal{E}_-[t], \\ q(t) - \dfrac{Q(t) B'(t) l^0}{\langle B'(t) l^0, Q(t) B'(t) l^0 \rangle^{\frac{1}{2}}}, & \text{if } x \notin \mathcal{E}_-[t]. \end{cases} \quad (4.22)$$

Such are the two approaches to the calculation of l^0.

In the second approach one may use several internal ellipsoids $\mathcal{E}_-^k[t]$ since they all have a common center $w^\star(t)$ for each t. The solution formula is as follows

$$U(t, x) = \begin{cases} q(t), & \text{if } x \in \bigcup_k \mathcal{E}_-^k[t], \\ q(t) - \dfrac{Q(t) B'(t) l^0}{\langle B'(t) l^0, Q(t) B'(t) l^0 \rangle^{\frac{1}{2}}}, & \text{if } x \notin \bigcup_k \mathcal{E}_-^k[t]. \end{cases} \quad (4.23)$$

In the calculation of reachability sets for phase coordinates x of large dimensions and controls of small dimensions the internal ellipsoids $\mathcal{E}_-^k[t]$ may turn out to be degenerate. A separate treatment of such situations is given below, in Sect. 4.4.3.

For the two-dimensional system

$$\dot{x}_1 = x_2, \quad \dot{x}_2 = u + 1, \quad \mathcal{M} = \mathcal{E}(m, M), \quad m = (1, 1)', \quad M = \frac{1}{2} I, \quad |u| \leq 1,$$

indicated here are the projections of the backward reach set $\mathcal{W}[\tau]$ for $\tau = 0, \vartheta = 3$ (Figs. 4.14, 4.15, 4.16, 4.17, 4.18, 4.19, 4.20, 4.21, and 4.22).

4.2 A Planar Motion Under Newton's Law

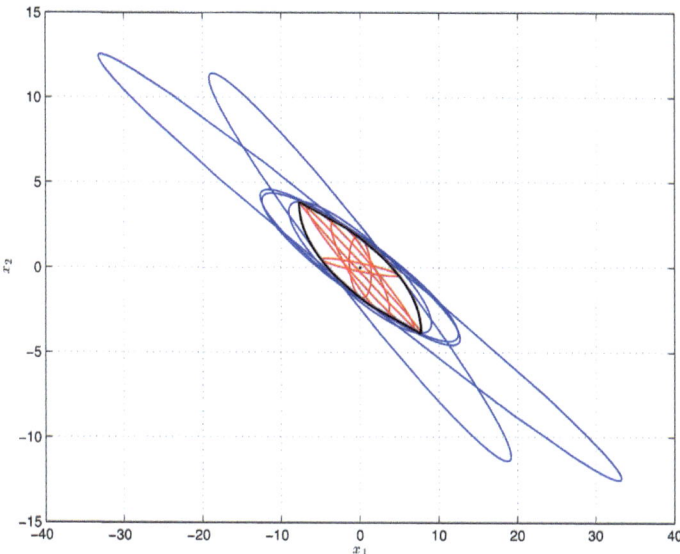

Fig. 4.14 $N = 5, DIM = 2$. Backward reach set—internal and external estimates

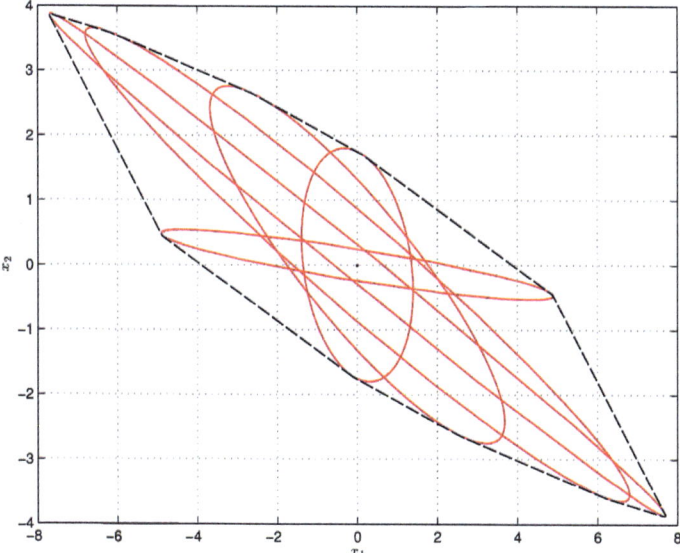

Fig. 4.15 $N = 5, DIM = 2$. Backward reach set—internal estimates and their convex hull

164   4 Solution Examples on Ellipsoidal Methods: Computation in High Dimensions

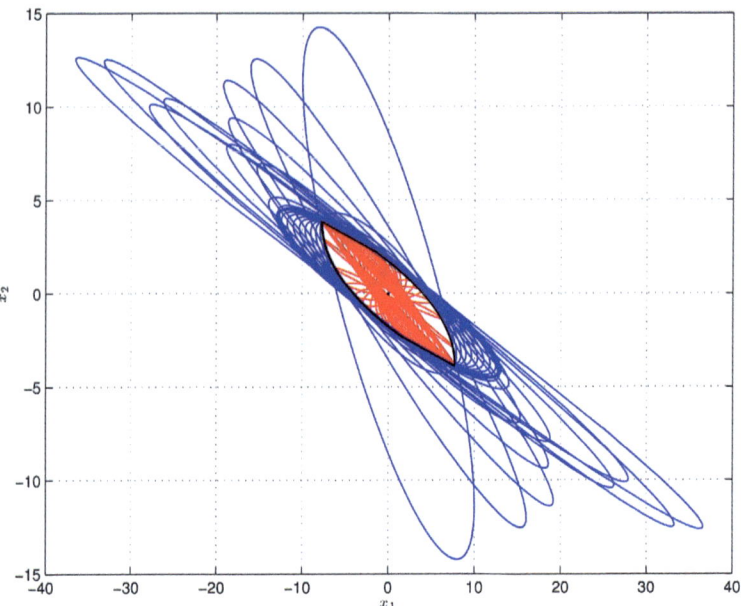

Fig. 4.16 $N = 30, DIM = 2$. Backward reach set—internal and external estimates

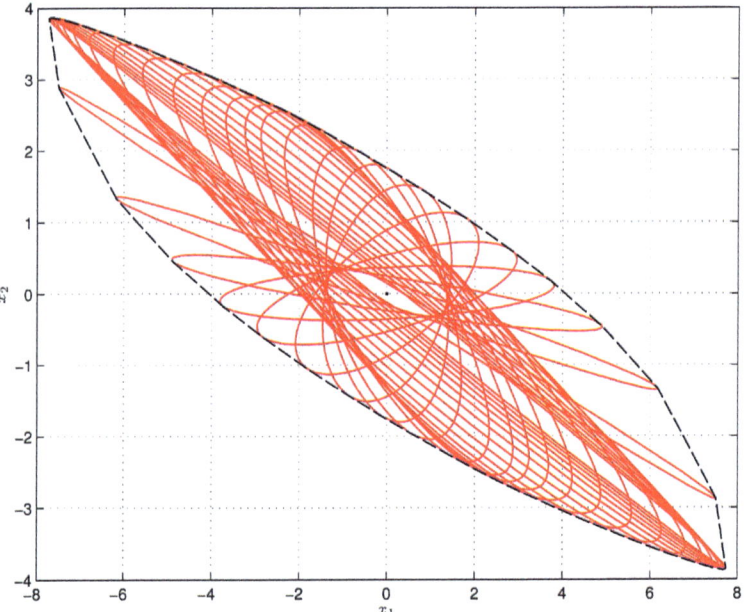

Fig. 4.17 $N = 30, DIM = 2$. Backward reach set—internal estimates and their convex hull

4.2 A Planar Motion Under Newton's Law

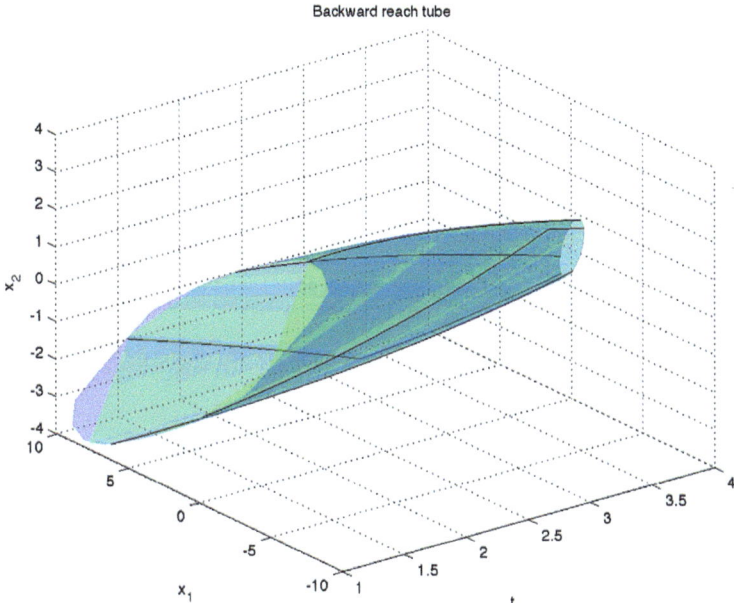

Fig. 4.18 $N = 5, Dim = 2$. Backward reach tube—internal and external estimates

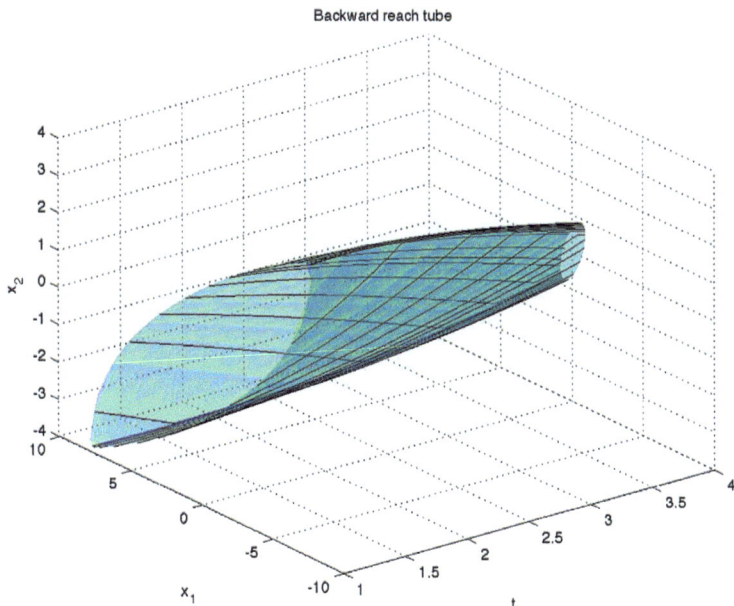

Fig. 4.19 $N = 15, DIM = 2$. Backward reach tube—coinciding internal and external estimates

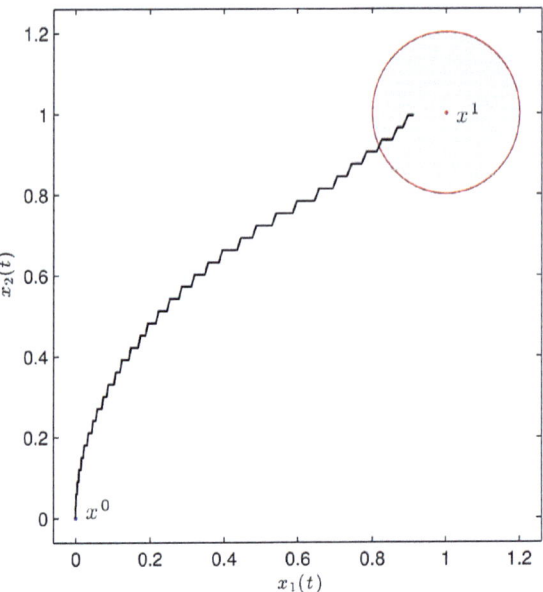

Fig. 4.20 $N = 5, Dim = 2$. Closed-loop chattering control—aiming method (b) in the ellipsoidal norm

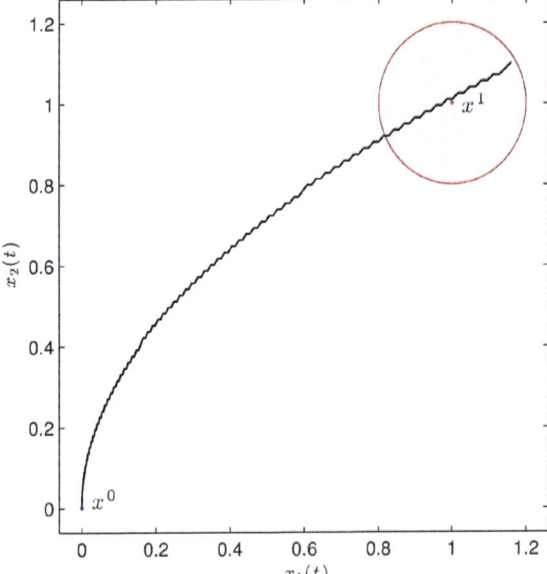

Fig. 4.21 $N = 15$, $Dim = 2$. Closed-loop chattering control—aiming in the Euclidean norm

4.2 A Planar Motion Under Newton's Law

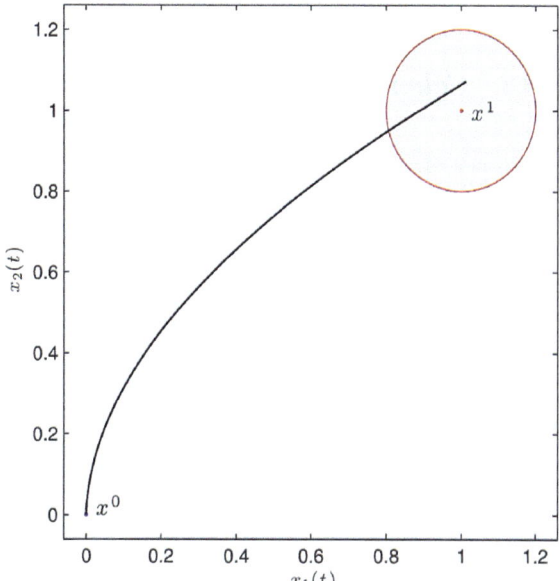

Fig. 4.22 $N = 15$, $DIM = 2$. Continuous closed-loop target control

Indicated further is the closed-loop controlled trajectory that reaches target $\mathcal{M}$ with aiming in the Euclidean metric (a) and in the metric of the approximating ellipsoid (b).

Aiming Method (c) for Continuous Control

In the previous figures the controls are of chattering type, ensuring sliding-type regimes when the trajectory fluctuates along the boundary of the backward reach set $\mathcal{W}[t]$. However, one may introduce a solution with continuous control, when having started from inside $\mathcal{W}[\tau]$ the trajectory does not reach the boundary of $\mathcal{W}[t]$, $t > \tau$. Such control is constructed as follows ($T = W_-^{-1}(t)$)

$$U_e^*(t,x) = \begin{cases} q(t), & \text{if } \langle x - w^\star(t), W_-^{-1}(t)(x - w^\star(t)) \rangle \leq 1, \\ & \text{otherwise} \\ q(t) - QB'(t)(x - w^\star(t)) \| Q^{1/2} B'(t)(x - w^\star(t)) \|^{-1}. \end{cases}$$

4.3 Damping Oscillations

4.3.1 Calming Down a Chain of Springs in Finite Time

Given is a vertical chain of a finite number n of suspended springs subjected to a vertical control force applied to the lower end of the chain. The chain also includes loads of given mass attached in between the springs at their lower ends and the masses of the springs are taken to be negligible relative to the loads. The upper end of the chain is rigidly fixed to a suspension (see Fig. 4.23).

The oscillations of the chain are then described by the following system of second-order ODEs ($i = 2, \ldots, j - 1, j + 1, \ldots, n - 1$):

$$\begin{cases} m_1 \ddot{w}_1 = k_2(w_2 - w_1) - k_1 w_1, \\ \ldots \\ m_i \ddot{w}_i = k_{i+1}(w_{i+1} - w_i) - k_i(w_i - w_{i-1}), \\ \ldots \\ m_j \ddot{w}_j = k_{j+1}(w_{j+1} - w_j) - k_j(w_j - w_{j-1}) + u_j(t), \\ \ldots \\ m_n \ddot{w}_n = -k_n(w_n - w_{n-1}) \end{cases} \qquad (4.24)$$

with $t > t_0 = 0$. Here n is the number of springs as numbered from top to bottom. The loads are numbered similarly, so that the i-th load is attached to the lower end of the i-th spring; w_i is the displacement of the i-th load from the equilibrium, m_i is

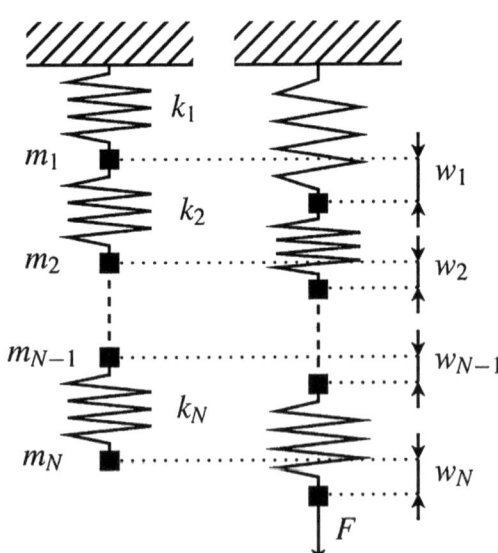

Fig. 4.23 An oscillating system

4.3 Damping Oscillations

the mass of the i-th load, k_i is the stiffness coefficient of the i-th spring. The gravity force enters (4.24) implicitly, through the lengths of the springs at the equilibrium.

The initial state of the chain at time $t_0 = 0$ is given by the displacements w_i^0 and the velocities of the loads $\dot{w}_i^0$, with

$$w_i(t_0) = w_i^0, \quad \dot{w}_i(t_0) = \dot{w}_i^0, \quad i = 1, \ldots, n.$$

The control $u = u_j(t)$ in this equation is indicated at the j-th load (which may be $j = n$) with its values confined to the interval $Q = [u_{\min}, u_{\max}]$. In particular, this constraint may be symmetric: $Q = [-\mu, \mu]$ or one-sided: directed only down ($Q = [0, \mu]$) or only up ($Q = [-\mu, 0]$).

We now rewrite system (4.24) in normal matrix form introducing the extended vector $x \in \mathbb{R}^{2n}$ such that $(x_1, \ldots, x_n)' = w$, $(x_{n+1}, \ldots, x_{2n})' = \dot{w}$. Then

$$\dot{x}(t) = Ax(t) + b^{(j)} u_j(t), \quad x(t_0) = x^0 = \begin{pmatrix} w^0 \\ \dot{w}^0 \end{pmatrix} \quad (4.25)$$

$$A = \begin{pmatrix} 0 & I \\ -M^{-1}K & 0 \end{pmatrix}, \quad b^{(j)'} = (0, \ldots, 0, b_{n+j}^{(j)}, 0, \ldots, 0), \quad b_{n+j}^{(j)} = m_j^{-1},$$

$$M = \begin{pmatrix} m_1 & & \\ & \ddots & \\ & & m_n \end{pmatrix},$$

$$K = \begin{pmatrix} k_1 + k_2 & -k_2 & & & \\ -k_2 & k_2 + k_3 & -k_3 & & \\ \ldots & \ddots & \ddots & \ddots & \ldots \\ & & -k_{n-1} & k_{n-1} + k_n & -k_n \\ & & & -k_n & k_n \end{pmatrix}.$$

If the controls are applied at several loads $j_1, \ldots, j_k$, then vector $b^{(j)}$ in 4.25 has to be substituted by matrix $B = (b^{(j_1)}, \ldots, b^{(j_k)})$ and the control will be k-dimensional: $u' = (u_{j_1}(t), \ldots, u_{j_k}(t))$.

Exercise 4.3.1. (i) Prove that system (4.25) is controllable if $j = 1$ or $j = n$.
(ii) Is the system controllable for any $j \neq 1$ and $j \neq n$?.
(iii) Find the necessary and sufficient conditions for controllability of system (4.25), when $m_i \equiv m$, $k_i \equiv k$ (m, k are the same for all $i = 1, \ldots, n$).

Examples of Solutions Through Ellipsoidal Methods

The emphasis of this book is not only to describe exact solutions but also to indicate effective numerical methods. Based on ellipsoidal approximations such methods rely on specially derived ordinary differential equations, indicated above in Chap. 3. We now present some examples of feedback control for a chain of springs. These are solved by schemes explained in Sect. 3.11.

Example 4.2. Chain of four springs with four loads with control at each load

Indicated are projections of the backward reachability set for system (4.24) with four springs and four loads of different masses $\{m_i\} = \{3, 1, 1.5, 5\}$ and different coefficients of stiffness $\{k_i\} = \{2, 1, 4, 1\}$.

We assume target set $\mathcal{M} = \mathcal{E}(0, M)$, $M = 0.1I$ and the control in (4.25) to be four-dimensional: $Bu = \{b^{(5)}u_1, \ldots, b^{(8)}u_4\}$ (applied to each load),

$$u \in \mathcal{E}(0, Q), \quad Q = \begin{pmatrix} 1 & 1 & 0 & 0 \\ 1 & 2 & 0 & 0 \\ 0 & 0 & 2 & 0 \\ 0 & 0 & 0 & 1 \end{pmatrix}.$$

The phase space for our system is of dimension 8, with unit orths of from $e'_1 = (1, 0, \ldots, 0) \in \mathcal{R}^8$ to $e'_8 = (0, \ldots, 0, 1) \in \mathcal{R}^8$. Our aim is to calculate the solvability set (backward reach set) $\mathcal{W}[t] = \mathcal{W}(t, \vartheta, \mathcal{M})$ with $\vartheta = 3$, seeking for the projections of eight-dimensional $\mathcal{W}[t]$ on two-dimensional planes $\{w_5, w_6\}$ and $\{w_7, w_8\}$ of system velocities. To find these projections we need the related external and internal ellipsoidal approximations that are calculated for good curves $l(t)$ along which they would be tight. The boundary conditions $l = l(\vartheta)$ for such curves are taken along a partition of unit circles on the planes e_5, e_6 and e_7, e_8 for the velocities. The evolution of these projections in time will reflect the dynamics of tube $\mathcal{W}[t]$. The result of intersecting approximating ellipsoids is given in Figs. 4.24, 4.25, 4.26, and 4.27. □

Example 4.3. Target control trajectories for chain of four springs with four loads: Control at each load

Indicated are the target controlled trajectories with four springs and four loads of different masses $\{m_i\} = \{5, 2, 4, 1\}$ and different coefficients of rigidity $\{k_i\} = \{2, 1, 1.5, 3\}$, taking $\mathcal{M} = \mathcal{E}(0, 0.1I)$. The aiming is done within scheme (b) of Sect. 3.11, in the metric of internal ellipsoids. The results are illustrated in Figs. 4.28, 4.29, 4.30, and 4.31 with boundary of target set in blue.

Example 4.4. Oscillating springs with two loads. Considered is a system with two loads and control applied to one of them.

4.3 Damping Oscillations

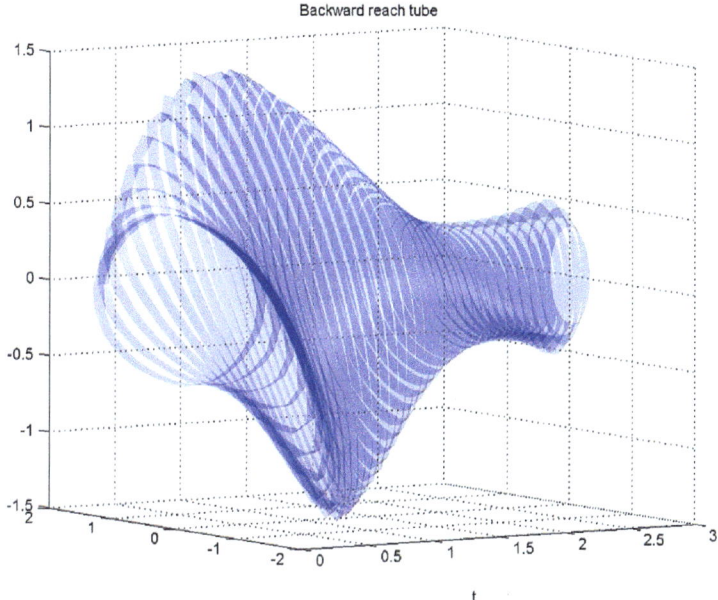

Fig. 4.24 $Dim = 8$. External approximations of backward reach set—coordinates x_5, x_6

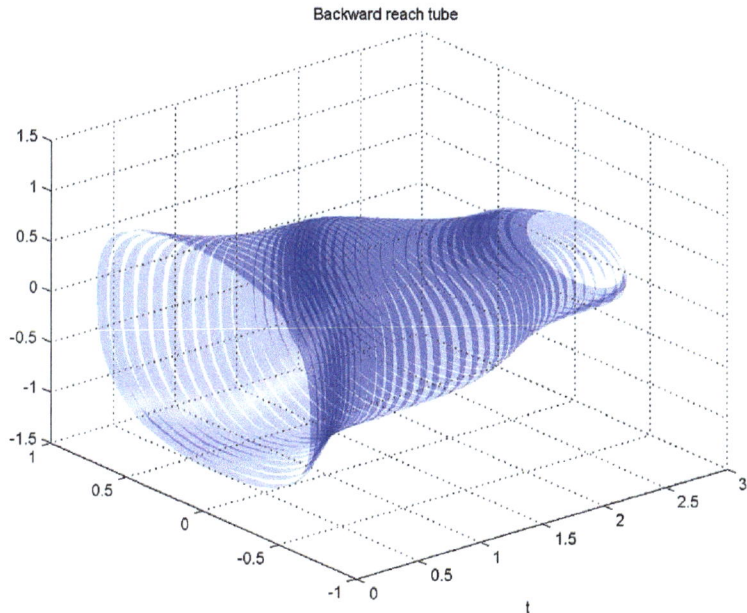

Fig. 4.25 $Dim = 8$. External approximations of backward reach set—coordinates x_7, x_8

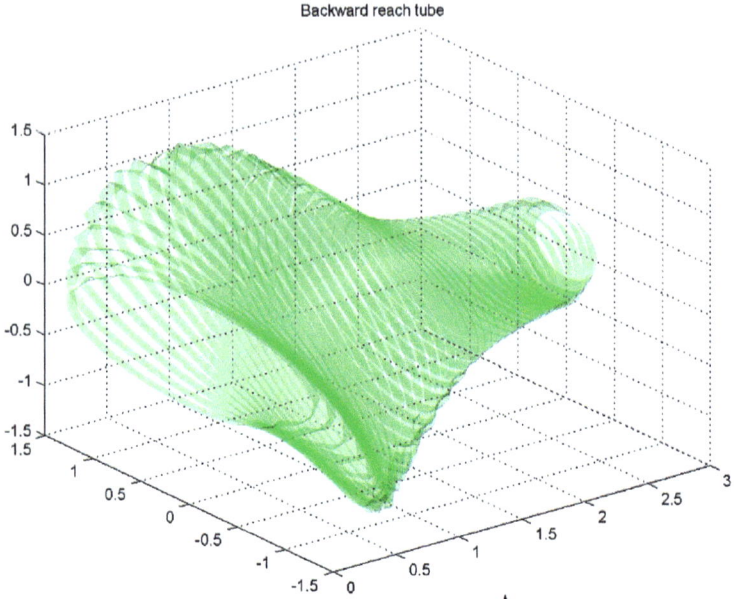

Fig. 4.26 $Dim = 8$. Internal approximations of backward reach set—coordinates x_5, x_6

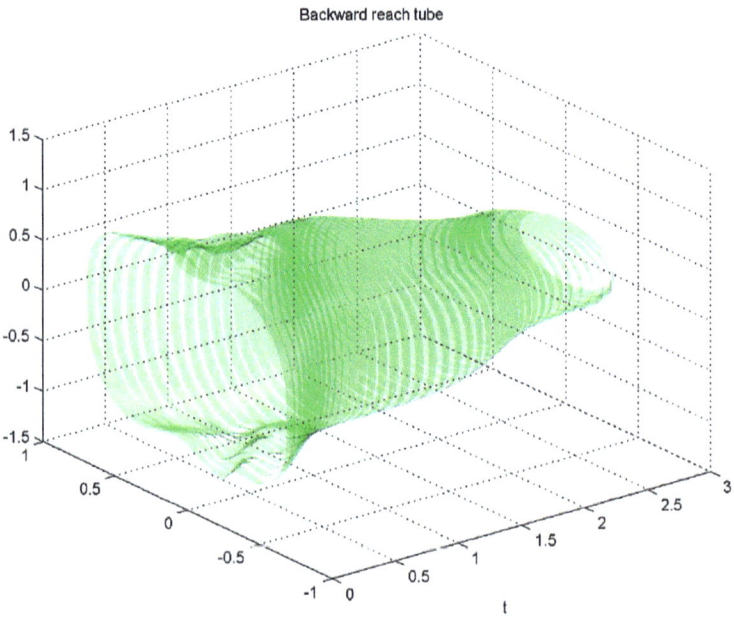

Fig. 4.27 $Dim = 8$. Internal approximations of backward reach set—coordinates x_7, x_8

4.3 Damping Oscillations

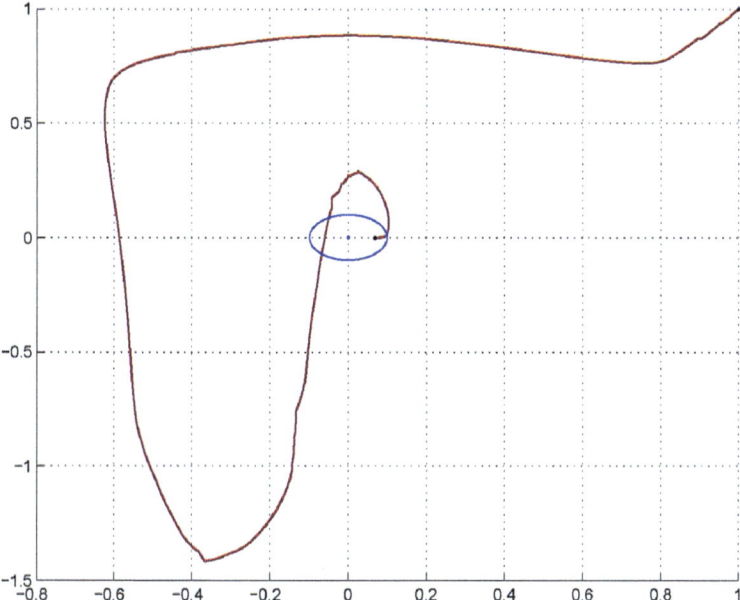

Fig. 4.28 $Dim = 8$. Projection of phase trajectory—position coordinates x_1, x_2

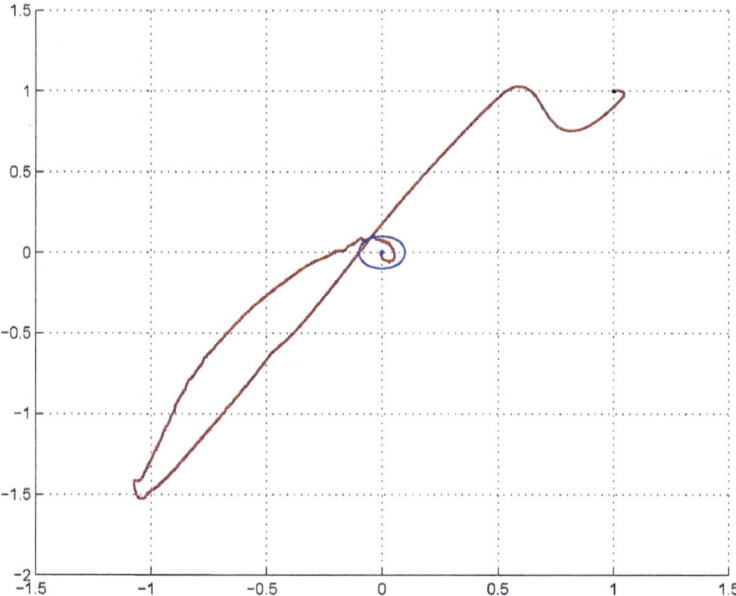

Fig. 4.29 $Dim = 8$. Projection of phase trajectory—position coordinates x_3, x_4

Fig. 4.30 $Dim = 8$. Projection of phase trajectory: velocities x_5, x_6

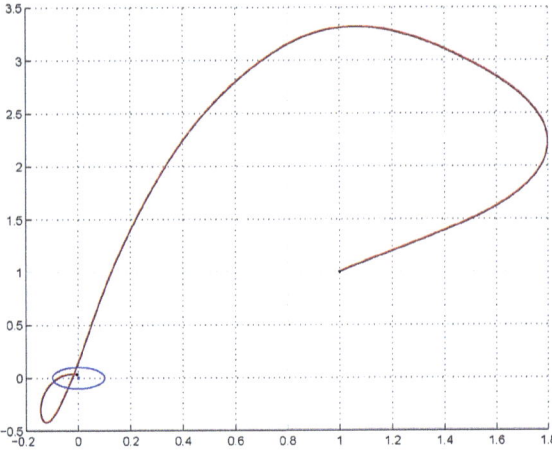

Fig. 4.31 $Dim = 8$. Projection of phase trajectory: velocities x_7, x_8

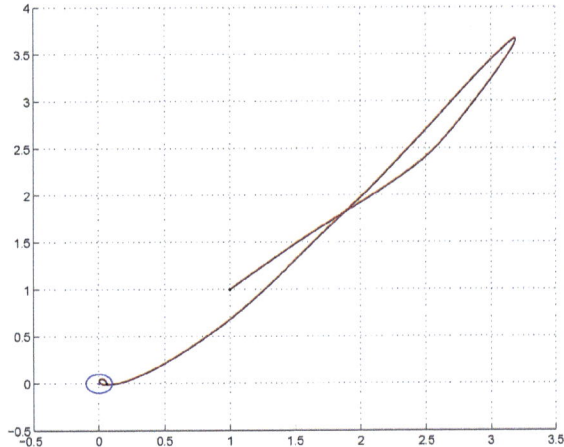

Calculated are:

- projections of approximations of backward reach set at final time by arrays of ellipsoids (their boundaries are marked blue for externals and green for internals; an accurate approximation through many ellipsoids indicates that the result is seen as almost the same).
- solutions to the problem of ellipsoidal control synthesis with pictures showing starting set is in green and boundary of target set in red.

(i) *Two loads. Control at upper load.*

Here, in system (4.24), the number of loads is $N = 2$, $m_1 = 2$, $m_2 = 2$, $k_1 = 1.3$, $k_2 = 2$, $w \in \mathbb{R}^2$, $t_0 = 0$, $t_1 = 7$. The control force $u \in [-1, 1]$ is applied only to the upper load. In system (4.25) $x \in \mathbb{R}^4$. The approximated backward reach tube, as calculated in backward variables $\bar{t} = 7 - t$, $t \in [0, 7]$, emanates at time $\bar{t} = 0$,

4.3 Damping Oscillations

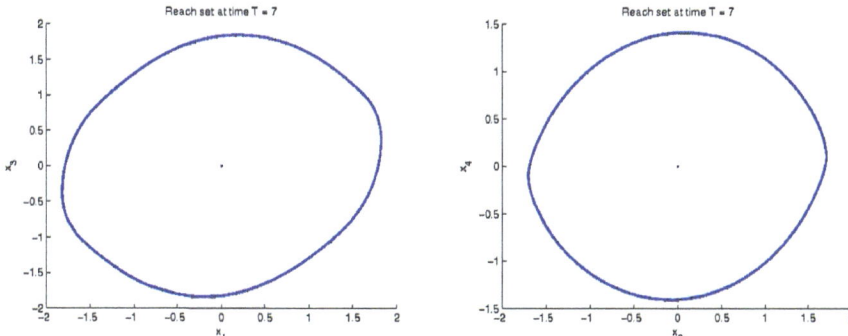

Fig. 4.32 Dim 4. Projections of approximation of backward reach set with control at upper load on two-dimensional planes in axes (position, velocity) for each load

($t = 7$), from target set M—a ball of radius 0.01 centered at 0, and terminates at $t = 0, T = \bar{t} = 7$. Reach sets at shown in backward time at $T = 7 (t = 0)$, see Fig. 4.32.

The controlled trajectory evolves from starting set X^0, centered at point $x(0) = x^0 = [0.5688, 1.0900, 0.8921, 0.2423]'$ to target set M—a ball of radius 0.1 centered at $m = 0$. It is shown in Figs. 4.33 and 4.34. □

(ii) *Two loads. Control at lower load.*

The system in same as in (i), but control is applied only at lower load.

The approximation of backward reach set is calculated as in previous case (i), within the same time interval, but with changed location of control. It is shown at final time in Fig. 4.35.

The starting set is centered at $x^0 = [1.4944, 1.9014, 0.2686, 1.2829]^T$ and terminal set is the same as in (i), see Figs. 4.36 and 4.37.

Example 4.5. Oscillating springs with four loads. Considered is a system with four loads and control applied to only one of them.

Calculated are:

- projections of backward reach set at final time (their boundaries are marked blue for externals and green for internals; an accurate solution indicates that they are seen as almost the same).
- solutions to the problem of ellipsoidal control synthesis with pictures showing starting set is in green and boundary of target set in red.

Control Applied at First (Upper) Load

Here, in system (4.24), the number of loads is $N = 4$, $m_1 = 1.5$, $m_2 = 2$, $m_3 = 1.1$, $m_4 = 4$, $k_1 = 3, k_2 = 1.4$, $k_3 = 2.1$, $k_4 = 1$, $w \in \mathbb{R}^4$, $t_0 = 0$, $t_1 = 10$. The control force $u \in [-1, 1]$ is applied only to the upper load. In system (4.25) $x \in \mathbb{R}^8$. The backward reach tube, as calculated in backward variables

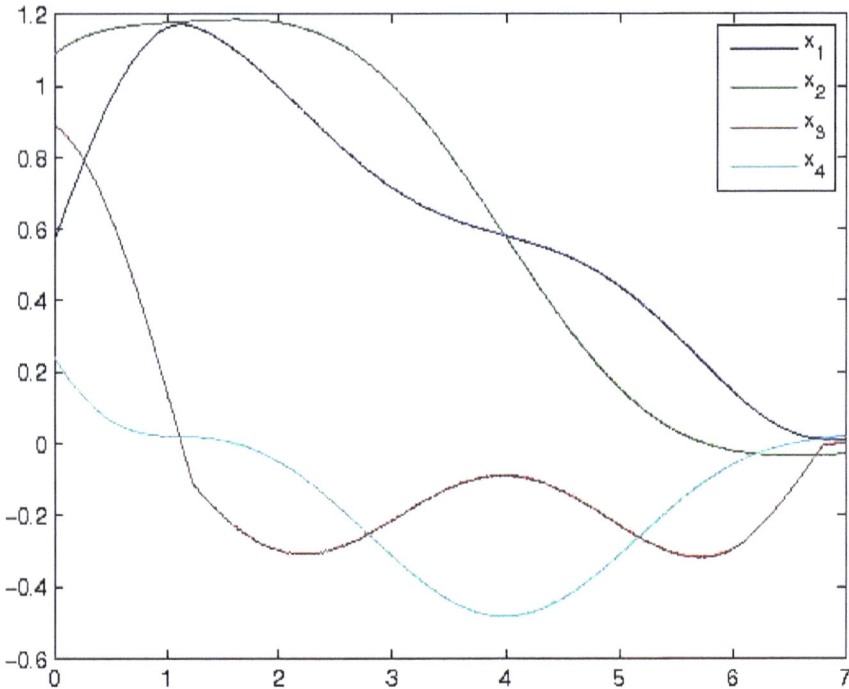

Fig. 4.33 Dim 4. The target controlled trajectories for system of two loads with control at upper load

$\bar{t} = 10 - t$, $t \in [0, 10]$, emanates at time $\bar{t} = 0$, $(t = 10)$, from target set $\mathcal{M}$—a ball of radius 0.01 centered at 0, and terminates at $t = 0, T = 10$. The projections of these reach sets at shown in backward time at $T = 10(t = 0)$, see Fig. 4.38.

The controlled trajectory evolves from starting set $\mathcal{X}^0$, centered at point $x(0) = x^0 = [-1.3691; 0.4844; 0.8367; -0.0432; 1.6817; 0.5451; -1.9908; 0.2277]'$ to target set $\mathcal{M}$—a ball of radius 0.1 centered at $m = 0$. It is shown in Figs. 4.39 and 4.40.

Control Applied at Fourth (Lowest) Load

Here the problem parameters are the same as with control at first (upper) load except that now it is applied to the fourth (lowest) load.

Projections of the reach set are shown in Fig. 4.41. The controlled trajectory now evolves from starting set $\mathcal{X}^0$, centered at point

$$x(0) = x^0$$
$$= [-0.6296; -1.9940; -2.6702; -3.6450; -0.1013; -0.1054; -0.2913; -0.6383]'$$

to the same target set $\mathcal{M}$—a ball of radius 0.1 centered at $\{0\}$. It is depicted in Figs. 4.42 and 4.43.

4.3 Damping Oscillations

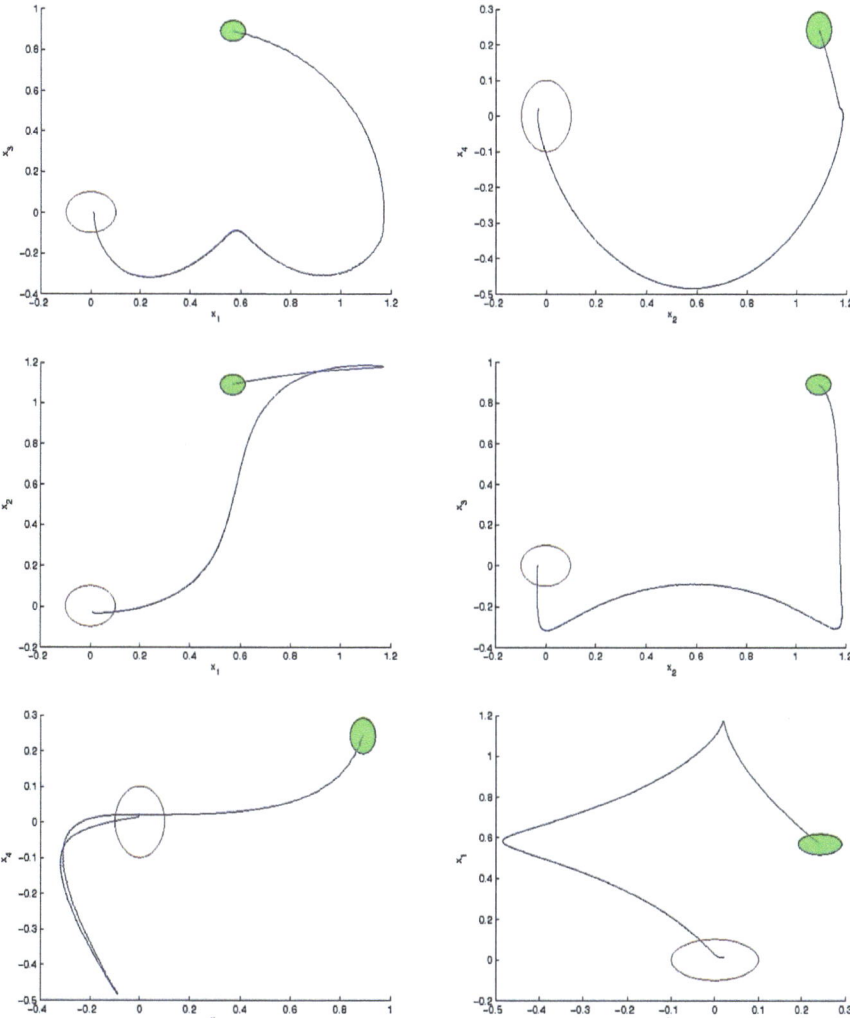

Fig. 4.34 Dim 4. Projections of phase trajectories of target controlled system with control at upper load

In order to calculate *exact solutions* to our control problems one needs to heavily rely on calculating fundamental matrix $G(t,s)$. Their calculation may also occur in other elements of the solution process. The calculation of such matrices for large dimensions is a special direction in numerical methods (see [89,90,92]). In the next lines we indicate an algorithmic method applicable to differential equations for the chain of springs.

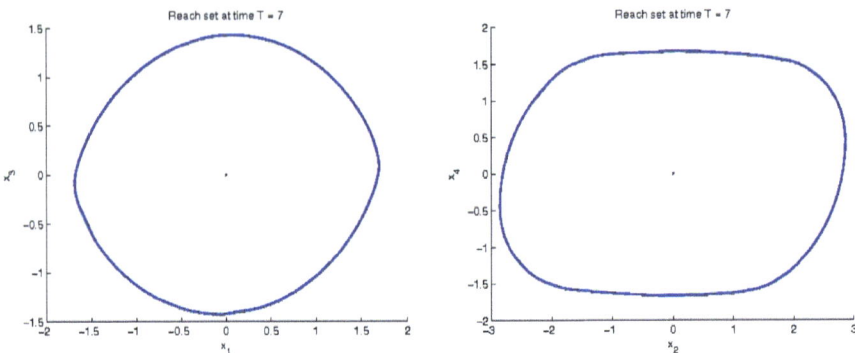

Fig. 4.35 Projections of approximation of backward reach set with control at lower load on two-dimensional planes in axes (position, velocity) for each load

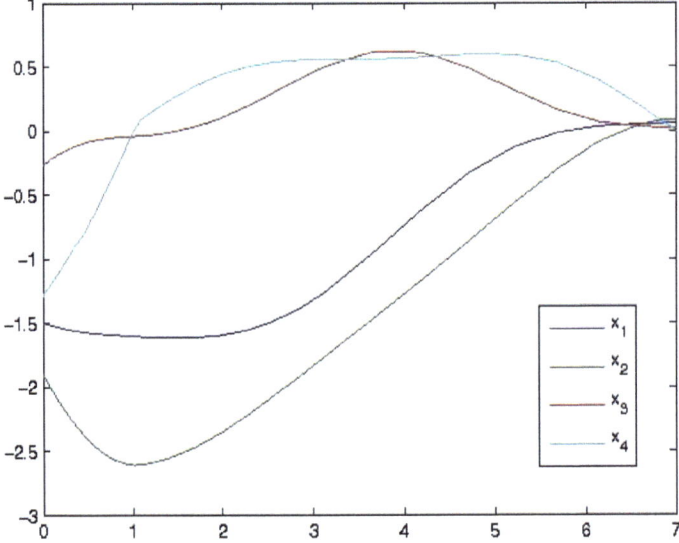

Fig. 4.36 The target controlled trajectories for system of two loads with control at lower load

Computing the Fundamental Matrix

For stationary systems matrix function $G(t,s) = \exp A(t-s)$ which is a matrix exponent. Consider calculation of such an exponent $X(t) = e^{At}$, assuming

$$A = \begin{pmatrix} 0 & I \\ -M^{-1}A_0 & 0 \end{pmatrix},$$

4.3 Damping Oscillations

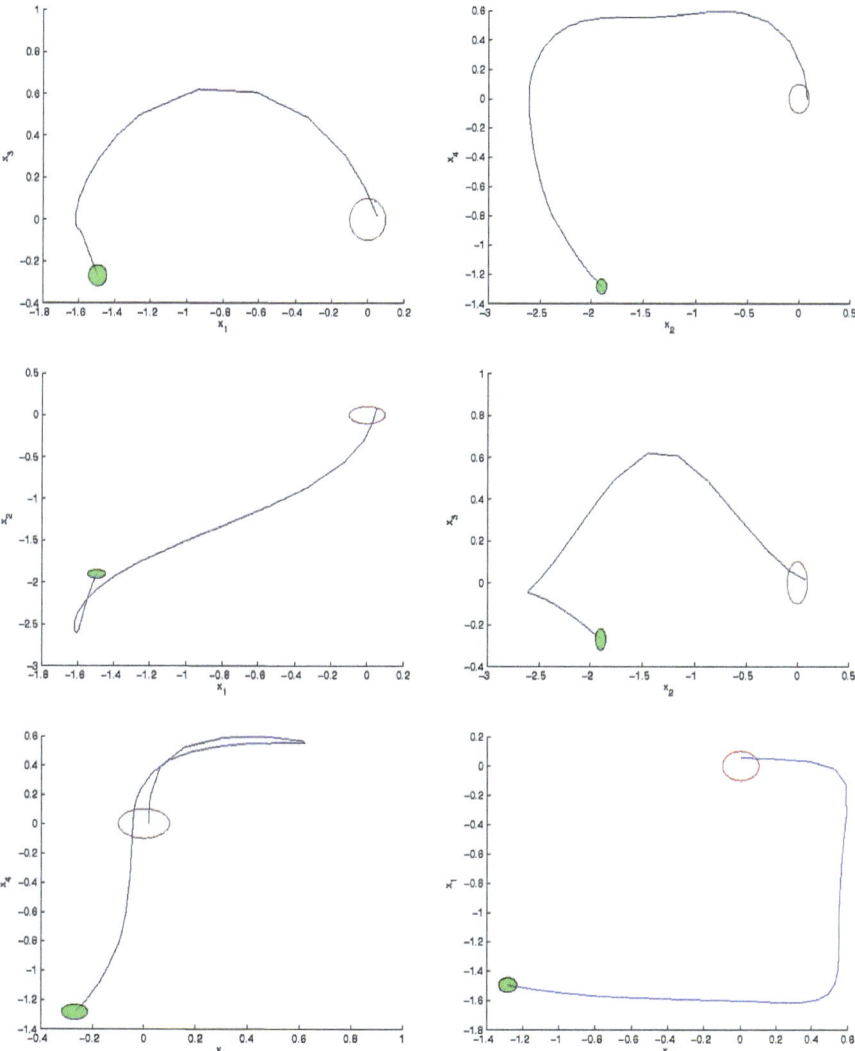

Fig. 4.37 Dim 4. Projections of phase trajectories of target controlled system with control at lower load

where matrix M is diagonal and matrix A_0 is symmetric and tri-diagonal (nonzero elements are allowed only on the main diagonal and those directly above and under).

Consider the powers of A. Having observed

$$A^2 = \begin{pmatrix} -M^{-1}A_0 & 0 \\ 0 & -M^{-1}A_0 \end{pmatrix},$$

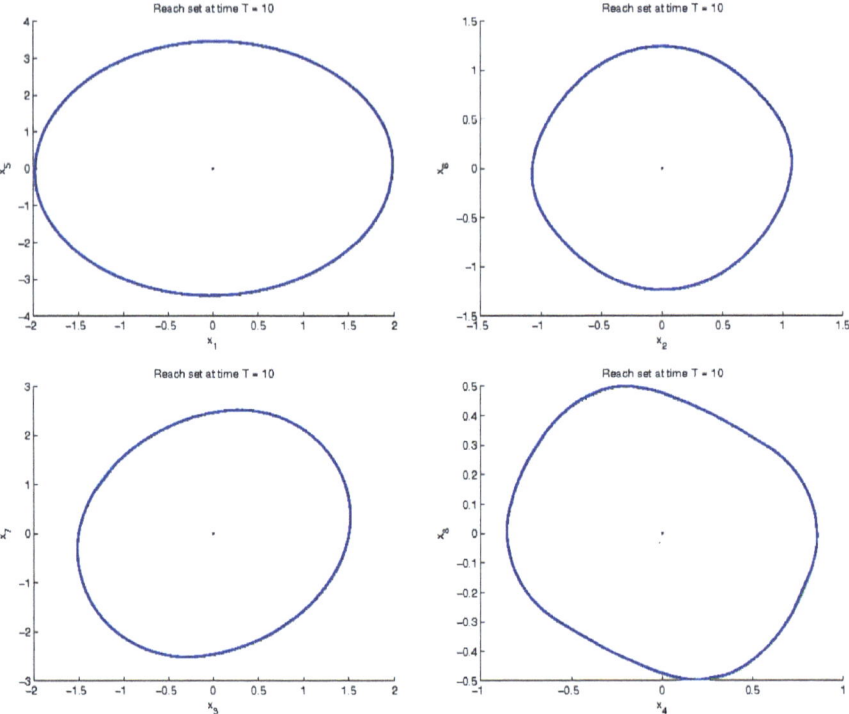

Fig. 4.38 Projections of approximation of backward reach set with control at first (upper) load on two-dimensional planes in axes (position, velocity) for each load

we further obtain

$$A^{2k} = \begin{pmatrix} (-1)^k (M^{-1} A_0)^k & 0 \\ 0 & (-1)^k (M^{-1} A_0)^k \end{pmatrix},$$

$$A^{2k+1} = \begin{pmatrix} 0 & (-1)^k (M^{-1} A_0)^k \\ (-1)^{k+1} (M^{-1} A_0)^{k+1} & 0 \end{pmatrix}.$$

Introducing new matrix $L = (M^{-\frac{1}{2}} A_0 M^{-\frac{1}{2}})$, we have

$$A^{2k} = \begin{pmatrix} (-1)^k M^{-\frac{1}{2}} L^{2k} M^{\frac{1}{2}} & 0 \\ 0 & (-1)^k M^{-\frac{1}{2}} L^{2k} M^{\frac{1}{2}} \end{pmatrix},$$

$$A^{2k+1} = \begin{pmatrix} 0 & (-1)^k M^{-\frac{1}{2}} L^{2k+1} L^{-1} M^{\frac{1}{2}} \\ (-1)^{k+1} M^{-\frac{1}{2}} L^{2k+1} L M^{\frac{1}{2}} & 0 \end{pmatrix}$$

4.3 Damping Oscillations

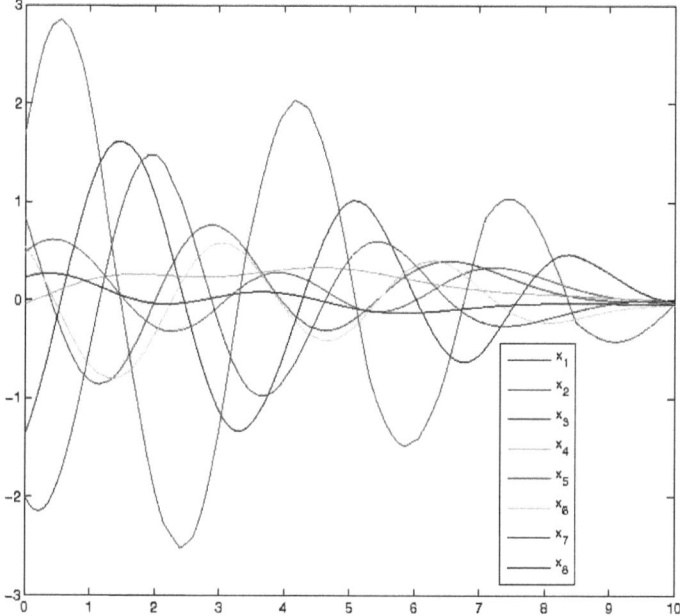

Fig. 4.39 The target controlled trajectories for system of four loads with control at first (upper) load

and further on

$$e^{At} = \sum_{k=0}^{\infty} \frac{1}{k!} (At)^k =$$

$$= \begin{pmatrix} M^{-\frac{1}{2}} & 0 \\ 0 & M^{-\frac{1}{2}} \end{pmatrix} \begin{pmatrix} \cos(Lt) & \sin(Lt)L^{-1} \\ -\sin(Lt)L & \cos(Lt) \end{pmatrix} \begin{pmatrix} M^{\frac{1}{2}} & 0 \\ 0 & M^{\frac{1}{2}} \end{pmatrix}. \quad (4.26)$$

The algorithm for calculating e^{At} is as follows:

1. Calculate symmetric matrix $L^2 = M^{-\frac{1}{2}} A_0 M^{-\frac{1}{2}}$.
2. Calculate singular decomposition for matrix $L^2 = UKV^T$, where matrices U, V are orthogonal (here $U = V$ due to symmetry of L^2) and K is diagonal. Since L^2 is a tri-diagonal matrix, one may use special algorithms effective for this specific class of matrices (with complexity of order $N \log^3 N$).
3. Calculate matrix $K^{\frac{1}{2}}$.
4. For each instant t calculate diagonal matrices $\sin(tK^{\frac{1}{2}})$, $\cos(tK^{\frac{1}{2}})$.
5. Calculate matrix $\cos(tL) = U \cos(tK^{\frac{1}{2}})V^T$, $\sin(tL) = U \sin(tK^{\frac{1}{2}})V^T$.
6. Calculate e^{At} using formula (4.26).

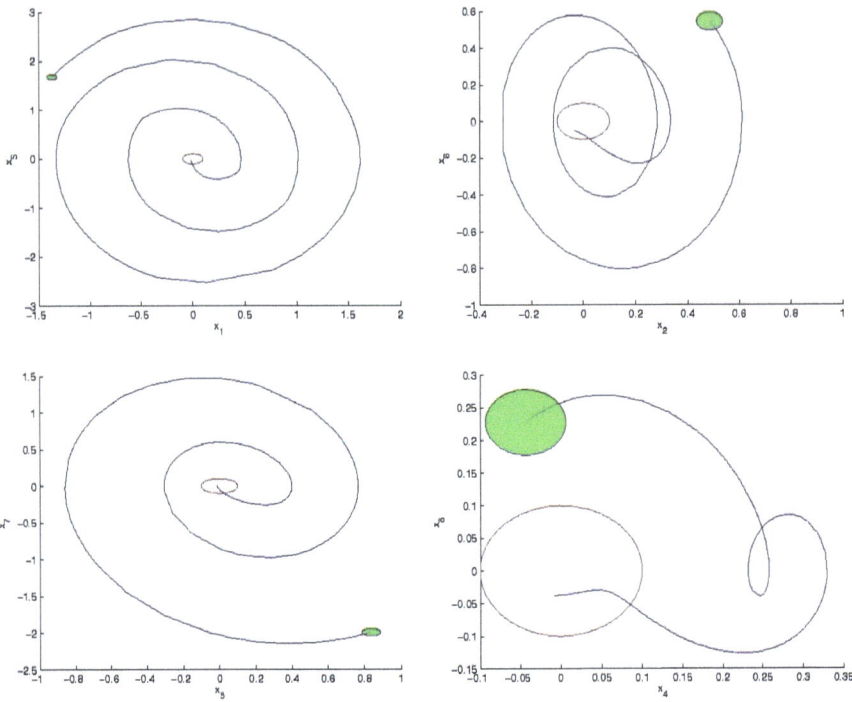

Fig. 4.40 Projections of phase trajectories of target controlled system with control at first (upper) load

The presented algorithm relies on the special structure of matrix A and proves to be far more effective than the general algorithms that do not rely on such structure. A simulation experiment indicates an eight-times more effective performance of the structure related algorithm as compared to the "ordinary" one.

4.4 Computation in High-Dimensional Systems. Degeneracy and Regularization[2]

4.4.1 Computation: The Problem of Degeneracy

As we have observed, the basic problem treated here is how to calculate ellipsoidal estimates for reachability tubes of a control system. Its solution is a crucial element for calculating feedback target control $\mathcal{U}(t, x)$ that steers the system trajectory from

---

[2] This section follows paper [58].

4.4 Computation in High-Dimensional Systems. Degeneracy and Regularization

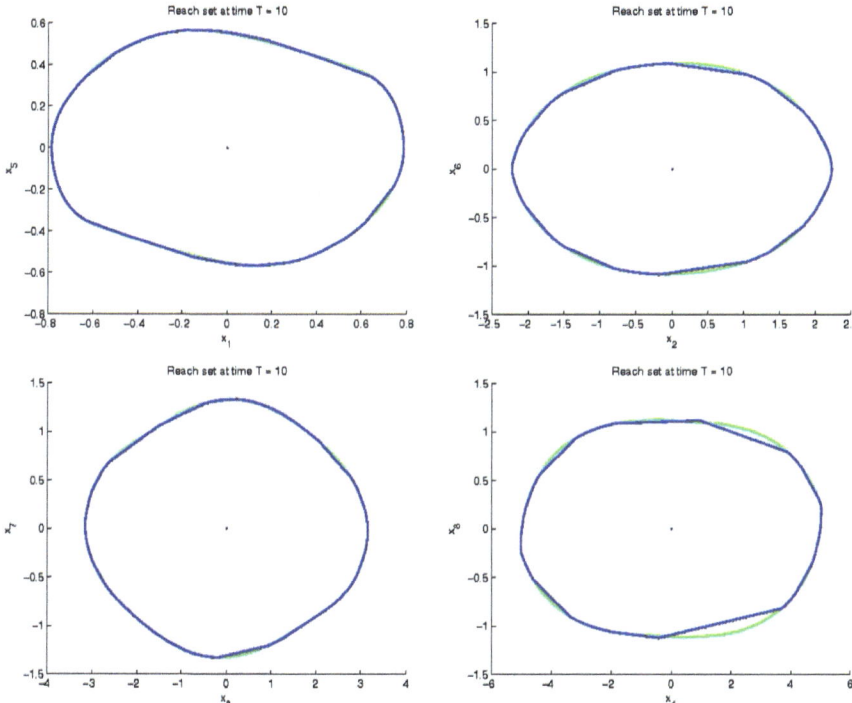

Fig. 4.41 Projections of approximations of backward reach set with control at fourth (lowest) load on two-dimensional plane in axes (position, velocity) for each load

given position $\{t_0, \mathcal{X}^0\}$ to given target set $\mathcal{M}$. Namely, as indicated in Sect. 2.6, to apply the aiming rule, one needs to calculate the *backward reachability* (solvability) *tube* $\mathcal{W}[t]$ for the investigated system. For linear-convex systems of type (3.1) this is done through ellipsoidal methods of Sect. 3.11 which proved successful for many problems. This route is especially useful for systems of higher dimensions where finding set $\mathcal{W}[t]$ directly is computationally too cumbersome. Thus we use its internal ellipsoidal approximations. Here we require that constraints in the problem are ellipsoidal: $\mathcal{X}^0 = \mathcal{E}(x^0, X^0)$, $\mathcal{Q}(t) = \mathcal{E}(q(t), Q(t))$. Otherwise sets $\mathcal{Q}(t)$ have to be approximated by their internal or external ellipsoids. Functions $q(t)$, $Q(t)$ for the constraints are assumed to be continuous.

With $\mathcal{W}[t]$ given, the desired control strategy $\mathcal{U} = U_e^*(t, x)$ is given by (3.136), (3.137). Following Sects. 3.9, 3.11, and (3.123), we observe that the internal ellipsoids are generated by the following ODEs:

$$\dot{w}^*(t) = A(t)w^*(t) + B(t)q(t) \qquad (4.27)$$

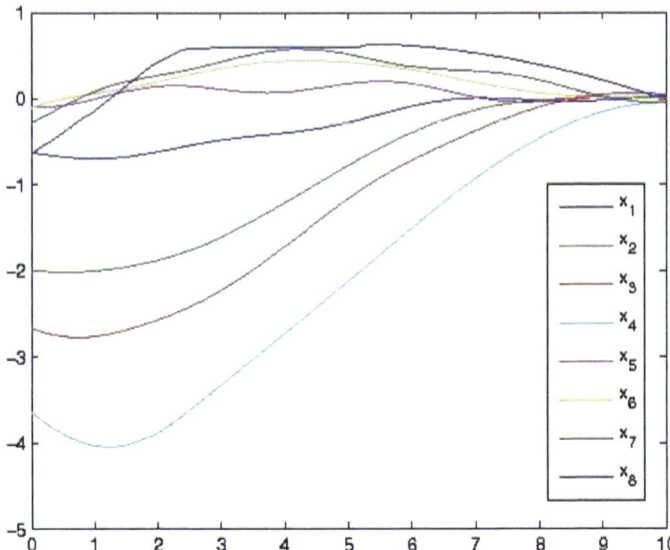

Fig. 4.42 The target controlled trajectories for system of four loads with control at fourth (lowest) load

with initial condition $w^\star(t_0) = m$, and

$$\dot{W}_-(t) = A(t)W_-(t) + W_-(t)A'(t) - \\ - [\mathbf{H}'(t)W_-(t) + W_-(t)\mathbf{H}(t)], \quad (4.28)$$

$$\mathbf{H}(t) = W_-^{-\frac{1}{2}}(t)S(t)Q_B^{\frac{1}{2}}(t),$$

with boundary condition $W_-(t_1) = M$. Here $S(t)$ is a parameterizing orthogonal matrix such that vectors $S(t)Q_B^{\frac{1}{2}}l(t)$ and $M^{\frac{1}{2}}l(t)$ are collinear and column $l(t)$ is a "good" curve—the solution to equation

$$\frac{dl(t)}{dt} = -A'(t)l(t), \quad l(t_0) = l.$$

Ellipsoidal approximations $\mathcal{E}_-^w[t] = \mathcal{E}(w^\star(t), W_-(t))$ are "tight" in the sense that they touch the exact reachability set in the direction of vector $l(t)$:

$$\rho(l(t) \mid \mathcal{E}_-^w[t]) = \rho(l(t) \mid X[t]).$$

The computation of external estimates is less difficult than of the internals which are therefore less investigated. However, the internals are crucial for designing feedback controls, so these difficulties, which increase with dimension, have to be coped with. We therefore concentrate on internals.

4.4 Computation in High-Dimensional Systems. Degeneracy and Regularization

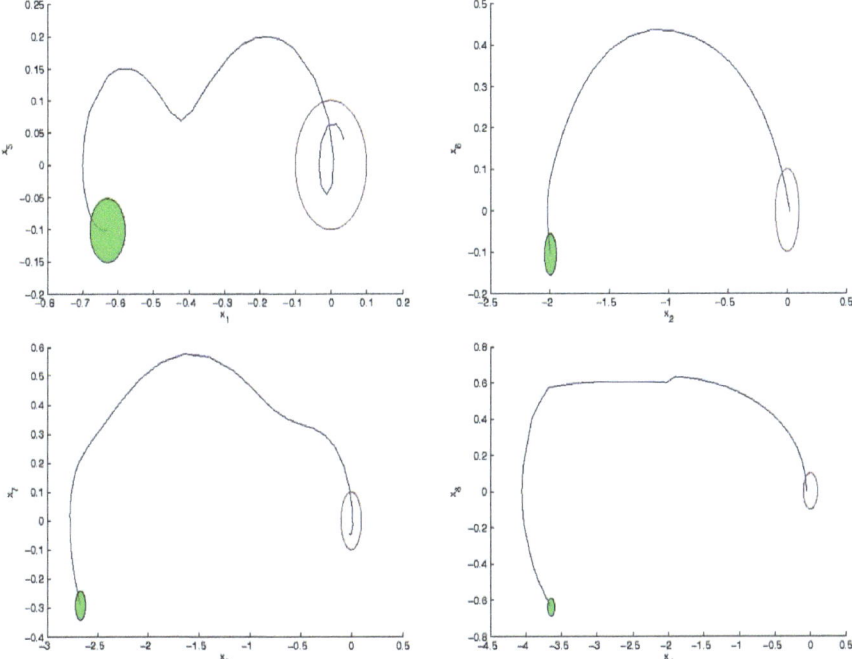

Fig. 4.43 Projections of phase trajectories of target controlled system with control at fourth (lowest) load

An attempt to apply established approximation formulas from [182] to oscillating systems of high dimensions [58] with scalar control ($n_u = 1$) revealed that the configuration matrix $W_-[t]$ of ellipsoid $\mathcal{E}_-^w[t]$ is *ill-conditioned* (namely, $n-1$ out of n existing semi-axes of $\mathcal{E}_-^w[t]$ are of length close to zero). This raises a serious issue for computations for the following reasons:

(a) errors caused by numerical integration of approximating ODEs may cause matrix $W_-[t]$ to have negative eigenvalues, which is unacceptable;
(b) the only information provided by a degenerate ellipsoid with a single positive axis is the value of support function in the direction of that axis. However, this value may be obtained by simpler calculations (not involving the solution of a matrix ODE).

We indicate some means of coping with these difficulties. For case (a) this will be to regularize the original procedures of Sect. 3.9. The idea is to calculate a set of ellipsoidal approximations which are then being "mixed" throughout the process. For case (b) the issue is that high-dimension systems require a polynomially increasing computational burden. This is met by introducing a new efficient method for calculating matrices $S(t)$.

Finally in this section we discuss a parallelized software implementation of the presented formulas. The given algorithms have proved to be effective for systems of ODEs of dimension up to 500.

4.4.2 Regularizing the Ellipsoidal Estimates

Numerical experiments with internal ellipsoidal estimates of the reachability tube for a high-order oscillating system of the (4.24) type (see [58]) with a scalar control lead to the following conclusion. If the terminal set $\mathcal{M}$ has small diameter, then the internal ellipsoidal estimates of the reach tube are close to degenerate. Namely, as time t increases, only one eigenvalue of matrix $W_-(t)$ grows. Other eigenvalues remain close to those of matrix M.

We start this section by demonstrating the cause of such degeneracy. To do this, we analyze the formula for the internal ellipsoidal estimate of the geometrical sum of two ellipsoids. After that we indicate how one copes with this degeneracy by weakening the requirements on the tightness of estimates. Finally we extend this approach to calculating the internal estimates of the reachability tube.

Degeneracy of the Sum of Degenerate Ellipsoids

Recall from Sect. 3.7 that the formula for an internal ellipsoidal estimate of the geometrical sum of m ellipsoids $\mathcal{E}(q_i, Q_i)$, $i = 1, \ldots, m$, tight in direction $\ell \in \mathbb{R}^n$:

$$\mathcal{E}(q_1, Q_1) + \cdots + \mathcal{E}(q_m, Q_m) \supseteq \mathcal{E}(q, Q), \tag{4.29}$$

$$q = \sum_{i=1}^{m} q_i, \quad Q = Q[m]' Q[m], \quad Q[m] = Q_1^{\frac{1}{2}} + S_2 Q_2^{\frac{1}{2}} + \cdots + S_m Q_m^{\frac{1}{2}},$$

where S_i are orthogonal matrices satisfying condition:

$Q_1^{\frac{1}{2}} \ell$ is collinear with $S_j Q_j^{\frac{1}{2}} \ell$, $j = 2, \ldots, m$. If any of the matrices Q_i, $i = 1, \ldots m$, is degenerate, relation (4.29) makes sense for directions ℓ such that $Q_i \ell \neq 0$, for all $i = 1, \ldots, m$.

Theorem 4.4.1. *Suppose* $\operatorname{rank} Q_i = r_i$ *and* $Q_i \ell \neq 0$, $i = 1, \ldots, m$. *Then the rank of matrix* $Q = Q[m]' Q[m]$ *is limited by* $r_1 + \cdots + r_m - (m - 1)$.

Proof. Denote column spaces of matrices $Q_1^{\frac{1}{2}}$ and $S_i Q_i^{\frac{1}{2}}$ as L_1 and L_i, respectively, $i = 2, \ldots, m$. Then the range $\operatorname{im} Q \subseteq \sum_{i=1}^{m} L_i$. By definition of matrices S_i the nonzero vector $l^{(1)} = Q_1^{\frac{1}{2}} \ell$ belongs to all subspaces L_i. Therefore the total number of linearly independent vectors in the image $\operatorname{im} Q$ cannot be greater than $r_1 + (r_2 - 1) + \cdots + (r_m - 1)$. ☐

4.4 Computation in High-Dimensional Systems. Degeneracy and Regularization

Corollary 4.4.1. *The set of rank 1 matrices is closed with respect to operation (4.29).*

This means a collection Q of matrices of rank 1, made according to (4.29), is also of rank 1.

Remark 4.4.1. The degeneracy of estimates is the result of their tightness (the requirement that they touch the exact set). This is not a property of the particular formula (4.29). Indeed, if rank $Q_i = 1$, then the summed ellipsoids may be presented as $\mathcal{E}_i = \mathcal{E}(q_i, Q_i) = \text{conv}\{q_i \pm a_i\}$, where vectors a_i are the only nonzero semi-axes of these ellipsoids. Then the exact sum $\mathcal{E}_1 + \cdots + \mathcal{E}_m$ will be a polyhedron conv$\{q_1 + \cdots + q_m \pm a_1 \pm \cdots \pm a_m\}$. And for almost all directions ℓ the only tight approximation will be one of the diagonals of this polyhedron. In this case formula (4.29) describes exactly those diagonals.

Remark 4.4.2. For some choices of orthogonal matrix S the dimension of the internal ellipsoidal estimate may be strictly less than the dimension of ellipsoids being added. For example, if $Q_1 = Q_2 = I$, $\ell = l^{(1)}$, $S = \text{diag}\{1, -1, -1, \ldots, -1\}$, then (4.29) gives $Q = \text{diag}\{1, 0, 0, \ldots, 0\}$.

Regularizing the Estimate for the Sum of Degenerate Ellipsoids

Applying results of the above to systems of type (4.27), (4.28), we indicate how several degenerate estimates may be combined (mixed) to get an estimate of higher dimension. This approach is based on the formula for internal ellipsoidal estimates of the convex hull for the union of ellipsoids [174].

Lemma 4.4.1. *If ellipsoids $\mathcal{E}_i^w = \mathcal{E}(w, W_-^{(i)})$, $i = 1, \ldots, m$, are internal estimates of a convex set X, then ellipsoid*

$$\mathcal{E}_\alpha^w = \mathcal{E}(w, W^\alpha), \quad W_-^\alpha = \sum_{i=1}^m \alpha_i W_-^{(i)}, \quad \alpha_i \geq 0, \quad \sum_{i=1}^m \alpha_i = 1, \quad (4.30)$$

will also be an internal estimate of X.

Proof. Since $\sqrt{t}$ is concave, we have

$$\rho(\ell \mid \mathcal{E}_\alpha^w) = \langle w, \ell \rangle + \left(\sum_{i=1}^m \alpha_i \langle \ell, W_-^{(i)} \ell \rangle \right)^{\frac{1}{2}} \quad (4.31)$$

$$\sum_{i=1}^m \alpha_i (\langle q_i, \ell \rangle + \langle \ell, W_-^{(i)} \ell \rangle^{\frac{1}{2}}) \leq \max_{i=1,\ldots,m} \rho(\ell \mid \mathcal{E}_-^w i),$$

hence

$$\mathcal{E}_\alpha^w \subseteq \operatorname{conv} \bigcup_{i=1}^m \mathcal{E}_i^w \subseteq X.$$

□

Theorem 4.4.2. *Suppose $w = 0$ and the dimension of L (the linear hull of $\mathcal{E}_i^w$) is r. Then if $\alpha_i > 0$, $i = 1, \ldots, m$, the following equality holds: the image of mapping W_-^α is $\operatorname{im} W_\alpha = L$. In particular, the matrix W_-^α is of rank r.*

Proof. This theorem means that

$$\operatorname{im}(\alpha_1 W_-^{(1)} + \cdots + \alpha_m W_-^{(m)}) = \operatorname{im} W_-^{(1)} + \cdots + \operatorname{im} W_-^{(m)}.$$

Due to the symmetricity of matrices $W_-^{(i)}$, the latter is equivalent in terms of matrix kernels to

$$\ker(\alpha_1 W_-^{(1)} + \cdots + \alpha_m W_-^{(m)}) = \ker W_-^{(1)} \cap \cdots \cap \ker W_-^{(m)}.$$

The inclusion of the right-hand side into the left is obvious. On the other hand, if

$$x \in \ker(\alpha_1 W_-^{(1)} + \cdots + \alpha_m W_-^{(m)}),$$

then taking the inner product of equality $\alpha_1 W_-^{(1)} + \cdots + \alpha_m W_-^{(m)} = 0$ by x, on both sides, we get

$$\alpha_1 \langle x, W_-^{(1)} x \rangle + \cdots + \alpha_m \langle x, W_-^{(m)} x \rangle = 0.$$

Since matrices $W_-^{(1)}$ are non-negative definite and α_i is positive, we observe that $\langle x, W_-^{(i)} x \rangle = 0$. Hence $x \in \ker W_-^{(1)}$. □

Note some properties of estimates $\mathcal{E}_\alpha^w$:

1. Suppose ellipsoidal approximation $\mathcal{E}_1^w$ is tight in direction ℓ, i.e. $\rho(\ell \mid \mathcal{E}_1^w) = \rho(\ell \mid X)$. Let us now estimate the difference between its support function and that of ellipsoid $\mathcal{E}_\alpha^w$ in the same direction. After some calculation we have:

$$\rho(\ell \mid \mathcal{E}_\alpha^w) \geq \langle w, \ell \rangle + \alpha_1 \langle \ell, W_-^{(1)} \ell \rangle^{\frac{1}{2}}$$

$$= \langle w, \ell \rangle + \langle \ell, W_-^{(1)} \ell \rangle^{\frac{1}{2}} \sqrt{1 - (1 - \alpha_1)} -$$

$$= \rho(\ell \mid X) - \frac{1}{2}(1 - \alpha_1) \langle \ell, W_-^{(1)} \ell \rangle^{\frac{1}{2}} + O((1 - \alpha_1)^2).$$

4.4 Computation in High-Dimensional Systems. Degeneracy and Regularization

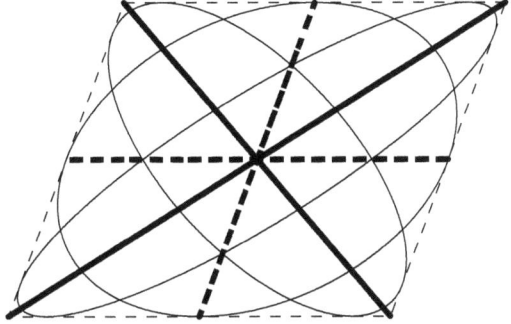

Fig. 4.44 Internal ellipsoidal approximations of sum of two ellipsoids

Therefore, the closer α_1 is to 1, the closer are ellipsoids $\mathcal{E}_\alpha^w$ to the tight internal ellipsoid along direction ℓ.

2. If all the ellipsoids $\mathcal{E}_i^w$, $i = 1, \ldots, m$, are tight approximations of X along the same direction ℓ, then in (4.31) an equality is true and ellipsoid $\mathcal{E}_\alpha^w$ is also a tight approximation of X along ℓ.

Example 4.6. Figure 4.44 shows ellipsoidal estimates of the sum of two degenerate ellipsoids. Original ellipsoids $\mathcal{E}_i$ are two line segments (thick dotted line), and their sum is a parallelogram (thin dotted line). Due to Theorem 4.4.1 tight approximations of the sum are also degenerate ellipsoids (line segments) shown by thick solid line. Regularized approximations with $\alpha_1 = \frac{1}{10}, \frac{1}{2}, \frac{9}{10}$ are presented with thin solid lines. These are non degenerate due to Theorem 4.4.2 and touch the parallelogram (note that they are tight in direction of normals to the sides of the parallelogram, see property 2). Besides that, for $\alpha = \frac{1}{10}$ and $\frac{9}{10}$ thanks to property 1 the support functions for the estimates are close to the support function of the parallelogram along corresponding directions.

Figure 4.45 shows internal ellipsoidal approximations of solvability set, as calculated by (4.31), for an oscillating system $\dot{x}_1 = x_2$, $\dot{x}_2 = -x_1 + u$ on time interval $[0, \pi]$ with parameter $\gamma = \frac{1}{20}$ (left) and $\frac{1}{2}$ (right). Exact (tight) approximations are degenerate ellipsoids (shown with thick lines).

4.4.3 Regularizing the Estimate for the Reachability Tube

We choose in (4.30) the following values of parameters α:

$$\alpha_1 = 1 - \sigma\gamma + \sigma\gamma\beta_i, \quad \alpha_i = \sigma\gamma\beta_i, \quad i = 2, \ldots, m,$$

where $\beta_i \geq 0$ and $\sum_{i=1}^m \beta_i = 1$, $\gamma \geq 0$.
Here σ is sufficiently small, ensuring $\alpha_1 > 0$. Then

$$\sigma^{-1}(W_-^\alpha - W_-^{(1)}) = \gamma\left(\sum_{i=1}^m \beta_i W_-^{(i)} - W_-^{(1)}\right).$$

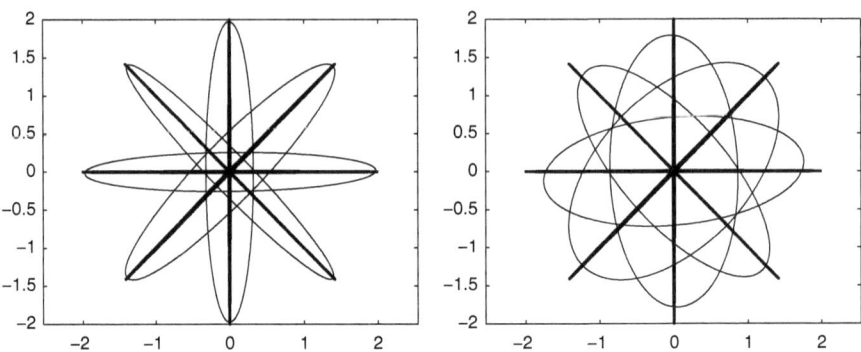

Fig. 4.45 Internal ellipsoidal estimates of the reachability set of a 2D system (*left*: $\gamma = \frac{1}{20}$, *right*: $\gamma = \frac{1}{2}$)

We use this result to mix m ellipsoidal estimates of the reachability tube as follows:

$$\dot{W}_-^{(i)} = A(t)W_-^{(i)} + W_-^{(i)}A'(t) - [\mathbf{H}'(t)W_-^{(i)} + W_-^{(i)}\mathbf{H}(t)] +$$

$$+ \gamma \left(\sum_{j=1}^{m} \beta_{ij} W_-^{(j)} - W_-^{(i)} \right), \quad W_-^{(i)}(t_1) = M,$$

with $\mathbf{H}(t) = (W_-^{(i)}(t))^{-\frac{1}{2}} S(t) Q_B^{\frac{1}{2}}(t)$.

Recall that $S_i(t)$ are arbitrary orthogonal matrices such that vectors $S_i(t)(B(t)Q(t)B'(t))^{\frac{1}{2}} l_i(t)$ and $(W_-^{(i)}(t))^{\frac{1}{2}} l_i(t)$ are directionally collinear. The column vector functions $l_i(t)$ are solutions to related adjoint systems

$$\frac{d l_i(t)}{dt} = -A'(t) l_i(t), \, l_i(t_0) = \ell_i.$$

Remark 4.4.3. Parameter $\gamma \geq 0$ controls the level of "mixed" approximations: the higher is γ, the greater is the impact of mixing. Parameters $\beta_{ij} \geq 0$ control the configuration of the mixture (namely, which ellipsoids are being mixed).

Remark 4.4.4. The choice of identical coefficients $\beta_{ij} = \hat{\beta}_j$ (in particular, $\beta_{ij} = \frac{1}{m}$) reduces the number of operations, since in this case the sum $\sum_{j=1}^{m} \beta_{ij} W_-^{(j)}(t) = \sum_{j=1}^{m} \hat{\beta}_j W_-^{(j)}(t)$ does not depend on j and is calculated only once for each time step.

Theorem 4.4.3. *Suppose matrix-valued solutions $W_-^{(i)}(t)$ to Eq. (4.28) are extendable to interval $[t_0, t_1]$, and are positive definite on this interval. Then the set-value function*

$$\mathcal{W}_-[t] = \text{conv} \bigcup_{i=1}^{m} \mathcal{E}_i^w[t] = \text{conv} \bigcup_{i=1}^{m} \mathcal{E}(w^*(t), W_-^{(i)}(t))$$

4.4 Computation in High-Dimensional Systems. Degeneracy and Regularization

satisfies for $t \in [t_0, t_1]$ the funnel equation

$$\lim_{\sigma \to 0+} \sigma^{-1} h_+((I - \sigma A(t))\mathcal{W}_-[t - \sigma], \mathcal{W}_-[t] - \sigma \mathcal{E}(0, Q_B(t))) = 0 \quad (4.32)$$

with boundary condition $\mathcal{W}_-[t_1] \subseteq \mathcal{M} = \mathcal{E}(m, M)$.

Without loss of generality consider $q(t) \equiv 0$, $m = 0$, and therefore, $w^*(t) \equiv 0$. Then, changing variables $z(t) = G(t_0, t)x(t)$, as in Sect. 1.1, we come to system (4.25) with $A(t) \equiv 0$.

The support function of the reach set $\mathcal{W}_-[t]$ is

$$\rho(\ell \mid \mathcal{W}_-[t]) = \max\{\langle \ell, W_-^{(i)}(t)\ell \rangle^{\frac{1}{2}} \mid i = 1, \ldots, m\}. \quad (4.33)$$

Let $\sigma > 0$ be sufficiently small, such that for $\delta \in [0, \sigma]$ the maximum for given direction ℓ is achieved at same $i = i_0$, i.e. $\rho(\ell \mid \mathcal{W}_-[t]) = \langle \ell, W_-^{(i_0)}(t)\ell \rangle^{\frac{1}{2}}$.

Assuming $\|\ell\| = 1$, the estimate for the support function of $\mathcal{W}_-[t - \sigma]$ is

$$\rho(\ell \mid \mathcal{W}_-[t - \sigma]) = \langle \ell, W_-^{(i_0)}(t - \sigma)\ell \rangle^{\frac{1}{2}} =$$

$$+ \langle \ell, W_-^{(i_0)}(t)\ell \rangle^{\frac{1}{2}} - \frac{\sigma}{2} \langle \ell, W_-^{(i_0)}(t)\ell \rangle^{-\frac{1}{2}} \langle \ell, \dot{W}_-^{(i_0)}(t)\ell \rangle^{\frac{1}{2}} + o(\sigma).$$

Omitting brackets (t) in the notations, we further estimate, due to (4.4.3),

$$\langle \ell, \dot{W}_-^{(i_0)}(t)\ell \rangle = -\langle \ell, (W_-^{(i_0)})^{\frac{1}{2}} SQ_B^{\frac{1}{2}} \ell \rangle - \langle \ell, (W_-^{(i_0)})^{\frac{1}{2}} SQ_B^{\frac{1}{2}} \ell \rangle +$$

$$+ \gamma \left(\sum_{j=1}^{m} \beta_{i_0 j} \langle \ell, W_-^{(j)} \ell \rangle - \langle \ell, W_-^{(i_0)} \ell \rangle \right) \geq -$$

$$-2 \|(W_-^{(i_0)})^{\frac{1}{2}} \ell\| \, \|S\| \, \|Q_B^{\frac{1}{2}} \ell\| \geq -2 \langle \ell, W_-^{(i_0)} \ell \rangle^{\frac{1}{2}} \langle \ell, Q_B \ell \rangle^{\frac{1}{2}}.$$

Here the multiplier of γ is non-positive, since for $i = i_0$ the expression $\langle \ell, W^{(i)}(t)\ell \rangle$, $(i = i_0)$ is at its maximum. Returning to the estimate of $\rho(\ell \mid \mathcal{W}_-[t - \sigma])$, we finally get

$$\rho(\ell \mid \mathcal{W}_-[t - \sigma]) - \rho(\ell \mid \mathcal{W}_-[t]) + \sigma \langle \ell, Q_B(t)\ell \rangle^{\frac{1}{2}} = o(\sigma),$$

which is equivalent to the funnel equation (4.32) of the above ($A(t) \equiv 0$).

Corollary 4.4.2. *Set-valued function $\mathcal{W}_-[t]$ is an internal estimate of the reachability tube $\mathcal{W}[t]$, and functions $W_-^{(i)}[t]$ are internal ellipsoidal estimates of $\mathcal{W}[t]$.*

This follows from the fact that the solvability set $\mathcal{W}[t]$ is the maximum solution to funnel equation (4.32) with respect to inclusion.

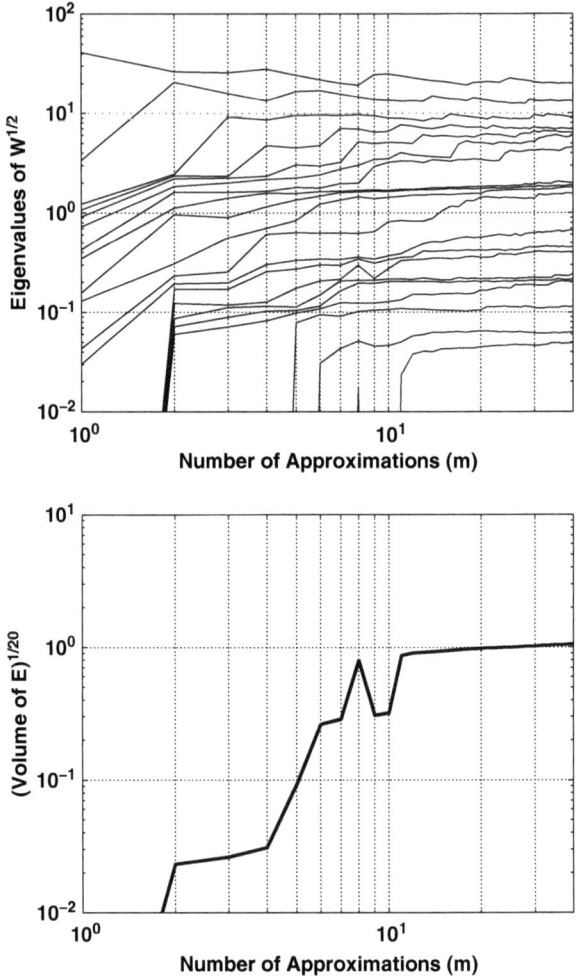

Fig. 4.46 The size of ellipsoidal estimates depending on the number of mixed estimates m (*top*: eigenvalues, *bottom*: volume in the power $\frac{1}{n}$)

Exercise 4.4.1. Prove formula (4.33).

Example 4.7. Figure 4.45 shows internal ellipsoidal estimates of the reachability set of an oscillating system $\dot{x}_1 = x_2$, $\dot{x}_2 = -x_1 + u$, within time interval $[0, \pi]$ with $\gamma = \frac{1}{20}$ and $\frac{1}{2}$. Here exact approximations are degenerate ellipsoids (shown with thick lines).

Exercise 4.4.2. Figure 4.46 demonstrates the dependence of the size of ellipsoidal estimates for a higher-order oscillating system on the system dimension (number of nodes $N = 10$, dimension $n = 2N = 20$). Shown on the top are 20 graphs of all eigenvalues of the matrix $(W_-^{(1)})^{\frac{1}{2}}$ (these eigenvalues are the sizes of semi-axes of the estimates). On the bottom is the volume of the estimates plotted in power $\frac{1}{n} = \frac{1}{20}$ (the geometrical mean of the axes).

Analyzing similar plots for a number of values of N, one may conclude that for robust computation of the reachability tube the number of approximations m should be chosen close to the system dimension n.

4.4.4 Efficient Computation of Orthogonal Matrix S

Equation (4.28) for the ellipsoidal approximation includes the operation of finding an orthogonal matrix $S = S(v_1, v_2) \in \mathbb{R}^{n \times n}$, such that Sv_2 is collinear with v_1 for some nonzero vectors $v_1, v_2 \in \mathbb{R}^n$.

Note that with $n \geq 2$ the matrix $S(v_1, v_2)$ is not unique (for $n = 2$ there are at least two such matrices, while for $n \geq 3$ there are infinitely many).

Function $S(v_1, v_2)$ should be defined as sufficiently smooth in variables v_1, v_2, so that integration schemes for higher-order ODEs could be applied to (4.28).

Matrix $S(v_1, v_2)$ may be calculated, for example, by computing the singular value decomposition of vectors v_1, v_2 and by multiplying the corresponding orthogonal matrices [132]. The computational burden for this procedure is of order $O(n^3)$, and continuous dependence of S on v_1, v_2 is not guaranteed.

The following theorem gives explicit formulas for calculating $S(v_1, v_2)$, which relax the burden to $O(n^2)$ operations and also ensure S to be sufficiently smooth.

Theorem 4.4.4. *Take some nonzero vectors $v_1, v_2 \in \mathbb{R}^n$. Then matrix $S \in \mathbb{R}^{n \times n}$ calculated as*

$$S = I + Q_1(\mathcal{S} - I)Q_1^T, \tag{4.34}$$

$$\mathcal{S} = \begin{pmatrix} c & s \\ -s & c \end{pmatrix}, \ c = \langle \hat{v}_1, \hat{v}_2 \rangle, \ s = \sqrt{1-c^2}, \ \hat{v}_i = \frac{v_i}{\|v_i\|}, \tag{4.35}$$

$$Q_1 = [q_1, q_2] \in \mathbb{R}^{n \times 2}, \tag{4.36}$$

$$q_1 = \hat{v}_1, \quad q_2 = \begin{cases} s^{-1}(\hat{v}_2 - c\hat{v}_1), & s \neq 0, \\ 0, & s = 0 \end{cases}$$

is orthogonal and satisfies the property of collinearity for Sv_2 and v_1.

Proof. Suppose v_1, v_2 are not collinear and pass to their normalized versions $\hat{v}_i = v_i \|v_i\|^{-1}$. Denote $\mathcal{L}_v = \mathcal{L}(v_1, v_2)$ to be the plane generated by vectors v_1 and v_2. We shall describe an orthogonal transformation S which is a rotation in the plane $\mathcal{L}_v$ that transforms $\hat{v}_2$ into $\hat{v}_1$. We also impose an additional requirement on the orthogonal complement $\mathcal{L}_v^\perp$ to $\mathcal{L}_v$ which is that the induced operator $S_v^\perp$ will be equal to identity.

Compose matrix $V = [\hat{v}_1, \hat{v}_2]$ and find its decomposition: $V = QR$, where matrix $Q \in \mathbb{R}^{n \times n}$ is orthogonal and matrix $R \in \mathbb{R}^{n \times 2}$ is upper triangular. Matrices Q and R may be written in block form as

$$Q = [Q_1, Q_2], \quad Q_1 \in \mathbb{R}^{n \times 2}, \quad Q_2 \in \mathbb{R}^{n \times (n-2)},$$

$$R = \begin{bmatrix} R_1 \\ O_{(n-2) \times 2} \end{bmatrix}, \quad R_1 = \begin{pmatrix} 1 & c \\ 0 & s \end{pmatrix}.$$

Note that the columns of Q_1 and Q_2 form orthonormal bases in plane $\mathcal{L}_v$ and its orthogonal complement $\mathcal{L}_v^{\perp}$, respectively. With an additional constraint $R_{11} > 0$, $R_{22} > 0$ matrix Q_1 is unique, while matrix Q_2 may be arbitrary, with orthonormal columns orthogonal to those of Q_1. The indicated relations may be regarded as the Gram—Schmidt orthogonalization procedure for finding matrix Q_1.

Set

$$S = Q \begin{bmatrix} S & O \\ O & I \end{bmatrix} Q'.$$

This matrix is orthogonal as the product of orthogonal matrices. It describes the composition of three transformations: a rotation of plane $\mathcal{L}_v$ to plane $\mathcal{L}(e_1, e_2)$, a rotation in that plane, and a return to original coordinates.

We now prove that S does not depend on Q_2. Indeed, $S = Q_1 S Q_1' + Q_2 Q_2'$. But from the equality $QQ' = Q_1 Q_1' + Q_2 Q_2' = I$ it follows that $Q_2 Q_2' = I - Q_1 Q_1'$, so we get relation (4.34), which does not contain Q_2.

Matrix S is also orthogonal when v_1 is collinear with v_2, i.e. $s = 0$. In this case $c = \pm 1$ where for $c = 1$ we have $S = I$ and for $c = -1$ we get $S = I - 2\hat{v}_1 \hat{v}_1'$, and check that $SS' = I$.

Now calculate $S\hat{v}_2$. From (4.36) we get $Q_1' \hat{v}_2 = (c, s)'$ for any value of s. We further have

$$S\hat{v}_2 = \hat{v}_2 + Q_1 \begin{pmatrix} c-1 & s \\ -s & c-1 \end{pmatrix} \begin{pmatrix} c \\ s \end{pmatrix} =$$

$$\hat{v}_2 + [\hat{v}_1, s^{-1}(\hat{v}_2 - c\hat{v}_1)] \begin{pmatrix} 1-c \\ -s \end{pmatrix} = \hat{v}_1.$$

For any vector $v \perp \hat{v}_1, \hat{v}_2$ we have $Q_1' v = 0$, hence $Sv = v$. □

Remark 4.4.5. The computational complexity in using (4.34)–(4.36) is of order $O(n^2)$. Moreover, the multiplication by matrix S is now performed with $O(n^2)$ operations instead of the usual $O(n^3)$.

4.4 Computation in High-Dimensional Systems. Degeneracy and Regularization 195

Remark 4.4.6. It may be checked directly that for nonzero vectors v_1, v_2 function $S = S(v_1, v_2)$ has continuous derivatives of any order everywhere except the cone generated by collinear v_1, v_2.

4.4.5 Parallel Computation

Here we discuss the application of parallel computations to the computation of internal ellipsoidal estimates and their application to finding feedback controls.

Computing the Ellipsoidal Estimates

In order to solve the ODE (4.4.3) numerically, using μ parallel processes, we decompose the index set $I = \{i = 1, \ldots, m\}$ into μ disjoint subsets I_k, such that $I = I_1 \cup \ldots \cup I_\mu$. Process k will calculate and store matrices $W_-^{(i)}$, $i \in I_k$. (To balance the load between the processes the cardinalities of I_k should be approximately identical, namely, of order m/μ.)

Direct integration of ODE (4.4.3) would lead to an excessive amount of data exchange between processes for computing the term $\sum_{i=1}^{m} \beta_{ij} W_-^{(j)}$. To avoid this, we combine (4.4.3) with (4.30). As a result each process solves its own ODE (4.4.3) with the above term replaced by $\sum_{i \in I_k} \beta_{ij} W_-^{(j)}$, where $\sum_{j \in I_k} \beta_{ij} = 1$. Approximations from different processes are mixed at prescribed discrete times by applying formula (4.30) with $\alpha_i = \frac{1}{m}$. As a result the amount of data exchange is reduced substantially.

Computing the Feedback Controls

For the feedback problem the exact "extremal aiming" control should be calculated as

$$\mathcal{U}_e(t, x) = \begin{cases} -Q(t) B'(t) l \langle l, Q_B(t) l \rangle^{-\frac{1}{2}} & \text{if } B'(t) l \neq 0; \\ \mathcal{E}(q(t), Q(t)), & \text{if } B'(t) l = 0, \end{cases}$$

where vector $l = l(t, x) = \partial V^-/\partial x$, $V^-(t, x) = d(G(t_1, t)x, G(t_1, t)\mathcal{W}_-[t])$ indicates the shortest path from x to $\mathcal{W}_-[t]$. Here set $\mathcal{W}_-[t]$ is the convex hull of a union of sets. So the calculation of vector l reduces to a computationally difficult problem of the max-min type, namely,

$$V^-(t, x) = \max_{\|q\| \leq 1} \min_{i=1,\ldots,m} \{\langle q, G(t_1, t)x \rangle - \rho(G'(t_1, t)q \mid \mathcal{E}_i^w[t])\} \qquad (4.37)$$

(especially so, since parameters of the sets are stored across multiple processes).

To cope with such a difficulty, we replace $V^-(t,x)$ by

$$\hat{V}^-(t,x) = d(G(t_1,t)x, G(t_1,t)\hat{W}_-[t]), \quad \hat{W}_-[t] = \bigcup_{i=1}^{m} \mathcal{E}_i^w[t] = \bigcup_{i=1}^{m} \mathcal{E}(w(t), W_-^{(i)}(t)),$$

which is equivalent to interchanging the maximum and minimum in (4.37)

$$\hat{V}^-(t,x) = \min_{i=1,\ldots,m} \max_{\|q\| \le 1} \{\langle q, G(t_1,t)x\rangle - \rho(G'(t_1,t)q \mid \mathcal{E}_i^w[t])\}. \tag{4.38}$$

This leads to a vector $l^* = l^*(t,x) = \partial V^-/\partial x = l_{i_0}^*$, where $i_0 \in 1,\ldots,m$ is the minimizers index and $l_{i_0}^*$ the maximizer in (4.38) for the fixed $i = i_0$.

Thus, to synthesize controls, each process locally finds an ellipsoid nearest to the on-line position. Then the process that is the nearest among such ellipsoids is chosen to calculate the control by using its own ellipsoid. The specified control is finally communicated to all the other processes.

Simulation Results

Numerical results were obtained in computing feedback controls for a chain of oscillating springs (4.24), [57], with the following parameters (here N is the number of links and $n = 2N$ is the system dimension):

- $N = 25$ ($n = 50$) for a system with disturbance, without matching condition;
- $N = 50$ ($n = 100$) for a system with one-directional scalar control ($u \in [0, \mu]$);
- $N = 50$ ($n = 100$) for an inhomogeneous oscillating system (the lower part is twice as heavy as the upper);
- $N = 100$ ($n = 200$) for a system with scalar control;
- $N = 250$ ($n = 500$) for a system with control of dimension N.

We do not compare these results with a non-parallel version of our code, since in the latter case the existing memory limitations usually prevent it from being run with large values of n.

Chapter 5
The Comparison Principle: Nonlinearity and Nonconvexity

Abstract This chapter introduces generalizations and applications of the presented approach prescribed earlier to nonlinear systems, nonconvex reachability sets and systems subjected to non-ellipsoidal constraints. The key element for these issues lies in the Comparison Principle for HJB equations which indicates schemes of approximating their complicated solutions by arrays of simpler procedures. Given along these lines is a deductive derivation of ellipsoidal calculus in contrast with previous inductive derivation.

Keywords Nonlinearity • Nonconvexity • Comparison principle • Unicycle • Boxes • Zonotopes • Ellipsoids

As indicated above, the solutions to many problems of control synthesis for systems described by ODEs reduce to an investigation of first-order PDEs—the HJB type—and their modifications [21, 24, 178, 198, 226, 248]. Similar equations may also be used for calculating forward and backward reachability sets for control systems with or without disturbances (the HJB equations).[1]

It is also well known that solutions to equations of the HJB type are rather difficult to calculate, and their respective algorithms are still being developed [221, 244]. However, for many problems, as those on reachability, on design of safety zones for motion planning or on verification of control algorithms, one may often be satisfied with approximate solutions that require a smaller computational burden and may be achieved through substituting original HJB equations by variational inequalities [23] due to certain *Comparison Principles* [48, 95, 149, 186]. This chapter indicates such comparison theorems which are also applicable to nonlinear systems for both smooth and nonsmooth solutions and to description of nonconvex sets. They also allow to derive ellipsoidal methods through *deductive procedures* in contrast with inductive approaches of Chap. 3. At the same time,

---

[1] Such considerations are also true for systems with uncertain (unknown but bounded) disturbances (the HJBI equation) [123, 176, 222], and for dynamic game-type problems. They involve more complicated equations of Hamilton–Jacobi–Bellman–Isaacs (HJBI) type, [18, 19, 102, 123]. [121, 150, 183]. These issues are mostly, except Chaps. 9 and 10, beyond the scope of the present volume.

when applied to linear-convex reachability problems, such estimates may lead to effective external and internal approximations of reach sets that converge to exact. This includes the treatment of controls with hard bounds generated not only by ellipsoids, but also by box-valued sets and symmetric polyhedrons (zonotopes).

5.1 The Comparison Principle for the HJB Equation

5.1.1 Principal Propositions for Comparison Principle

In this section we deal with ordinary systems without disturbances. Consider first the nonlinear equation

$$\dot{x} = f(t, x, u), \ t \in [t_0, \vartheta] = T_\vartheta \tag{5.1}$$

which coincides with Eq. (1.1) whose properties are described in the first lines of Sect. 1.1.

As indicated in Sect. 2.3, given *initial set* X^0 (for $t = t_0$) and a *target set* M (for $t = \vartheta$), it makes sense to construct *forward* $X[t] = X(t; t_0, X^0)$, $\forall t \geq t_0$), and *backward* $W[t] = W(t; \vartheta, M), t \leq \vartheta$, reachability tubes for system (5.1), emanating from set-valued *positions* $\{t_0, X^0\}$ and $\{\vartheta, M\}$, respectively [121, 178].

Introduce notations

$$\mathcal{H}(t, x, p) = \max\{< p, f(t, x, u) > | u \in \mathcal{P}(t)\}.$$

It is well known (see Sect. 2.3 and also [145, 178, 198]) that the solution $V(t, x)$ of the respective "forward" HJB equation of type

$$V_t + \mathcal{H}(t, x, V_x) = 0, \ V(t_0, x) = d(x(, X^0), \tag{5.2}$$

allows to calculate $X[t] = X(t, t_0, X^0)$ as *the level set*:

$$X[\tau] = \{x : V(\tau, x) \leq 0\}.$$

The last property is independent of whether V was obtained as a classical or a generalized solution of Eq. (5.2).

Thus the exact description of set $X[\tau]$ in general requires to solve the first-order PDE (5.2). Such a problem may be rather difficult to solve as the reachability sets for nonlinear systems may turn out to have a very peculiar form (see [225, 255]).

We shall therefore seek for the upper and lower estimates of functions $V(t, x)$, and as a consequence, also the external and internal estimates of sets $X[\tau]$. This move is motivated by a desire to describe the estimates through relations simpler than those for the exact solution.

5.1 The Comparison Principle for the HJB Equation

Assumption 5.1.1. *There exist functions $H(t, x, p)$ and functions $w^+(t,x) \in C_1[T_\vartheta]$, $\mu(t) \in L_1$, which satisfy the inequalities*

$$\mathcal{H}(t, x, p) \leq H(t, x, p), \quad \forall \{t, x, p\}, \tag{5.3}$$

$$w_t^+ + H(t, x, w_x^+) \leq \mu(t). \tag{5.4}$$

Theorem 5.1.1. *Let functions $H(t, x, p), w^+(t, x), \mu(t)$ satisfy Assumption 5.1.1. Then the following estimate is true*

$$X[t] \subseteq X_+[t], \tag{5.5}$$

where

$$X_+[t] = \{x : w^+(t, x) \leq \int_{t_0}^t \mu(s) ds + \max\{w^+(t_0, x) | x \in X^0\}\}. \tag{5.6}$$

Let there exist a pair $x^{0*} \in X^0$, $u^*(t) \in \mathcal{P}(t)$, $t \geq t_0$, such that the respective trajectory $x^*(t) \in X[t]$. Then

$$dw^+(t, x)/dt|_{x=x^*(t)} = w_t^+(t, x^*) + \langle w_x^+(t, x^*), f(t, x^*, u^*) \rangle \leq$$

$$\leq w_t^+(t, x^*) + \mathcal{H}(t, x^*, w_x^+) \leq w_t^+(t, x^*) + H(t, x^*, w_x^+) \leq \mu(t).$$

The last relations imply

$$dw^+(t, x)/dt|_{x=x^*(t)} \leq \mu(t).$$

Integrating this inequality from t_0 to t, we have

$$w^+(t, x^*(t)) \leq$$

$$\leq \int_{t_0}^t \mu(s) ds + w^+(t_0, x^*(t_0)) \leq \int_{t_0}^t \mu(s) ds + \max\{w^+(t_0, x) | x \in X^0\},$$

which means $x^*(\tau) \in X_+[\tau]$ and the theorem is proved.

Recall that when function $V(t, x)$ is nondifferentiable, Eq. (5.2) is written down only as a formal symbolic notation and its solution should be considered in a generalized "viscosity" sense [16, 50, 80] or equivalent "minimax" sense [247] or "proximal" sense [48].

Similar theorems are true for *backward reachability sets* $W[t]$ (or, in other words the "weakly invariant sets" relative target set $\mathcal{M}$ or the "solvability sets" in terms of [174]). Namely, if $W(t, \vartheta, \mathcal{M}) = W[t]$ is a backward reachability set from *target set*—a compact $\mathcal{M} \subseteq \mathbb{R}^n$, then

$$W[\tau] = \{x : V^{(b)}(\tau, x) \leq 0\},$$

where $V^{(b)}(t, x)$ is the solution (classical or generalized) of the "backward" HJB equation

$$V_t^{(b)} - \mathcal{H}(t, x, -V_x^{(b)}) = 0, \quad V^{(b)}(\vartheta, x) = d^2(x, \mathcal{M}). \tag{5.7}$$

which, in its turn, is generated by the problem

$$V^{(b)}(\tau, x) = \min\{d^2(x[\vartheta], \mathcal{M}) \mid x[\tau] = x, \ u(\cdot, \cdot) \in \mathcal{U}_C\}. \tag{5.8}$$

Here $\mathcal{U}_C$ is the class of closed-loop controls in which the last problem does have an optimal closed-loop solution $u^0(\cdot, \cdot)$ which generates an *optimal trajectory* $x^0[\vartheta] = x^0(\vartheta; \tau, x)$.

Assumption 5.1.2. *There exist functions* $H(t, x, p)$, $w^{(b+)}(t, x) \in C_1$, $v(t) \in L_1$, *which satisfy the inequalities*

$$\mathcal{H}(t, x, p) \leq H(t, x, p), \quad \forall \{t, x, p\},$$

$$w_t^{(b+)} - H(t, x, -w_x^{(b+)}) - v(t) \geq 0,$$

$$w^{(b+)}(\vartheta, x) \leq V^{(b)}(\vartheta, x), \quad \forall x.$$

Denote

$$\mathcal{H}^{(b)}(t, x, p) = \min\{\langle p, f(t, x, u)\rangle \mid u \in \mathcal{P}(t)\} = -\mathcal{H}(t, x, -p).$$

Then, under the last assumption, we have

$$dw^{(b+)}(t, x)/dt - v(t) \geq w_t^{(b+)} - \mathcal{H}^{(b)}(t, x, w_x^{(b+)}) \geq w_t^{(b+)} - H(t, x, -w_x^{(b+)}) - v(t) \geq 0.$$

Integrating this inequality from τ to ϑ along an optimal trajectory $x^0[t]$ which starts at x and ends at $x^0[\vartheta]$, generating the value $V^{(b)}(\tau, x)$, (as introduced in the previous lines), we come to relations

$$w(\vartheta, x^0[\vartheta]) \geq w^{(b+)}(\tau, x) + \int_\tau^\vartheta v(t)dt,$$

where $w^{(b+)}(\vartheta, x^0[\vartheta]) \leq V^{(b)}(\vartheta, x^0[\vartheta])$ according to the last condition of Assumption 5.1.2.

However, along the optimal trajectory $x^0[t]$, $t \in [\tau, \vartheta]$, we have

$$V^{(b)}(\tau, x) = V^{(b)}(t, x^0[t]) = V^{(b)}(\vartheta, x^0[\vartheta]). \tag{5.9}$$

(see, for example, [48, Sects. 7.5–7.7]). This brings us to the next theorem.

5.1 The Comparison Principle for the HJB Equation

Theorem 5.1.2. *Let functions* $H(t, x, p), w^{(b+)}(t, x), v(t)$ *satisfy Assumption 5.1.2. Then there exists a lower estimate for* $V^{(b)}(\tau, x)$:

$$w^{(b+)}(\tau, x) + \int_\tau^\vartheta v(t)dt \leq V^{(b)}(\tau, x),$$

and an external estimate for $\mathcal{W}[\tau]$:

$$\mathcal{W}[\tau] \subseteq W_+[\tau],$$

where

$$W_+[\tau] = \{x : w^{(b+)}(\tau, x) + \int_\tau^\vartheta v(t)dt \leq 0\}.$$

Exercise 5.1.1. (a) Prove a proposition similar to Theorem 5.1.1 for the backward reach set $\mathcal{W}[\tau]$.
(b) Prove a proposition similar to Theorem 5.1.2 for the forward reach set $\mathcal{X}[\tau]$.

Let us now pass to the discussion of internal estimates for the reachability sets and the related HJB equations. We shall consider the internal estimates for backward reachability sets $\mathcal{W}[t] = \mathcal{W}(t, x, \mathcal{M})$. As in the above, here we do not necessarily require differentiability of the value function $V^{(b)}(t, x)$.

Let function $V^{(b)}(t, x)$ of Eq. (5.7) be continuous in all the variables, being a generalized (viscosity) solution of this equation (in particular, it may also turn out to be classical).

Consider the next assumption.

Assumption 5.1.3. *There exist function* $h(t, x, p)$ *and function* $w^-(t, x) \in C_1$, *which satisfy the inequalities*

$$\mathcal{H}(t, x, p) \geq h(t, x, p), \quad \forall \{t, x, p\},$$

$$w_t^-(t, x) - h(t, x, -w_x^-(t, x)) \leq 0,$$

$$w^-(\vartheta, x) \geq V^{(b)}(\vartheta, x).$$

Under Assumption 5.1.3 we have

$$w_t^- - \mathcal{H}(t, x, -w_x^-) \leq w_t^- - h(t, x, -w_x^-) \leq 0.$$

Integrating the last inequality from τ to ϑ, along some *optimal trajectory* $x^0[s] = x^0(s, \tau, x)$ which ends in $x^0[\vartheta]$, we come to

$$w^-(\vartheta, x^0[\vartheta]) \leq w^-(\tau, x)$$

and further, due to third condition, to

$$V(\vartheta, x^0[\vartheta]) \leq w^-(\tau, x). \tag{5.10}$$

However $V(\vartheta, x^0[\vartheta]) = V(\tau, x)$, according to (5.9). Comparing this with (5.10), we come to the next proposition.

Theorem 5.1.3. *Under Assumption 5.1.3 the following estimate is true*

$$V^{(b)}(\tau, x) \leq w^-(\tau, x).$$

Denoting $W^-[t] = \{x : w^-(t, x) \leq 0\}$ and using the last inequality, we come to the conclusion.

Corollary 5.1.1. *Under Assumption 5.1.3 the next inclusion is true*

$$W^-[\tau] \subseteq \mathcal{W}[\tau]. \tag{5.11}$$

We shall further observe, in the linear-convex case, how comparison theorems may be used for obtaining ellipsoidal estimates of reachability sets.

5.1.2 A Deductive Approach to Ellipsoidal Calculus

Let us now return to linear systems of type

$$\dot{x} = A(t)x + B(t)u, \tag{5.12}$$

under an ellipsoidal constraint on control u and a similar one on target set $\mathcal{M}$, namely,

$$u \in \mathcal{P}(t) = \mathcal{E}(p_u(t), P(t)), \quad \mathcal{M} = \mathcal{E}(m, M), \ t \in [t_0, \vartheta], \tag{5.13}$$

where $P'(t) = P(t) > 0, M' = M > 0$. As usual, a nondegenerate ellipsoid is given here by the inequality

$$\mathcal{E}(p_u, P) = \{u : \langle u - p_u, P^{-1}(u - p_u) \rangle \leq 1\},$$

with its support function being

$$\rho(l \mid \mathcal{E}(p_u, P)) = \max\{\langle l, p \rangle | p \in \mathcal{E}(p_u, P)\} = \langle l, p_u \rangle + \langle l, Pl \rangle^{1/2}.$$

5.1 The Comparison Principle for the HJB Equation

Then

$$-\mathcal{H}(t, x, -p) = \min\{\langle p, f(t, x, u)\rangle \mid u \in \mathcal{P}(t)\} = \mathcal{H}^{(b)}(t, x, p)$$

$$= \langle p, A(t)x\rangle + \min_u\{\langle p, B(t)u\rangle \mid u \in \mathcal{E}(p_u(t), P(t))\} =$$

$$= \langle p, A(t)x + B(t)p_u(t)\rangle - \langle p, B(t)P(t)B'(t)p\rangle^{1/2},$$

and with these relations the HJB equation (5.7), namely,

$$V_t^{(b)} + \mathcal{H}^{(b)}(t, x, V_x^{(b)}) = 0, \quad V^{(b)}(\vartheta, x) = d^2(x, \mathcal{M}). \tag{5.14}$$

acquires the form specific for a linear system.

In order to get the lower bound for $V^{(b)}$ and the external bound for set $\mathcal{W}[t]$, we apply Theorem 5.1.2 and Corollary 5.1.1. We further use the next inequalities

$$\langle p, B(t)P(t)B'(t)p\rangle^{1/2} \leq \gamma^2(t) + (4\gamma^2(t))^{-1}\langle p, B(t)P(t)B'(t)p\rangle,$$

where $\gamma^2(t) > 0$ is arbitrary. Here an equality is attained with

$$\gamma^2(t) = (1/2)\langle p, B(t)P(t)B'(t)P(t)\rangle^{1/2}.$$

Under Assumption 5.1.2, as applied to system (5.1), the function $w(t, x)$ is taken to be quadratic:

$$w(t, x) = \langle x - x^*, K(t)(x - x^*)\rangle - 1, \tag{5.15}$$

where $K(t) = K'(t) > 0$ is differentiable. Then we have

$$dw(t, x)/dt = w_t + \mathcal{H}^{(b)}(t, x, w_x) =$$

$$= w_t + \langle w_x, A(t)x + B(t)p_u(t)\rangle - \langle w_x, B(t)P(t)B'(t)w_x\rangle^{1/2} \geq$$

$$\geq w_t + \langle w_x, A(t)x + B(t)p_u(t)\rangle - \gamma^2(t) - (4\gamma^2(t))^{-1}\langle w_x, B(t)P(t)B'(t)w_x\rangle.$$

Substituting $w(t, x) = \langle x - x^*(t), K(t)(x - x^*(t))\rangle - 1$, we further obtain

$$w_t + \mathcal{H}^{(b)}(t, x, w_x) \geq \{\langle x - x^*(t), \dot{K}(t)(x - x^*(t))\rangle - 2\langle \dot{x}^*(t), K(t)(x - x^*(t))\rangle +$$

$$+ 2\langle K(t)(x - x^*(t)), A(t)x + B(t)p_u(t)\rangle -$$

$$- (\gamma^2(t))^{-1}\langle K(x - x^*(t)), B(t)P(t)B'(t)K(x - x^*(t))\rangle\} - \gamma^2(t). \tag{5.16}$$

In order that satisfy this scheme we shall demand that the expression in curly brackets of (5.16) is equal to zero for all $x - x^*$:

$$\langle x - x^*(t), \dot{K}(t)(x - x^*(t)) \rangle - 2\langle \dot{x}^*(t), K(t)(x - x^*(t)) \rangle +$$
$$+ 2\langle K(t)(x - x^*(t)), A(t)(x - x^*(t)) + A(t)x^*(t) + B(t)p_u(t) \rangle -$$
$$- (\gamma^2(t))^{-1} \langle K(t)(x - x^*(t)), B(t)P(t)B'(t)K(t)(x - x^*(t)) \rangle = 0$$

Equalizing with zero the terms with multipliers of second order in $x - x^*$, then those of first order in the same variable, we observe that the last equality will be fulfilled if and only if the following equations are true

$$\dot{K} + KA(t) + A'(t)K - (\gamma^2(t))^{-1} KB(t)P(t)B'(t)K = 0, \qquad (5.17)$$

$$\dot{x}^* = A(t)x^*(t) + B(t)p_u(t), \qquad (5.18)$$

with boundary conditions

$$K(\vartheta) = M^{-1}, \quad x^*(\vartheta) = m. \qquad (5.19)$$

Under Eqs. (5.17), (5.18) relation (5.16) yields inequality

$$dw/dt \geq -\gamma^2(t).$$

Integrating it from τ to ϑ, along an optimal trajectory $x^0[t]$, $t \in [\tau, \vartheta]$, which starts at $x[\tau] = x$ and ends at $x[\vartheta]$, generating $V^{(b)}$, we come to condition

$$w(\tau, x) \leq w(\vartheta, x) + \int_\tau^\vartheta \gamma^2(t)dt, \qquad (5.20)$$

where $w(t, x) = \langle x - x^*(t), K(t)(x - x^*(t)) \rangle - 1$, and $K(t), x^*(t)$ are defined through Eqs. (5.17), (5.18) with

$$w(\vartheta, x) = \langle x - x^*(\vartheta), M^{-1}(x - x^*(\vartheta)) \rangle.$$

Summarizing the above, we come to the proposition

Theorem 5.1.4. *(i) Function*

$$\mathbf{w}^0(t, x | \gamma(\cdot)) = w(t, x) - \int_t^\vartheta \gamma^2(s)ds$$

5.1 The Comparison Principle for the HJB Equation

is a lower bound of $V^{(b)}(t,x)$:

$$\mathbf{w}^0(t,x|\gamma^2(\cdot)) \le V^{(b)}(t,x). \tag{5.21}$$

(ii) The inclusion

$$\mathcal{W}[t] \subseteq \mathcal{W}_+[t], \tag{5.22}$$

is true, where

$$W_+[t] = \left\{ x : \langle x - x^*(t), K(t)(x - x^*(t)) \rangle \le 1 + \int_t^\vartheta \gamma^2(s)ds \right\}.$$

We further transform Eq. (5.17) to new variables, substituting K for $\mathcal{K}_+$ according to relations

$$\mathcal{K} = K^{-1}, \quad \dot{\mathcal{K}} = -\mathcal{K}\dot{K}\mathcal{K}, \quad \mathcal{K}_+(t) = \left(1 + \int_t^\vartheta \gamma^2(s)ds\right)\mathcal{K}(t)$$

We then come to equations

$$\dot{\mathcal{K}}_+ = A(t)\mathcal{K}_+ + \mathcal{K}_+ A'(t) - \pi(t)\mathcal{K}_+ - (\pi(t))^{-1}B(t)P(t)B'(t), \tag{5.23}$$

and boundary condition $\mathcal{K}_+(\vartheta) = M$.
Here

$$\pi(t) = \gamma^2(t)\left(1 + \int_t^\vartheta \gamma^2(s)ds\right)^{-1}.$$

The formulas derived here allow the next conclusion.

Theorem 5.1.5. *The following inclusion (external estimate) is true*

$$\mathcal{W}[t] \subseteq \mathcal{W}_+[t],$$

where

$$\mathcal{W}_+[t] = \mathcal{E}(x^*(t), \mathcal{K}_+(t)) = \{x : \langle x - x^*(t), \mathcal{K}_+^{-1}(t)(x - x^*(t)) \rangle \le 1\},$$

whatever be the function $\pi(t) > 0$.

Remark 5.1.1. The last relation, namely, the external ellipsoidal approximation of the backward reach set $\mathcal{W}[t]$ was derived through a **deductive scheme**, from the HJBI equations. Recall that similar results were achieved earlier, in Chap. 3, through an **inductive scheme** (see also [174, 180]), without applying the Hamiltonian formalism. There the structure of differential equations indicated in Theorem 3.3.1

for the configuration matrices of the approximating "inductive" ellipsoids is similar to (5.23) for matrices $\mathcal{K}_+$ of the "deductive" external ellipsoids derived in this section.

We now pass to internal ellipsoidal approximations of sets $\mathcal{W}[\tau]$. Starting with Eq. (5.14), where

$$\mathcal{H}^{(b)}(t,x,p) = \langle p, A(t)x + B(t)p_u(t)\rangle - \langle p, B(t)P(t)B'(t)p\rangle^{1/2},$$

we shall apply Theorem 5.1.3 and Corollary 5.1.1. In doing this we shall use the inequality

$$\langle p, B(t)P(t)B'(t)p\rangle^{1/2} \geq \langle T(t)p, T(t)p\rangle^{-1/2}\langle p, S(t)(B(t)P(t)B'(t))^{1/2}p\rangle,$$

where $p \in \mathbb{R}^n$ is arbitrary and $S(t)$ is any continuous matrix function, whose values $S(t)$ are orthogonal matrices, namely, $S(t)S'(t) = I$ and $T(t) = T'(t)$ is a matrix function with $T(t)p \neq 0$.

Here an equality is attained under coordinated collinearity of vectors $S(t)(B(t)P(t)B'(t))^{1/2}p$ and $T(t)p$.

As in the "external" case, under Assumption 5.1.3, when applied to system (5.12), function $w^-(t,x)$ is taken quadratic: $w^-(t,x) = \langle x - x^*, K_-(t)(x - x^*)\rangle - 1$, with $K_-(t) = (K_-)'(t) > 0$ and $x^*(t)$ differentiable. Taken $T(t) = (K_-)^{-1}$, we then have

$$w_t^- + \mathcal{H}^{(b)}(t,x,w_x^-) = w_t^- + \langle w_x^-, A(t)x + B(t)p_u(t)\rangle - \langle w_x, B(t)P(t)B'(t)w_x^-\rangle^{1/2} \leq$$

$$\leq w_t^- + \langle w_x^-, A(t)x + B(t)p_u(t)\rangle -$$

$$-\langle (K_-)^{-1}(t)w_x^-, (K_-)^{-1}(t)w_x^-\rangle^{-1/2}\langle (K_-)^{-1}(t)w_x^-, S(t)(B(t)P(t)B'(t))^{1/2}w_x^-\rangle.$$

Note that in the domain

$$D(r) = \{t, x : \langle x - x^*(t), K_-(t)(x - x^*(t))\rangle < r^2, t \in [\tau, \vartheta]\}$$

the following condition is true ($P_B(t) = B(t)P(t)B'(t)$)

$$w_t^- + \mathcal{H}^{(b)}(t,x,w_x^-) \leq \Big\{\langle x - x^*(t), \dot{K}_-(t)(x - x^*(t))\rangle$$

$$-2\langle \dot{x}^*(t), K_-(t)(x - x^*(t))\rangle + 2\langle K_-(t)(x - x^*(t)), A(t)x + B(t)p_u(t)\rangle -$$

$$- 2r^{-1}\langle x - x^*(t), S(t)P_B^{1/2}(t)K_-(t)(x - x^*(t))\rangle\Big\}. \tag{5.24}$$

5.1 The Comparison Principle for the HJB Equation

The last relation will reflect the scheme of Theorem 5.1.4 with r sufficiently large. Then, in order to satisfy this scheme, it will be sufficient that the expression in curly brackets will be equal to zero in $D(r)$. Demanding this equality, we have

$$\langle x - x^*(t), \dot{K}_-(t)(x - x^*(t))\rangle - 2\langle \dot{x}^*(t), K_-(t)(x - x^*(t))\rangle +$$

$$+ 2\langle K_-(t)(x - x^*(t)), A(t)(x - x^*(t)) + A(t)x^*(t) + B(t)p_u(t)\rangle$$

$$- 2r^{-1}\langle x - x^*(t), S(t)P_B^{1/2}(t)K_-(t)(x - x^*(t))\rangle = 0$$

for all $\{t, x \in D(r)\}$. Equalizing with zero the terms with multipliers of second order in $x - x^*$, then those of first order in the same variable, we observe that the last equality will be fulfilled if and only if the next equation

$$\dot{K}_- + (K_- A(t) + A'(t)K_- - r^{-1}(K_- P_B^{1/2} S'(t) + S(t)P_B^{1/2}(t)K_-) = 0, \quad (5.25)$$

is true together with Eq. (5.18) for $x^*(t)$ and with boundary conditions

$$K_-(\vartheta) = M^{-1}, \quad x^*(\vartheta) = m. \quad (5.26)$$

Under (5.25), (5.18), (5.26) relation (5.24) yields inequality

$$dw^-/dt \le 0,$$

integrating which from τ to ϑ, along an optimal trajectory $x^0[t]$, $t \in [\tau, \vartheta]$, which starts at $x[\tau] = x$, we come to condition

$$w^-(\tau, x) \ge w(\vartheta, x^0[\vartheta]) = V^{(b)}(\vartheta, x^0[\vartheta]) = V^{(b)}(t, x), \quad (5.27)$$

where $w(t, x) = \langle x - x^*(t), K_-(t)(x - x^*(t))\rangle - 1$.

Summarizing the above, we come to the proposition

Theorem 5.1.6. *Function $w^-(t, x)$ is an upper bound of $V^{(b)}(t, x)$:*

$$w^-(t, x) \ge V^{(b)}(t, x).$$

The inclusion

$$\mathcal{W}[t] \supseteq W_-[t], \quad (5.28)$$

is true, where

$$W_-[t] = \{x : \langle x - x^*(t), K_-(t)(x - x^*(t))\rangle \le 1\}.$$

Transforming Eq. (5.28) to new variables, with

$$\mathcal{K}_- = K^{-1}, \quad \dot{\mathcal{K}}_- = -\mathcal{K}_- \dot{K}_- \mathcal{K}_-,$$

we come to equations

$$\dot{\mathcal{K}}_- = A(t)\mathcal{K}_- + \mathcal{K}_- A'(t) - r^{-1}(t)(\mathcal{K}_- S(t) P_B^{1/2}(t) + P_B^{1/2}(t) S'(t) \mathcal{K}_-), \quad (5.29)$$

and boundary condition $\mathcal{K}_-(\vartheta) = M$. Then

$$W_-[t] = \{x : \langle x - x^*(t), (\mathcal{K}_-(t))^{-1}(x - x^*(t)) \rangle \leq 1\}. \quad (5.30)$$

Here $r(t)$ is a tuning parameter which may be selected on-line, as a function of $\|\mathcal{K}_-(t)\|$. Then the structure of Eq. (5.29) will be of the type we had in (3.109).

Note that the *centers* $x^*(t)$ of the indicated internal ellipsoids *will be the same as for external estimates*, being in both cases described by Eq. (5.18).

Remark 5.1.2. By an appropriate selection of parameterizers $\pi(t), S(t)$ for Eqs. (5.23), (5.29), similar to those of Sects. 3.3 and 3.8, one may obtain for each "good" direction $l \in \mathbb{R}^n$ of Assumption 3.3.1 an ellipsoid which is "tight" along this direction. Namely, for each l one may indicate such parameters $\pi(\cdot), S(\cdot)$ for which equalities

$$\rho(l | \mathcal{W}[t]) = \rho(l | W_+[t]), \quad \rho(l | \mathcal{W}[t]) = \rho(l | W_-[t])$$

would be true.

5.2 Calculation of Nonconvex Reachability Sets

Consider the reachability set $\mathbf{X}[\Theta] = X(\Theta, t_0, X^0)$ of Definition 2.3.1 *within an interval* $\Theta = [\alpha, \beta]$. As indicated in Sect. 2.7

$$\mathbf{X}[\Theta] = \bigcup \{X[t] \mid t \in \Theta \}.$$

We shall now calculate $\mathbf{X}[\Theta]$ through ellipsoidal approximations restricting ourselves to the "external" case. The "internal" solutions could be designed within a similar scheme.

5.2 Calculation of Nonconvex Reachability Sets

Recall that due to Theorem 3.3.2 and Comparison in Principle as in Sect. 5.1.2, we may write[2]:

$$X[t] = \bigcap_l \{\mathcal{E}(x^*(t), X_+^l[t]) \mid \langle l, l \rangle = 1\}$$

and equivalently $X[t] = \{x : V^0(t, x) \leq 1\}$. where

$$V^0(t, x) = \max_l \{V_+^0(t, x, l) \mid \langle l, l \rangle \leq 1\},$$

$$V_+^0(t, x, l) = \langle x - x^*(t), (X_+^l[t])^{-1}(x - x^*(t)) \rangle. \tag{5.31}$$

Then we have

$$\mathbf{X}[\Theta] = \bigcup \{X[t] \mid t \in \Theta\} = \{x : \mathbf{V}^0(\Theta, x) \leq 1\}, \tag{5.32}$$

where

$$\mathbf{V}^0(\Theta, x) = \min_t \{V_+^0(t, x) \mid t \in \Theta\} = \min_t \max_l \{V_+^0(t, x, l) \mid \langle l, l \rangle = 1, \, t \in \Theta\}. \tag{5.33}$$

This leads to

Theorem 5.2.1. *The reachability set* $\mathbf{X}[\Theta]$ *within the time interval* $\Theta = [\alpha, \beta]$ *may be described through (5.31)–(5.33).*

According to the theory of minimax problems (see [62]), we have

$$\mathbf{V}^0(\Theta, x) = \min_t \max_l \{V_+^0(t, x, l) \mid \langle l, l \rangle = 1, \, t \in \Theta\} \geq$$

$$\geq \max_l \min_t \{V_+^0(t, x, l) \mid \langle l, l \rangle = 1, \, t \in \Theta\}. \tag{5.34}$$

Then, in terms of sets, the relation for $\mathbf{X}[\Theta]$ may be expressed as

$$\mathbf{X}[\Theta] = \bigcup_t \bigcap_l \{\mathcal{E}(x^*(t), X_+^l[t]) \mid \langle l, l \rangle = 1, t \in \Theta\} \subseteq$$

$$\subseteq \bigcap_l \bigcup_t \{\mathcal{E}(x^*(t), X_+^l[t]) \mid \langle l, l \rangle = 1, t \in \Theta\}, \tag{5.35}$$

where $x^*[t], X_+^l[t]$ are defined by Theorems 3.3.1, 3.3.2 (Eqs. (3.31), (3.32)).

---

[2] In Sect. 5.1.2 we calculated the backward reach set $\mathcal{W}[t]$ due to value function $V^{(b)}(t, x)$. Here, in a similar way, we calculate the forward reach set $X[t]$ due to value function $V^0(t, x)$.

Recalling notation $E_+[t,l] = \mathcal{E}(x^*(t), X_+^l[t])$, introduce

$$X[t] = \bigcap_l \{E_+[t,l] \mid \langle l,l\rangle = 1\}, \quad X[\Theta,l] = \bigcup_t \{E_+[t,l] \mid t \in \Theta\}.$$

Then, in terms of sets we have the following proposition

Theorem 5.2.2. *The next relations are true*

$$X[\Theta] = \bigcup_t \{X[t] \mid t \in \Theta\} \subseteq \bigcap_l \{X[\Theta,l] \mid \langle l,l\rangle = 1\}.$$

Remark 5.2.1. The conditions for *equality of sets*

$$\bigcup_t \bigcap_l \{\cdot\} = \bigcap_l \bigcup_t \{\cdot\}$$

would be the set-valued analogy of the theorem on existence of a *saddle point* in l, t ($\min_t \max_l V_+(t,x,l) = \max_l \min_t V_+(t,x,l)$) for *function* $V_+(t,x,l)$.

We further we observe

$$\bigcup_t \bigcap_{l(\cdot)} \{\mathcal{E}(x^*(t), X_+^l[t]) \mid l(\cdot) \in SS_g(\cdot), t \in \Theta\} \subseteq$$

$$\bigcap_{l(\cdot)} \bigcup_t \{\mathcal{E}(x^*(t), X_+^l[t]) \mid l(\cdot) \in SS_g(\cdot), t \in \Theta\}, \quad (5.36)$$

where $SS_g(\cdot)$ is the set of all good curves $\|l(t)\| = 1, t \in \Theta$, normalized according to Sect. 3.3, to lie on a unit sphere.

Exercise 5.2.1. Indicate conditions when the operation of inclusion ($\subseteq$) in (5.36) may be substituted by an equality.

Exercise 5.2.2. Indicate internal ellipsoidal approximations for set $\mathbf{X}[\Theta]$.

Example 5.8.

Consider system

$$\dot{x}_1 = x_2 + u + r\cos\varphi t, \quad \dot{x}_2 = -\omega^2 t + r\sin\varphi t, \quad (5.37)$$

with $t \in [0, \tau]$, $|u| \leq \mu$, $x(t_0) \in \mathcal{X}_0$.

We illustrate the calculation of $\mathbf{X}[\Theta]$, $\Theta = [0, \tau]$, and its upper estimate $X_+^l[\Theta]$, assuming $\tau = 1$, $\omega = 1$, $\mu = 1$, $\mathcal{X}_0 = x_c + \mathcal{B}(0)$, $x_c' = (1,1)$.[3]

---

[3] The illustrations for this example were done by Zakroischikov.

5.2 Calculation of Nonconvex Reachability Sets

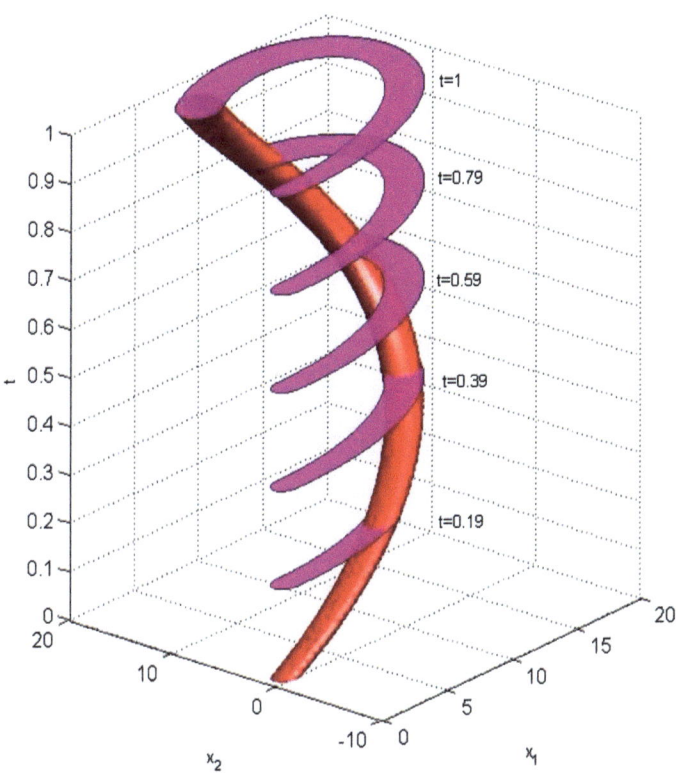

Fig. 5.1 The set $\mathcal{X}_+[\Theta]$ (in *red*) and its covering by intersection of nonconvex sets $\mathcal{X}_+^l[\Theta]$

Shown in Fig. 5.1 is the set $\mathcal{X}_+[\Theta]$ (in red) and its covering by intersection of nonconvex sets $\mathcal{X}_+^l[\Theta]$, calculated along good curves $l(t)$. An example of such covering

$$\mathcal{X}_+[\Theta \mid l^*, l^{**}] = \mathcal{X}_+[\Theta, l^*] \bigcap \mathcal{X}_+[\Theta, l^{**}]$$

by only two such sets calculated from good curves generated by $l^*(\tau) = (-1, 0)$, $l^{**}(\tau) = (-0.54, 0.83)$ is indicated in Fig. 5.2.
Figures 5.3 and 5.4 show set $\mathbf{X}[\Theta]$ and its external approximation by an ellipsoidal tube.

Fig. 5.2 Covering of $\mathcal{X}_+[\Theta]$ by only two sets from $\mathcal{X}_+^l[\Theta]$

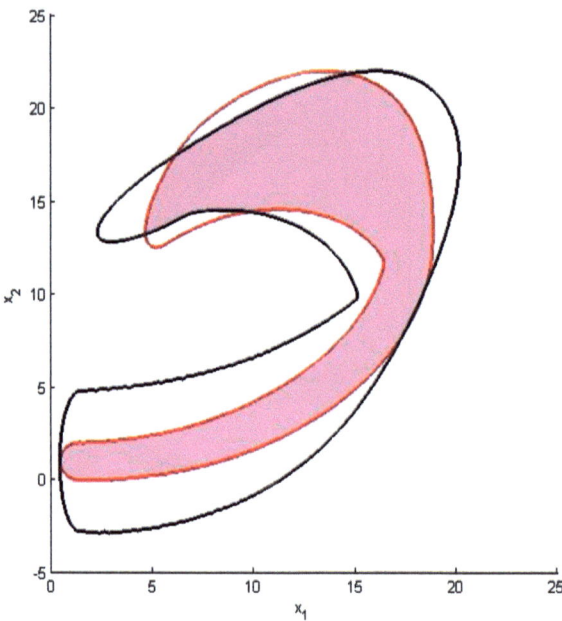

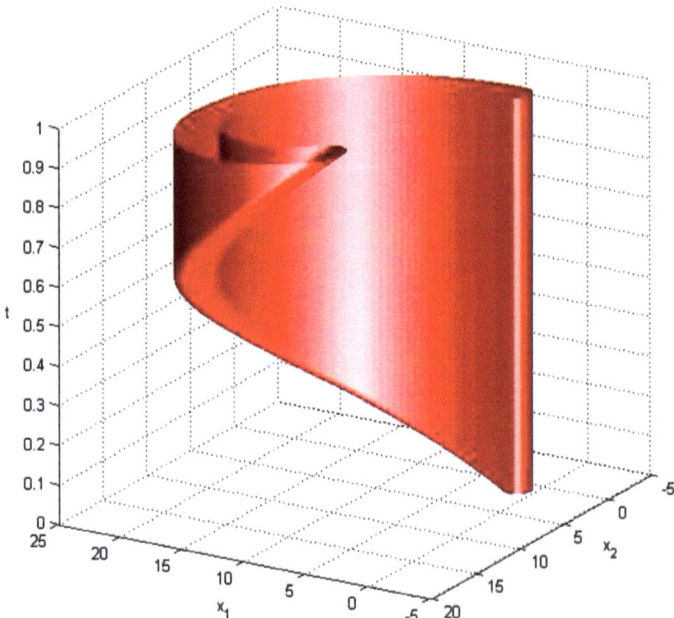

Fig. 5.3 Set $\mathbf{X}[\Theta]$

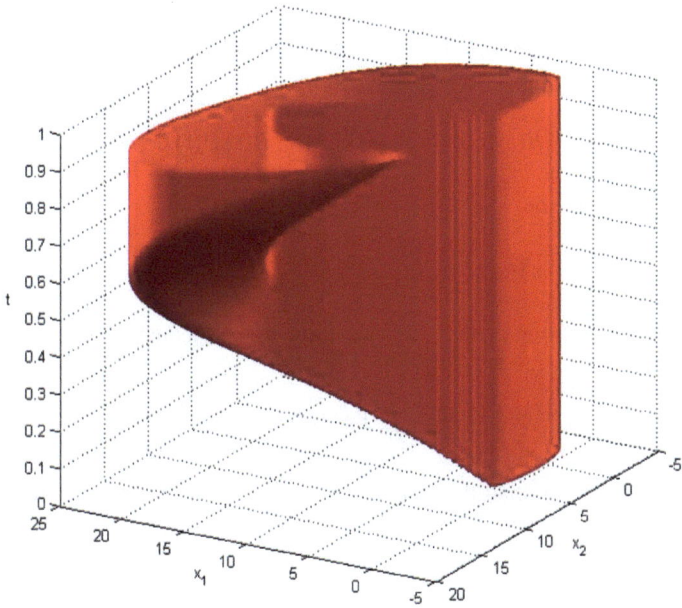

Fig. 5.4 External approximation of set $\mathbf{X}[\Theta]$

5.3 Applications of Comparison Principle

5.3.1 Forward Reachability

Here we briefly indicate a result similar to Theorem 5.1.1 of Sect. 5.1 for a forward reachability set. Take linear system (5.12), with constraints on $x(t_0), u(t)$ given by nondegenerate ellipsoids:

$$x(t_0) \in \mathcal{E}(x^0, X^0), \quad u(t) \in \mathcal{E}(p_u(t), P(t)), \tag{5.38}$$

Then

$$\mathcal{H}(t, x, p) = \langle p, A(t)x + B(t)p_u(t)\rangle + \langle p, B(t)P(t)B'(t)p\rangle^{1/2} \leq \tag{5.39}$$

$$\leq \langle p, A(t)x + p_u(t)\rangle + \mu(t) + (4\mu(t))^{-1}(\langle p, B(t)P(t)B'(t)p\rangle = H(t, p, x)$$

for any $\mu(t) > 0$, with equality attained if

$$\mu(t) = \mu_e(t, p) = \frac{1}{2}\langle p, B(t)P(t)B'(t)p\rangle^{1/2}.$$

Let us now look for $w(t, x)$ as a quadratic form, $w(t, x) = \langle (x - x^*(t), K(t)(x - x^*(t)) \rangle - 1$, where $K(t) = K'(t)$ is differentiable and requires that w satisfies the PDE

$$w_t + \langle w_x, A(t)x + B(t)p_u \rangle + (4\mu(t))^{-1} \langle w_x, B(t)P(t)B'(t)w_x \rangle = 0 \quad (5.40)$$

with boundary condition

$$w(t_0, x) \equiv \langle x - x^0, (X^0)^{-1}(x - x^0) \rangle. \quad (5.41)$$

Then, after an integration from t_0 to t, we come to the inequality

$$w(t, x(t)) \leq w(t_0, x(t_0)) + \int_{t_0}^t \mu(s)ds \leq 1 + \int_{t_0}^t \mu(s)ds. \quad (5.42)$$

Here $K(t)$ may be obtained through a standard procedure of solving the resulting Riccati equation

$$\dot{K} = -KA(t) - A'(t)K - \frac{1}{\mu(t)}(K, B(t)P(t)B'(t)K), \quad (5.43)$$

with equation

$$\dot{x}^* = A(t)x^* + B(t)p_u(t).$$

where $K(t_0) = (X^0)^{-1}$, $x^*(t_0) = x^0$. Transformations

$$\dot{\mathcal{K}} = -\mathcal{K}\dot{K}\mathcal{K}, \quad \mathcal{K} = K^{-1}, \quad \mathcal{K}_+(t) = \left(1 + \int_{t_0}^t \mu(s)ds\right)\mathcal{K}(t),$$

convert into

$$\dot{\mathcal{K}}_+ = A(t)\mathcal{K}_+ + \mathcal{K}_+A'(t) + \frac{1}{\pi(t)}B(t)Q(t)B'(t) + \pi(t)\mathcal{K}_+, \quad \mathcal{K}_+(t_0) = (X^0)^{-1},$$

$$(5.44)$$

Here

$$\pi(t) = \mu(t)/(1 + \int_{t_0}^t \mu(s)ds).$$

These equations now coincide with those of [21, 23, 24].

The level sets for function $w(t, x) = w_\pi(t, x)$ are ellipsoids $\mathcal{E}(x^*(t), \mathcal{K}_+(t))$, that depend on parameterizing functions π, namely,

$$\{x : w_\pi(t, x) \leq 1\} = \mathcal{E}(x^*(t), \mathcal{K}_+(t)) = \mathcal{E}_\pi^+[t]. \quad (5.45)$$

5.3 Applications of Comparison Principle

Selecting k such functions $\pi_i(\cdot)$, we come to the estimate

$$X[t] \subseteq \cap\{\mathcal{E}^+_{\pi_i}[t] \mid 1 \leq i \leq k\}. \tag{5.46}$$

An appropriate selection of $\pi_i(\cdot)$ ensures that ellipsoids $\mathcal{E}^+_{\pi_i}[t]$ are tight in the sense that there is no other external ellipsoid of type (5.44) that could be squeezed in between $\mathcal{E}^+_{\pi_i}[t]$ and $X[t]$.

5.3.2 Systems with Hamiltonians Independent of the State

For a system of type

$$\dot{x} = f(t, u), \quad u \in \mathcal{P}(t), \tag{5.47}$$

with $\mathcal{P}(t)$ compact and f continuous in t, x, find

$$V(\tau, x) = \min_u \{\varphi(x(t_0)) \mid x(\tau) = x\}. \tag{5.48}$$

Here are two simple examples of such systems:

$$\dot{x}_i = \langle u, L_i(t)u \rangle, \quad i = 1, \ldots, n, \quad \langle u, Pu \rangle \leq 1,$$

where $P = P' > 0$ and $L_i(t)$ are $n \times n$ matrices, and

$$\dot{x}_i = \sin(u_i - \alpha_i(t)), \quad u_i \in [-\mu_i, \mu_i], \quad i = 1, \ldots, n.$$

Function $V(t, x)$ could be sought for through equation

$$V_t + H(t, V_x) = 0, \quad V(t_0, x) = \varphi(x), \tag{5.49}$$

with Hamiltonian

$$H(t, p) = \max\{\langle p, f(t, u) \rangle \mid u \in \mathcal{P}(t)\},$$

and solution $V(t, x)$ taken, if necessary, as a viscosity solution. We shall indicate an explicit formula for this function.

Assumption 5.3.1. *(i) Function $\varphi(x)$ is proper, convex.*
(ii) Its conjugate $\varphi^*(l)$ is strictly convex and $0 \in \text{int}(\text{Dom}\varphi)$, so that $\varphi^*(l) \to \infty$, as $\langle l, l \rangle \to \infty$, [237].

Due to a theorem of Lyapunov on the range of a vector measure (see [3, 202])[4] a vector $x(t_0)$ of type

$$x(t_0) = x + Q(\tau), \quad Q(\tau) = -\int_{t_0}^{\tau} f(t, \mathcal{P}(t)) dt,$$

ranges over $x + Q(\tau)$, where $Q(\tau)$ is a convex compact set.

Then[5]

$$V(\tau, x) = \min_{q}\{\varphi(q) \,|\, q \in (x+Q(\tau))\} = \min_{q} \sup_{l}\{\langle l, q\rangle - \varphi^*(l) \,|\, l \in \mathbb{R}^n, q \in Q(\tau)\}$$

$$= \sup_{l}\{\langle l, x\rangle - \rho(-l \,|\, Q(\tau)) - \varphi^*(l)\}.$$

From here it follows that $V(\tau, x)$ is closed, convex in x, being the conjugate of a closed convex function $V^*(\tau, x) = \rho(l \,|\, -Q(\tau)) + \varphi^*(l)$.

Lemma 5.3.1. *Under Assumption 5.3.1 on $\varphi(x)$ the following equality is true:*

$$V(\tau, x) = V^{**}(\tau, x). \tag{5.50}$$

We further have

$$V^*(\tau, l) = \sup_{x}\{\langle l, x\rangle - V(\tau, x)\}$$

$$= \sup_{x}\{\langle l, x\rangle - \min_{u(\cdot)} \max_{p}\{\langle p, x(t_0)\rangle - \varphi^*(p)\}$$

$$= \sup_{x}\{\langle l, x\rangle - \min_{u(\cdot)} \max_{p}\{\langle p, x - \int_{t_0}^{\tau} f(t, u(t))dt\rangle - \varphi^*(p)\}\}$$

$$= \max_{u(\cdot)} \sup_{x}\{\langle l, x\rangle - \max_{p}\{\langle p, x - \int_{t_0}^{\tau} f(t, u(t))dt\rangle - \varphi^*(p)\}\}$$

$$= \max_{u(\cdot)} \sup_{x} \min_{p}\{\langle l - p, x\rangle + \int_{t_0}^{\tau} \langle p, f(t, u(t))\rangle dt + \varphi^*(p)\}$$

$$= \max_{u(\cdot)} \min_{p} \sup_{x}\{\langle l - p, x\rangle + \int_{t_0}^{\tau} \langle p, f(t, u(t))\rangle dt + \varphi^*(p)\}$$

---

[4] Do not confuse famous Lyapunov (1911–1973) with celebrated A.M. Lyapunov (1856–1918), founder of modern stability theory.

[5] Here and below the conjugates of V are taken only in the second variables with τ fixed.

5.3 Applications of Comparison Principle

$$= \max_{u(\cdot)} \{ \int_{t_0}^{\tau} \langle l, f(t, u(t)) \rangle dt + \varphi^*(l) \}$$

$$= \int_{t_0}^{\tau} \mathcal{H}(t, l) dt + \varphi^*(l).$$

Therefore, due to (5.50),

$$V(\tau, x) = V^{**}(\tau, x) = \max_{l} \{ \langle l, x \rangle - \int_{t_0}^{\tau} \mathcal{H}(t, l) dt - \varphi^*(l) \}.$$

Assumption 5.3.1 allowed to interchange the sup and min operations above by applying the minmax theorem of Ky Fan [72]. Also used was the property

$$\max_{u(\cdot)} \int_{t_0}^{\tau} \langle l, f(t, u(t)) \rangle dt = \int_{t_0}^{\tau} \max \{ f(t, u) \mid u \in \mathcal{P}(t) \} dt,$$

indicated in [238, Chap. 14, Sect. F]. The above results in the next proposition.

Theorem 5.3.1. *Under Assumption 5.3.1 the following relation is true:*

$$V(\tau, x) = \max_{p} \{ \langle p, x \rangle - \int_{t_0}^{\tau} \mathcal{H}(t, p) dt - \varphi^*(p) \}. \quad (5.51)$$

With H independent of both t, x the last relation is a formula of the *Lax–Hopf type*, [17, 247]. The requirements of Assumption 5.3.1 are clearly satisfied by function $\varphi(x) = d^2(x, X^0)$ with X^0 convex and compact.

Exercise 5.3.1. For system

$$\dot{x}_1 = \sin u, \quad \dot{x}_2 = \cos u, \quad u \in [-\pi, \pi], \quad t \in [t_0, \tau], \quad (5.52)$$

find value function

$$V(\tau, x) = \min \{ ||x(t_0) - x^0|| \mid x(\tau) = x \}, \quad x^0 \in X^0.$$

Here the Hamiltonian is

$$\mathcal{H}(t, p) = \max_{u} \{ p_1 \sin u + p_2 \cos u \mid u \in [-\pi, \pi] \} = \max_{\alpha} \{ (p_1^2 + p_2^2)^{1/2} \sin(u + \alpha) \},$$

where $\alpha = \arccos(p_1(p_1^2 + p_2^2)^{-1/2})$. This gives $\mathcal{H}(t, p) = (p_1^2 + p_2^2)^{1/2}$. We also have

$$\varphi^*(p) = I(\mathcal{B}(0)) + \langle p, x^0 \rangle,$$

where $I(\mathcal{B}(0))$ is the indicator function for a Euclidean unit ball.

Following (5.52), we have

$$V(\tau, x) = \max_{p}\{\langle p, x - x^0\rangle - (\tau - t_0)(p_1^2 + p_2^2)^{1/2} \mid p_1^2 + p_2^2 = 1\}$$

Calculating the maximum, we come to the maximizers

$$p_1^0 = (x_1 - x_1^0)((x_1 - x_1^0)^2 + (x_2 - x_2^0)^2)^{-1/2}, \quad p_2^0 = (x_2 - x_2^0)((x_1 - x_1^0)^2 + (x_2 - x_2^0)^2)^{-1/2}.$$

This yields $V(\tau, x) = ((x_1 - x_1^0)^2 + (x_2 - x_2^0)^2)^{1/2} - \tau$.
We may now check that $V(\tau, x)$ is a solution to the HJB equation (5.49) which is

$$V_t + \max_u \{V_{x1} \sin u + V_{x2} \cos u\} = 0, \quad u \in [-\pi, \pi].$$

Here $V_t(\tau, x) = -1$, $V_x(\tau, x) = (p_1^0, p_2^0)' = (\cos \varphi, \sin \varphi)'$, so that $H(t, p^0) = 1$. Hence this equation is satisfied together with its boundary condition. We finally calculate the reachability set for system (5.52) which is

$$X[\tau] = \{x : V(\tau, x) \le 0\} = \{x : (x_1 - x_1^0)^2 + (x_2 - x_2^0)^2 \le \tau^2\}.$$

5.3.3 A Bilinear System

Consider the two-dimensional nonlinear system

$$\dot{x}_1 = u_1 x_1 - a_{12} x_1 \log x_2, \quad \dot{x}_2 = u_2 x_2 - a_{21} x_2 \log x_1, \tag{5.53}$$

with positive coefficients a_{ij}, $i, j = 1, 2$ and a hard bound

$$u \in \mathcal{P} = \{p : p_{11}^{-2} u_1^2 + p_{22}^{-2} u_2^2 \le 1\} \tag{5.54}$$

on the control $u = (u_1, u_2)'$ with $p = (p_{11}, p_{22})'$, $p_{ii} > 0$.

Let us look for the reach set $X(t, t_0, x^0)$ for this equation, starting from point $x^0 = (x_1^0, x_2^0)$, where $x_i^0 > 0$, $i = 1, 2$. Then the corresponding forward HJB equation is

$$V_t + \max_u \{V_{x_1}(u_1 x_1 - a_{12} x_1 \log x_2) + V_{x_2}(u_2 x_2 - a_{21} x_2 \log x_1) | u \in \mathcal{E}\} = 0,$$

where $\mathcal{E}$ is the ellipsoid defined by (5.54). This further gives

$$V_t + (p_{11}(V_{x_1})^2 x_1 + p_{22}(V_{x_2})^2 x_2)^{\frac{1}{2}} - (a_{12} x_1 \log x_2 V_{x_1} + a_{21} x_2 \log x_1 V_{x_2}) = 0, \tag{5.55}$$

5.3 Applications of Comparison Principle

which has to be considered with boundary condition

$$V(t_0, x) = (\log x - \log x^0, \log x - \log x^0)^2. \tag{5.56}$$

Rather than solving the last equation, let us transform Eq. (5.53) to another system of coordinates (system (5.53) may be linearized). Namely, taking $z_i = \log x_i$, $z_i^0 = \log x_i^0$, we have

$$\dot{z}_1 = -a_{12}z_2 + u_1, \quad \dot{z}_2 = -a_{21}z_1 + u_2, \quad z(0) = z^0, \tag{5.57}$$

under the same constraint (5.54). The reach set $Z(\tau, 0, z^0)$ from point z^0 for this system may be written down using relations similar to (3.32), (2.68), with changes in the definition of value function $V(t, x)$ from $d^2(z(t_0), z^0)$ to $d(z(t_0), z^0)$.
This leads to the value function

$$V^*(\tau, z) = \max\{\langle l, z \rangle - \Phi(\tau, z^0, l) \mid \langle l, l \rangle \le 1\},$$

and further on, in view of Lemma 2.3.1, to a system of inequalities $(l, z \in \mathbb{R}^2)$

$$Z(\tau, 0, z^0) = \{z : l_1 z_1 + l_2 z_2 \le \Phi(\tau, z_1^0, z_2^0, l_1, l_2)\}, \forall l \in \mathbb{R}^2,$$

where

$$\Phi(\tau, z_1^0, z_2^0, l_1, l_2)$$

$$= \int_0^\tau \Big((l_1 \cosh \alpha(\tau - s) - l_2 \beta^{-1} \sinh \alpha(\tau - s))^2 p_{11}$$

$$+ (-l_1 \beta^{-1} \sinh \alpha(\tau - s) + l_2 \cosh \alpha(\tau - s))^2 p_{22} \Big)^{1/2} ds$$

$$+ (l_1 \cosh \alpha\tau - l_2 \beta^{-1} \sinh \alpha\tau)z_1^0 + (-l_1 \beta \sinh \alpha\tau + l_2 \cosh \alpha\tau)z_2^0,$$

where $\alpha = (a_{12}a_{21})^{\frac{1}{2}}$, $\beta = (a_{12}/a_{21})^{\frac{1}{2}}$. Returning to the initial coordinates x, we observe that the boundary $\partial X(\tau, 0, x^0)$ of the exact reach set $X(\tau, 0, x^0)$ is given by equation

$$\max_l\{(l_1 \log x_1 + l_2 \log x_2) - \Phi(\tau, \log x_1^0, \log x_2^0, l_1, l_2) | l : \langle l, l \rangle \le 1\} = 0 \tag{5.58}$$

in the variables x_1, x_2, where $x_i^0 = \exp z_i^0$. The solution to (5.55), (5.56) then happens to be $V(\tau, x) = V^*(\tau, \log x_1, \log x_2)$.

Rather than calculating the exact $V(\tau, x)$ and $X(\tau, 0, x^0)$, it may be simpler to approximate these. To explore this option, let us first approximate function $V^*(\tau, z)$ by quadratic functions and set $Z(\tau, 0, z^0)$ by ellipsoids.

To approximate the convex compact reach set $Z(\tau, 0, z^0))$ by external ellipsoids $\mathcal{E}(z^\star(t), Z_+(t))$ we may apply Eq. (5.44) where $\mathcal{K}_+$ is substituted for Z_+. We have

$$\dot{Z}_+ = AZ_+ Z_+ A' + \frac{1}{\pi(t)} Q(t) + \pi(t) Z_+, \quad Z_+(t_0) = 0, \tag{5.59}$$

and

$$\dot{z}^\star = Az^\star, \quad z^\star(t_0) = z^0,$$

where

$$A = \begin{pmatrix} 0 & -a_{12} \\ -a_{21} & 0 \end{pmatrix}, \quad P = \begin{pmatrix} 1 & 0 \\ 0 & 1 \end{pmatrix}.$$

In order to ensure the property of tightness for ellipsoids $\mathcal{E}(z^\star(t), Z_+(t))$, and therefore an equality in (5.46), we have to select the parameterizing functions $\pi(t)$ as follows (see Chap. 3):

$$\pi(t) = \langle l, F(t) Q F'(t) l \rangle^{\frac{1}{2}} \langle l, Z_+(t) l \rangle^{-\frac{1}{2}}. \tag{5.60}$$

Here

$$F(t) = \exp(-At) = \begin{pmatrix} \cosh \alpha t & \beta \sinh \alpha t \\ \beta^{-1} \sinh \alpha t & \cosh \alpha t \end{pmatrix}.$$

The reach tube $Z[t] = Z(t, 0, z^0)$ will now be exactly described by the formula (see Chap. 3)

$$z^\star(t, l) = z^\star + Z_+[t] F(t) l \langle l, F'(t) Z_+[t] F(t) l \rangle^{-\frac{1}{2}}, \tag{5.61}$$

its surface being totally covered by the parameterized family of nonintersecting curves $z^\star(t, l)$ with parameter $l \in \mathbb{R}^2$ and $t \geq 0$. The equality

$$Z[t] = \cap \{\mathcal{E}(z^\star(t), Z_+(t)) | \pi(\cdot)\}$$

will be true, where each ellipsoid

$$\mathcal{E}(z^\star(t), Z_+(t)) = \{z : \Sigma_{i,j=1}^2 \langle z_i - z_i^\star[t], Z_{+ij}[t](z_j - z_j^\star[t]) \rangle \leq 1\}.$$

Here Z_{+ij} are the coefficients of Z_+. Returning to the original coordinates x and taking $x_i^\star(t) = \exp z_i^\star(t)$, we observe that ellipsoids $\mathcal{E}(z^\star(t), Z_+(t))$ transform into star-shaped nonconvex sets

$$S[t] = \{x : \Sigma_{i,j=1}^2 \langle \log x_i - \log x_i^\star[t], Z_{+ij}[t](\log x_j - \log x_j^\star[t]) \rangle \leq 1\},$$

5.3 Applications of Comparison Principle

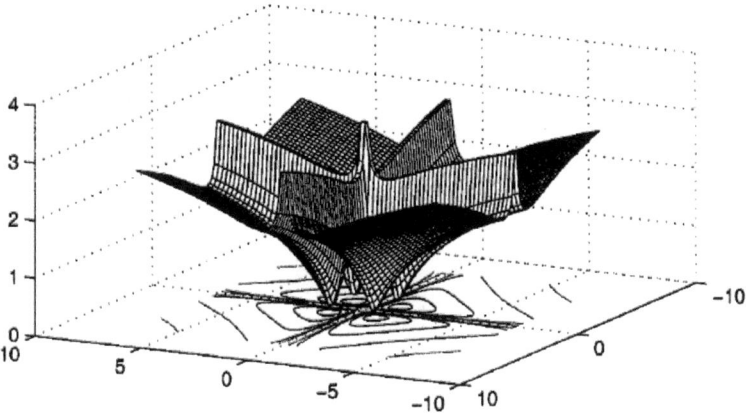

Fig. 5.5 Cost at time $t = 0$

so that now

$$X[t] = \cap \{S[t] | \pi(\cdot)\}. \tag{5.62}$$

The surface of the tube $X[t] = X(t, 0, x^0)$ will be totally covered by the family of nonintersecting curves $(i = 1, 2)$

$$x_i^*(t, l) = \exp(\log(x_i^*[t] + (Z_+[t]F(t)l)_i \langle l, F'(t)Z_+[t]F(t)l \rangle^{-\frac{1}{2}}),$$

where $(h)_i$ stands for the i-th coordinate of vector h.

It is clear from the text that solutions to this example all lie *within the first quadrant* $\{x_1 > 0, x_2 > 0\}$. They are illustrated in Figs. 5.5, 5.6, 5.7, 5.8, 5.9, and 5.10. However, these figures are actually produced for a more general system

$$|\dot{x}_1| = u_1 |x_1| - a_{12}|x_1| \log |x_2|, \ |\dot{x}_2| = u_2 |x_2| - a_{21}|x_2| \log |x_1|, \ |x_1|(0) > 0, |x_2|(0) > 0,$$

which includes Eq. (5.53), if considered within the domain $\{x_1 > 0, x_2 > 0\}$. The pictures for the quadrants $\{x_1 < 0, x_2 > 0\}, \{x_1 < 0, x_2 < 0\}, \{x_1 > 0, x_2 < 0\}$ are symmetrical with the one for $\{x_1 > 0, x_2 > 0\}$, relative to either the coordinate axes or the origin.

Here Figs. 5.5, 5.6, and 5.7 illustrate the exact value function $V(\tau, x)$ for $\tau = 0$, $\tau = 0.5$, $\tau = 1$, respectively. Figure 5.8 shows four reach tubes $X[t]$ each of which originates in one of the four coordinate quadrants. Figures 5.9 and 5.10 show the cuts of the value function (for each of the four reach tubes) surrounded by the boundaries of sets $S[t]$—the preimages of ellipsoidal surfaces, for $\tau = 0.5, \tau = 1$.[6]

---

[6]This example was worked out by O.L. Chucha.

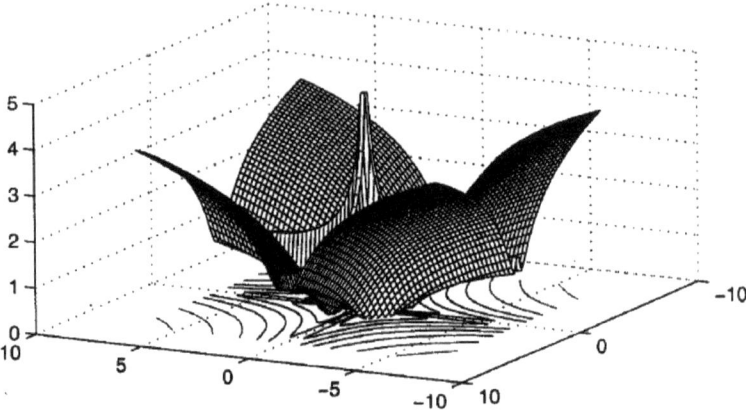

Fig. 5.6 Cost at time t = 0.5

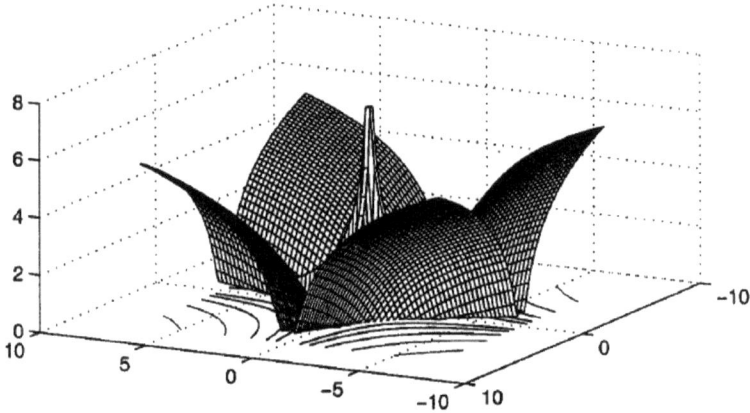

Fig. 5.7 Cost at time t = 1

5.3.4 External Ellipsoids for the Unicycle: Reachability

It is well known that reachability sets for nonlinear systems are typically nonconvex and may have a peculiar form. But in some cases it may suffice to have ellipsoidal estimates of their convex hulls instead. This may be given by an intersection of an array of ellipsoids. Explained here is an example of such a convexifying approximation. Consider the nonlinear system

5.3 Applications of Comparison Principle 223

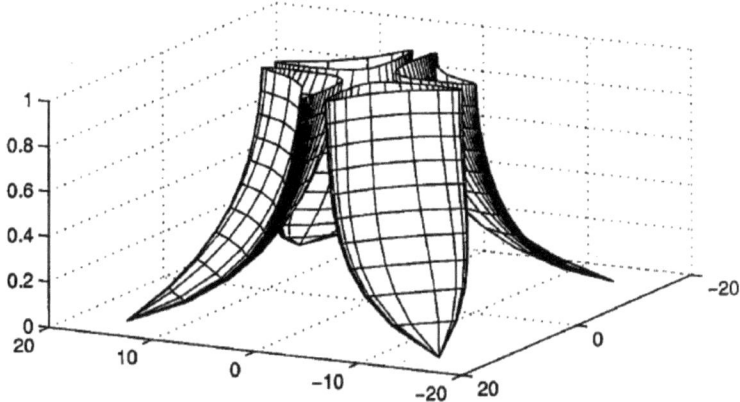

Fig. 5.8 Reach tube at time t=0

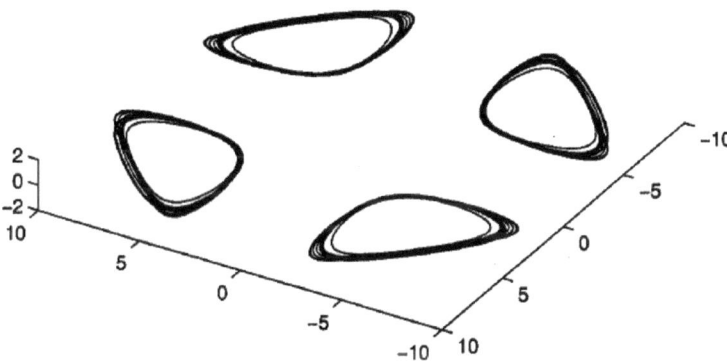

Fig. 5.9 Estimate at time t=0.5

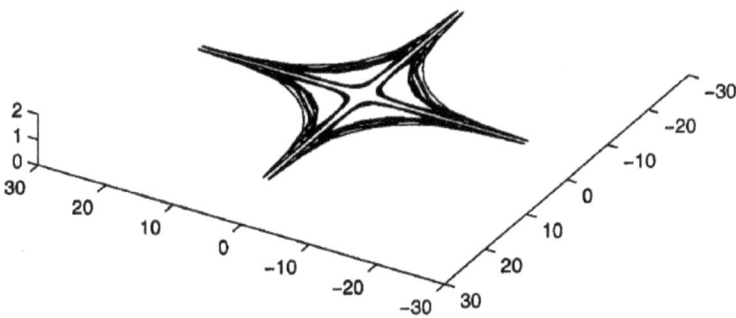

Fig. 5.10 Estimate at time t=1

$$\dot{x}_1 = x_2,$$
$$\dot{x}_2 = r \cos x_5,$$
$$\dot{x}_3 = x_4, \qquad (5.63)$$
$$\dot{x}_4 = r \sin x_5,$$
$$\dot{x}_5 = \alpha u,$$

which describes a controlled planar motion for the "unicycle."

Here u is the control, $r = \text{const} > 0$, $\alpha = \text{const} > 0$, $x^{(0)} = x(t_0) \in X^0 = \mathcal{E}(x^0, X^0)$.

The domain $\mathcal{U}(\cdot)$ of admissible controls u is

$$\mathcal{U}(\cdot) = L^\infty([t_0, t]) \cap \{u : u(t) \in U = [-1, 1]., \ a.e\}. \qquad (5.64)$$

As before, let $X[t] = X(t, t_0, X^0)$ stand for the reachability set of system (5.63):

$$X(t, t_0, X^0) = \{x \mid \exists u(\cdot) \in \mathcal{U}(\cdot) \ : \ x(t_0) = x^{(0)} \in X^0 : x(t, t_0, x^{(0)}) = x\}.$$

Problem 5.3.1. Find a family of external ellipsoidal estimates for the reachability set $X[t]$ of system (5.63) such that for each $t \in [t_0, t_1]$ it touches the convex hull conv$X[t]$ of $X[t]$ at a certain point $\bar{x}(t)$.

To solve this problem we apply Sect. 5.1 of this chapter.

Using the Comparison Principle

Following Sects. 2.1 and 2.3, we introduce the value function:

$$V(t, x) = \inf_{u(\cdot)} \{d\left(x[t_0], X^0\right) \mid x[t] = x\},$$

so that

$$X[t] = \{x \mid V(t, x) \leq 0\}$$

Then the related HJB equation for $V(t, x)$ is as follows:

$$V_t + \max_{|u| \leq 1} \{V_{x_1} x_2 + V_{x_3} x_4 + r V_{x_2} \cos x_5 + r V_{x_4} \sin x_5 + \alpha u V_{x_5}\} = 0. \qquad (5.65)$$

Solving the internal problem, we have

$$V_t + \mathcal{H}(t, x, V_x) = 0, \quad \mathcal{H}(t, x, p) = p_1 x_2 + p_3 x_4 + r p_2 \cos x_5 + r p_4 \sin x_5 + \alpha |p_5| = 0. \qquad (5.66)$$

5.3 Applications of Comparison Principle

To find the necessary approximation for $\text{conv}\,X[t]$ we apply the comparison principle (Theorem 5.1.1). Let $w^+(t,x) \in C_1$, $\mu(t) \in L_1$, $\mathcal{H}(t,x,p)$ satisfy conditions

$$\mathcal{H}(t,x,p) \le H(t,x,p), \quad \forall\, \{t,x,p\},$$

$$w_t^+(t,x) + H\left(t, x, w_x^+(t,x)\right) \le \mu(t),\ \forall t \ge t_0. \tag{5.67}$$

Then the next estimate is true.

$$X[t] \subseteq X_+[t], \tag{5.68}$$

where

$$X_+[t] = \left\{ x \,\bigg|\, w^+(t,x) \le \int_{t_0}^t \mu(\tau)d\tau + \max_{x \in X^0} w^+(t_0, x) \right\}. \tag{5.69}$$

We now parameterize the family of functions $w^+(t,x)$ by the pair $\{\bar{x}(\cdot), \bar{p}(\cdot)\}$ which is the solution to the *characteristic ODE system* for our HJB PDE equation (5.66) [51]. This ODE is written as

$$\begin{aligned}
\dot{x}_1 &= x_2, & \dot{p}_1 &= 0, \\
\dot{x}_2 &= r\cos x_5, & \dot{p}_2 &= -p_1, \\
\dot{x}_3 &= x_4, & \dot{p}_3 &= 0, \\
\dot{x}_4 &= r\sin x_5, & \dot{p}_4 &= -p_2, \\
\dot{x}_5 &= \alpha \bar{u}, & \dot{p}_5 &= r(p_4 \cos x_5 - p_2 \sin x_5),
\end{aligned}$$

where $\bar{u} = \text{sgn}(p_5)$ with $p_5 \ne 0$ and $\bar{u} \in [-1, 1]$ with $p_5 = 0$. We will look for such ellipsoidal approximations $X_+[t]$ of reachability set $X[t]$, that for each t would touch $\text{conv}\,X[t]$ at point $\bar{x}(t)$. We further omit the argument "t" in $\bar{x}(t)$, $\bar{p}(t)$, denoting them as $\bar{x}$, $\bar{p}$.

For arbitrary $h > 0$, λ_1, λ_2 the next inequality is true

$$p_2 \cos x_5 + p_4 \sin x_5 \le \frac{1}{2}h^{-1} \langle p, \left(e_2 e_2' + e_4 e_4'\right) p\rangle +$$

$$+ h^{-1}\langle \lambda_1 e_2 + \lambda_2 e_4, p\rangle + \lambda_1 \cos x_5 + \lambda_2 \sin x_5 + \gamma.$$

Here $\lambda_1 = \bar{p}_2 - r_1 \cos \bar{x}_5$, $\lambda_2 = \bar{p}_4 - r_1 \sin \bar{x}_5$, $\gamma = \frac{1}{2}h + \frac{1}{2}h^{-1}(\lambda_1^2 + \lambda_2^2)$, $[e_1, e_2, e_3, e_4, e_5] = I$, with equality reached at $(\bar{x}, \bar{p})$. Also true is inequality

$$\cos \phi \le a(\phi - \bar{\phi})^2 + 2b(\phi - \bar{\phi}) + c,$$

where

$$a = \begin{cases} -\sin\bar{\phi}/(2(\bar{\phi}-\pi)), & \bar{\phi} \neq \pi \\ \frac{1}{2}, & \bar{\phi} = \pi \end{cases}, \quad b = a(\bar{\phi}-\pi), \quad c = \cos\bar{\phi}, \quad \bar{\phi} \in [0, 2\pi]$$

with equality at $\phi = \bar{\phi}$. Upper estimates for functions $-\cos\phi$, $\sin\phi$, $-\sin\phi$ may be obtained similarly. Applying them, we come to

$$p_2 \cos x_5 + p_4 \sin x_5 \leq \frac{1}{2}h^{-1}\langle p, (e_2 e'_2 + e_4 e'_4)\,p\rangle + h^{-1}\langle \lambda_1 e_2 + \lambda_2 e_4, p\rangle +$$

$$+ \langle x - \bar{x}, (\lambda_1 a_1 + \lambda_2 a_2)e_5 e'_5 (x-\bar{x})\rangle + 2\langle \lambda_1 b_1 + \lambda_2 b_2)e_5, x - \bar{x}\rangle + \lambda_1 c_1 + \lambda_2 c_2 + \gamma,$$

with equality at $x = \bar{x}$, $p = \bar{p}$.

Using this relation, we find an upper estimate $H(t, x, p)$ for $\mathcal{H}(t, x, p)$:

$$\mathcal{H}(t, x, p) = p_1 x_2 + p_3 x_4 + r p_2 \cos x_5 + r p_4 \sin x_5 + \alpha |p_5| \leq$$

$$\leq \langle p, A(x - x^*)\rangle + \langle p, Bp\rangle + \langle x - x^*, C(x - x^*)\rangle +$$

$$+ 2\langle f, x - x^*\rangle + 2\langle C(x^* - \bar{x}), x - x^*\rangle + \langle g, p\rangle + \langle Ax^*, p\rangle + \mu = H(t, x, p),$$

with equality reached at $x = \bar{x}$, $p = \bar{p}$ for any function $x^* = x^*(t)$. Here

$$A = e_1 e'_2 + e_3 e'_4, \quad B = \frac{1}{2}rh^{-1}(e_2 e'_2 + e_4 e'_4) + \frac{1}{2|\bar{p}_5|}\alpha e_5 e'_5, \quad C = r(\lambda_1 a_1 + \lambda_2 a_2)e_5 e'_5$$

$$f = r(\lambda_1 b_1 + \lambda_2 b_2)e_5, \quad g = rh^{-1}(\lambda_1 e_2 + \lambda_2 e_4),$$

$$\mu = \frac{1}{2}\alpha|\bar{p}_5| + \langle x^* - \bar{x}, C(x^* - \bar{x})\rangle + 2\langle f, x^* - \bar{x}\rangle + r(\lambda_1 c_1 + \lambda_2 c_2 + \gamma).$$

Now, for obtaining an ellipsoidal estimate we take $w^+(t, x)$ as a quadratic form:

$$w^+(t, x) = \langle x - x^*(t), K(t)(x - x^*(t))\rangle$$

and write down for functions $w^+(t, x)$ and $\mathcal{H}(t, x, p)$ the inequality (5.67):[7]

$$\langle x - x^*(t), \dot{K}(t)(x - x^*(t))\rangle - 2\langle x - x^*(t), K(t)\dot{x}^*(t)\rangle +$$

$$+ \langle x - x^*(t), (K(t)A + A'K(t) + 4B(t) + C(t))(x - x^*(t))\rangle +$$

$$+ 2\langle (K(t)g(t) + C(t)(x^* - \bar{x}) + KAx^* + f(t), x - x^*(t)\rangle + \mu(t) \leq \mu(t).$$

---

[7]This example was worked out by V.V.Sinyakov.

5.3 Applications of Comparison Principle

In order that this inequality would be true *for any* $(x - x^*(t))$ it is sufficient to have the next equations satisfied:

$$\dot{K}(t) + K(t)A + A'K(t) + 4KB(t)K + C(t) = 0, \tag{5.70}$$

$$\dot{x}^*(t) = (A + K^{-1}(t)C(t))x^*(t) + K^{-1}(t)(f(t) - C(t)\bar{x}(t)) + g(t). \tag{5.71}$$

From initial condition

$$\max_{x \in X^0} w^+(t_0, x) = 1$$

we now find such conditions for Eqs. (5.70), (5.71), namely:

$$K(t_0) = \left(X^0\right)^{-1}, \quad x^*(t_0) = x^0.$$

Substituting K for its inverse $\mathcal{K}$ as

$$\mathcal{K}(t) = K^{-1}(t), \quad \dot{\mathcal{K}}(t) = -\mathcal{K}(t)\dot{K}(t)\mathcal{K}(t),$$

then multiplying both parts of (5.70) by matrix $K^{-1}(t)$ we get

$$\dot{\mathcal{K}}(t) = A\mathcal{K}(t) + \mathcal{K}(t)A' + \mathcal{K}(t)C(t)\mathcal{K}(t) + 4B(t), \quad \mathcal{K}(t_0) = X^0,$$

$$\dot{x}^*(t) = (A + \mathcal{K}(t)C(t))x^*(t) + \mathcal{K}(t)(f(t) - C(t)\bar{x}(t)) + g(t), \quad x^*(t_0) = x^0.$$

If under specific selection of $\{\bar{x}(\cdot), \bar{p}(\cdot)\}$ the solution to the last equations exists, then we may introduce notation

$$\mathcal{K}_+(t) = \left(1 + \int_{t_0}^{t} \mu(s)ds\right)\mathcal{K}(t),$$

coming to the inclusion

$$X[t] \subseteq X_+[t] = \mathcal{E}(x^*(t), \mathcal{K}_+(t)) = \left\{x : \langle x - x^*(t), \mathcal{K}_+^{-1}(t)(x - x^*(t))\rangle \leq 1\right\},$$

where $X_+[t]$ touches $X[t]$ at point $\bar{x}[t]$ (Figs. 5.11 and 5.12).

228 5 The Comparison Principle: Nonlinearity and Nonconvexity

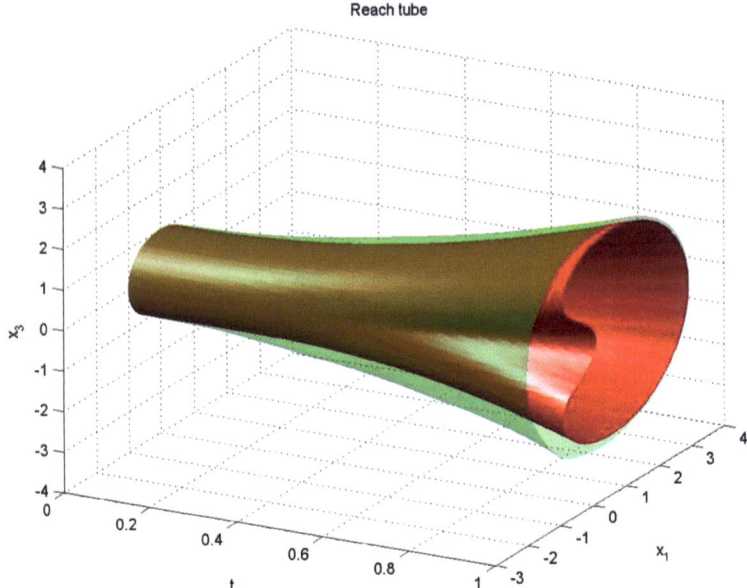

Fig. 5.11 The external ellipsoidal-valued estimate for the reachability tube (1 ellipsoid)

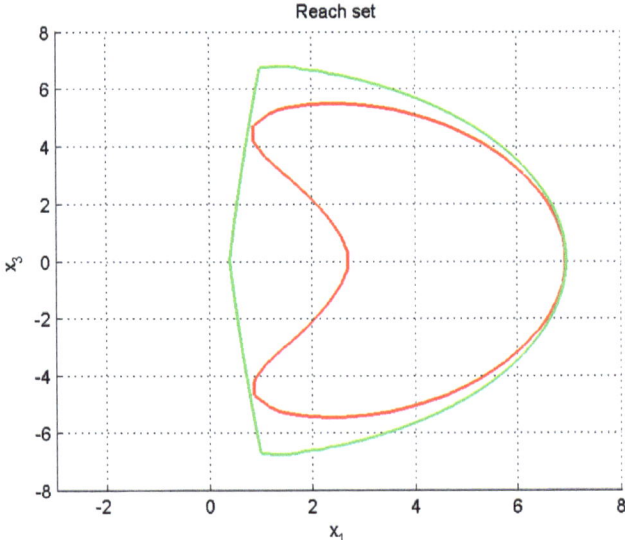

Fig. 5.12 The external ellipsoidal estimate for the reachability set at time $t_1 = 1$ (3 ellipsoids)

5.4 Ellipsoidal Methods for Non-ellipsoidal Constraints

In the previous texts the main types of hard bounds on system parameters were taken as ellipsoidal-valued. However one of the conventional types of such bounds are also polyhedrons which led to specific algorithms based, for example, on linear programming and related constructions. A detailed theory of approximating reachability sets and control solutions for linear-convex systems by parallelotopes, including the rectangular case (boxes), was developed by Kostousova [115]. However, in this section, we indicate that symmetrical polyhedrons may be well treated by ellipsoidal approximations, with reasonable computational burden, especially if one involves parallel calculation.

5.4.1 Degenerate Ellipsoids: Box-Valued Constraints

Let us now assume that system (5.12) is subjected to hard bounds of the "box" type, namely,

$$u(t) \in \mathcal{P}(t), \quad x(t_0) \in \mathcal{X}^0,$$

where

$$\mathcal{P}(t) = \{u \in \mathbb{R}^m : |u_i - u_i^0(t)| \leq \mu_i(t)\}, \quad \mu_i(t) \geq 0, \tag{5.72}$$

$$\mathcal{X}^0 = \{x \in \mathbb{R}^n : |x_j - x_j^0| \leq \nu_j\}, \quad \nu_j \geq 0, \; i = \{1,\ldots,m\}, \; j = \{1,\ldots,n\},$$

and $u_i^0(t), x_j^0$ are given.

Will it be possible to use ellipsoidal approximations for the respective reach sets now, that $\mathcal{P}, \mathcal{X}^0$ are not ellipsoids? To demonstrate that this is indeed possible, we proceed as follows.

Let us define a box $\mathcal{P}$ with center p as $\mathcal{P} = \mathcal{B}(p, P)$ where $P = \{p^{(1)}, \ldots, p^{(m)}\}$ is an array of m vectors ("directions") $p^{(i)}$ such that

$$\mathcal{B}(p, P) = \{x : x = p + \sum_{i=1}^{m} p^{(i)}\alpha_i, \; \alpha_i \in [-1, 1]\}.$$

Then box $\mathcal{P}(t)$ of (5.72) may be presented as $\mathcal{P}(t) = \mathcal{B}(u^0(t), P(t))$, where $P = \{p^{(1)}, \ldots, p^{(m)}\}$, with $p^{(i)} = \mathbf{e}^{(i)}\mu_i(t)$ and $\mathbf{e}^{(i)}$ is a unit orth oriented along the axis $0x_i$. Box $\mathcal{B}(u^0(t), P(t))$ is a rectangular parallelepiped. A linear transformation T of box $\mathcal{B}(p, P)$ will give

$$T\mathcal{B}(p, P) = \mathcal{B}(Tp, TP).$$

Thus, in general, box $\mathcal{B}(Tu^0(t), TP(t))$ will not be rectangular. Let us now approximate a box by a family of ellipsoids. Taking set $\mathcal{B}(0, P)$, we may present it as the sum of m degenerate ellipsoids $\mathcal{E}(0, Q_{ii})$, where

$$Q_{ii} = \mathbf{e}^{(i)}\mathbf{e}^{(i)'}q_{ii}, \quad q_{ii} = \mu_i^2.$$

Here Q_{ii} is an $m \times m$ diagonal matrix with diagonal elements $q_{kk} = 0$, $k \neq i$, $q_{ii} = \mu_i^2$ (its only nonzero element is $q_{ii} = \mu_i^2$).

Then

$$\mathcal{B}(0, P) = \sum_{i=1}^{m} \mathcal{E}(0, Q_{ii}) \subseteq \mathcal{E}(0, Q(p)),$$

where $p = \{p_1, \ldots, p_m\}$ and

$$Q(p) = \left(\sum_{i=1}^{m} p_i\right)\left(\sum_{i=1}^{m} p_i^{-1} Q_{ii}\right), \quad p_i > 0. \tag{5.73}$$

Remark 5.4.1. These relations were usually applied to nondegenerate ellipsoids, (see, [174], and Sect. 2.7 of this book). However, on applying Lemma 3.2.1 of [174], one may observe that *they are also true for the degenerate case.*

Indeed, given vector $l \in \mathbb{R}^m$, take $p = \{p_1, \ldots, p_m\}$ as

$$p_i = |l_i|\mu_i \text{ if } l_i \neq 0, \quad p_i = \epsilon m^{-1}||l|| \text{ if } l_i = 0, \quad ||l||^2 = \sum_{i=1}^{m} l_i^2 \tag{5.74}$$

Here $|l_i| = \sqrt{l_i^2}$, so

$$\rho(l|\mathcal{E}(0, Q(p))) = \langle l, Q(p)l \rangle^{1/2} \leq \sum_{i=1}^{m} |l_i|\mu_i + \epsilon||l||, \quad \sum_{i=1}^{m} |l_i|\mu_i = \rho(l|\mathcal{B}(0, P))$$
$$\tag{5.75}$$

and

$$\rho(l|Q(p)) - \rho(l|\mathcal{B}(0, P)) \leq \epsilon||l||. \tag{5.76}$$

Note that with $\varepsilon > 0$ matrix $Q(p)$ is always nondegenerate. However, if we allow $\epsilon = 0$, then, with p taken as in (5.74), relation (5.76) will turn into an equality. Now let m^* be the number of nonzero coordinates of vector p in $Q(p)$.

5.4 Ellipsoidal Methods for Non-ellipsoidal Constraints

Then we observe that with $m = m^*$ the matrix $Q(p)$ is nondegenerate. However, with $m^* < m$, $Q(p)$ will turn out to be degenerate and set $\mathcal{E}(0, Q(p))$ will be an elliptical cylinder.[8]

Theorem 5.4.1. *(i) An external ellipsoidal approximation*

$$\mathcal{B}(0, P) \subseteq \mathcal{E}(0, Q(p))$$

is ensured by ellipsoid $\mathcal{E}(0, Q(p))$, where $Q(p)$ is given by (5.73).
(ii) With p selected due to (5.74), the inequality (5.76) will be true and if one takes $\epsilon = 0$ in (5.74), then (5.76) turns into an equality. However, with $\epsilon = 0$, $m^* < m$ the ellipsoid $\mathcal{E}(0, Q(p))$ becomes degenerate (an elliptical cylinder).

A similar approximation is true for box $\mathcal{B}(0, X) = X^0$.

Lemma 5.4.1. *Under a linear transformation T we have*

$$T\mathcal{B}(0, \mathbf{Q}) = \mathcal{B}(0, T\mathbf{Q}) \subseteq \mathcal{E}(0, TQ(p)T') \tag{5.77}$$

This follows directly from the above.

5.4.2 Integrals of Box-Valued Functions

Consider a set-valued integral

$$\int_{t_0}^{\tau} \mathcal{B}(0, B(t)P(t))dt \tag{5.78}$$

and a partition $\sigma[N]$ similar to the one of Sect. 2.2. As before, the $m \times m$ diagonal matrix $P(t)$ and $n \times m$ matrix $B(t)$ are continuous.
Then

$$\int_{t_0}^{\tau} \mathcal{B}(0, B(t)P(t))dt = \lim \sum_{i=1}^{N} \sum_{j=1}^{m} \mathcal{E}(0, B(t_j)Q_{jj}(t_j)B'(t_j))\sigma_i$$

with $N \to \infty$ we have $\sigma[N] = \max\{\sigma_i\} \to 0$. Applying again the formula for the external ellipsoidal approximation of the sum of ellipsoids, we have

---

[8] In the next subsection related to zonotopes we indicate a regularization procedure that ensures a numerical procedure that copes with such degeneracy. This is done by substituting cylinders for far-stretched ellipsoids along degenerate coordinates.

$$\sum_{i=1}^{N}\sum_{j=1}^{m} \mathcal{E}(0, B(t_j)Q_{jj}(t_j)B'(t_j))\sigma_i \subseteq \mathcal{E}(0, X_{+N}(p_N[\cdot])),$$

$$X_{+N}(p_N(\cdot)) = \left(\sum_{i=1}^{N}\sum_{j=1}^{m} p_{jj}(t_i)\right)\left(\sum_{i=1}^{N}\sum_{j=1}^{m} p_{jj}^{-1}(t_i)Q_{jj}(t_i)\right), \quad p_{jj}(t_i) > 0.$$

Here $p_N[\cdot] = \{p_{jj}(t_i) \mid j = 1, \ldots, m, \ i = 1, \ldots, N\}$.

Taking $p_{jj}(t_i)$ to be the values of continuous functions $p_{jj}(t)$, $j = 1, \ldots, m$, and passing to the limit in the previous relation, with $N \to \infty$, $\sigma[N] \to 0$, we come to the next conclusion.

Lemma 5.4.2. *The following inclusion is true*

$$\int_{t_0}^{\tau} \mathcal{B}(0, B(t)P(t))dt \subseteq \mathcal{E}(0, X_{+}(\tau, p[\cdot])) \qquad (5.79)$$

$$X_{+}(p[\cdot]) = \sum_{j=1}^{n}\left(\int_{t_0}^{\tau} p_{jj}(t)dt\right)\left(\sum_{j=1}^{n}\int_{t_0}^{\tau} p_{jj}^{-1}(t)B(t)Q_{jj}(t)B'(t)dt\right)$$

for any continuous functions $p_{jj}(t) > 0$.

Here $p[\cdot] = \{p_{jj}(\cdot) \mid j = 1, \ldots, m, \ t \in [t_0, \tau]\}$. For a nonrectangular box $T(t)\mathcal{B}(0, P(t)) = \mathcal{B}(0, T(t)P(t))$ and a nonzero box $T_0\mathcal{B}(0, X^0) = \mathcal{B}(0, T_0 X^0)$ in a similar way we have

Theorem 5.4.2. *The following inclusion is true*

$$X[\tau] = \mathcal{B}(0, T_0 X^0) + \int_{t_0}^{\tau} \mathcal{B}(0, T(t)B(t)P(t))dt \subseteq \mathcal{E}(0, X_{+}(\tau, p[\cdot])) \qquad (5.80)$$

where

$$X_{+}(\tau, p[\cdot]) = \left(\sum_{k=1}^{n} p_{kk}^{(0)} + \sum_{j=1}^{m}\int_{t_0}^{\tau} p_{jj}(t)dt\right) \times \qquad (5.81)$$

$$\times \left(\sum_{j=1}^{n} p_{kk}^{(0)-1} T_0 X_{kk}^0 T_0' + \sum_{j=1}^{m}\int_{t_0}^{\tau} p_{jj}^{-1}(t)T(t)B(t)Q_{jj}(t)B'(t)T'(t)dt\right).$$

Here $p[\cdot] = \{p_{kk}^{(0)}, p_{jj}(\cdot) \mid k = 1, \ldots, n; \ j = 1, \ldots, m, \ t \in [t_0, \tau]\}$.

5.4 Ellipsoidal Methods for Non-ellipsoidal Constraints

In order that an equality

$$\rho(l \mid X[\tau]) = \rho(l|\mathcal{B}(0, T_0 X^0)) + \int_{t_0}^{\tau} \rho(l|\mathcal{B}(0, T(t)B(t)P(t)))dt = \rho(l|\mathcal{E}(0, X_+(\tau, p[\cdot]))) \tag{5.82}$$

would be possible for a given $l \in \mathbb{R}^n$, we would formally have to choose $X_+(\tau, p[\cdot])$ taking

$$p_{jj}(t) = \langle l, T(t)B(t)Q_{jj}(t)B'(T)T'(t)l \rangle^{1/2}, \quad p_{kk}^{(0)} = \langle l, T_0 X_{kk}^0 T_0' l \rangle^{1/2}. \tag{5.83}$$

But a nondegenerate matrix $X_+(\tau, p[\cdot])$ would be possible only if $p_{jj}(t) \neq 0$ almost everywhere and $p_{kk}^{(0)} \neq 0$. The equality is then checked by direct calculation.

Lemma 5.4.3. *In order that for a given $l \in \mathbb{R}^n$ there would be an equality (5.82), it is necessary and sufficient that $p_{jj}(t)$, $p_{kk}^{(0)}$ would be selected according to (5.83) and both of the conditions $p_{jj}(t) \neq 0$ almost everywhere and $p_{kk}^{(0)} \neq 0$ would be true.*

Otherwise, either an equality (5.82) will still be ensured, but with a degenerate $\mathcal{E}(0, X_+(\tau, p[\cdot]))$, or, for any ϵ given in advance, an inequality

$$\rho(l|\mathcal{E}(0, X_+(\tau, p[\cdot]))) - \rho(l|\mathcal{B}(0, T_0 X^0)) - \int_{t_0}^{\tau} \rho(l|\mathcal{B}(0, T(t)B(t)P(t)))dt \leq \epsilon \|l\| \tag{5.84}$$

may be ensured with a nondegenerate $\mathcal{E}(0, X_+(\tau, p[\cdot]))$. This may be done by selecting

$$p[\cdot] = p^\epsilon[\cdot] = \{p_{kk}^{(0)}, p_{jj}(\cdot) \mid k = 1, \ldots, n; \ j = 1, \ldots, m, \ t \in [t_0, \tau,]\}$$

as

$$p_{jj}^{(\epsilon)}(t) = \langle l, T(t)B(t)Q_{jj}(t)B'(t)T'(t)l \rangle^{1/2} + \frac{\epsilon \|l\|}{2m(\tau - t)},$$

$$p_{kk}^{(0\epsilon)} = \langle l, T_0 X_{kk}^0 T_0' l \rangle^{1/2} + \frac{\epsilon \|l\|}{2n}.$$

It may be useful to know when $p_{jj}(t) \neq 0$ almost everywhere.

Lemma 5.4.4. *In order that $p_{jj}(t) = \langle l, T(t)B(t)Q_{jj}(t)B'(t)T'(t)l \rangle^{1/2} \neq 0$ almost everywhere, for all $l \in \mathbb{R}^n$, it is necessary and sufficient that functions $T(t)B(t)\mathbf{e}^{(j)}$ would be linearly independent. (The j-th column of $T(t)B(t)$ would consist of linearly independent functions.)*

This follows from the definition of linearly independent functions. Note that with $\epsilon = 0$ we have

$$\left(\sum_{k=1}^{n} p_{kk}^{(0)} + \sum_{j=1}^{m} \int_{t_0}^{\tau} p_{jj}(t)dt\right) = \langle l, X_{+}(\tau, p[\cdot])l\rangle^{1/2}, \qquad (5.85)$$

$$\sum_{k=1}^{n} p_{kk}^{(0)} = \langle l, T_0 X^0 T_0'l\rangle^{1/2} \sum_{k=1}^{n} \langle l, T_0 X_{kk}^0 T_0'l\rangle^{-1/2}.$$

The parameters of the ellipsoid $\mathcal{E}(0, X_{+}(\tau, p[\cdot]))$ may be expressed through a differential equation. Taking $X_{+}[\tau] = X_{+}(\tau, p[\cdot])$ and differentiating it in τ, we get

$$\dot{X}_{+}[\tau] = \left(\sum_{j=1}^{m} p_{jj}(\tau)\right)\left(\sum_{k=1}^{n} p_{kk}^{(0)}\right)^{-1} T_0 X_{kk}^0 T_0'$$

$$+ \sum_{j=1}^{m} \int_{t_0}^{\tau} p_{jj}^{-1}(t) T(t) B(t) Q_{jj}(t) B'(t) T'(t) dt\right)$$

$$+ \left(\sum_{k=1}^{n} p_{kk}^{(0)} + \sum_{j=1}^{m} \int_{t_0}^{\tau} p_{jj}(t)dt\right)\left(\sum_{j=1}^{m} p_{jj}^{-1}(\tau) T(\tau) B(\tau) Q_{jj}(\tau) B'(\tau) T'(\tau)\right)$$

Denoting

$$\pi_j(\tau) = p_{jj}(\tau)\left(\sum_{k=1}^{n} p_{kk}^{(0)} + \sum_{j=1}^{m} \int_{t_0}^{\tau} p_{jj}(t)dt\right)^{-1} = \mu_j |l_j| \langle l, X_{+}[\tau]l\rangle^{-1/2},$$

$$\pi_k^{(0)} = p_{kk}^{(0)}\left(\sum_{k=1}^{n} p_{kk}^{(0)}\right)^{-1}, \quad X_{+}^0 = \langle l, T_0 X^0 T_0'l\rangle^{1/2} \sum_{k=1}^{n} \langle l, T_0 X_{kk}^0 T_0'l\rangle^{-1/2} T_0 X_{kk}^0 T_0'$$

and rearranging the coefficients similarly to Sect. 3.3, we come to

$$\dot{X}_{+}[\tau] = \left(\sum_{j=1}^{m} \pi_j(\tau)\right) X_{+}[\tau] + \sum_{j=1}^{m} \pi_j^{-1}(\tau) T(\tau) B(\tau) Q_{jj}(\tau) B'(\tau) T'(\tau), \quad X_{+}[t_0] = X_{+}^0.$$
(5.86)

Remark 5.4.2. If boxes $\mathcal{B}(p(t), P(t))$, $\mathcal{B}(x^0, X^0)$ have nonzero centers $p(t)$, then all the previous relations hold, with centers changed from 0 to $x^0(t)$, where

$$\dot{x}^0 = B(t)p(t), \quad x(t_0) = x^0,$$

so that $\mathcal{E}(0, X_+(\tau, p[\cdot]))$ turns into $\mathcal{E}(x^0(\tau), X_+(\tau, p[\cdot]))$.

Theorem 5.4.3. *(i) The matrix $X_+(\tau, p[\cdot])$ of the external ellipsoid $\mathcal{E}(x^0(\tau), X_+(\tau, p[\cdot]))$ that ensures inclusion (5.80) satisfies the differential equation and initial condition (5.86).*
(ii) The ellipsoid $\mathcal{E}(x^0(\tau), X_+(\tau, p[\cdot]))$ ensures the equality (5.82) ($\mathcal{E}(x^0(\tau), X_+(\tau, p[\cdot]))$ touches set $X[\tau]$ of (5.80) along direction l), if parameters $p[\cdot]$ are selected as in (5.83) and $\pi_j(t)$ are defined for all $t \in [t_0, \tau]$, with $\pi_k^{(0)} \neq 0$. Then $\mathcal{E}(x^0(\tau), X_+(\tau, p[\cdot]))$ is nondegenerate for any l.
(iii) In order that $\mathcal{E}(x^0(\tau), X_+(\tau, p[\cdot]))$ would be nondegenerate for all l (with box $\mathcal{B}(x^0, X^0) = \{x^0\}$ being a singleton), it suffices that functions $T(t)B(t)e^{(j)}$ would be linearly independent on $[t_0, t_1]$.
(iv) In general, for any given ϵ, selecting $\pi_j^{(\epsilon)}(t)$, $\pi_k^{(0\epsilon)}$ similarly to $\pi_j(t), \pi_k^{(0)}$, but with $p_{jj}(t)$, $p_{kk}^{(0)}$ substituted for $p_{jj}^{(\epsilon)}(t)$, $p_{kk}^{(0\epsilon)}$, one is able to ensure the inequality (5.84).

We may now proceed with the approximation of reach sets for system (5.12).

5.4.3 Reachability Tubes for Box-Valued Constraints: External Approximations

Consider system (5.12) under box-valued constraints (5.72). Its reach set will be $\mathcal{X}^*[t] = G(t, t_0)\mathcal{X}[t]$, where

$$\mathcal{X}[t] = \mathcal{B}(x^0, X^0) + \int_{t_0}^{t} G(t_0, s) B(s) \mathcal{B}(u^0(s), P(s)) ds. \tag{5.87}$$

Let us first apply the results of the previous section to the approximation of $\mathcal{X}[t]$. Taking $T(s) = G(t_0, s)$, $T_0 = I$, we have

$$\mathcal{X}[t] \subseteq \mathcal{E}(G(t_0, t)x^0(t), X_+[t])),$$

where

$$\dot{X}_+[t] = \left(\sum_{j=1}^{m} \pi_j(t) \right) X_+[t] + \sum_{j=1}^{m} \pi_j^{-1}(\tau) G(t_0, t) B(t) Q_{jj}(t) B'(t) G'(t_0, t),$$

$$\tag{5.88}$$

with initial condition

$$X_+[t_0] = \sum_{k=1}^{n} (\pi_k^{(0)})^{-1} X_{kk}^0, \tag{5.89}$$

and with $x^0(t)$ evolving due to equation

$$\dot{x}^0 = A(t)x^0 + B(t)u^0(t), \quad x(t_0) = x^0. \tag{5.90}$$

Further on, denoting $X_+^*[t] = G(t,t_0)X_+[t]G'(t,t_0)$, we obtain

$$X_+^*[t] \subseteq \mathcal{E}(x^0(t), G(t,t_0)X_+[t]G'(t,t_0)) = \mathcal{E}(x^0(t), X_+^*[t]), \tag{5.91}$$

where now $X_+^0 = X_+^*[t_0]$

$$\dot{X}_+^*[t] = A(t)X_+^* + X_+^*A'(t) + \left(\sum_{j=1}^m \pi_j(t)\right)X_+[t] + \sum_{j=1}^m \pi_j^{-1}(t)B(t)Q_{jj}(t)B'(t). \tag{5.92}$$

Theorem 5.4.4. *The inclusion (5.91) is true, whatever be the parameters $\pi_j(t) > 0$, $\pi_k > 0$ of Eq. (5.92).*

Let us now presume that Assumption 3.3.1 of Chap. 3 is fulfilled, so that vector function $l(t)$ along which we would like to ensure the tightness property is taken as $l(t) = G(t_0, t)l$, $l \in \mathbb{R}^n$. Then, following the schemes of Sect. 3.8 of Chap. 3, we come to the next results.

Theorem 5.4.5. *Under Assumption 3.3.1 of Chap. 3, in order that equality*

$$\rho(l|\mathcal{X}[t]) = \rho(l|\mathcal{E}(x^0(t), X_+(t, p[\cdot]))$$

would be true for a given "direction" l, the external ellipsoids $\mathcal{E}(x^0(t), X_+(t, p[\cdot]))$ should be taken with

$$\pi_j(t) = \frac{\langle l, G(t_0,t)B(t)Q_{jj}(t)B'(t)G'(t_0,t)l\rangle^{1/2}}{\langle l, X_+^*[t]l\rangle^{1/2}}, \tag{5.93}$$

$$X_+^{*0} = \left(\sum_{k=1}^n v_k|l_k|\right)\left(\sum_{k=1}^n (v_k|l_k|)^{-1}X_{kk}^0\right), \tag{5.94}$$

provided $\pi_j(t) > 0$ almost everywhere and $|l_k| > 0$, $k = \{1, \ldots, m\}$.

Otherwise, the ellipsoids $\mathcal{E}(x^0(t), X_+(t, p[\cdot]))$ have to be taken for any given $\epsilon > 0$ with parameters

$$\pi_j^\epsilon(t) = \frac{\langle l, G(t_0,t)B(t)Q_{jj}(t)B'(t)G'(t_0,t)l\rangle^{1/2}}{\langle l, X_+^*[t]l\rangle^{1/2} + \epsilon\|l\|}, \tag{5.95}$$

$$X_+^{*0\epsilon} = \left(\sum_{k=1}^n v_k|l_k| + \epsilon\|l\|\right)\left(\sum_{j=1}^m (v_j|l_j|)^{-1}X_{jj}^0\right), \tag{5.96}$$

5.4 Ellipsoidal Methods for Non-ellipsoidal Constraints

instead of $\pi_j(t)$, X_+^{*0}. The inequality

$$\rho(l \mid \mathcal{E}(x^0(t), X_+(t, p[\cdot]))) - \rho(l \mid X[t]) \leq \epsilon \|l\|$$

will then be true.

The exact reach set $\mathcal{X}^*[t]$ is the sum of two sets:

$$\mathcal{X}^*[t] = \mathcal{X}_0^*[t] + \mathcal{X}_u^*[t],$$

where $\mathcal{X}_0^*[t] = G(t, t_0)\mathcal{B}(x^0, X^0)$ is a (nonrectangular) box and

$$\mathcal{X}_u^*[t] = \int_{t_0}^{t} G(t,s) B(s) \mathcal{B}(u^0, P(s)) ds,$$

is a convex compact set. Set $\mathcal{X}_0^*[t]$ cannot be exactly approximated by nondegenerate ellipsoids (see Fig. 5.13), while set $\mathcal{X}_u^*[t]$ may be represented exactly by nondegenerate ellipsoids under condition (iii) of Theorem 5.4.3. Let us reformulate this condition.

Theorem 5.4.6. *In order that for any $l \in \mathbb{R}^n$ an equality*

$$\rho(l \mid \mathcal{X}_u^*[t]) = \rho(l \mid \mathcal{E}(0, X_+^*[t]))$$

would be possible for an appropriately selected ellipsoid $\mathcal{E}(0, X_+^*[t])$, it is sufficient that the pair $\{A(t), e^k\}$ would be completely controllable for any $k = 1, \ldots, m$. Then $X_+^*[t]$ will be correctly defined when described by Eq. (5.92), with parameters $\pi_j^*(t)$ and initial condition $X_+^*[t_0] = X_+^{*0} = 0$, selected due to (5.93), (5.94).

This follows from the definition of complete controllability (see Sect. 1.3). Under such condition the boundary of set $X_+^*[t]$ will have no faces ("platforms") and the set can be totally described by "tight" ellipsoids, as in Chap. 3, Sects. 3.7 and 3.9 (see Fig. 3.1).

Finally, a parametric presentation of set $\mathcal{X}^*[t]$, similar to (5.92), (5.93) can be produced. Then

$$x^*(t) = x^0(t) + X_+^*[t] l^* \langle l^*, X_+^*[t] l^* \rangle^{1/2}, \tag{5.97}$$

with

$$x^*(t_0) = x^0 + X_+^0 l \langle l, X_+^0 l \rangle^{1/2}, \quad l^*(t) = G(t_0, t))' l \tag{5.98}$$

with either $\pi_j(t)$, π_k^0 or $\pi_j^\epsilon(t)$, $\pi_k^{(0\epsilon)}$ selected as indicated in Theorem 5.4.5. This results in an array of external ellipsoids of either type $\mathcal{E}(x^0(t), X_+^*[\tau])$ which yields an equality similar to (5.82), namely

$$\rho(l^* \mid X[\tau]) = \rho(l^* \mid \mathcal{E}(x^0, X_+^*[\tau]) = \langle l^*, x^*(\tau) \rangle, \tag{5.99}$$

or type $\mathcal{E}(x^0(t), X_+^{*\epsilon}[t])$, which yields

$$\rho(l^*|\mathcal{E}(0, X_+^*(\tau, p[\cdot])) - \rho(l^* \mid X[\tau]) \leq \epsilon \|l^*\| \qquad (5.100)$$

similar to (5.84). Recall that here

$$\rho(l^* \mid X[\tau]) = \rho(l^*|\mathcal{B}(0, T_0 X^0)) - \int_{t_0}^{\tau} \rho(l^*|\mathcal{B}(0, T(t)B(t)P(t)))dt$$

In Figs. 5.13 and 5.14, we return to Example 3.6 of Chap. 3. Here external ellipsoidal approximations are used to construct reachability sets emanated from a box $\mathcal{X}^0 = X[t_0] = \{x : |x_i| \leq 1, \quad i = 1, 2\}$ rather than from an ellipsoid, as in Chap. 3, Sect. 3.3. Calculations are made due to relations (5.90), (5.92), and Theorems 5.4.4, 5.4.5, with $m = 1, n = 2$.[9]

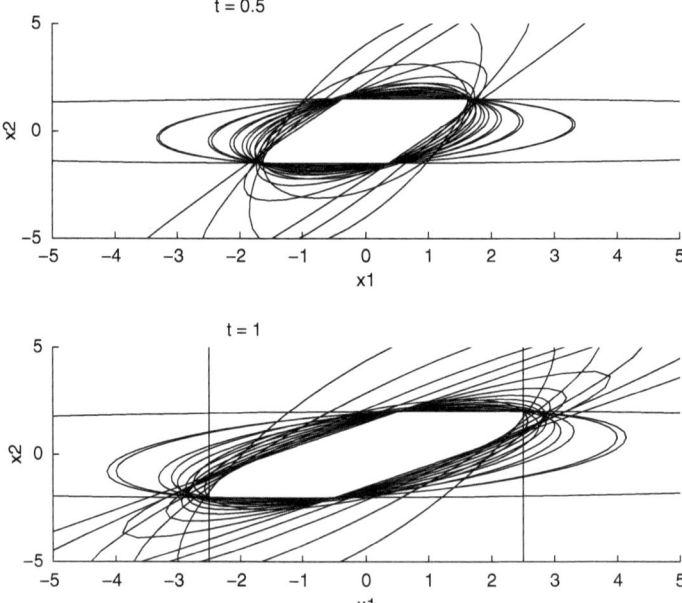

Fig. 5.13 Approximations of reachability sets emanating from a box

---

[9]Illustrations to this example were worked out by M. Kirillin.

5.4 Ellipsoidal Methods for Non-ellipsoidal Constraints

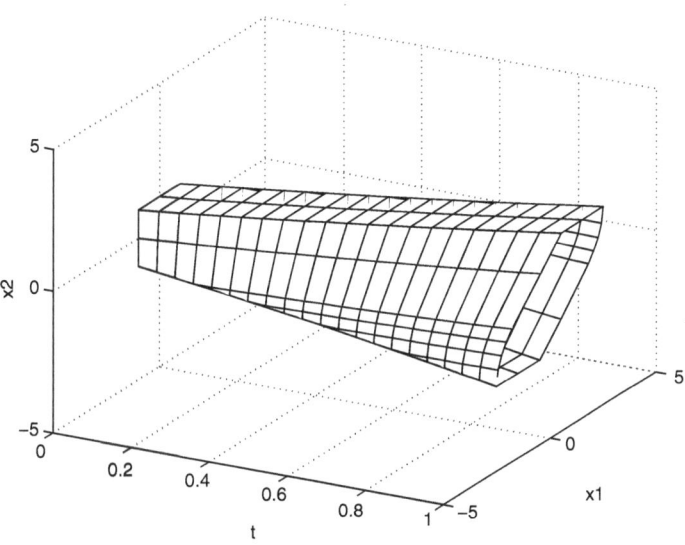

Fig. 5.14 Trajectory tube emanating from a box

5.4.4 Reach Tubes for Box-Valued Constraints: Internal Approximations

Following Remark 5.4.1, we recall that the results of Chap. 3, Sects. 3.2–3.4, 3.7, and 3.8 are all also true for degenerate ellipsoids. We may therefore directly apply them to box-valued constraints

$$\mathcal{P}(t) = \mathcal{B}(u^0(t), P(t)), \quad \mathcal{X}^0 = \mathcal{B}(x^0, X^0),$$

using relations (1.37), in view of inclusions

$$\mathcal{B}(u^0(t), P(t)) = \sum_{j=1}^{m} \mathcal{E}(u^0(t), Q_{ii}(t)), \quad \mathcal{B}(x^0, X^0) = \sum_{k=1}^{n} \mathcal{E}(x^0, X_{kk}),$$

where

$$Q_{ii}(t) = \mathbf{e}^{(i)}\mathbf{e}^{(i)'}q_{ii}, \quad q_{ii} = \mu_i^2, \quad X_{kk}^0 = \mathbf{e}^{(k)}\mathbf{e}^{(k)'}x_{kk}^0, \quad x_{kk}^0 = v_k^2,$$

and $\mathbf{e}^{(i)}, \mathbf{e}^{(k)}$ are unit orths in the respective spaces $\mathbb{R}^m, \mathbb{R}^n$. This leads to the following statements.

Theorem 5.4.7. *(i) An ellipsoid $\mathcal{E}(x^0(t), X_-(t))$ that satisfies relations*

$$X_-(t) = Y'(t)Y(t), \quad Y'(t) = G(t,t_0)X^*(t), \qquad (5.101)$$

where

$$X_-^*(t) = \left(\sum_{k=1}^{n} (X_{kk}^0)^{1/2} S'_{0k} + \int_{t_0}^{t} \sum_{j=1}^{n} G(t_0,s)(B(s)Q_{jj}(s)B'(s))^{1/2} S'_j(s) ds \right)$$

and S_{0k}, S_j are any orthogonal matrices of dimensions $n \times n$, $(S_{0k} S'_{0k} = I$, $SS_j S'_j = I)$ is an internal ellipsoidal approximation for the reach set $X[t]$ of (5.87).
(ii) In order that for a given "direction" l the equality

$$\rho(l \mid X[t]) = \rho(l \mid \mathcal{E}(x^0(t), X_-^*(t)))$$

would be true, it is necessary an sufficient that there would exist a vector $d \in \mathbb{R}^n$, such that the equalities

$$S_{0k} X_{kk}^{1/2} l = \lambda_{0k} d, \quad S_j(B(s)Q_{jj}(s)B'(s))^{1/2} G'(t_0,s) l = \lambda_j(s) d, \qquad (5.102)$$

$$k = 1, \ldots, n, \quad j = 1, \ldots, m, \quad s \in [t_0, t],$$

would be true for some scalars $\lambda_{0k} > 0$ $\lambda_j(s) > 0$.
(iii) The function $x^0(t)$ is the same as for external approximations, and is given by (5.90).

We may now express relations for $X_-(t)$ through differential equations similar to those of Sect. 3.9.

Differentiating $X_-(t)$ and using the previous relations, we come to the proposition.

Theorem 5.4.8. *Matrix $X_-(t)$ of the ellipsoid*

$$\mathcal{E}(x^0(t), X_-(t)) \subseteq X[t],$$

satisfies the equation

$$\dot{X}_-^* = A(t)X_- + X_- A'(t) + Y'(t)S'(t)(B(t)Q_{jj}(t)B'(t))^{1/2} +$$

$$+ (B(t)Q_{jj}(t)B'(t))^{1/2} S(t)Y(t)$$

5.5 Ellipsoidal Methods for Zonotopes

with initial condition

$$X_-(t_0) = \Big(\sum_{k=1}^{n}(X_{kk}^0)^{1/2}S'_{0k}\Big)'\Big(\sum_{k=1}^{n}(X_{kk}^0)^{1/2}S'_{0k}\Big),$$

where

$$\dot{Y}' = A(t)Y' + \sum_{j=1}^{m}(B(t)Q_{jj}B'(t))^{1/2}S'_j(t),$$

$$Y'(t_0) = \sum_{k=1}^{n}(X_{kk}^0)^{1/2}S'_{0k}.$$

and S_{0k}, S_j are orthogonal matrices of dimensions $n \times n$, ($S_{0k}S'_{0k} = I$, $S_j S'_j = I$).

In order that equality $\rho(l \mid X[t]) = \rho(l \mid \mathcal{E}(x^0(t), X_-^*(t)))$ would be true, it is necessary and sufficient that relations of type (5.102) would be satisfied.

5.5 Ellipsoidal Methods for Zonotopes

5.5.1 *Zonotopes*

In this section we consider hard bounds on control u and initial set $X(t_0) = X^0$ in the form of *symmetric polyhedrons* also known as *zonotopes*, denoted here as

$$\mathcal{P}(t) = \mathcal{Z}(p(t), P(t)), \quad X^0 = \mathcal{Z}(x^0, X^0) \tag{5.103}$$

and described as follows.

Definition 5.5.1. A zonotope is understood to be a symmetric polyhedron of type

$$\mathcal{Z}(p, P) = \{x : x = p + \sum_{i=1}^{m} p^i \alpha_i, \; \alpha_i \in [-1, 1]\}.$$

where $p \in \mathbb{R}^m$ is its center and $P = \{p^1, \ldots, p^m\}$ is an array of m vectors ("directions") p^i which define its configuration.
The support function for $\mathcal{Z}$ is

$$\rho(l \mid \mathcal{Z}(p, P)) = \langle p, l \rangle + \sum_{i=1}^{m} |\langle p^i, l \rangle|.$$

If the interior int $\mathcal{Z}(p, P) \neq \emptyset$, then zonotope $\mathcal{Z}(p, P)$ is said to be *nondegenerate*.

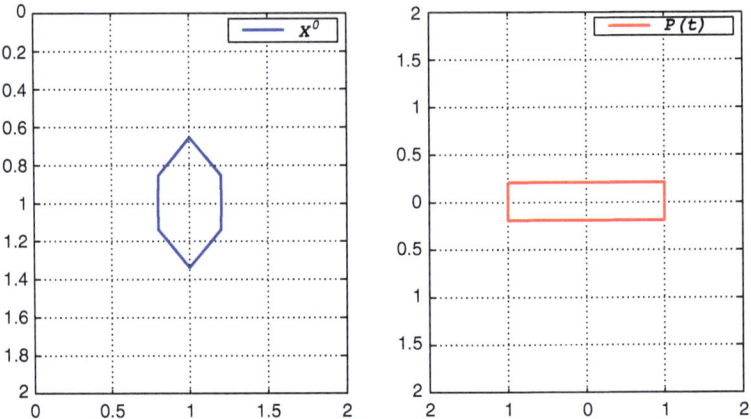

Fig. 5.15 Hard bounds on control u and initial set X^0 for Example 5.9

Examples of zonotopes used below in Example 5.9 are given in Fig. 5.15. Note that a particular case for the zonotope is a box which is initially defined by a set of orthogonal vectors p^i.

Definition 5.5.2. An ellipsoid $\mathcal{E}(q, Q)$ with support function

$$\rho(l|\mathcal{E}(q, Q)) = \langle l, q \rangle + \langle l, Ql \rangle^{1/2}.$$

is said to be degenerate if its matrix $Q = Q' \geq 0$ is degenerate.

Similar to Sect. 3.6 we observe that zonotope $\mathcal{Z}(0, P)$ may be presented as the sum of m degenerate ellipsoids, namely, $\mathcal{Z}(0, P) = \sum_{i=1}^{m} \mathcal{E}(0, P_{ii})$, where $P_{ii} = p^i p^{i'}$, $i = 1, \ldots, m$.

Then the hard bounds on u, $x(t_0)$ may be described as

$$u(t) \in p(t) + \sum_{j=1}^{m_p} \mathcal{E}(0, P_{jj}(t)), \quad x(t_0) \in x^0 + \sum_{i=1}^{m_0} \mathcal{E}(0, X_{ii}^0), \qquad (5.104)$$

Here $P_{jj}(t) = p^j(t) p^{j'}(t)$, $X_{ii}^0 = x^i x^{i'}$

In order to calculate reachability sets for zonotopes one may follow the schemes of the previous section as taken for boxes. These may be repeated for zonotopes as a useful exercise. We formulate the main task beginning with internal approximation.

5.5.2 Internal Ellipsoidal Tubes for a Zonotope

Problem 5.5.1. (i) For the reachability set $X[t] = X(t, t_0, X^0)$ of system (5.12), under constraints 5.103 specify the family $\mathbf{E} = \{\mathcal{E}_-\}$ of internal ellipsoidal approximations such that

$$\operatorname{conv}\left\{\bigcup_{\mathbf{E}} \mathcal{E}_- \mid \mathcal{E}_- \in \mathbf{E}\right\} = X[t].$$

(ii) Indicate a subfamily of tight recurrent approximations $\mathcal{E}_-^{l^*}[t] \subseteq X[t]$, that touch the reachability set along "good" curves" ensuring equalities

$$\rho(l^*(t)|\mathcal{E}_-^{l^*}[t]) = \rho(l^*(t)|X[t]), \quad \operatorname{conv} \bigcup_{l^*} \mathcal{E}_-^{l^*}[t] = X[t].$$

Recall that such a tight approximation is unique if it is selected among the class $\mathbf{E}_-$, in the sense of Definitions 3.2.3, 3.7.3 of Chap. 3. Namely $\mathcal{E}_-[t]$ is unique if for any ellipsoid $\mathcal{E}$ that satisfies $\mathcal{E}_-[t] \subseteq \mathcal{E} \subseteq X[t]$ it follows that $\mathcal{E} = \mathcal{E}_-[t]$.

Following the reasoning of Sects. 5.4.3 and 5.4.4, but applying it our zonotopes $Z(p(t), P(t)), Z(x^0, X^0)$, we come to a conclusion similar to Theorem 5.4.8.

Theorem 5.5.1. (i) The reachability set $X[t]$ allows to have the following internal approximation

$$\mathcal{E}(x_-(t), X_-(t)) \subseteq G(t, t_0) Z(x^0, X^0) + \int_{t_0}^t G(t, s) B(s) Z(p(s), P(s)) ds = X[t],$$

where $X_-(t) = (X_-^*(t))' X_-^*(t)$ and $x_-(t)$ satisfy relations

$$X_-^*(t) = \sum_{i=1}^{m_0} S_{0,i}(X_{ii}^0)^{\frac{1}{2}} G'(t, t_0) + \int_{t_0}^t \sum_{j=1}^{m_p} S_j(s)(B(s) P_{jj}(s) B'(s))^{\frac{1}{2}} G'(t, s) ds,$$
(5.105)

$$x_-(t) = G(t, t_0) x^0 + \int_{t_0}^t G(t, s) B(s) p(s) ds,$$
(5.106)

for any orthogonal matrices $S_{0,i}$, $i = 1, \ldots, m_0$, and continuous orthogonal-valued matrix functions $S_j(t)$, $j = 1, \ldots, m_p$.

(ii) At fixed time t the ellipsoid $\mathcal{E}(x_-(t), X_-(t))$ touches the surface of the reachability set $X(t, t_0, Z(x^0, X^0))$ along direction $l(t) = l^*(t)$, $l(t_0) = l^0$, namely,

$$\rho(l^*|X(t, t_0, Z(x^0, X^0))) = \rho(l^*|\mathcal{E}(x_-(t), X_-(t))) = \langle l^*, x_-(t) \rangle + \langle l^*, X_-(t) l^* \rangle^{1/2},$$
(5.107)

if only the time-varying parameters $S_{0,i}^t$, $S_j^t(\cdot)$ are selected such that

$$S_{0,i}^t(X_{ii}^0)^{\frac{1}{2}}G'(t,t_0)l^0 = \lambda_{0,j}^t p^t, \quad S_j^t(s)(B(s)P_{jj}(s)B'(s))^{\frac{1}{2}}G'(t,s)l^* = \lambda_j^t(s)p^t, \tag{5.108}$$

$$s \in [t_0, t], \quad i = 1, \ldots, m_0, \quad j = 1, \ldots, m_p,$$

for certain scalar functions $\lambda_j^t(\cdot) > 0$, $\lambda_{0,j}^t > 0$, $j = 1, \ldots, m_p$, and functions $p^t \in \mathbb{R}^n$, $p^t \neq 0$.

Differentiating relations (5.105), (5.106), and applying equality

$$\dot{X}_-(\tau) = (\dot{X}_-^*(\tau))' X_-^*(\tau) + (X_-^*(\tau))' \dot{X}_-^*(\tau),$$

we come to differential equations for the dynamics of approximating ellipsoids $\mathcal{E}(x_-(t), X_-(t))$.

$$\begin{cases} \dot{X}_-(\tau) = A(\tau)X_-(\tau) + X_-(\tau)A'(\tau) + \\ + (X_-^*(\tau))'(\sum_{j=1}^{m_p} S_j^t(\tau)(B(\tau)P_{jj}(\tau)B'(\tau))^{\frac{1}{2}}) + (\sum_{j=1}^{m_p}(B(\tau)P_{jj}(\tau)B'(\tau))^{\frac{1}{2}} S_j^{t'}(\tau))X_-^*(\tau), \\ X_-(t_0) = (\sum_{i=1}^{m_0} S_{0,i}^t(X_{ii}^0)^{\frac{1}{2}})'(\sum_{i=1}^{m_0} S_{0,i}^t(X_{ii}^0)^{\frac{1}{2}}), \\ \dot{x}_-(\tau) = A(t)x_-(\tau) + B(\tau)p(\tau), \quad x_-(t_0) = x_0. \end{cases} \tag{5.109}$$

Among these ellipsoidal-valued tubes we now have to specify those that ensure *tight approximations*.

We now as usually (see Sects. 3.3 and 3.9) select a "good" curve as a solution to equation

$$\dot{l}^*(t) = -A'(t)l^*(t), \quad l^*(t_0) = l^0$$

for a certain direction l^0, so that $l^*(t) = G'(t_0, t)l^0$.

For curve $l^*(t)$ we then find the ellipsoidal tube $\mathcal{E}(x_-(t), X_-^*(t))$ designed for $l = l^*(t)$, such that for each $t \geq t_0$ we have the equality

$$\rho(l^*(t)|\mathcal{X}(t, t_0, \mathcal{Z}(x^0, X^0)) = \rho(l^*(t)|\mathcal{E}(x_-(t), X_-^*(t)) = \langle l^*(t), x^*(t) \rangle. \tag{5.110}$$

In order to satisfy this equality it is necessary to ensure, due to Theorem 5.5.1 (ii), that orthogonal matrices $S_{0,i}^t$, $S_j^t(\cdot)$, $i = 1, \ldots, m_0$, $j = 1, \ldots, m_p$ would satisfy conditions (5.108) for certain $\lambda_{0,i}^t > 0$, $\lambda_j^t(\cdot) > 0$, $p^t \in \mathbb{R}^n$, $p^t \neq 0$. Substituting in these conditions $l^*(t)$, we arrive at

$$S_{0,i}^t(X_{ii}^0)^{\frac{1}{2}}l^0 = \lambda_{0,i}^t p^t, \quad S_j^t(s)B(s)(P_{jj}(s))^{\frac{1}{2}}l^*(s) = \lambda_j^t(s)p^t, \tag{5.111}$$

$$s \in [t_0, t], \quad i = 1, \ldots, m_0, \quad j = 1, \ldots, m_p.$$

Note that for a good curve $l^*(s)$ functions $S_{0,i}^t$, $S_j^t(s)$, $\lambda_{0,i}^t$, $\lambda_j^t(s)$, p^t, the last relations do not depend on t, so further on the upper index t or the functions may be deleted.

Summarizing the above, we conclude the following

Theorem 5.5.2. (i) *For the reachability tube* $X[t] = X(t, t_0, Z(x^0), X^0)$ *there exists an internal approximation*

$$\mathcal{E}(x_-(t), X_-(t)) \subseteq X(t, t_0, Z(x^0, X^0)), \quad t \geq t_0,$$

where $X_-(t) = (X_-^*(t))'(X_-^*(t))$ *and* $x_-(t)$ *satisfy differential equations (5.109) and* $S_{0,i}$, $S_j(\cdot)$, $i = 1, \ldots, m_0$, $j = 1, \ldots, m_p$, *are orthogonal matrices that evolve continuously in time.*

(ii) *The ellipsoidal tube* $\mathcal{E}(x_-(t), X_-^*(t))$ *touches the boundary* $\partial X[t]$ *of tube* $X[t]$ *from the inside, along direction* $l^*(t) = G'(t_0, t)l^0$, *ensuring equalities (5.107) whenever relations (5.111) turn out to be true.*

The last theorem gives the solution to Problem 5.5.1 described through recurrent relations. The internal ellipsoidal tube $\{\mathcal{E}(x_-(t), X_-^*(t)), t \geq t_0\}$ touches the reachability tube $\{X(t, t_0, \mathcal{M}(x^0, X^0)), t \geq t_0\}$ along the curve $l^*(t) = G'(t_0, t)l^0$, ensuring for any l^0 the equalities

$$(l^*(t), x^{(l*)}(t)) = \rho(l^*(t) | X[t]) = \rho(l^*(t) | \mathcal{E}(x_-(t), X_-^*(t))), \tag{5.112}$$

where

$$x^{(l*)}(t) = x_-(t) + X_-^*(t)l^*(\langle l^*, X_-^*(t)l^* \rangle)^{-1/2}, \tag{5.113}$$

These relations represent the maximum principle and the ellipsoidal maximal principle which hold simultaneously at $x = x^{(l*)}(t)$, $l(t) = l^*(t)$. They also indicate that at each time t point $x^{(l*)}(t)$ lies on the support plane to $X[t]$ generated by vector $l^*(t)$ with equalities (5.112) reached at $x^{(l*)}(t)$.

Remark 5.5.1. (a) Note that in formula (5.113) vector $x^{(l*)}(t)$ does not depend on the length of $l^*(t)$. Then, denoting $x^{(l*)}(t) = x[t, l^*(t)]$, we may consider the surface $\Xi_-[t, l^0] = x[t, l^*(t)]$, where $l^*(t)$ depends on l^0, which together with t serve as a parameter of dimension $n + 1$. Hence, for t^* given, we have the n-dimensional surface

$$\Xi_-[t^*, \cdot] = \bigcup \{x[t^*, l_0] \mid \|l^0\| = 1\}$$

and

$$\Xi[t^*, \cdot] = \partial X[t^*] = \bigcup \{x[t^*, l^0] \mid \|l^0\| = 1\},$$

where ∂X is the boundary of set X. Then $X[t] = \text{conv}\{x[t^*, l^0] \mid \|l^0\| = 1\}$ is the convex hull of $\Xi_-[t^*, \cdot]$.

(b) Also note that in Eqs. (5.109), (5.111) matrix $X_*^*(t)$ depends on the selection of good curve $l = l^*(t)$ and was marked as $X_*^*(t)$, while $x_-(t)$ does not depend on l^* and *is the same for all internal and external ellipsoids within classes* $\mathbf{E}_-, \mathbf{E}_+$. Now, if we take a finite number k of good directions $l_i^*(t)$, $i = 1, \ldots, k$, then, for a selected $l = l_i^*(t)$, we shall mark matrix $X_*^*(t) = X_-^i(t)$. We further introduce set $\Xi_k[t] = \bigcup_{i=1}^{k} \mathcal{E}(x_-(t, X_-^i(t),] \subset X[t]$ which is a "star-shaped" set and such that its convex hull $\text{conv}\{\Xi_k[t]\} \subset X[t[$ is a convex internal approximation of $X[t]$. Sets $\Xi_k[t]$ may be effectively used in related numerical procedures.

Example 5.9. Consider system

$$\begin{cases} \dot{x}_1 = x_2 + u_1 \\ \dot{x}_2 = -\omega^2 x_1 + u_2 \end{cases},$$

$$u \in \mathcal{Z}(p, P)$$

$$x(0) \in \mathcal{Z}(x_0, X_0)$$

where sets $\mathcal{Z}(x_0, X_0)$, $\mathcal{Z}(p, P)$ are shown in Fig. 5.15.

In Fig. 5.16 we demonstrate the exact reachability tube with one of its tight internal ellipsoidal approximations. This tube was constructed as the union of an array of such tight approximations. In Fig. 5.17 we depict a reachability set for a zonotope with some of its tight internal ellipsoids at time $t = 1$.[10]

5.5.3 External Ellipsoidal Tubes for a Zonotope

Problem 5.5.2. (i) For the reachability set $X[t] = X(t, t_0, X^0)$ of system (5.12), under constraints 5.103 specify the family $\mathbf{E} = \{\mathcal{E}+\}$ of external ellipsoidal approximations such that for a given $\varepsilon > 0$ we have

$$h(X[t], \bigcap\{\mathcal{E}_+ \mid \mathcal{E}_+ \in \mathbf{E}\}) \leq \varepsilon, \quad X[t] \subseteq \bigcap\{\mathcal{E}_+ \mid \mathcal{E}_+ \in \mathbf{E}\}.$$

(ii) Indicate a subfamily of tight recurrent approximations $\mathcal{E}_+^l[t] \supseteq X[t]$, that touch the reachability set along "good" curves ensuring equality

$$\rho(l(t)|\mathcal{E}_+^l[t]) \leq \rho(l(t)|X[t]) + \varepsilon \|l(t)\|, \quad h(X[t], \bigcap_{\|l\| \leq 1} \mathcal{E}_+^l[t]) \leq \varepsilon.$$

---

[10]Examples illustrated in Figs. 5.16, 5.17, 5.18, and 5.19 of this section were worked out by M. Kirillin.

5.5 Ellipsoidal Methods for Zonotopes

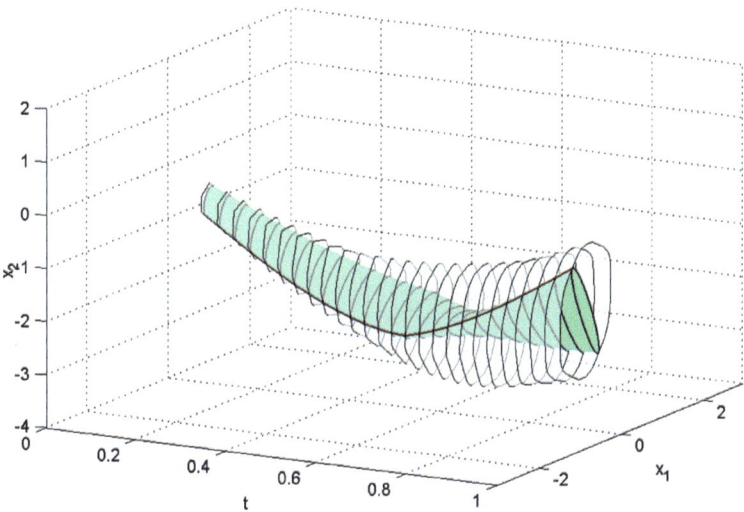

Fig. 5.16 Reach tube and one tight internal approximation

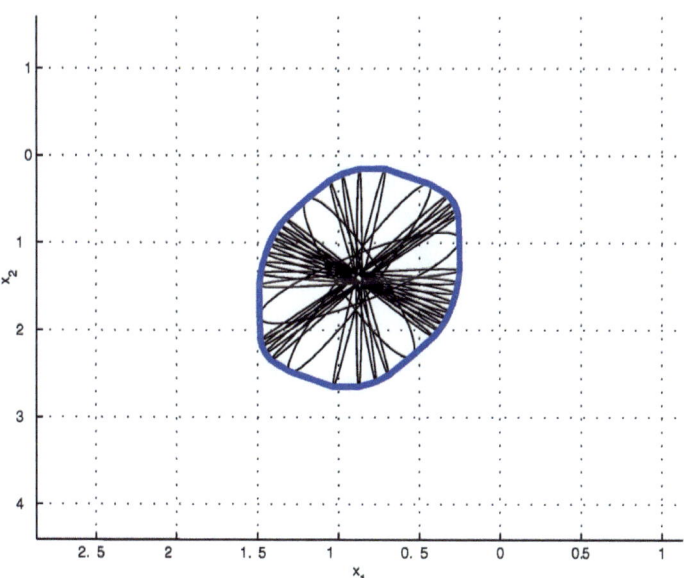

Fig. 5.17 The reachability set with its internal approximations

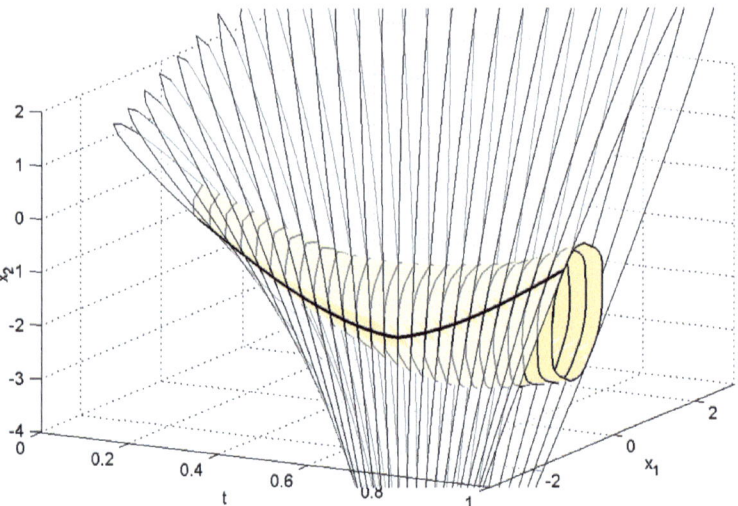

Fig. 5.18 Reachability tube $\mathcal{X}_\varepsilon^z[t]$ with one tight external approximation

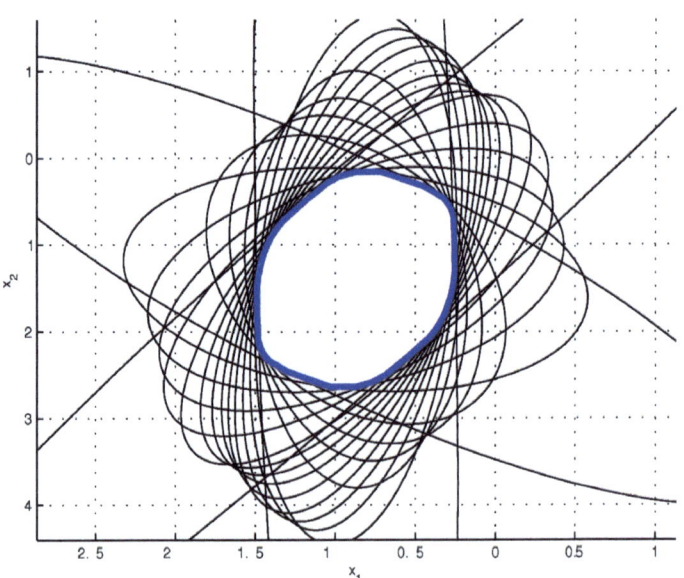

Fig. 5.19 Reachability set with its tight external approximations

5.5 Ellipsoidal Methods for Zonotopes

To solve this problem we first define some neighborhoods of a zonotope.

Definition 5.5.3. For a zonotope $Z(0, P) = \sum_{i=1}^{m} \mathcal{E}(0, P_{ii})$, with degenerate matrices P_{ii} denote
$P = \sum_{i=1}^{m} P_{ii}$ and further

$$P_{ii}^\varepsilon = P_{ii} + \frac{\varepsilon}{m} I, \quad P^\varepsilon = \sum_{i=1}^{m} P_{ii}^\varepsilon = P + \varepsilon I.$$

Then the ε-neighborhood of $Z(0, P)$ is defined as

$$Z_\varepsilon(0, P) = Z(0, P^\varepsilon) = \sum_{i=1}^{m} \mathcal{E}(0, P_{ii}^\varepsilon) \supset Z(0, P)$$

Here with $X^0 = \sum_{i=1}^{m} X_{ii}^0$, $X^0 + \varepsilon I = X^{0\varepsilon}$, we have the ε-neighborhood of $Z(0, X^0)$ denoted as

$$Z_\varepsilon(0, X^0) = Z(0, X^{0\varepsilon}) = \sum_{i=1}^{m} \mathcal{E}(0, X_{ii}^{0\varepsilon}) \supset Z(0, X^0)$$

To approximate reachability sets $X[t] = X(t, t_0, Z^0)$ we introduce constraints on $u, x(t_0)$

$$u(t) \in \mathcal{P}_\varepsilon = Z(p(t), P^\varepsilon(t)), \quad x(t_0) \in X_\varepsilon^0 = Z(x^0, X^{0\varepsilon}) \quad (5.114)$$

Then the reachability set of system (5.12) under such constraints will be denoted as

$$X_\varepsilon^z[t] = x^*(t) + G(t, t_0) Z(0, X^{0\varepsilon}) + \int_{t_0}^{t} G(t, s) B(s) Z(0, P^\varepsilon(s)) ds, \quad (5.115)$$

$$x^*(t) = G(t, t_0) x^0 + \int_{t_0}^{t} G(t, s) B(s) p(s) ds. \quad (5.116)$$

Exact reach set $X[t]$ will be defined through (5.115) where $X^{0\varepsilon}, P^\varepsilon(s)$ are substituted by $X^0, P(s)$.

Lemma 5.5.1. *For any $\varepsilon > 0$ we have*
(i) the inclusion $X[t] \subset X_\varepsilon^z[t]$,
(ii) the inequality $h(X[t], X_\varepsilon^z[t]) \leq k(t)\varepsilon$, where

$$k(t) = \sup \left\{ \|G'(t, t_0)l\|^{1/2} + \int_{t_0}^{t} \|B(s)G'(t, s)l\|^{1/2} ds \,\bigg|\, l \in \mathcal{B}(0) \right\}.$$

Exercise 5.5.1. Prove Lemma 5.5.1

We now calculate the external ellipsoidal ε-approximation to $\mathcal{X}_\varepsilon^z[t]$.

Remark 5.5.2. Since $\mathcal{Z}(p, P) = p + \mathcal{Z}(0, P)$, the centers of the zonotopes are not involved in ellipsoidal approximation which is the same as when $x^0 = 0$, $p(t) = 0$, which also yields $x(t) = 0$. The approximation is then the sum of $x^*(t)$ and the external $\mathcal{E}_+[t]$ under $x(t) = 0$.

Using schemes of Sect. 3.3 with those of Sect. 5.4 of this chapter, we come to the next assertion[11]

Theorem 5.5.3. *(i) For the ε-neighborhoods of $X[t]$ there exist an external ellipsoidal approximation*

$$X[t] \subseteq \mathcal{X}_\varepsilon[t]^z \subseteq \mathcal{E}(x^*(t), X_+^\varepsilon(t)), \tag{5.117}$$

where $x^*(t)$, is given by (5.116) and $X_+^\varepsilon(t)$ satisfies equations

$$\begin{cases} \dot{X}_+^\varepsilon(t) = A(t)X_+^\varepsilon(t) + X_+^\varepsilon A'(t) + \\ + \sum_{j=1}^{m_p} \pi_j(t)X_+^\varepsilon(t) + \sum_{j=1}^{m_p} (\pi_j(t))^{-1} B(t) P_{jj}^\varepsilon(t) B'(t), \\ X_+^\varepsilon(t_0) = (\sum_{i=1}^{m_0} p_{0,i})(\sum_{i=1}^{m_0} (p_{0,i})^{-1} X_{ii}^{0\varepsilon}), \end{cases} \tag{5.118}$$

for any positive $p_{0,i} > 0$, $i = 1, \ldots, m_0$, and any continuous functions $\pi_j(t) > 0$, $j = 1, \ldots, m_p$.

(ii) At each time t ellipsoid $\mathcal{E}(x^*(t), X_+^\varepsilon(t))$ touches the ε-neighborhood $\mathcal{X}_\varepsilon^z[t]$ along direction $l^*(t) = G'(t_0, t)l^0$, so that relation

$$\rho(l^*(t)|\mathcal{X}_\varepsilon^z[t]) = \rho(l^*(t)|\mathcal{E}(x^*(t), X_+^\varepsilon(t)), \quad t \geq t_0, \tag{5.119}$$

is true as long as the next conditions are satisfied

$$p_{0,i} = \langle l, X_{ii}^{0\varepsilon} l \rangle^{\frac{1}{2}}, \quad i = 1, \ldots, m_0,$$

$$\pi_j(s) = \langle l^*(s), B(s) P_{jj}^\varepsilon(s) B'(s) l^*(s) \rangle^{1/2}, s \in [t_0, t], \quad j = 1, \ldots, m_p.$$

Relation (5.119) yields the next property: set $\mathcal{X}_\varepsilon^z[t]$ may be presented as the following intersection of approximating ellipsoids, namely

$$\mathcal{X}_\varepsilon^z[t] = \cap \{\mathcal{E}(x^*(t), X_+^\varepsilon(t)) \mid p_{0,i} > 0, \pi_j(\cdot) > 0\}. \tag{5.120}$$

---

[11] Note that designing external approximations of reachability sets $X[t]$ we have to apply them to nondegenerate ε-neighborhoods of ellipsoids $\mathcal{E}_{ii}$ instead of exact $\mathcal{E}_{ii}$, as in Sect. 5.4.2. This is because all ellipsoids $\mathcal{E}_{ii}$ are degenerate, and their approximations may also turn out to be such. But since we need all externals to be nondegenerate, this will be guaranteed by applying our scheme to nondegenerate neighborhoods of $\mathcal{E}_{ii}$.

5.5 Ellipsoidal Methods for Zonotopes

For a fixed vector $l^0 \in \mathbb{R}^n$ let $x_l^*(t)$ be the related good curve along which ellipsoid $\mathcal{E}(x^*(t), X_+^\varepsilon(t))$ touches the tube $\mathcal{X}_\varepsilon^z[t]$, ensuring equalities

$$\langle l^*(t), x_l^*(t) \rangle = \rho(l^*(t) | \mathcal{X}_\varepsilon^z[t]) = \rho(l^*(t) | \mathcal{E}(x^*(t), X_+^\varepsilon(t))). \tag{5.121}$$

Then point $x_l^*(t)$ lies on the boundaries of both $\mathcal{E}(x^*(t), X_+^\varepsilon(t)))$ and $\mathcal{X}_\varepsilon^z[t]$, satisfying an ellipsoidal maximum principle

$$\langle l^*(t), x_l^*(t) \rangle = \max\{\langle l^*(t), x \rangle \mid x \in \mathcal{E}(x^*(t), X_+^\varepsilon(t))\}, \tag{5.122}$$

along trajectory

$$x_l^*(t) = x^*(t) + \frac{X_+^\varepsilon(t) l^*(t)}{\langle l^*(t), X_+^\varepsilon(t) l^*(t) \rangle^{1/2}}.$$

Remark 5.5.3. Similar to Remark 5.5.1(a) we may introduce a two-parameter surface $\Xi[t, l^0]$ but with array of functions $\{l^*[t]\}$ calculated for $\mathcal{X}_\varepsilon(t)$ rather than for $\mathcal{X}[t]$, leading to some similar conclusions. However, one of the differences is that with number of ellipsoids that form zonotope $Z(x^0, X^0)$ being $m^0 < n$, the matrix $X^0 = \sum_{i=1}^{m^0} X_{ii}^0$ will be degenerate, and in $\mathbb{R}^n$ set $Z(x^0, X^0)$ will be a "zonotopic cylinder."

Example 5.10. Is the same as Example 5.9. Here Fig. 5.18 indicates the ε-neighborhood $\mathcal{X}_\varepsilon^z[t]$ for the exact zonotopic reachability tube $\mathcal{X}[t]$ calculated as an intersection of its tight external approximations one of which is shown touching $\mathcal{X}_\varepsilon^z[t]$ along a good curve $l^*(t)$. Figure 5.19 demonstrates the zonotopic cross-section of the ε-neighborhood $\mathcal{X}_\varepsilon^z[t]$ at fixed time $t = 1$, together with an array of its tight ellipsoidal approximations.

Chapter 6
Impulse Controls and Double Constraints

Abstract In the first section of this chapter we deal with the problem of feedback impulse control in the class of generalized inputs that may involve delta functions and discontinuous trajectories in the state space. Such feedback controls are not physically realizable. The second section thus treats the problem of feedback control under double constraints: both hard bounds and integral bounds. Such solutions are then used for approximating impulse controls by bounded "ordinary" functions.

Keywords Bounded variation • Impulsive inputs • δ-Function • Closed-loop control • Value function • Variational inequality • Double constraints

Control inputs in the form of impulses of the δ-function type have been studied since the conception of control theory, being motivated by designing space trajectories, automation, biomedical issues, and also problems in economics (see [23, 42, 43]). However, impulse control problems were initially studied mostly as open-loop solutions, [120, 219] with detours to closed-loop strategies being rather rare [23]. In the first section we describe the Dynamic Programming approach to problems of feedback impulse control for any finite-dimensional linear system. This is reached through *variational inequalities* that propagate HJB techniques to impulse control. So, we begin with problems of open-loop control, then concentrate on closed-loop solutions emphasizing their specifics and the type of HJB equation that describes them. The solutions arrive as δ-functions. However such functions are ideal mathematical elements which require physically realizable approximation by "ordinary" functions. This is done in the second section, where we solve the problem *with double constraints*—joint hard instantaneous and soft L_1—integral bounds on the controls. The achieved solution then yields realizable approximations of impulsive feedback inputs. The contents of this chapter rely on investigations [55, 152, 154].

6.1 The Problem of Impulse Controls

Considering impulse control problems in finite time, it could at first seem reasonable, say, in system (1.2), to minimize the norm $\|u(\cdot)\|$ of the control, taking it in the space L_1. But due to the specifics of this space the minimum may not be attained within its elements. It is therefore natural to apply the next scheme.

Consider the following generalization of the Mayer-Bolza problem in classical Calculus of Variations (see [33]).

Problem 6.1.1. Minimize the functional

$$J(u(\cdot)) = \mathrm{Var}\, U(\cdot) + \varphi(x(\vartheta + 0)) \to \inf, \tag{6.1}$$

over controls $U(\cdot)$, due to system

$$dx(t) = A(t)x(t)dt + B(t)dU(t), \quad t \in [t_0, \vartheta], \tag{6.2}$$

with continuous coefficients $A(t)$, $B(t)$, under restriction

$$x(t_0 - 0) = x^0. \tag{6.3}$$

Here $\varphi(x)$ is a convex function, $\mathrm{Var}\, U(\cdot)$ stands for the *total variation* of function $U(\cdot)$ over the interval $[t_0, \vartheta+0]$, where $U(\cdot) \in \mathbf{V}^p[t_0, \vartheta]$ is the space of vector-valued functions of bounded variation. The *generalized control* $U(t)$ attains its values in $\mathbb{R}^p$ which means that each component $U_i(t)$, $(i = 1, \ldots, p)$, assumed *left-continuous*, is a function of bounded variation on $[t_0, \vartheta + 0]$ and $B(t) \in \mathbb{R}^{n \times p}$ are taken to be continuous. The minimum over $U(\cdot)$ will now be attained.[1]

Equation (6.2) with condition (6.3) is a symbolic relation for

$$x(t) = G(t, t_0)x^0 + \int_{t_0}^t G(t, \xi)B(\xi)dU(\xi), \tag{6.4}$$

where the last term in the right-hand side is a Stieltjes or Lebesgue–Stieltjes integral (see [234]). The terminal time is fixed and the terminal cost function $\varphi(x) : \mathbb{R}^n \to \mathbb{R} \cup \{\infty\}$ is closed and convex.

A special selection of $\varphi(x) = I(x \,|x^{(1)})$ yields

Problem 6.1.2. Steer $x(t)$ from point $x^0 = x(t_0)$ to point $x^{(1)} = x(\vartheta)$ with minimal variation of control $U(t)$:

$$\mathrm{Var}\{U(t) \mid t \in [t_0, \vartheta + 0]\} = \mathrm{Var}\, U(\cdot) \to \inf, \tag{6.5}$$

---

[1] In this book we give a concise description of impulse controls that are confined only to δ-functions, but not their derivatives. A general theory of impulse control that also involves derivatives of δ-functions is beyond the scope of this book and is presented in [152, 154]. Such theory leads to the description of "fast" or "ultra fast" control solutions achieved on a quantum level and in "nano"-time.

6.1 The Problem of Impulse Controls

due to system (6.3), under boundary conditions

$$x(t_0 - 0) = x^{(0)}, \quad x(\vartheta + 0) = x^{(1)}. \tag{6.6}$$

Our main interest in this book lies in closed-loop control, but prior to that we start with open-loop solutions presenting results in terms of the above. Such type of problems in terms of the moment problem was indicated in Sect. 1.3 and addressed in detail in [120, 219].

6.1.1 Open-Loop Impulse Control: The Value Function

Define

$$V(t_0, x^0) = \min\{J(u(\cdot)) \mid x(t_0 - 0) = x^{(0)}\}$$

as the value function for Problem 6.1.1. We shall also use notation $V(t_0, x^0) = V(t_0, x \mid \vartheta + 0, \varphi(\cdot))$, emphasizing the dependence on $\{\vartheta, \varphi(\cdot)\}$.

Let us start by minimizing $V_1(t_0, x^0) = V(t_0, x^0 \mid \vartheta, I(\cdot \mid x^{(1)}))$. Then we first find conditions for the solvability of boundary-value problem (6.6) under constraint $\mathrm{Var} U(\cdot) \leq \mu$, with μ given.

Since

$$\langle l, x(\vartheta + 0) \rangle = \langle l, x^{(1)} \rangle \leq$$

$$\langle l, G(\vartheta, t_0) x^0 \rangle + \max \left\{ \int_{t_0}^{\vartheta + 0} \langle l, G(\vartheta, \xi) B(\xi) dU(\xi) \rangle \mid \mathrm{Var} U(\cdot) \leq \mu \right\},$$

and since the conjugate space for vector-valued continuous functions $C^p[t_0, \vartheta]$ is $\mathbf{V}^p[t_0, \vartheta]$, then treating functions $B'(t)G(\vartheta, t)l$ as elements of $C^p[t_0, \vartheta]$, we have

$$\langle l, x^{(1)} \rangle \leq \langle l, G(\vartheta, t_0) x^0 \rangle + \mu \|B'(\cdot)G'(\vartheta, \cdot)l\|_{C[t_0, \vartheta]}, \tag{6.7}$$

whatever be $l \in \mathbb{R}^n$.

Here

$$\|B'(\cdot)G'(\vartheta, \cdot)l\|_{C[t_0, \vartheta]} = \max_t\{\|B'(t)G'(\vartheta, t)l\| \mid t \in [t_0, \vartheta]\} = \|l\|_V,$$

and $\|l\|$ in the braces is the Euclidean norm. We further drop the upper index p in $C^p, \mathbf{V}^p$ while keeping the dimension p.

Denote $c(t_0, x^{(1)}) = x^{(1)} - G(\vartheta, t_0)x^{(0)}$. Moving the first term in the right-hand side of (6.7) to the left, then dividing both sides by $\|B'(\cdot)G'(\vartheta, \cdot)l\|_{C[t_0, \vartheta]}$, we come to the next proposition.

Theorem 6.1.1. *Problem 6.2 of steering $x(t)$ from $x^0 = x(t_0)$ to $x^{(1)} = x(\vartheta + 0)$ under constraint $\mathrm{Var}U(\cdot|[t_0, \vartheta]) \leq \mu$ is solvable iff (6.7) is true for all $l \in \mathbb{R}^n$. The optimal control $U^0(t)$ for this problem is of minimal variation*

$$\mathrm{Var}U^0(\cdot) = \mu^0 = \sup\left\{ \frac{\langle l, c(t_0, x^{(1)}) \rangle}{\|B'(\cdot)G'(\vartheta, \cdot)l\|_{C[t_0, \vartheta]}} \,\middle|\, l \in \mathbb{R}^n, \right\}. \qquad (6.8)$$

Note that the criteria of controllability in the class of functions $U(\cdot) \in \mathbf{V}^p[t_0, \vartheta]$ for system (6.2) are the same as in Sect. 1.3.

Lemma 6.1.1. *In (6.8) the maximum (supremum over l) is attainable if system (6.2) is strongly completely controllable. (Assumption 1.5.2 is true.)*

Indeed, if $\mu^0 \neq 0$ and Assumption 1.5.2 requiring strong complete controllability is true, then $\|l\|_V$ defines a finite-dimensional norm in $\mathbb{R}^n$ (check this property (!)) and (6.8) is equivalent to

$$\mu^0 = \max\{\langle l, x^{(1)} - G(\vartheta, t_0)x^0 \rangle \mid \|l\|_V \leq 1\} = \|x^{(1)} - G(\vartheta, t_0)x^0\|_V^*, \qquad (6.9)$$

where $\|x\|_V^*$ is the finite-dimensional norm conjugate to $\|l\|_V$.

Exercise 6.1.1. Find the norm $\|x\|_V^*$.

Remark 6.1.1. Note that under Assumption 1.5.2 we have $\mu^0 = \mu^0(t_0, x^0; \vartheta, x^{(1)}) < \infty$.

Let l^0 be the maximizer in (6.9). Then from (6.7), (6.8) we observe

$$\mu^0 = \max\left\{ \int_{t_0}^{\vartheta+0} \langle l^0, G(\vartheta, \xi)B(\xi)dU(\xi) \rangle \,\middle|\, \mathrm{Var}U(\cdot) \leq \mu^0 \right\} =$$

$$= \int_{t_0}^{\vartheta+0} \langle l^0, G(\vartheta, \xi)B(\xi)dU^0(\xi) \rangle. \qquad (6.10)$$

This is *the maximum principle for impulse controls* which is given *in integral form*.

Denote $\psi^0[\xi] = \psi(\xi, \vartheta, l) = l'G(\vartheta, \xi)$. Then we can rewrite (6.10) as

$$\mu^0 = \max\left\{ \int_{t_0}^{\vartheta+0} \psi^0[\xi]B(\xi)dU(\xi) \,\middle|\, \mathrm{Var}U(\cdot) \leq \mu^0 \right\} = \int_{t_0}^{\vartheta+0} \psi^0[\xi]B(\xi)dU^0(\xi). \qquad (6.11)$$

Theorem 6.1.2. *The optimal impulse control $U^0(\cdot)$ that minimizes variation $\mathrm{Var}U(\cdot)$ satisfies **the maximum principle** (6.11) under $l = l^0$—the maximizer of (6.9). With $\mu^0 > 0$ and under Assumption 1.5.2 it is also sufficient for the optimality of $U^0(\cdot)$.*

6.1 The Problem of Impulse Controls

The proof of sufficiency is similar to Sect. 1.5.

Remark 6.1.2. The optimal control $U^0(\cdot)$ for Problem 6.1.1 may not be unique.

Denote $\mathcal{T}(l) = \arg\max_t \{\|B'(t)G'(\vartheta,t)l\| \mid t \in [t_0, \vartheta]\}$ which is the set of points τ where

$$\|B'(\tau)G'(\vartheta,\tau)l\| = \max_t\{\|B'(t)G'(\vartheta,t)l\| \mid t \in [t_0, \vartheta]\}.$$

Lemma 6.1.2. *(i) Among optimal controls $U^0(\cdot)$ there always exists one of type*

$$U(t) : \begin{cases} U(t) = const, & \tau_{i-1} < t \leq \tau_i, \, i = \overline{1, m+1}, \\ U(\tau_i + 0) - U(\tau_i) = \alpha_i, & i = \overline{1, m}. \end{cases}$$

where $\alpha \neq 0$ each of the instants $\tau_i \in \mathcal{T}(l^0)$, so that

$$\frac{dU^0[t]}{dt} = \sum_{i=1}^m p^{(i)} \delta(t - \tau_i), \tag{6.12}$$

where

$$t_0 \leq \tau_1 < \tau_2 < \ldots < \tau_k \leq \vartheta, \; p^{(i)} \in \mathbb{R}^p,$$

$$\mathrm{Var} U^0(\cdot) = \sum_{i=1}^m |p^{(i)}| = \mu^0;$$

(ii) the number of such instants τ_i is $m \leq n$ where n is the system dimension.

Exercise 6.1.2. Prove Lemma 6.1.2.[2]

A particular case arises under scalar control U.

Assumption 6.1.1. *Coefficients $A, B = const$, $B = b$, $|A| \neq 0$, $\mathrm{rank}[AB, A^2B, \ldots, A^nB] = n$.*

Lemma 6.1.3. *Under Assumption 6.1.1 points $\tau_i \in \mathcal{T}(l^0)$ are isolated.*

This fact facilitates the calculation of $U^0(t)$.

Exercise 6.1.3. Prove Lemma 6.1.3.

Example 6.1.1. Consider the motion of a ball which rolls along a planar curve as in Fig. 6.1. The ball may be controlled by striking it along the direction of the motion

---

[2] The proof of this Lemma may be found in [120, 219].

Fig. 6.1 Motion of a ball along a planar curve

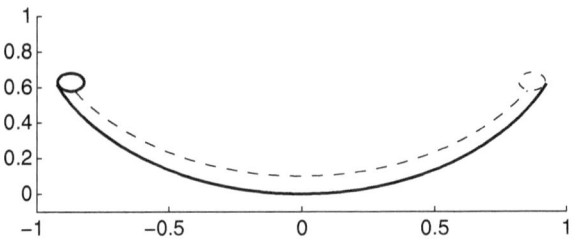

with positive or negative impulse (a strike), so as to calm down the motion to a total stop at given finite time. The problem is to find the total number of strikes with their intensities and directions.

The equation for the motion of the controlled ball is

$$\ddot{s} + \omega^2 s = u, \tag{6.13}$$

where s is the length from the origin along the curve with $t \in [0, \theta]$. The stopping rule is

$$\begin{cases} s(\vartheta) = 0, \\ \dot{s}(\vartheta) = 0. \end{cases}$$

Exercise 6.1.4. 1. What should be the curve that yields equation of motion without friction to be (6.13)?
2. Write down the equation of motion for the ball when the curve is a circular arc.

Write down (6.13) as

$$\begin{cases} dx_1 = x_2 dt, \\ dx_2 = -\omega^2 x_1 dt + U dt, \end{cases}$$

where $u = dU/dt$, $\text{Var} U \leq \mu$, $t \in [0, \pi/\omega]$, $x(0) = [x_1^0, x_2^0]'$.

Thus we have

$$A = \begin{pmatrix} 0 & 1 \\ -\omega^2 & 0 \end{pmatrix}, \quad B = \begin{pmatrix} 0 \\ 1 \end{pmatrix},$$

and the fundamental matrix for the homogeneous system is

$$G(\pi/\omega, t) = \begin{pmatrix} -\cos(\omega t) & \frac{1}{\omega}\sin(\omega t) \\ -\omega\sin(\omega t) & -\cos(\omega t) \end{pmatrix},$$

6.1 The Problem of Impulse Controls

Hence

$$G(\pi/\omega, t)B = \begin{pmatrix} \frac{1}{\omega}\sin(\omega t) \\ -\cos(\omega t) \end{pmatrix},$$

with

$$x^{(1)} = x(\pi/\omega) = \begin{pmatrix} 0 \\ 0 \end{pmatrix}, \quad x^{(0)} = x(0) = \begin{pmatrix} 1 \\ 0 \end{pmatrix},$$

Then

$$c = x^{(1)} - X(\pi/\omega, 0)x^{(0)} = \begin{pmatrix} -1 \\ 0 \end{pmatrix}.$$

Using (6.9) and making an equivalent substitution $\frac{l_1}{\omega} \to l_1, -l_2 \to l_2$, we come to

$$\mu_0 = \max_l \frac{-l_1\omega}{\max_{0 \le t \le \frac{\pi}{\omega}} |l_1 \sin(\omega t) + l_2 \cos(\omega t)|}.$$

Introducing $\varphi = \arccos\left(l_1/\sqrt{l_1^2 + l_2^2}\right)$ we get

$$l_1 \sin(\omega t) + l_2 \cos(\omega t) = \sqrt{l_1^2 + l_2^2} \sin(\omega t + \varphi),$$

so that

$$\mu_0 = \max_l \frac{-l_1\omega}{\sqrt{l_1^2 + l_2^2} \max_{0 \le t \le \pi/\omega} |\sin(\omega t + \varphi)|}. \tag{6.14}$$

Note that the right hand side of (6.14) depends only on the ratio $\varphi = \text{arctg}(l_2/l_1)$. Then, since $l_1 > 0$ is obviously infeasible, we may take $l_1 = -1$, arriving at the minmax problem of finding

$$\min_{l_2} \sqrt{1 + l_2^2} \max_{0 \le t \le \pi/\omega} |\sin(\omega t + \varphi)|,$$

Its solution gives $l_2^0 = 0$, $\mu_0 = \omega$ with only one extremum $t = \pi/2\omega$. Then the desired control is

$$u = u^0 = \alpha\delta\left(t - \frac{\pi}{2\omega}\right).$$

The constant $\alpha = -\omega$ may be found from condition

$$\langle l^0, c \rangle = \int_{t_0}^{\vartheta+0} l^{0\prime} G(\vartheta, t) B(t) dU^0(t),$$

which follows from (6.9), (6.11).

So we have

$$\int_0^{\pi/\omega} \sin(\omega t) u^0 dt = -\omega, \quad \int_0^{\pi/\omega} \cos(\omega t) u^0 dt \implies \alpha \sin \pi/2 = -\omega,$$

The specified control $u^0 = dU^0/dt$ ensures $\text{Var}U(\cdot) = \min$. It is also a *time-optimal control* for starting point $x^{(0)}$ and $x^{(1)} = 0$ (prove that).

Exercise 6.1.5. 1. Indicate the set of solutions for this example when the time interval is $\left[0, \frac{2\pi}{\omega}\right]$, the criterion is $\text{Var}U(\cdot) = \min$ and c is arbitrary.
2. Find the time optimal solution for any starting point $\{t, x\}$ and terminal point $x(t_{optimal}) = 0$.

We now move to solutions in the class of *closed-loop* (feedback) controls. This requires us to develop a Dynamic Programming approach within this new class.

6.1.2 Closed-Loop Impulse Control: The HJB Variational Inequality

Returning to Problem 6.1.1 with arbitrary starting position, we shall look for the value function $V(t, x)$, assuming $\|l\|_V$ *is a norm*. (As indicated above this is ensured if system (6.2) is strongly completely controllable.) Recall that $\mu^0 = \mu^0(t, x; \vartheta, x^{(1)})$.

We have

$$V(t, x) =$$

$$\inf\{\mu^0(t, x; \vartheta, x^{(1)}) + \varphi(x^{(1)})\} = \inf\{\|x^{(1)} - G(\vartheta, t) - x\|_V^* + \varphi(x^{(1)}) \mid x^{(1)} \in \mathbb{R}^n\}. \quad (6.15)$$

Denote

$$\Phi(x^{(1)}) = \|c(t, x^{(1)})\|_V^* + \varphi(x^{(1)}), \quad c(t, x^{(1)}) = x^{(1)} - G(\vartheta, t)x,$$

6.1 The Problem of Impulse Controls

and suppose $\inf\{\Phi(x^{(1)})|x^{(1)} \in \mathbb{R}^n\} > -\infty$. Then, since $\|c(t, x^{(1)})\|_V^*$ is a norm, we have

$$\Phi^*(0) = \max\{-\Phi(x^{(1)})|x^{(1)} \in \mathbb{R}^n\} < \infty.$$

Hence $0 \in \text{int Dom}\Phi^*$ which indicates that the level sets of $\Phi(x)$ are compact (closed and bounded). Then clearly, $V(t, x) < \infty$, the "infimum" in $x^{(1)}$ in the above is attainable (let it be point x^*) and can be substituted for "minimum." This, together with Lemma 6.1.2 brings us to the next conclusion.

Theorem 6.1.3. *With $V(t, x) < \infty$ there exists a point x^* where*

$$\min\{\Phi(x^{(1)})|x^{(1)} \in \mathbb{R}^n\} = \mu^0(t, x; \vartheta, x^*) + \varphi(x^*).$$

For every starting position $\{t, x\}$ there exists an open-loop control $U^0(\cdot|t, x)$ of type (6.12) which steers system (6.1) from $\{t, x\}$ to $\{\vartheta, x^*\}$ thus solving Problem 6.1.1.

In order to calculate the conjugate of $V(t, x)$ in the second variable we apply formula

$$V(t, x) = \min_{x^{(1)}}\{\max_l\{\langle l, G(\vartheta, t)x - x^{(1)}\rangle + \varphi(x^{(1)}) \mid \|l\|_V \leq 1, x^{(1)} \in \mathbb{R}^n\},$$

where the min and max are interchangeable (check this property). Changing the order of these operations and using the definition of convex conjugate functions we have

$$V(t, x) = \max_l \min_{x^{(1)}}\{\langle l, G(\vartheta, t)x - x^{(1)}\rangle + \varphi(x^{(1)}) \mid x^{(1)} \in \mathbb{R}^n, \|l\|_V \leq 1\} =$$

$$= \max_l\{\langle l, G(\vartheta, t)x\rangle - \max_{x^{(1)}}\{\langle l, x^{(1)}\rangle - \varphi(x^{(1)}) \mid x^{(1)} \in \mathbb{R}^n\} \mid \|l\|_V \leq 1\} =$$

$$= \max_l\{\langle G'(\vartheta, t)l, x\rangle - \varphi^*(l) - I(l|\mathcal{B}_V[t, \vartheta])\}, \qquad (6.16)$$

with

$$\mathcal{B}_V[t, \vartheta] = \{l : \|l\|_V \leq 1\} \text{ and } \|l\|_V = \max_\xi\{\|B'(\xi)G'(\vartheta, \xi)l\| \mid \xi \in [t, \vartheta]\},$$

where $\|l\|$ is the Euclidean norm.

This is a unit ball in $\mathbb{R}^n$ whose definition depends on system parameters and the interval $[t, \vartheta]$.

The last formula yields the next result.

Theorem 6.1.4. *The value function $V(t,x)$ is convex in x and its conjugate in the second variable is*

$$V^*(t,l) = \varphi^*(G'(t,\vartheta)l) - I(G'(t,\vartheta)l \,|\, \mathcal{B}_V[t,\vartheta]). \tag{6.17}$$

Exercise 6.1.6. Check whether $V^{**}(t,x) = V(t,x)$.

Exercise 6.1.7. Prove Theorem 6.1.4 without Assumption 1.5.2 on complete controllability of system (6.2). Then $\|l\|_V$ will be only a seminorm rather than a norm in $\mathbb{R}^n$.

Theorem 6.1.5. *The value function $V(t, x \mid \vartheta + 0, \varphi(\cdot))$ satisfies **the principle of optimality** as a semigroup property: for any $\tau \in [t, \vartheta]$ the next relation is true*

$$V(t,x) = V(t, x \mid \vartheta + 0, \varphi(\cdot)) = V(t, x \mid \tau, V(\tau + 0, \cdot \mid \vartheta + 0, \varphi(\cdot))). \tag{6.18}$$

Proof. If at time τ there is no jump, then $V(\tau, x(\tau)) = V(\tau + 0, x(\tau + 0))$. So, by applying formulas (6.16), (6.17) we have:

$$V(\tau, x\mid\vartheta + 0, \varphi(\cdot)) = \max_l\{\langle l, G(\vartheta, \tau)x\rangle - \varphi^*(l) - I(l\mid\mathcal{B}_V[\tau,\vartheta]) \mid l \in \mathbb{R}^n\},$$

$$V(t, x\mid\tau, V(\tau,\cdot)) = \max_l\{\langle l, G(\tau,t)x\rangle - V^*(\tau,l) - I(l \mid \mathcal{B}_V[t,\tau]) \mid l \in \mathbb{R}^n\} =$$

$$= \max_l\{\langle G'(\tau,\vartheta)l, G(\vartheta,t)x\rangle - \varphi^*(G'(\tau,\vartheta)l) - I(G'(\tau,\vartheta)l\mid\mathcal{B}_V[\tau,\vartheta])$$

$$-I(G'(\tau,\vartheta)l\mid\mathcal{B}_V[t,\tau]) \mid l \in \mathbb{R}^n\}.$$

After substituting $G'(\tau,\vartheta)l = \lambda$ we get

$$V(t,x\mid\tau, V(\tau,\cdot)) = V(t,x\mid\tau, V(\tau,\cdot\mid\vartheta + 0, \varphi(\cdot))) =$$

$$= \max_\lambda\{\langle \lambda, G(\vartheta,t)x\rangle - \varphi^*(\lambda) - I(\lambda \mid \lambda \in \mathcal{B}_V[t,\vartheta])\} = V(t,x\mid\vartheta + 0, \varphi(\cdot)).$$

In the last line we presume $\mathcal{B}_V[t,\vartheta] = \{\lambda : \|\lambda\|_V \leq 1\}$. □

Now, if at time τ there is jump $x(\tau + 0) - x(\tau) \neq 0$, we have

$$V(\tau, x(\tau + 0)) = V(\tau + 0, x(\tau + 0)) = V(\vartheta + 0, \varphi(\cdot)),$$

$$V(t,x) = V(\tau, x(\tau+0)) = V(\tau, x(\tau)) + \alpha(\tau); \quad \alpha(\tau) = V(\tau, x(\tau+0)) - V(\tau, x(\tau)).$$

Using these relations with previous type of reasoning, we come to (6.18).

Remark 6.1.3. For the specific problem of this section note that the **inequality**

$$V(\vartheta, x\mid\vartheta, \varphi(\cdot)) \leq \varphi(x)$$

6.1 The Problem of Impulse Controls

is true, since

$$V(\vartheta, x|\vartheta, \varphi(\cdot)) =$$

$$= \max_l \{\langle l, x \rangle - \varphi^*(l) \mid \|B'(\vartheta)l\| \leq 1\} \leq \max_l \{\langle l, x \rangle - \varphi^*(l) \mid l \in \mathbb{R}^n\} \leq \varphi(x).$$

This inequality may even be a strict one. Thus, with $\varphi(x) = I(x|\{x^{(1)}\})$, $B(\vartheta) = I$, we have $V(\vartheta, x|\vartheta, \varphi(\cdot)) = \max_l \{\langle l, x - x^{(1)} \rangle | \|l\| \leq 1\} = \|x - x^{(1)}\| < \varphi(x)$.

The principle of optimality may now be used for deriving an analogy of the HJB equation. However, here we have to count on two types of points $\tau \in [t, \vartheta]$, namely, those, where there is no control ($U(t) = const$, with $t \in [\tau - \varepsilon, \tau + \varepsilon]$ for some $\varepsilon > 0$) and those where there is an impulse $u^0 \delta(t - \tau)$.

In the first case take $t = \tau, \sigma \in [0, \varepsilon)$. Then we come to inequality

$$v(\tau) = V(\tau, x) \leq V(\tau + \sigma, G(\tau + \sigma, \tau)x) = v(\tau + \sigma),$$

where $v(\sigma + \tau)$ is nondecreasing due to the principle of optimality. Hence $V(\tau, x)$ is directionally differentiable at $\{\tau, x\}$ along $\{1, A(\tau)x\}$ and here

$$V'(\tau, x; 1, A(\tau)x) = dv(\sigma)/d\sigma \,|_{\sigma=0} \geq 0.$$

With V totally differentiable we have

$$H_1(\tau, x, V_t, V_x) = \frac{dV(\tau, x)}{d\tau} = V_t(\tau, x(\tau)) + \langle V_x(\tau, x(\tau)), A(\tau)x \rangle \geq 0. \quad (6.19)$$

Further on, in the absence of differentiability the scalar product should be interpreted as the directional derivative.

In the second case there is a jump at $t = \tau$. Then τ does not change and we take $V_t(\tau, x(\tau)) = 0$. Here denote the jump as $u = \alpha h$, where the unit vector h defines *the direction* of the jump and $\alpha > 0$—the *size* of the jump.

Now we have

$$H_2(\tau, x, 0, V_x) = \alpha \min_h \{\langle V_x(\tau, x), B(\tau)h \rangle + \|h\| \mid \|h\| = 1\} = 0, \quad (6.20)$$

where the minimizer h^0 of unit length gives us the direction of the jump. Clearly, here the multiplier $\alpha > 0$ may be omitted.

The size α of the jump is obtained from condition

$$H_1(\tau + 0, x(\tau + 0), V_t(\tau + 0, x(\tau + 0)), V_x(\tau + 0, x_+(\tau + 0))) = 0. \quad (6.21)$$

Here $x(\tau + 0) = x(\tau) + B(\tau)u^0$, $u^0 = \alpha h^0$, $\alpha > 0$. With $\varphi(x) \equiv 0$ it can also arrive from the calculation $\alpha = V(\tau + 0, x(\tau + 0)) - V(\tau, x(\tau))$.

We have thus found that $V(t, x)$ must satisfy the relation (a variational inequality)

$$\mathcal{H}(t, x, V_t, V_x) = \min\{H_1(t, x, V_t, V_x), H_2(t, x, V_t, V_x)\} = 0, \qquad (6.22)$$

under boundary condition

$$V(\vartheta, x) = V(\vartheta, x; \vartheta, \varphi(\vartheta)). \qquad (6.23)$$

If necessary, $V(t, x)$ may be always treated as a generalized (viscosity-type) solution (see [16]).

Denote $\mathbf{q} = V_t$, $\mathbf{p} = V_x$. The last conclusions are now summarized as follows.

Theorem 6.1.6. *The value $V(t, x)$ function is a solution (classical or generalized) to the HJB-type Eq. (6.22) under boundary condition (6.23), where*

$$H_1(t, x, \mathbf{q}, \mathbf{p}) = \mathbf{q} + \langle \mathbf{p}, A(t)x \rangle, \qquad H_2(t, x, \mathbf{q}, \mathbf{p}) = \min_h\{\langle \mathbf{p}, B(t)h \rangle + \|h\| \mid \|h\| = 1\}.$$

Denote $\mathcal{H}^0 = \arg\min_h\{\langle \mathbf{p}, B(t)h \rangle + \|h\| \mid \|h\| = 1\}$.

Theorem 6.1.7. *The synthesizing control $\mathcal{U}^0(t, x)$ has the following form:*

(i) *if $H_2(t, x, V_t, V_x) > 0$, then $\mathcal{U}^0(t, x) = \{0\}$;*
(ii) *if $H_2(t, x, V_t, V_x) = 0$, then $\mathcal{U}^0(t, x) = \{u^0 = \alpha h : h \in \mathcal{H}^0\}$,*
with (ii) taken under $H_2(t, x(t+0), V_t(t, x(t+0)), V_x(t, x(t+0))) = 0$, and with $H_1(t, x(t+0), V_t(t, x(t+0)), V_x(t, x(t+0))) = 0$, $x(t+0) = x + \alpha B(t)h$.

Related assertions are proved by direct calculation. These theorems are illustrated next.

Example 6.1.2. Consider the problem of minimizing $\text{Var}\, U(\cdot)$ due to equation

$$dx = b(t)dU, \quad x(0) = x, \quad x(2) = 0, \quad b(t) = 1 - (1-t)^2.$$

Here $x \in \mathbb{R}$, $t \in [0, 2]$. Following Theorems 6.1.6, 6.1.7, we have

$$V(t, x) =$$

$$= \max_l \left\{ \frac{\langle l, -x \rangle}{\max_\xi |b(\xi)l|} \,\middle|\, \xi \in [t, 2], l \in \mathbb{R}, l \neq 0 \right\} = \frac{|x|}{\max_\xi\{b(\xi) | \xi \in [t, 2]\}}$$

$$= \begin{cases} |x|b^{-1}(1), & t \leq 1, \\ |x|b^{-1}(t), & t > 1. \end{cases}$$

6.1 The Problem of Impulse Controls

Let $\tau = 1$. Consider the following cases:

- With $t < \tau$ we have $V_t(t,x) = 0, V_x(t,x) = \text{sign} x \, b^{-1}(1)$, so that $H_1(t,x,V_t,V_x) = V_t(t,x) = 0$ and

$$H_2(t,x,V_t,V_x) = 1 - |b(t)V_x| = 1 - \frac{b(t)}{b(1)} = (1-t)^2 > 0.$$

There are no jumps at these points.
- With $t > \tau$ we have

$$V_t(t,x) = \frac{2(t-1)|x|}{t^2(2-t)^2}, \quad V_x(t,x) = \frac{\text{sign}(x)}{b(t)}.$$

Here $H_1(t,x,V_t,V_x) = V_t(t,x) > 0$ with $x \neq 0$, while $H_1(t,x,V_t,V_x) = 0$ with $x = 0$ and

$$H_2(t,x,V_t(t,x),V_x(t,x)) = 1 - |b(t)V_x| = 1 - b(t)b^{-1}(t) = 0, \; \forall x.$$

Therefore, if we start moving at $t^* > \tau$, then there must be an immediate jump to zero with $u^0 = -x(t^*)\text{sign} x(t^*) b^{-1}(t) |\delta(t-t^*)$.
- With $t = \tau$ we have

$$b(\tau) = 1, \; V(\tau,x) = |x|b^{-1}(\tau) = |x|, \; V_t(\tau,x) = 0, V_x(\tau,x) = \text{sign} x,$$

while

$$H_2(\tau+0, x(\tau+0), V_t(\tau+0, x(\tau+0)), V_x(\tau+0, x(\tau+0))) =$$
$$\min_h \{\langle \text{sign}(x), h\rangle + 1\} = h^0 + 1 = 0,$$

and

$$V(\tau, x(\tau)) - V(\tau+0, x(\tau+0)) = \alpha = |x|,$$
$$H_1(\tau+0, x(\tau+0), V_t(\tau+0, x(\tau+0)), V_x(\tau+0, x(\tau+0))) = 0.$$

There is a jump $u^0 = -x \text{sign} x \delta(t-\tau)$.

Finally, the synthesized system now looks like

$$dx/dt = -b(t)x\delta(t-\tau) \text{ or } dx = -b(t)x \, dU^0(t,\tau), \quad (6.24)$$

where $U^0(t,\tau) = \mathbf{1}_\tau(t)$. (Recall that $\mathbf{1}_\tau(t) = 0$, if $t \in [0,\tau]$ and $\mathbf{1}_\tau(t) = 1$, if $t \in (\tau, 2]$.)

The solution to symbolic Eq. (6.24) is given by the Lebesgue–Stieltjes integral equation (see [234])

$$x(t) = x - \int_0^t b(t)x(t)\,dU^0(t,\tau),$$

where $x(t)$ is a left-continuous function. □

It now just remains to interpret the synthesized equation in the general case. This will be done at the end of next section after discussing approximate solutions to impulse controls.

6.2 Realizable Approximation of Impulse Controls

Impulse controls introduced above are "ideal" elements. In order to allow a physical realization of the proposed schemes we introduce the next "realistic" scheme. Here the original control will be subjected to *double constraints* where in the previous problem there is an additional hard bound on the norm of the control vector, while the bounding parameter may tend to infinity.

6.2.1 The Realistic Approximation Problem

Problem 6.2.1. Consider Problem 6.1.1 under additional constraint $u(t) \in \mathcal{B}_\mu(0)$, so that there arrives the next problem: find

$$J(t_0, x^0 | u(\cdot)) = \int_{t_0}^{\vartheta} \|u(t)\|\,dt + \varphi(x(\vartheta)) \to \inf, \quad (6.25)$$

$$\dot{x}(t) = A(t)x(t) + B(t)u(t), \quad t \in [t_0, t_1], \quad (6.26)$$

$$x(t_0) = x^0, \quad \|u(t)\| \le \mu. \quad (6.27)$$

The solution to this problem exists since the set of all admissible controls $u(\cdot)$ is now weakly compact in $L_2([t_0, t_1]; \mathbb{R}^p)$ and convex, while the cost functional $J(u(\cdot))$ is weakly lower semicontinuous and convex.

The value function for the last problem is

$$V(t, x, \mu) = \min_u \{J(t, x | u(\cdot)) \mid t, x\}.$$

Using the schemes of Chap. 2 (see Sect. 2.3.1), we come to the next proposition.

6.2 Realizable Approximation of Impulse Controls

Theorem 6.2.1. *The value function $V(t, x)$ is a solution to the HJB equation*

$$\frac{\partial V(t,x,\mu)}{\partial t} + \min \left\{ \left\langle \frac{\partial V(t,x,\mu)}{\partial x}, A(t)x(t) + B(t)u \right\rangle + \|u\| \;\Big|\; \|u\| \leq \mu \right\} = 0, \tag{6.28}$$

with boundary condition

$$V(t, x, \mu) = \varphi(x). \tag{6.29}$$

As we shall see, $V(t, x, \mu)$ is directionally differentiable at each position $\{t, x\}$, so that it satisfies Eq. (6.28) for all t.

Function $V(t, x, \mu)$ may be calculated through methods of convex analysis, similar to the schemes of Sects. 2.4 and 2.6. Indeed, after applying the minmax theorem of [70], and a permutation of the minimum in u with the integral, we have

$$V(t, x, \mu) =$$

$$= \min_u \max_l \left\{ \int_t^\vartheta (\|u(\xi)\| + \langle l, G(\vartheta, \xi) B(\xi) u(\xi) \rangle) d\xi + \langle l, G(\vartheta, t) x \rangle \right.$$

$$\left. - \varphi^*(l) \;\Big|\; l \in \mathbb{R}^n, \|u\| \leq \mu \right\} =$$

$$= \max_l \{\phi(t, x, l) + \langle l, G(\vartheta, t) x \rangle - \varphi^*(l)\} \tag{6.30}$$

where

$$\phi(t, x, l) = \int_t^\vartheta \min_u \{\langle l, G(\vartheta, s) B(s) u \rangle + \|u\| \;|\; \|u\| \leq \mu\} ds.$$

Here *the subproblem*

$$h(s, l) = \min_u \{\psi[s] B(s) u + \|u\| \;|\; \|u\| \leq \mu\},$$

with $\psi[s] = l' G(\vartheta, s)$, has the solution

$$h(s, l) = \begin{cases} 0 & \text{if } \|\psi[s] B(s)\| \leq 1 \\ \mu(1 - \|\psi[s] B(s)\|) & \text{if } \|\psi[s] B(s)\| > 1, \end{cases}$$

This yields

Lemma 6.2.1. *The value function*

$$V(t, x, \mu) = \max_l \{\langle l, G(\vartheta, t) x \rangle - \mu \int_t^\vartheta \{\|\psi[s] B(s)\| - 1\}_+ ds - \varphi^*(l)\} =$$

$$= \max_l \{\psi[t] x + \phi(t, x, l) - \varphi^*(l) \;|\; l \in \mathbb{R}^n\},$$

where

$$\{\|\psi[s]B(s)\| - 1\}_+ = \begin{cases} \|\psi[s]B(s)\| - 1\|, & \text{if } \|\psi[s]B(s)\| - 1 > 0 \\ 0, & \text{if } \|\psi[s]B(s)\| - 1 \le 0. \end{cases}$$

Exercise 6.2.1. (a) Check that function $V(t, x, \mu)$ satisfies Eqs. (6.28), (6.29).
(b) Prove that with $\mu \to \infty$ function $V(t, x, \mu)$ converges pointwise to $V(t, x)$ of (6.16).
(c) Prove that with $\mu \to \infty$ the HJB Eqs. (6.28), (6.29) converges to Eqs. (6.22), (6.23).

Lemma 6.2.2. *The conjugate function to $V(t, x, \mu)$ in the second variable is*

$$V^*(t, l, \mu) = \mu \int_t^\vartheta \{\|l'G(t,s)B(s)\| - 1\}_+ ds + \varphi^*(l'G(t,\vartheta)). \qquad (6.31)$$

The last formula is derived through direct calculation.

The control strategy $u^0(t, x, \mu)$ for Problem 6.2.1 is obtained from the next subproblem

$$\mathcal{U}^0(t, x, \mu) = \arg\min\left\{\left\langle\frac{\partial V(t, x, \mu)}{\partial x}, B(t)u\right\rangle + \|u\|\right\}.$$

Denote $B'(t)V_x(t, x, \mu) = c(t, x, \mu)$. Then this gives

$$\mathcal{U}^0(t, x, \mu) = \begin{cases} 0, & \text{if } \|c(t, x, \mu)\| < 1, \\ [0, -\mu c(t, x, \mu)], & \text{if } \|c(t, x, \mu)\| = 1, \\ -\mu c(t, x, \mu)/\|c(t, x, \mu)\|, & \text{if } \|c(t, x, \mu)\| > 1. \end{cases}$$

Theorem 6.2.2. *The closed-loop solution to Problem 6.2.1 is given by $\mathcal{U}^0(t, x, \mu)$ and the optimal trajectories emanating from $\{t, x\}$ satisfy the differential inclusion*

$$\frac{dx}{ds} \in A(s)x + B(s)\mathcal{U}^0(s, x, \mu). \qquad (6.32)$$

Here set-valued function $\mathcal{U}^0(s, x, \mu)$ is upper semicontinuous in $\{s, x\}$. Therefore a solution to (6.32) exists.

Suppose $x_\mu[s]$ is a solution to (6.32) emanating from $\{t, x\}$. Denote $u^0(s, x_\mu[s], \mu) \in \mathcal{U}^0(s, x_\mu[s], \mu)$ to be a measurable selector of $\mathcal{U}^0(s, x_\mu[s], \mu)$ and take continuous function $U^0(s, x_\mu[s], \mu)$ to be such that $dU^0(s, x[s], \mu) = u^0(s, x[s], \mu)ds$ a.e. Then $x_\mu[s]$ will also be a solution to the Stieltjes integral equation

$$x_\mu[s] = G(s, t)x + \int_t^s G(s, \xi)B(\xi)dU^0(\xi, x_\mu[\xi], \mu).$$

6.2 Realizable Approximation of Impulse Controls

With $\mu \to \infty$ the approximate control $U^0(s, x_\mu[s], \mu)$ weakly converges to the ideal $\mathcal{U}^0(s, x[s])$, so that with $\mu \to \infty$

$$\int_t^s G(s, \xi) B(\xi) dU^0(\xi, x_\mu[\xi], \mu) \to \int_t^s G(s, \xi) B(\xi) d\,\mathcal{U}^0(\xi, x[\xi]).$$

where $U^0(\xi) = \mathcal{U}^0(\xi, x(\xi))$ is a piecewise-constant function. The pointwise limits of trajectories $x_\mu[s]$ will be the optimal trajectories $x[s] = x^0[s]$ for the original Problem 6.1.1.

Another version of formalizing the limit transition from approximating Problem 6.2.1 to Problem 6.1.1 is as follows.

6.2.2 The Approximating Motions

We indicate the approximation of a delta impulse $u(t, t^*, x) = \alpha(t^*, x)\delta(t - t^*)$ by a rectangular pulse (a tower) of lower base $t \in [t^*, t^* + \theta(t^*, x, \mu)]$ and height $u(t, x, \mu) = \mu$, $t \in [t^*, t^* + \theta(t^*, x, \mu)]$ with $\mu\theta(t^*, x, \mu) = \alpha = constant \geq 0$.

Definition 6.2.1. The pair of functions $\Upsilon = \{u(t, x; \mu), \theta(t, x; \mu)\}$ (*magnitude* and *duration*), is said to be *a feedback control strategy* for Problem 6.2.1 if

$$u(t, x; \mu) \in \partial \mathcal{B}_\mu(0) \cup \{0\}, \quad \text{and} \quad u(t, x; \mu) \to u_\infty(t, x), \quad (\mu \to \infty)$$

$$\theta(t, x; \mu) \geq 0, \quad \text{and} \quad \mu\theta(t, x; \mu) \equiv \alpha(t, x) \geq 0, \quad (\mu \to \infty).$$

The component $u(t, x; \mu)$ is the approximating control impulse which is issued on the interval $[t, t + \theta(t, x; \mu)]$. Note that $\theta \to 0$ as $\mu \to \infty$, and in the limit one has a delta-function of direction $u_\infty(t, x)$ and intensity $\alpha(t, x)$ as the ideal impulse control.

Definition 6.2.2. Fix a control strategy Υ, number $\mu > 0$ and a partition $t_0 = \tau_0 < \tau_1 < \ldots < \tau_s = t_1$ of interval $[t_0, t_1]$. An **approximating motion** of system (6.28) is the solution to the differential equation

$$\tau_i = \tau_{i-1} + \theta(\tau_{i-1}, x_\Delta(\tau_{i-1}); \mu),$$

$$\dot{x}_\Delta(\tau) = \mu B(\tau) u(\tau_{i-1}, x_\Delta(\tau_{i-1}); \mu), \quad \tau_{i-1} \leq \tau < \tau_i.$$

Number $\sigma = \max_i \{\tau_i - \tau_{i-1}\}$ is the *diameter* of the partition.

Definition 6.2.3. A **constructive motion** of system (6.2) under feedback control Υ is a piecewise continuous function $x(t)$, which is the pointwise limit of approximating motions $x_\Delta(t)$ as $\mu \to \infty$ and $\sigma \to 0$.

Due to [75] the approximating differential inclusion (6.32) under feedback strategy $\mathcal{U}^0(s, x, \mu)$ does have a solution, a realizable constructive approximating motion. And thus the ends meet.

An illustrative explanation to the indicated procedure is given by the following situation. Suppose the on-line position of system (6.1) is $\{t, x\}$, and from (6.22) it follows that control $U^0(\cdot)$ has a jump $\bar{h}\delta(t - \bar{t})$. Then the feedback strategy pair $\bar{u} = u(\bar{t}, \bar{x}; \mu)$, $\bar{\theta} = \theta(\bar{t}, \bar{x}; \mu)$ is to be chosen such that

$$B(t)\bar{h} = \mu \left[\int_{\bar{t}}^{\bar{t}+\bar{\theta}} B(t) dt \right] \bar{u}.$$

In the limit this yields $u(\bar{t}, \bar{x}; \mu) \to \bar{h}/\|\bar{h}\|$ and $\mu\theta(\bar{t}, \bar{x}; \mu) \to \|\bar{h}\|$ with $\mu \to \infty$.

That is, an impulse $\bar{h}\delta(t - \bar{t})$ of intensity $\alpha = \bar{h}$ is approximated by a platform of height (magnitude) μ, direction $\bar{h}$, and width (duration) $\mu^{-1}\|\bar{h}\|$.

Example 6.1.3. We now deal with the same 2-dimensional system as in Example 6.1.1, but looking for the solution in the form of feedback control strategies. This will be done by specifying certain *impulse domains* which serve as regions where the instantaneous impulses are activated. They are in a certain sense similar to switching surfaces for bang-bang control.

We consider system

$$\begin{cases} dx_1 = x_2 dt, \\ dx_2 = -x_1 dt + dU, \end{cases}$$

where impulses arrive as generalized derivatives of $u = dU/dt$ and, $t \in [0, \pi/2]$, $x(t) = [x_1, x_2]'$. Our aim will be to minimize $\text{Var} U(\xi)$, $\xi \in [t-0, \vartheta+0]$, under $x(\vartheta) = 0$, $\vartheta = \pi/2$.

Following (6.8), (6.16), we have

$$\mu^0(t, x) = V(t, x) = \max\{\langle l, -G(\pi/2, t)x \rangle \mid \|l\|_V \leq 1\}.$$

where

$$G(\pi/2, t) = \begin{pmatrix} \sin t & \cos t \\ -\cos t & \sin t \end{pmatrix},$$

and

$$\|l\|_V = \max\{|\langle l, G(\pi/2, \xi)b \rangle| \mid \xi \in [t, \pi/2]\} =$$

$$= \sqrt{l_1^2 + l_2^2} \max\{|\cos(\xi - \varphi)| \mid \xi \in [t, \pi/2]\}$$

where $b = [0, 1]'$ and $\varphi = \arcsin l_2 (l_1^2 + l_2^2)^{-1/2}$, $\varphi \in [-\pi/2, \pi/2]$.

6.2 Realizable Approximation of Impulse Controls

Further assuming $t = 0$, we have. $\xi \in [0, \pi/2]$ and $\cos t = 1, \sin t = 0$. This gives

$$\max |\cos(\xi - \varphi)| = \begin{cases} 1, & \vartheta \in (0, \pi/2), \\ \max\{|\sin \varphi|, |\cos \varphi|\}, & \text{otherwise}, \end{cases}$$

and

$$\|l\|_V = \begin{cases} \sqrt{l_1^2 + l_2^2} & l_1 l_2 > 0, \\ \max\{|l_1|, |l_2|\} & \text{otherwise}. \end{cases}$$

Then

$$\mu^0(0, x) = \|z\| = \max\{\langle l, z \rangle \mid_V^* \|l\|_V \leq 1\} = \begin{cases} \sqrt{z_1^2 + z_2^2} & z_1 z_2 > 0, \\ \max\{|z_1| + |z_2|\} & z_1 z_2 < 0, \end{cases}$$

where $z = G(\pi/2, 0)x = \{x_2, -x_1\}$. The maximizer in the last problem is

$$l^0 = \begin{cases} (1, -1) & z_1 > 0, z_2 < 0, \\ (-1, 1) & z_1, 0, z_2 > 0 \\ z/\|z\| & z_1 z_2 > 0, \end{cases}$$

and the impulses arrive at $\tau = \text{argmax}\{|b'G'(\pi, \xi)l^0| \mid \xi \in [0, \pi/2]\}$ which here, at $z_1 z_2 < 0$, are $\tau_1 = 0, \tau_2 = \pi/2$. Then the optimal synthesizing strategy $u^0(t, x)$ at point $\{t \in (-0, \vartheta], x\}$, for $\alpha_1 > 0, \alpha_2 > 0$, is

$$u(t, x) = -\alpha_1 \delta(t - 0) + \alpha_2 \delta(t - \pi/2), \quad \alpha_1, \alpha_2 \geq 0, \quad \alpha_1 + \alpha_2 = \mu^0,$$

Checking this with terminal condition $x(\vartheta) = 0$, we finally get $\alpha_1 = x_2, \alpha_2 = x_1$.

This solution is illustrated in the following figures. Namely, Fig. 6.2 demonstrates the general picture of ideal feedback solutions from position $\{(t - 0), x\}$ with optimal discontinuous trajectories with jumps of the velocities x_2 shown in Fig. 6.3. The realizable solution under double constraint is indicated in Fig. 6.4. Here the plane is divided into four domains three of which, D_0, D_μ, and $D_{-\mu}$ correspond, respectively, to controls $u = 0, u = \mu, u = -\mu$. The fourth external domain $D_\emptyset$ indicates initial states from which the solution to problem of reaching final state $x = 0$ under constraint μ is impossible (not solvable). Figure 6.5 illustrates convergence of solutions from case Fig. 6.4 to case Fig. 6.3 with $\mu \to \infty$.

272   6 Impulse Controls and Double Constraints

Fig. 6.2 General scheme of feedback impulse control

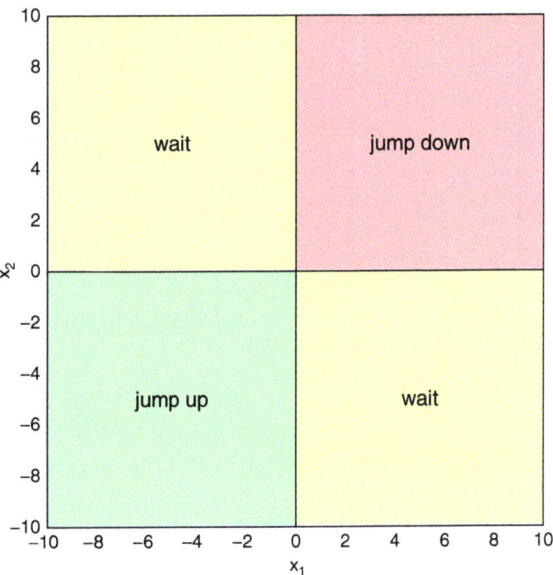

Fig. 6.3 Ideal control trajectories

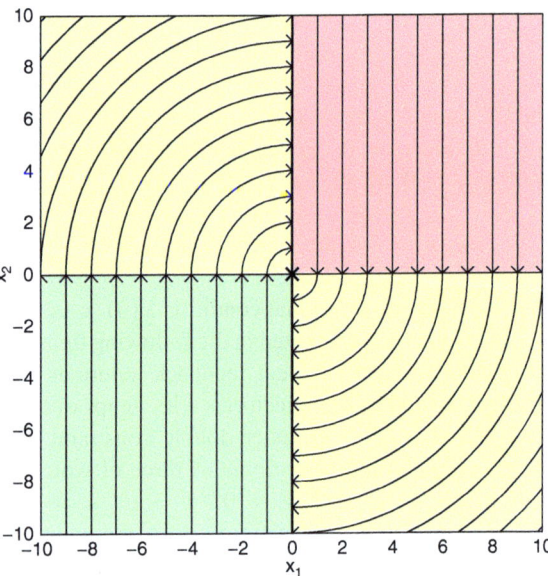

6.2 Realizable Approximation of Impulse Controls

Fig. 6.4 Double-constraint control trajectories

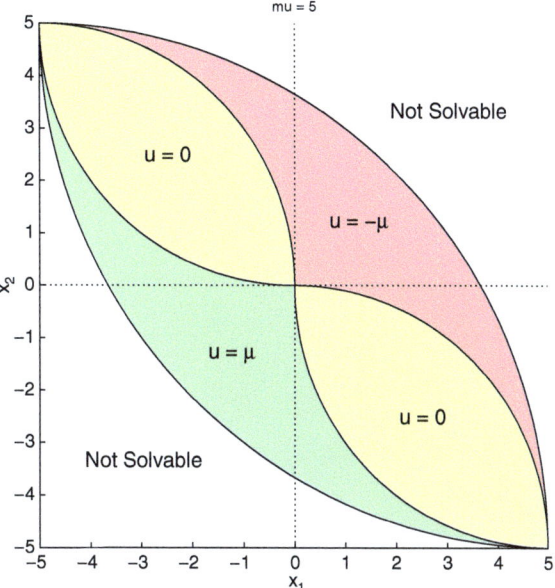

Fig. 6.5 The convergence of double-constraint controls to the impulse control

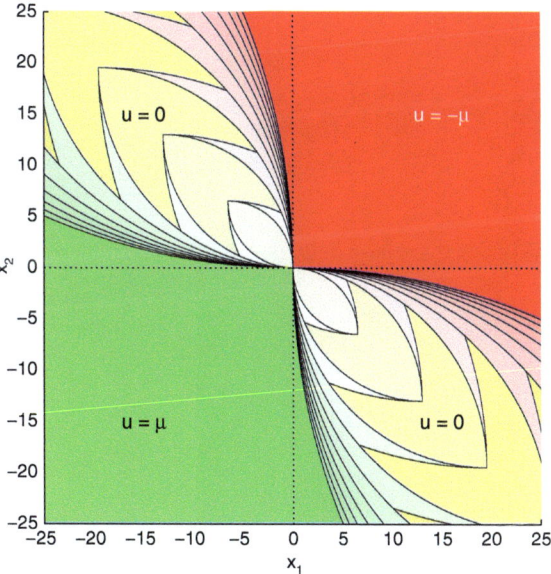

Chapter 7
Dynamics and Control Under State Constraints

Abstract The topics of this chapter are problems of reachability and system dynamics under state constraints in the form of reach tubes. Indicated are general approaches based on the Hamiltonian formalism and a related Comparison Principle. Further emphasis is on the dynamics of linear systems under hard bounds on the controls and system trajectories. A detailed solution is presented based on ellipsoidal approximations of bounded trajectory tubes.

Keywords State constraints • Reach tubes • Comparison Principle • Hamiltonian approach • Linear-convex systems • Ellipsoidal techniques

The issue of control under state constraints is at the heart of many applied problems. An abundant literature on such issues is published under the notion of "viability theory" [5, 6]. The specifics of this book lies in general treatment of such problems under time-dependent state constraints through Hamiltonian techniques with related Comparison Principle, while the linear-convex case is covered in detail through ellipsoidal approximations. Such a move provides a natural merger of theory with appropriate computation schemes. The material of this chapter uses results of papers [158, 185]. Its topics are also a useful preparation for further consideration of problems on feedback control under state constraints and external obstacles addressed in the next chapter.

7.1 State-Constrained Reachability and Feedback

This section deals with general theory of reach sets and tubes under hard bounds on controls, and *time-dependent geometric (hard) constraints on the states*. The solution relations are given here in Hamiltonian terms, while the linear-convex case is described through tight external approximations by parametrized ellipsoid-valued tubes. The result is an exact parametric representation of state-constrained reachability tubes through families of external ellipsoidal tubes.

7.1.1 The Reachability Problem Under State Constraints

We begin with the general nonlinear case. Similar to (1.1) the controlled system is thus described by the same differential equation, with same properties, namely

$$\dot{x} = f(t, x, u), \quad t_0 \leq t \leq \tau, \tag{7.1}$$

in which the state $x \in \mathbb{R}^n$ and the control $u \in \mathbb{R}^p$ are restricted by

$$u(t) \in \mathcal{P}(t), \quad x(t) \in \mathcal{Y}(t), \tag{7.2}$$

for all $t \geq t_0$. Here $\mathcal{P}(t), \mathcal{Y}(t)$ are compact set-valued functions in $\mathbb{R}^p, \mathbb{R}^n$, respectively, continuous in the Hausdorff metric. The initial state is restricted by $x(t_0) \in \mathcal{X}^0$, a compact subset of $\mathbb{R}^n$. The function $f(t, x, u)$ is assumed to ensure uniqueness and uniform extension of solutions to any finite interval of time for any $x(t_0) = x^0$, $u(t) \in \mathcal{P}(t)$, $t \geq t_0$.

Definition 7.1.1. Given set-valued *positions* $\{t_0, \mathcal{X}^0\}$, $\mathcal{X}^0 \cap \mathcal{Y}(t_0) \neq \emptyset$, the *reach set* (or "attainability domain") $X(\tau, t_0, \mathcal{X}^0)$ *at time* $\tau > t_0$ is the set

$$X[\tau] = X(\tau, t_0, \mathcal{X}^0)$$

of all states $x[\tau] = x(\tau, t_0, x^0)$, $x^0 = x(t_0) \in X(t_0) = \mathcal{X}^0$, that can be reached at time τ by system (7.1), from some $x^0 \in \mathcal{X}^0$, using all possible controls u that ensure constraints (7.2). The set-valued function $\tau \mapsto X[\tau] = X(\tau, t_0, \mathcal{X}^0)$ is the *reach tube* from $\{t_0, \mathcal{X}^0\}$.

The *Basic Problem* is simply stated:

Problem 7.1.1. Calculate the reach sets $X(\tau, t_0, \mathcal{X}^0)$, $\tau \geq t_0$.

A fairly general approach is to relate reach sets to an optimization problem [178]. This could be done by calculating certain *value functions* that may be selected in several ways. Consider first the value function

$$V(\tau, x) = \min_{u} \left\{ d^2(x(t_0), \mathcal{X}^0) + \int_{t_0}^{\tau} d^2(x(t), \mathcal{Y}(t)) dt \right\}, \tag{7.3}$$

under the restriction $x(\tau) = x$. Here the minimum is over all measurable functions (controls) $u(t) \in \mathcal{P}(t)$, function $x(t), (t \geq t_0)$, is the corresponding trajectory of (7.1), and

$$d^2(x, X) = \min\{\langle x - z, x - z \rangle | z \in X\}$$

is the square of the distance function $d(x, X)$.

7.1 State-Constrained Reachability and Feedback

Lemma 7.1.1. *The following relation is true:*

$$X(\tau, t_0, X^0) = \{x : V(\tau, x) \leq 0\}.$$

This follows from the definition of the reach set $X(\tau, t_0, X^0)$, which is thus a *level set* of $V(\tau, x)$.

To state the important semigroup property of the value function, we extend definition (7.3) to more general boundary conditions, namely,

$$V(\tau, x \mid t_0, V(t_0, \cdot)) =$$

$$= \min_u \left\{ V(t_0, x(t_0)) + \int_{t_0}^{\tau} d^2(x(t), \mathcal{Y}(t)) dt \mid u(\cdot) \in \mathcal{P}(\cdot) \right\}, \quad (7.4)$$

under the restriction $x(\tau) = x$. Function $V(t_0, \cdot)$ defines a given boundary condition. In (7.2) this condition is $V(t_0, x) = d^2(x, X^0)$.

Theorem 7.1.1. *The value function $V(\tau, x)$ satisfies the principle of optimality, which has the semigroup form*

$$V(\tau, x \mid t_0, V(t_0, \cdot)) = V(\tau, x \mid t, V(t, \cdot \mid t_0, V(t_0, \cdot))), \quad t_0 \leq t \leq \tau. \quad (7.5)$$

This property is established through conventional arguments similar to Sect. 2.1, that also yield similar properties for reach sets. The solution of the reachability problem can be cast in the form of a solution of the "forward" HJB equation that follows from (7.5). To develop this approach we further assume that the function $V(t, x)$ is *continuously differentiable*.

A standard procedure, as before, then yields

$$V_t(t, x) + \max_u \{\langle V_x(t, x), f(t, x, u)\rangle - d^2(x, \mathcal{Y}(t)) \mid u \in \mathcal{P}(t)\} = 0, \quad t_0 \leq t \leq \tau, \quad (7.6)$$

with boundary condition $V(t_0, x) = d^2(x, X^0)$. Here V_t, V_x stand for the partial derivatives of $V(t, x)$. Note that the term $d^2(x, \mathcal{Y}(t)) \neq 0$ only outside the state constraint $\mathcal{Y}(t)$.

An alternative scheme relies on the value function

$$V(\tau, x) = \min_u \{\varphi_0(t_0, x(t_0))\}. \quad (7.7)$$

The minimum in (7.7) is over all $u(t) \in \mathcal{P}(t)$, $t \in [t_0, \tau]$, under restrictions $x(\tau) = x$ and $\varphi(t, x) \leq 1$, $t \in [t_0, \tau]$. We will use this formulation for the case when

$$\varphi_0(t_0, x) = \langle x - x^0, (X^0)^{-1}(x - x^0)\rangle^{1/2}, \quad X^0 = X^{0\prime} > 0,$$

$$\varphi(t, x) = \langle x - y(t), (Y(t))^{-1}(x - y(t))\rangle^{1/2}, \quad Y(t) = Y'(t) > 0, \quad (7.8)$$

so that the initial condition and the state constraints are ellipsoidal:

$$x(t_0) \in \mathcal{E}(x^0, X^0) = \{x : \varphi_0(t_0, x) \leq 1\},$$

$$x(t) \in \mathcal{E}(y(t), Y(t)) = \{x : \varphi(t, x(t)) \leq 1, \ t \in [t_0, \tau]\}.$$

Once $x(\tau, t_0, x(t_0)) = x(\tau) = x$ is fixed, one may figure out whether $x \in X(\tau, t_0, X^0)$ by looking at the value $V(\tau, x)$. Then the respective vector $x(t_0) \in \mathcal{E}(x^0, X^0)$ iff there exists a control $u(t)$, $t \in [t_0, \tau]$, that ensures $V(\tau, x) \leq 1$, so that $x(\tau) \in X(\tau, t_0, X^0)$ iff $V(\tau, x) \leq 1$.

Therefore the reach set at time τ is

$$X(\tau, t_0, X^0) = \{x \mid V(\tau, x) \leq 1\},$$

which is thus a level set of $V(\tau, x)$, and the previously stated *semigroup property* of the value function (the principle of optimality) holds again.

The solution of the reachability problem can again be cast in the form of a solution of the forward HJB equation—now somewhat different from (7.6). Again assume the functions $V(t, x)$ and $\varphi(t, x)$ are *continuously differentiable*.

Denote

$$\mathcal{H}(t, x, V, u) = V_t(t, x) + \langle V_x(t, x), f(t, x, u) \rangle. \tag{7.9}$$

Lemma 7.1.2. *The formal HJB equation is*

$$\max_{u \in \mathcal{P}(t)} \mathcal{H}(t, x, V, u) = 0 \ \text{if} \ \varphi(t, x) < 1, \tag{7.10}$$

and

$$\max_{u \in \mathcal{P}_s(t)} \mathcal{H}(t, x, V, u) = 0 \ \text{if} \ \varphi(t, x) = 1, \tag{7.11}$$

with

$$\mathcal{P}_s(t, x) = \mathcal{P}(t) \cap \{u \mid \mathcal{H}(t, x, \varphi, u) \leq 0\}, \tag{7.12}$$

and with boundary condition

$$V(t_0, x) = \varphi_0(t_0, x). \tag{7.13}$$

We sketch a proof of Lemma 7.1.2.

Together with $\varphi(t, x) < 1$, the principle of optimality (7.5), with $t_0 = \tau - \sigma$, gives

$$0 = \min_u \{V(\tau - \sigma, x(\tau - \sigma)) - V(\tau, x) \mid u \in \mathcal{P}(t), \ t \in [\tau - \sigma, \tau]\} \tag{7.14}$$

7.1 State-Constrained Reachability and Feedback

or

$$\max_{u \in \mathcal{P}(t)} \mathcal{H}(t, x, V, u) = \mathcal{H}(t, x, V, u^0) = 0. \tag{7.15}$$

With $\varphi(t, x) = 1$ we apply the same principle but only through those controls that do not allow the trajectory to move outside the state constraint. These are

$$u(t, x) \in \mathcal{P}_s(t, x) = \{u \in \mathcal{P}(t) \cap \{u \mid \mathcal{H}(t, x, \varphi, u) \leq 0\}\}.$$

Let $u^0(t, x)$ be the optimal control for the trajectory $x^0(t)$ that emanates from $x(\tau) = x$ and minimizes $\varphi_0(t_0, x(t_0))$ under constraints $\varphi(t, x(t)) \leq 1$, $t \in [t_0, \tau]$.

Note that with $\varphi(t, x^0(t)) = 1$ we have two cases: either $\mathcal{H}(t, x^0(t), \varphi, u^0) = 0$, which means the related optimal trajectory runs along the state constraint boundary, or

$$d\varphi(t, x^0)(t))/dt|_{u=u^0} = \mathcal{H}(t, x^0(t), \varphi, u^0) < 0,$$

so the optimal trajectory departs from the boundary, and for $\sigma > 0$ we have $\varphi(t + \sigma, x^0(t + \sigma)) < 1$. Relations $\mathcal{H}(t, x^0(t), \varphi, u^0) \equiv 0$ and (7.15) allow one to find the control u^0 along the state constraint boundary.

If all the operations in (7.6), (7.10), (7.11) result in smooth functions, then these equations may have a classical solution [51]. Otherwise (7.6), (7.7) form a symbolic generalized HJB equation, which has to be described in terms of subdifferentials, Dini derivatives, or their equivalents. However, the typical situation is that V is not differentiable and as said before, in Chap. 5, the treatment of (7.6), (7.7) then involves the notion of a generalized "viscosity" solution for these equations or their equivalents [50, 80, 247]. One approach is to use the method of characteristics as developed for this type of equation (see [51]) and modified for the nonsmooth case [248]. But it is a fairly complicated procedure, especially in the nonsmooth case for which the method requires additional refinement. Another approach is to look for the solution through level set methods [221, 244]. However, for the specific "linear-convex" problems of this book the function $V(t, x)$ is convex in x, hence directionally differentiable in any direction $\{(1, x)\}$, which proves it to be a generalized solution through classical technique (see [16, 123]). Moreover, in this case an effective ellipsoidal technique may be applied, which allows one to bypass the calculation of solutions to the HJB equation.

The Evolution Equation

The next question is: on which evolution equation would the multivalued mapping $\mathcal{X}[t] = \mathcal{X}(t, \cdot) = \mathcal{X}(t, \cdot | \mathcal{X}(t_0, \cdot))$ under state constraint be satisfied? An answer is in papers [158, 164], where this was suggested to be taken as *the integral funnel equation* for differential inclusion

$$\dot{x} = \in f(t, x, \mathcal{P}(t)), \quad x(t_0) \in X^0, \tag{7.16}$$

under constraint

$$x \in \mathcal{Y}(t), \ t \in [t_0, \vartheta].$$

Here is one of such equations (see [174] and also [164])

$$\lim_{\sigma \to +0} \frac{1}{\sigma} h_+ \left(Z[t+\sigma], \bigcup \left\{ x + \sigma(f(t, x, \mathcal{P}(t)) \middle| x \in Z[t] \cap \mathcal{Y}(t) \right\} \right) = 0, \ Z(t_0) = X^0, \tag{7.17}$$

where h_+ is *the Hausdorff semidistance*:

$$h_+(X', X'') = \min\{\varepsilon : X' \subseteq X'' + \varepsilon \mathcal{B}(0)\}$$

and $\mathcal{B}(0)$ is a unit ball in $\mathbb{R}^n$. The solution $Z[t]$ to this evolution equation is a multivalued function, with $Z[t_0] = X^0$, which satisfies (7.17) almost everywhere. As a rule, this solution is not unique. However, there exists a solution $X_m[t]$, sought for among these $Z(\cdot)$, which coincides with the realization of function $X(t, \cdot)$ and is the *inclusion-maximal* among all solutions $Z[t]$ to (7.17). Namely, $X[t] \supseteq Z[t]$, where $Z[t]$ is any solution that starts from $Z(t_0) = X^0$, see [174]. Note that Eq. (7.17) makes sense for any upper semi-continuous set-valued functions $\mathcal{Y}(t)$. Hence in equations of type (7.17) we may allow these functions to be right or left-continuous.

However, there is a more rigorous version of the funnel equation (see [158]), namely,

$$\lim_{\sigma \to +0} \frac{1}{\sigma} h_+ \left(Z[t+\sigma], \bigcup \left\{ x + \sigma(f(t, x, \mathcal{P}(t)) \middle| x \in Z[t] \cap \mathcal{Y}(t+\sigma) \right\} \right) = 0, \ Z(t_0) = X^0. \tag{7.18}$$

This version demands that support function $f(l, t) = \rho(l \mid \mathcal{Y}(t))$ would be differentiable in t and its partial derivative $\partial f(l, t)/\partial dt$ would be continuous. Then the set-valued solution $Z[t]$ satisfies (7.18) everywhere and will be unique.

Theorem 7.1.2. *(i) The solution $Z[t] = Z(t, t_0, X^0)$ to differential inclusion (7.16) satisfies the funnel equation (7.17) almost everywhere and its inclusion-maximal solution $Z[t] = X_m[t]$ is unique.*
(ii) The solution $Z[t] = Z(t, t_0, X^0)$ to differential inclusion (7.16) satisfies the funnel equation (7.18) everywhere and its solution $Z[t] = X[t]$ is unique.

In the sequel we always presume the following.

Assumption 7.1.1. *Functions $V(\tau, x)$ are continuous in all the variables, with nonempty compact zero level-sets that have nonempty interiors of full dimension.*

Such level sets

$$\mathcal{X}[\tau] = \{x : V(\tau, x) \leq 0\} \neq \emptyset$$

may be gotten due to the last assumption for Lemma 7.1.2. Such are those that solve our problem for linear systems with continuous coefficients and convex constraints continuous in time. The indicated class of function $V(\tau, \cdot)$ will be denoted as $\mathcal{K}_V$.

Remark 7.1.1. The variety $\mathcal{K}_V$ may be considered as a metric space with metric:

$$d(V'(\tau, \cdot), V''(\tau, \cdot)) = h(\mathcal{X}', \mathcal{X}''),$$

where $\mathcal{X}', \mathcal{X}''$—zero level-sets of V', V'', and the Hausdorff distance

$$h(\mathcal{X}, \mathcal{Z}) = \max\{h_+(\mathcal{X}, \mathcal{Z}), h_+(\mathcal{Z}, \mathcal{X})\},$$

where $h_+(\mathcal{X}, \mathcal{Z})$ is the Hausdorff semidistance.

This variety $\mathcal{K}_V$ may also be endowed with another metric, introducing $d(V'(\tau, \cdot), V''(\tau, \cdot))$ as the distance in the space $C_r[t_0, \tau]$ (of r-dimensional continuous functions) between respective functions $y'(t)$, $y''(t)$ ($t \in [t_0, \tau]$) of (7.8) or (7.19), which generated $V'(\tau, \cdot), V''(\tau, \cdot)$ due to Eqs. (7.9)–(7.13). Such a definition may be propagated only to functions $y(\cdot) \in Y(\cdot)$. Similar metrics may be also considered on the variety of functions $V^{(1)}(\tau, x)$.

A detailed discussion of possible metrics for spaces of functions similar to those generated by value functions of this chapter is given in monograph [98, Chap. 4, Appendix C.5].

Sets $\mathcal{X}[t]$ may thus be found, as in systems without state constraints, either through value functions $V(t, x)$, from their level sets, or through evolution Eqs. (7.17), (7.18). However, since direct calculation of these solutions is not simple, we remind that it may suffice to avoid exact solutions to related HJB equations under state constraints. We may hence apply the *comparison principle* which gives upper and lower approximations to value function $V(t, x)$ for HJB equation (7.6). Henceforth we may also obtain external and internal ellipsoidal estimates of sets $\mathcal{X}$. Such issues are described in the next sections.

7.1.2 Comparison Principle Under State Constraints

We shall now indicate the specifics of applying the comparison principle of Sect. 5.1 when the HJB equation is derived under state constraints. We further work with system (7.1), (7.2), but here the state constraint $\mathcal{Y}(t)$ will be presented as

$$\mathcal{Y}(t) = \{x : y(t) - g(t, x) \in \mathcal{R}(t)\}, \ t \in [t_0, \tau], \tag{7.19}$$

where values y, g lie in $\mathbb{R}^r$, $r \leq n$, while properties of functions $y(t), g(t,x)$ as same as those of $x(t), f(t,x)$, respectively, and set-valued function $\mathcal{R}(t)$ is taken to be convex, compact-valued, and Hausdorff-continuous. The Comparison Principle is further treated within two versions.

Version-A

We first use the approach used for Lemma 7.1.2 relying on relations of type (7.3)–(7.6) and modifying them in view of (7.19). Then introducing notation

$$\mathbf{H}(t,x,p) = \max\{\langle p, f(t,x,u)\rangle | u \in \mathcal{P}(t)\} - d^2(y(t) - g(t,x), \mathcal{R}(t)),$$

we observe, as in Sects. 2.3 and 2.4 (see also [178, 198]), that the solution $V(t,x)$ of the corresponding "forward" HJB equation

$$V_t + \mathbf{H}(t,x,V_x) = 0, \quad V(t_0,x) = d^2(x, \mathcal{X}^0) \qquad (7.20)$$

allows to calculate $X[t] = X(t, t_0, \mathcal{X}^0)$ as the level set $X[\tau] = \{x : V(\tau, x) \leq 0\}$. Here again this property is independent of whether V is a classical or a generalized solution to Eq. (7.20).

To obtain approximations of $V(t,x)$ we proceed as follows.

Assumption 7.1.2. *Given are functions $H^+(t,x,p), w^+(t,x) \in C_1$ and $\mu(t) \in L_1$, which satisfy the inequalities*

$$\mathbf{H}(t,x,p) \leq H^+(t,x,p), \quad \forall \{t,x,p\}, \qquad (7.21)$$

$$w_t^+ + H^+(t,x,w_x) \leq \mu(t), \qquad (7.22)$$

$$w^+(t_0,x) \leq V(t_0,x). \qquad (7.23)$$

Theorem 7.1.3. *Suppose $H^+(t,x,p), w^+(t,x), \mu(t)$ satisfy Assumption 7.1.2. Then the following estimate for the reachability set $X[t]$ is true:*

$$X[t] \subseteq X_+[t], \qquad (7.24)$$

where

$$X_+[t] = \left\{ x : w^+(t,x) \leq \int_{t_0}^t \mu(s)ds + V(t_0, x(t_0)) \right\}. \qquad (7.25)$$

Consider a pair $x^* \in X[t]$, $u^*(s) \in \mathcal{P}(s)$, $s \in [t_0, t]$, so that the corresponding trajectory $x^*(s, t, x^* | u^*(\cdot)) = x^*[s] \in X[s]$. Then

$$dw^+(t,x)/dt|_{x=x^*(t)}$$
$$= w_t^+(t,x^*) + \langle w_x^+(t,x^*), f(t,x^*,u^*)\rangle - d^2(y(t) - g(t,x^*(t)), \mathcal{R}(t))$$
$$\leq w_t^+(t,x^*) + \mathbf{H}(t,x^*,w_x^+) \leq w_t^+(t,x^*) + H^+(t,x^*,w_x^+) \leq \mu(t),$$

7.1 State-Constrained Reachability and Feedback

hence

$$dw^+(s, x)/ds|_{x=x^*(s)} \leq \mu(s).$$

Integrating this inequality from t_0 to t gives

$$w^+(t, x) \leq \int_{t_0}^t \mu(s)ds + w^+(t_0, x(t_0)) \leq \int_{t_0}^t \mu(s)ds + V(t_0, x(t_0))\},$$

which implies $x^*(t) \in X_+[t]$ and the theorem is proved.

We now move to the discussion of internal estimates for sets $X[t]$. As in the above, we do not necessarily require differentiability of the value function $V(t, x)$. Consider the next assumption.

Assumption 7.1.3. *Given are functions $H^-(t, x, p) \in C$ and $w^-(t, x) \in C_1$, which satisfy the inequalities*

$$(i) \quad \mathbf{H}(t, x, p) \geq H^-(t, x, p), \quad \forall \{t, x, p\},$$

$$(ii) \quad w_t^-(t, x) + H^-(t, x, w_x^-(t, x)) \geq 0,$$

$$(iii) \quad w^-(t_0, x) \geq V(t_0, x).$$

Theorem 7.1.4. *Suppose functions $H^-(t, x, p), w^-(t, x)$ satisfy Assumption 7.1.3. Then the following upper estimate for $V(t, x)$ is true:*

$$V(\tau, x) \leq w^-(\tau, x). \tag{7.26}$$

Indeed, with t, x given, consider an optimal trajectory $x(s)$ for problem (7.1), (7.2) with (7.19), $t_0 \leq s \leq t$, under condition $x(t) = x$. The definition of $V(t, x)$ yields

$$V(t, x) = V(t_0, x^0) + \int_{t_0}^t d^2(y(s) - g(s, x(s)), \mathcal{R}(s))ds.$$

Differentiating w^- along this optimal trajectory and having in view Assumption 7.1.3, we get

$$dw^-(s, x)/ds|_{x=x(s)} = w_s^-(s, x(s)) + \langle w_x^-(s, x(s)), f(s, x(s), u(s)) \rangle$$

$$= w_s^-(s, x(s)) + \mathbf{H}(s, x(s), w_x^-(s, x(s))) + d^2(y(s) - g(s, x(s)), \mathcal{R}(s))$$

$$\geq w_s^-(s, x(s)) + H^-(s, x(s), w_x^-(s, x(s))) + d^2(y(s) - g(s, x(s)), \mathcal{R}(s))$$

$$\geq d^2(y(s) - g(s, x(s)), \mathcal{R}(s)).$$

Integrating the last equality on $[t_0, t]$, we further have

$$w^-(t, x) \geq w(t_0, x^0) + \int_{t_0}^{t} d^2(y(s) - g(s, x(s)), \mathcal{R}(s))ds$$

$$\geq V(t_0, x^0) + \int_{t_0}^{t} d^2(y(s) - g(s, x(s)), \mathcal{R}(s))ds = V(t, x).$$

and the theorem is proved.

Denoting $X^-[t] = \{x : w^-(t, x) \leq 0\}$, $t \geq t_0$, and using the last inequality, we further come to the conclusion

Corollary 7.1.1. *Under Assumption 7.1.3 the following inclusion is true:*

$$X^-[t] \subseteq X[t]. \tag{7.27}$$

This statement is similar to those of 7.5–7.7 from [48].

Version-B

We now rely on the approach used for Lemma 7.1.1. After introducing notation

$$\mathbf{H}^1(t, x, p, p_0, \lambda) = \max\{\langle p, f(t, x, v)\rangle - \lambda \langle p_0, f(t, x, v)\rangle \,|\, v \in \mathcal{P}(t)\},$$

the corresponding "forward" HJB equation (7.7) transforms into

$$V_t^{(1)} + \lambda \theta_t + \mathbf{H}^1(t, x, V_x, \theta_x, \lambda) = 0, \quad V(t_0, x) = d(x, X^0) \tag{7.28}$$

where the multiplier λ is to be found from condition

$$d\theta(t, x)/dt \,|_{(7.1)} = \theta_t + \langle \theta_x, f(t, x, v)\rangle \leq 0, \tag{7.29}$$

which is the total derivative of constraint function θ along trajectories of system (7.1).

This allows one to calculate $X[t] = X(t, t_0, X^0)$ as the level set $X[\tau] = \{x : V^{(1)}(\tau, x) \leq 1\}$. This property is independent of whether $V^{(1)}$ is a classical or a generalized solution to Eq. (7.29). The comparison theorems and the respective proofs for Version-B are similar to Version-A.

Remark 7.1.2. In this subsection we dealt with reach sets that are forward. For the design of state-constrained feedback controls we may need to use backward reach sets. These are designed similarly and discussed with detail within related schemes of Sect. 5.1.

We now pass to the class of linear systems with convex constraints where the previous results may be developed with greater detail.

7.1.3 Linear-Convex Systems

Consider the linear system

$$\dot{x} = A(t)x + B(t)u, \quad t_0 \le t \le t_1, \tag{7.30}$$

in which $A(t), B(t)$ are continuous, and the system is completely controllable (see [109]). The control set $\mathcal{P}(t)$ is a nondegenerate ellipsoid, $\mathcal{P}(t) = \mathcal{E}(p(t), P(t))$, and

$$\mathcal{E}(p(t), P(t)) = \{u \mid \langle u - p(t), P^{-1}(t)(u - p(t)) \rangle \le 1\}, \tag{7.31}$$

with $p(t) \in \mathbb{R}^p$ (the center of the ellipsoid) and the symmetric positive definite matrix function $P(t) \in \mathbb{R}^{p \times p}$ (the shape matrix of the ellipsoid) continuous in t. The *support function* of the ellipsoid is

$$\rho(l \mid \mathcal{E}(p(t), P(t))) = \max\{\langle l, u \rangle) \mid u \in \mathcal{E}(p(t), P(t))\} = \langle l, p(t) \rangle + \langle l, P(t)l \rangle^{1/2}.$$

The *state constraint* is

$$x(t) \in \mathcal{Y}(t), \quad t \in [t_0, t_1], \tag{7.32}$$

in which $\mathcal{Y}(t)$ is an ellipsoidal-valued function $\mathcal{Y}(t) = \mathcal{E}(y(t), Y(t))$, $Y(t) = Y'(t) > 0$, $Y(t) \in \mathbb{R}^{n \times n}$, and with $y(t), Y(t)$ absolutely continuous. Lastly, $\mathcal{X}^0 = \mathcal{E}(x^0, X^0)$ is also an ellipsoid.

In [174, 181] it is also assumed that the constraints on the controls and initial values are ellipsoidal, which makes the explanations more transparent. However, these methods are applicable to box-valued constraints as well, see Sect. 5.4.3 and also [160, 182].

Lemma 7.1.3. *For the linear system (7.30), with convex-valued restrictions (7.31), (7.32) the reach set $X(\tau, t_0, \mathcal{X}^0)$ at time τ is convex and compact.*

The problem of calculating the reach set is now formulated as follows.

Problem 7.1.2. *Calculate the support functions $\rho(l \mid X(\tau, t_0, \mathcal{X}^0))$, $l \in \mathbb{R}^n$.*

This is equivalent to solving the following optimal control problem with state constraints:

$$\rho(l \mid X(\tau, t_0, \mathcal{X}^0)) = \max \langle l, x(\tau) \rangle$$

subject to $x(t_0) \in \mathcal{X}^0$, $u(t) \in \mathcal{P}(t)$, $x(t) \in \mathcal{Y}(t)$, $t_0 \le t \le \tau$.

For each vector l the solution to the optimal control problem is attained at the terminal point $x^{(0)}(\tau)$ of an *optimal trajectory* $x^{(0)}(t)$, $t \in [t_0, \tau]$, which starts from a point $x^{(0)}(t_0) \in X^0$ determined throughout the solution process.

We will solve Problem 7.1.2 by calculating the value function (7.4), namely,

$$V(\tau, x) = \min_u \{d(x(t_0), \mathcal{E}(x^0, X^0)) \mid x(\tau) = x,$$

$$u(t) \in \mathcal{P}(t), \ x(t) \in \mathcal{E}(y(t), Y(t)), \ t \in [t_0, \tau]\},$$

for system (7.30), using the techniques of convex analysis as in paper [166].

To handle the state constraints $X[\tau]$, one usually imposes the following *constraint qualification*.

Assumption 7.1.4. *There exist a control* $u(t) \in \mathcal{P}(t)$, $t \in [t_0, \tau]$, $x^0 \in X^0$, *and* $\varepsilon > 0$, *such that the trajectory* $x[t] = x(t, t_0, x^0) = x(t, t_0, x^0 \mid u(\cdot))$ *satisfies*

$$x(t, t_0, x^0) + \varepsilon \mathcal{B}(0) \subset \mathcal{Y}(t), \ t \in [t_0, \tau].$$

Here $\mathcal{B}(0) = \{z : (z, z) \leq 1, \ z \in \mathbb{R}^n\}$. Under Assumption 7.1.4 this gives

$$V(\tau, x) = \sup\{\langle l, x \rangle - \Psi(\tau, t_0, l) \mid l\},$$

where

$$\Psi(\tau, t_0, l) = \min\{\Psi(\tau, t_0, l, \Lambda(\cdot)) \mid \Lambda(\cdot)\}, \qquad (7.33)$$

$$\Psi(\tau, t_0, l, \Lambda(\cdot)) = s(t_0, \tau, l \mid \Lambda(\cdot))x^0 + (s(t_0, \tau, l \mid \Lambda(\cdot))X^0 s'(t_0, \tau, l \mid \Lambda(\cdot)))^{1/2}$$

$$+ \int_{t_0}^{\tau} \left(s(t, \tau, l \mid \Lambda(\cdot))B(t)p(t) + (s(t, \tau, l \mid \Lambda(\cdot))B(t)P(t)B'(t)s'(t, \tau, l \mid \Lambda(\cdot)))^{1/2} \right) dt$$

$$+ \int_{t_0}^{\tau} \left(d\Lambda(t)y(t) + (d\Lambda(t)Y(t)d\Lambda'(t))^{1/2} \right).$$

Here, $s(t, \tau, l \mid \Lambda(\cdot))$, $t \leq \tau$, is the *row-vector* solution to the adjoint equation

$$ds = -sA(t)dt + d\Lambda(t), \ s(\tau) = l', \qquad (7.34)$$

and $\Lambda(\cdot) \in \mathbf{V}_n[t_0, \tau]$—the space of n-dimensional functions of bounded variation on $[t_0, \tau]$. We also used the notation

$$\int_{t_0}^{\tau} (d\Lambda(\cdot)Y(t)d\Lambda'(\cdot))^{1/2} = \max \left(\int_{t_0}^{\tau} d\Lambda(t)h(t) \mid h(t) \in \mathcal{E}(0, Y(t)), t \in [t_0, \tau] \right).$$

This is the maximum of a Stieltjes integral over continuous functions $h(t) \in \mathcal{E}(0, Y(t))$. The minimum Ψ over $\Lambda(\cdot)$ is reached because of Assumption 7.1.4.

A direct consequence of formula (7.33) is the solution to Problem 7.1.2.

7.1 State-Constrained Reachability and Feedback

Theorem 7.1.5. *Under Assumption 7.1.4, the support function $\rho(l \mid X[\tau])$ is given by*

$$\rho(l \mid X[\tau]) = \Psi(\tau, t_0, l) = \min\{\Psi(\tau, t_0, l, \Lambda(\cdot)) \mid \Lambda(\cdot) \in \mathbf{V}_n[t_0, \tau]\}. \quad (7.35)$$

Let $\Lambda^{(0)}(\cdot) \in \arg\min\{\Psi(\tau, t_0, l, \Lambda(\cdot)) \mid \Lambda(\cdot) \in \mathbf{V}_n[t_0, \tau]\}$ be the minimizer of (7.14) and let $s^{(0)}[t] = s(t, \tau, l \mid \Lambda^{(0)}(\cdot))$ be the solution to (7.30) with $\Lambda(\cdot) = \Lambda^{(0)}(\cdot)$.

The application of formula (7.34) indicates a new approach to the next proposition. Alternate derivations of this approach can be found in [137, 166, pp. 116–121].

Theorem 7.1.6 (The "Standard" Maximum Principle Under State Constraints). *For Problem 7.1.2 under Assumption 7.1.4, the optimal control $u^{(0)}(t)$, initial condition $x^{(0)}$, and trajectory $x^{(0)}[t] = x(t, t_0, x^{(0)} \mid u^{(0)}(\cdot))$ must satisfy the "maximum principle"*

$$s^{(0)}[t]B(t)u^{(0)}(t) = \max\{s^{(0)}[t]B(t)u \mid u \in \mathcal{E}(p(t), P(t))\}, \quad t_0 \leq t \leq \tau, \quad (7.36)$$

and the "maximum conditions"

$$\int_{t_0}^{\tau} (d\Lambda^{(0)}(t)y(t) + (d\Lambda^{(0)}(t)Y(t)d\Lambda^{(0)'}(t))^{1/2}) = \int_{t_0}^{\tau} d\Lambda^{(0)}(t)x^{(0)}[t] \quad (7.37)$$

$$= \max\left\{\int_{t_0}^{\tau} (d\Lambda^{(0)}(t)z(t)) \mid z(t) \in \mathcal{E}(y(t), Y(t))\right\},$$

$$s^{(0)}[t_0]x^{(0)}[t_0] = s^{(0)}[t_0]x^0 + (s^{(0)}[t_0], X^0 s^{(0)}[t_0])^{1/2} =$$

$$\max\{s^{(0)}[t_0]x \mid x \in \mathcal{E}(x^0, X^0)\}. \quad (7.38)$$

Here we note that the minimum over $\Lambda(\cdot) \in \mathbf{V}_n[t_0, \tau]$ in (7.33) may be replaced by the minimum over the pair $\{M(t), \lambda(t)\}$, with $d\Lambda(t) = l'M(t)d\lambda(t)$, in which the $n \times n$ matrix $M(t)$ is continuous, and $\lambda(\cdot) \in \mathbf{V}_1[t_0, \tau]$ is a scalar function of bounded variation. Moreover, $M(t)$ may be chosen within a compact set C_0 of continuous functions. This new form of the multiplier $d\Lambda(t)$ is a result of combining the earlier schemes of [96, 158]. We summarize this result as follows.

Lemma 7.1.4. *The multiplier $\Lambda^{(0)}(t)$ allows the representation*

$$d\Lambda^{(0)}(t) = l'M^{(0)}(t)d\lambda^{(0)}(t), \quad (7.39)$$

with $M^{(0)}(\cdot) \in C_0$, $\lambda^{(0)}(\cdot) \in \mathbf{V}_1[t_0, \tau]$.

Denote by $S_M(t, \tau) = S(t, \tau \mid M(\cdot), \lambda(\cdot))$ the solution of the matrix equation

$$dS = -SA(t)dt + M(t)d\lambda(t), \quad S(\tau) = I. \quad (7.40)$$

This is a *symbolic expression* for the linear differential equation. Under *permutability* of $A(t)$ with its integral, namely, under condition

$$A(t) \int_\tau^t A(s)ds = \int_\tau^t A(s)ds\, A(t)$$

the solution to (7.40) for fixed τ is

$$S_M(t,\tau) = \exp\left(\int_t^\tau A(s)ds\right) - \int_t^\tau \left(\exp \int_n^s A(\xi)d\xi\right) M(s)d\lambda(s),$$

in which the second integral is a Stieltjes integral. In the absence of the permutation property the solution is expressed through expanding a matrix series of the *Peano* type [32]. Note that with $A(t) \equiv const$ the permutability property is always true.

As in the scheme of [158], the single-valued functional $\Psi(\tau, t_0, l, \Lambda(\cdot))$ may be substituted with the set-valued integral

$$R(\tau, t_0, M(\cdot), \lambda(\cdot), \mathcal{X}^0) = S(t_0, \tau \mid M(\cdot), \lambda(\cdot))\mathcal{E}(x^0, X^0) \qquad (7.41)$$

$$+ \int_{t_0}^\tau S(t,\tau \mid M(\cdot), \lambda(\cdot))B(t)\mathcal{E}(p(t), P(t))dt + \int_{t_0}^\tau d\lambda(t)M(t)\mathcal{E}(y(t), Y(t))dt,$$

which yields the next result whose analytical form differs from [158].

Lemma 7.1.5. *The following equalities hold:*

$$\mathcal{X}[\tau] = \cap\{R(\tau, t_0, M(\cdot), \lambda(\cdot), \mathcal{X}^0) \mid M(\cdot), \lambda(\cdot)\} =$$

$$\cap\{R(\tau, t_0, M(\cdot), \lambda(\cdot), \mathcal{X}^0) \mid M(\cdot), \lambda \equiv t\},$$

for $M(\cdot) \in C_0, \lambda \in \mathbf{V}_1[t_0, \tau]$.

Note that $l'M(t)d\lambda(t) = d\Lambda(\cdot)$ is the Lagrange multiplier responsible for the state constraint, with $M(t)$ continuous in t, and $\lambda(t)$, with bounded variation, responsible for the jumps of the multiplier $\Lambda(\cdot)$. Such properties of the multipliers are due to the linearity of the system and the type of the constraints on u, x, taken in this section (see also [96, 97]). In general, the multipliers responsible for the state constraints may be represented through a measure of general type and therefore may contain singular components. This will not be the case, however, *due to the next assumption which is now required*.

Assumption 7.1.5. *The multipliers $\Lambda^{(0)}(t), \lambda^{(0)}(t)$ do not have singular components.*

This assumption holds, for example, for piecewise absolutely continuous functions, with finite number of jumps.

Lemma 7.1.6. *Assumption 7.1.5 holds if the support function $\rho(q \mid \mathcal{Y}(t))$ is absolutely continuous in t, uniformly in q, $\langle q, q \rangle \leq 1$, and for each vector l the optimal trajectory for Problem 7.1.2 visits the boundary $\partial \mathcal{Y}(t)$ for a finite number of intervals.*

Note that Lemma 7.1.5 gives the reach set at time τ as an intersection of sets $R(\tau, t_0, M(\cdot), \lambda(\cdot), \mathcal{X}^0)$ parametrized by functions $M(\cdot) \in \mathcal{C}_0$. Lemma 7.1.5 may be used to calculate the reach set $\mathcal{X}[\tau]$ *for any fixed time* τ.

However, on many occasions our objective is to recursively calculate the whole tube $\mathcal{X}[\tau], \tau \geq t_0$. This means that while solving Problem 7.1.2 for increasing values of τ, we want a procedure that does not require one to solve the whole problem "afresh" for each new value of τ, but allows the use of the solutions for previous values.

There is a difficulty here, namely, given $M(\cdot), \lambda(\cdot)$, one may observe that in general, $R(\tau, t_0, M(\cdot), \lambda(\cdot), \mathcal{X}^0) \neq R(\tau, t, M(\cdot), \lambda(\cdot), R(t, t_0, M(\cdot), \lambda(\cdot), \mathcal{X}^0))$, which means R given by (7.41) does not satisfy the semigroup property. Therefore Theorem 7.1.5 and Lemmas 7.1.4 and 7.1.5 have to be modified to meet the recursion requirements. Such a move would yield a modified version of the maximum principle for linear systems under state constraints that eliminates the last difficulty.

7.2 State-Constrained Control: Computation

In this section we emphasize computational approaches. We start with a modified maximum principle. Then present an ellipsoidal technique to calculate reach sets for linear systems with constraints on the control and state. The suggested scheme introduces parametrized families of tight ellipsoidal-valued tubes that approximate the exact reach tube from above, touching the tube along specially selected "good" curves of Sect. 2.3 that cover the entire exact tube. This leads to recursive relations that compared to other approaches, simplify calculations. The proofs that rely on the mentioned "recurrent" version of the maximum principle under state constraints are followed by an example. Finally given is a description of specific properties used for proving the assertions stated in Sects. 7.1, 7.2 and crucial for easing the computational burden in calculation.

7.2.1 The Modified Maximum Principle

To achieve the desired results, in the following sections we shall use some properties related to the structure of the optimal controls and the state constraints. These properties and additional assumptions are summarized below in Sect. 7.2.1 and are typical of the problems under discussion. They are borrowed from previous investigations.

Starting with Assumptions 7.1.4 and 7.1.5 and presuming the conditions of Lemma 7.1.6 are true, let us first restrict Λ in (7.39) to satisfy the relation $d\Lambda(t) = l'M(t)d\lambda(t)$, in which $\lambda(t)$ is *absolutely continuous*. This means $d\Lambda(t) = l'M(t)\eta(t)dt$ and $\eta(t) = d\lambda(t)/dt \equiv 0$ for $t \in T_Y = \{t \mid x_l^{(0)}(t) \in \text{int}\mathcal{Y}(t)\}$, when $x_l^{(0)}(t)$ is the optimal trajectory for Problem 7.1.2, for the given l.

Denoting $L(t) = S^{-1}(t, \tau \mid M(\cdot), \eta(\cdot))M(t)\eta(t)$, we may replace

$$dS/dt = -SA(t) + M(t)\eta(t), \quad S(\tau, \tau \mid M(\cdot), l(\cdot)) = I, \tag{7.42}$$

whose solution is $S(t, \tau \mid M(\cdot), l(\cdot))$, with

$$dS_L/dt = -S_L(A(t) - L(t)), \quad S_L(\tau, \tau) = I,$$

whose solution is $S_L(t, \tau)$. Then $S(t, \tau \mid M(\cdot), l(\cdot)) \equiv S_L(t, \tau)$, $t \in [t_0, \tau]$.

In this case, $\mathcal{R}(\tau, t_0, M(\cdot), \eta(\cdot), \mathcal{X}^0)$ transforms into

$$\mathcal{R}(\tau, t_0, L(\cdot)) = S_L(t_0, \tau)\mathcal{E}(x^0, X^0) + \int_{t_0}^{\tau} S_L(t, \tau)(\mathcal{E}(B(t)p(t), B(t)P(t)B'(t))$$

$$+ L(t)\mathcal{E}(y(t), Y(t)))dt = \mathcal{X}(\tau, t_0, L(\cdot), \mathcal{X}^0) = \mathcal{X}_L[\tau],$$

and $\mathcal{X}_L[\tau]$ turns out to be the solution to the differential inclusion

$$\dot{x}(t) \in (A(t) - L(t))x + L(t)\mathcal{E}(y(t), Y(t)) + \mathcal{E}(B(t)p(t), B(t)P(t)B'(t)), \tag{7.43}$$

$$t \geq t_0, \quad x^0 \in \mathcal{E}(x^0, X^0). \tag{7.44}$$

Here also the compact set $\mathcal{C}_0$ of functions $M(t)$ transforms into a compact set $\mathcal{C}_{00}$ of functions $L(t)$.

We thus arrive at the following important property, which is similar to those proved in [158].

Lemma 7.2.1. *The reach set $\mathcal{X}[\tau]$ is the intersection*

$$\mathcal{X}[\tau] = \cap \{\mathcal{X}_L(\tau, t_0, \mathcal{E}(x^0, X^0)) \mid L(\cdot)\} \tag{7.45}$$

of the "cuts" or "cross-sections" $\mathcal{X}_L(\tau, t_0, \mathcal{E}(x^0, X^0))$ of the reach tubes (solution tubes) $\mathcal{X}_L(\cdot) = \{\mathcal{X}_L[t] : t \geq t_0\}$ to the differential inclusion (7.43), (7.44). The intersection is over all $L(\cdot) \in \mathcal{C}_{00}$, a compact set of continuous matrix functions $L(t), t \in [t_0, \tau]$.

A further calculation using convex analysis yields the next formula.

Theorem 7.2.1. *The support function*

$$\rho(l \mid \mathcal{X}(\tau, t_0, \mathcal{E}(x^0, X^0))) = \inf\{\rho(l \mid \mathcal{X}_L(\tau, t_0, \mathcal{E}(x^0, X^0))) \mid L(\cdot)\}, \tag{7.46}$$

7.2 State-Constrained Control: Computation

with

$$\rho(l \mid X_L(\tau, t_0, \mathcal{E}(x^0, X^0))) = \langle l, x_L^\star[\tau] \rangle + \langle l, G_L(\tau, t_0) X^0 G_L'(\tau, t_0) l \rangle^{1/2} \quad (7.47)$$

$$+ \int_{t_0}^\tau \langle l, G_L(\tau, s) B(s) P(s) B'(s) G_L'(\tau, s) l \rangle^{1/2} ds$$

$$+ \int_{t_0}^\tau \langle l, G_L(\tau, s) L(s) Y(s) L'(s) G_L'(\tau, s) l \rangle^{1/2} ds.$$

Here

$$x_L^\star[\tau] = G_L(\tau, t_0) x^0 + \int_{t_0}^\tau G_L(\tau, s)(B(s) p(s) + L(s) y(s)) ds, \quad (7.48)$$

and $G_L(\tau, s) = S_L(s, \tau)$ is the transition matrix for the homogeneous system

$$\dot{x} = (A(t) - L(t)) x, \quad G_L(t, t) = I, \quad L(\cdot) \in \mathcal{C}_{00}.$$

The significance of the last result is that in this specific problem the support function of the intersection (7.45) is equal to the *pointwise infimum* (7.46) of the support functions rather than to their *infimal convolution* as given by general theory [237].

It is not unimportant to specify when the infimum in (7.46) is attained; that is, it is actually a minimum. Indeed, it may happen that for a given l, the minimum over $L(\cdot)$ is in the class $\mathcal{C}_{00}$ (this, for example, is the case when $\lambda(t)$ turns out to be absolutely continuous, as in the above). But to ensure the minimum is always reached, we have to broaden the class of functions $L(\cdot)$.

To illustrate how to continue the procedure we will assume the following.

Assumption 7.2.1. *For each $l \in \mathbb{R}^n$, the optimal trajectory $x_l^0(t)$ of Problem 7.1.2 visits the boundary $\partial \mathcal{Y}(t)$ only during one time interval $[t_1, t_2]$, $t_0 \le t_1 \le t_2 \le \tau$.*

(The case of finite or countable collection of such intervals is treated in a similar way.)

Then, instead of the product $M(t) \eta(t)$, in (7.42) we must deal with multipliers of the form $\mathcal{M}(t) = M(t) \eta(t) + M_1 \delta(t - t_1) + M_2 \delta(t - t_2)$, where M_1, M_2 are $n \times n$ matrices.

By introducing a new multiplier $\mathcal{L}(t) = L(t) + L_1 \delta(t - t_1) + L_2 \delta(t - t_2)$ under transformation $\mathcal{L}(t) = S^{-1}(t, \tau \mid \mathcal{M}(\cdot)) \mathcal{M}(t)$, we shall match the formulas for $R(\tau, t_0, \mathcal{M}(\cdot))$ and its transformed version $\mathcal{R}(\tau, t_0, \mathcal{L}(\cdot))$.

Applying the schemes of [96, 97, 158], to the specific case of this paper, it is possible to rewrite the preceding assertions.

Lemma 7.2.2. *The support function*

$$\rho(l \mid X[\tau]) = \rho(l \mid X(\tau, t_0, \mathcal{E}(x^0, X^0))) = \min\{\rho(l \mid X_L(\tau, t_0, \mathcal{E}(x^0, X^0)) \mid L(\cdot)\}, \quad (7.49)$$

with

$$\rho(l \mid X_L(\tau, t_0, \mathcal{E}(x^0, X^0))) = \Phi(l, L(\cdot), \tau, t_0), \quad (7.50)$$

and

$$\Phi(l, L(\cdot), \tau, t_0) = \langle l, x_*[t] \rangle + \langle l, \mathcal{G}(\tau, t_0) X^0 \mathcal{G}'(\tau, t_0) l \rangle^{1/2} \quad (7.51)$$

$$+ \int_{t_0}^{\tau} \langle l, \mathcal{G}(\tau, s) B(s) P(s) B'(s) \mathcal{G}'(\tau, s) l \rangle^{1/2} ds$$

$$+ \int_{t_0}^{\tau} \langle l, \mathcal{G}(\tau, s) L(s) Y(s) L'(s) \mathcal{G}'(\tau, s) l \rangle^{1/2} ds$$

$$+ \int_{t_0}^{\tau} \langle l, \mathcal{G}(\tau, s) L_1 Y(s) L_1' \mathcal{G}'(\tau, s) l \rangle^{1/2} d\chi(s, t_1)$$

$$+ \int_{t_0}^{\tau} \langle l, \mathcal{G}(\tau, s) L_2 Y(s) L_2' \mathcal{G}'(\tau, s) l \rangle^{1/2} d\chi(s, t_2).$$

Here $\mathcal{G}(\tau, s)$ is the transition matrix for the homogeneous system

$$dx = (A(t) - L(t))x\,dt - L_1 x\,d\chi(t, t_1) - L_2 x\,d\chi(t, t_2),$$

so that

$$x(t) = \mathcal{G}(\tau, t_0) x^0,$$

and $\chi(s, t')$ is the step function,

$$\chi(s, t') \equiv 0, s < t'; \ \chi(s, t') \equiv 1, s \geq t', \ d\chi(s, t')/ds = \delta(s - t').$$

The vector $x_*[\tau]$ in (7.48), (7.51) may be described by equation

$$dx_*[t] = ((A(t) - L(t))x_*[t] + B(t)p(t) + L(t)y(t))dt,$$

$$- L_1(x_*[t] - y(t))d\chi(t, t_1) - L_2(x_*[t] - y(t))d\chi(t, t_2), x_*[t_0] = x^0. \quad (7.52)$$

Remark 7.2.1. Note that in the transition function $\mathcal{G}(t, s)$ we have the difference of a Riemann integral and a Riemann–Stieltjes integral. On the other hand, the last

7.2 State-Constrained Control: Computation

two integrals in (7.51) should be interpreted as Lebesgue–Stieltjes integrals [234]. This does not cause additional difficulty since the multipliers L_1, L_2 are among the optimizers in (7.47).

A result similar to (7.45) is all the more true for the sets $\mathcal{X}_L(\tau, t_0, \mathcal{E}(x^0, X^0))$.

Corollary 7.2.1. *The following intersection holds:*

$$X[\tau] = X(\tau, t_0, \mathcal{E}(x^0, X^0)) = \cap \{\mathcal{X}_L(\tau, t_0, \mathcal{E}(x^0, X^0)) \mid L(\cdot)\}.$$

The difference between Lemmas 7.2.1 and 7.2.2 is that in the former it is not guaranteed that the boundary $\partial X[\tau]$ is touched at each point by one of the intersecting sets $\mathcal{X}_L(\tau, t_0, \mathcal{E}(x^0, X^0))$, whereas in the latter the boundary $\partial X[\tau]$ is indeed touched at each point by one of the sets $\mathcal{X}_L(\tau, t_0, \mathcal{E}(x^0, X^0))$. This is because the minimum in (7.47) is attained.

Under Assumptions 7.1.4, 7.1.5, 7.2.1, and Assumption 7.2.3 given below, in Sect. 7.2.5, the reasoning of the above leads to the next result.

Lemma 7.2.3. *The reach set $X[t] = X(t, t_0, \mathcal{E}(x^0, X^0))$ is a convex compact set in $\mathbb{R}^n$ which evolves continuously in t.*

The boundary of the reach set $X[t]$ has an important characterization. Consider a point x^* on the boundary $\partial X[\tau]$ of the reach set $X[\tau] = X(\tau, t_0, \mathcal{E}(x^0, X^0))$.[1] Then there exists a *support vector* z^* such that

$$\langle z^*, x^* \rangle = \rho(z^* \mid X[\tau]).$$

Let $L^0(\cdot)$ be the minimizer for the problem (see (7.47)):

$$\rho(z^* \mid X[\tau]) = \min\{\Phi(z^*, L(\cdot), \tau, t_0) \mid L(\cdot)\} = \Phi(z^*, L^0(\cdot), \tau, t_0). \quad (7.53)$$

Then the control $u = u^*(t)$, the initial state $x(t_0) = x^{*0} \in \mathcal{E}(x^0, X^0)$, and the corresponding trajectory $x^*(t)$, along which system (7.30) is transferred from state $x^*(t_0) = x^{*0}$ to $x(\tau) = x^*$, are specified by the following "modified maximum principle".

Theorem 7.2.2 (The Modified Maximum Principle Under State Constraints). *For Problem 7.1.2 suppose state x^* is such that*

$$\langle z^*, x^* \rangle = \rho(z^* \mid X[\tau]).$$

Then the control $u^*(t)$, which steers the system (7.30) from $x^*(t_0) = x^{*0}$ to $x^*(\tau) = x^*$ under constraints $u(t) \in \mathcal{E}(p(t), P(t))$, $x(t) \in \mathcal{E}(y(t), Y(t))$ while ensuring

$$\langle z^*, x^* \rangle = \max\{\langle z^*, x \rangle \mid x \in X[\tau]\},$$

---
[1] The boundary $\partial X[\tau]$ of $X[\tau]$ may be defined as the set $\partial X[\tau] = X[\tau] \setminus \text{int} X[\tau]$. Under the controllability assumption, $X[\tau]$ has a nonempty interior, $\text{int} X[\tau] \neq \emptyset$, $\tau > t_0$.

satisfies the following pointwise "maximum principle" for the control ($s \in [t_0, \tau]$):

$$z^{*'}\mathcal{G}^0(\tau,s)B(s), u^*(s) = \max_u \{z^{*'}\mathcal{G}^0(\tau,s)B(s)u \mid u \in \mathcal{E}(p(s), P(s))\} \quad (7.54)$$

$$= \langle z^*, \mathcal{G}^{0'}(\tau,s)B(s)p(s)\rangle + \langle z^*, \mathcal{G}^0(\tau,s)B(s)P(s)B'(s)\mathcal{G}^0(\tau,s)z^*\rangle^{1/2}, \ s \in [t_0, \tau],$$

and the "maximum conditions" for the system trajectory (pointwise),

$$z^{*'}\mathcal{G}^0(\tau,s)\mathcal{L}^0(s)x^*(s) = \max_p \{z^{*'}\mathcal{G}^0(\tau,s)\mathcal{L}^0(s)p \mid p \in \mathcal{E}(y(s), Y(s))\} \quad (7.55)$$

$$= z^{*'}\mathcal{G}^0(\tau,s)\mathcal{L}^0(s)y(s) + \langle z^*, \mathcal{G}^0(\tau,s)\mathcal{L}^0(s)Y(s)\mathcal{L}^{0'}(s)\mathcal{G}^{0'}(\tau,s)z^*\rangle^{1/2},$$

and the initial state,

$$\langle z^*, \mathcal{G}^0(\tau,t_0)x^{*0}\rangle = \max\{\langle z^*, x\rangle \mid x \in \mathcal{G}^0(\tau,t_0)\mathcal{E}(x^0, X^0)\} = \rho(z^* \mid \mathcal{G}^0(\tau,t_0)\mathcal{E}(x^0, X^0))$$
$$(7.56)$$
$$= \langle z^*, \mathcal{G}^0(\tau,t_0)x^0\rangle + \langle z^*, \mathcal{G}^0(\tau,t_0)X^0\mathcal{G}^{0'}(\tau,t_0)z^*\rangle^{1/2}.$$

Here $\mathcal{G}^0(\tau, s)$ stands for the matrix function $\mathcal{G}(\tau, s)$ taken for $\mathcal{L}^0(s)$—the minimizer of problem (7.53).

The function $h(\tau, s) = z_*'\mathcal{G}^0(\tau, s)B(s)$ is taken right-continuous.[2]

Remark 7.2.2. Suppose we want to solve Problem 7.1.2, seeking $\rho(l^*(t) \mid X[t])$ along a curve $l^*(t)$, $t > t_0$. Then, taking $l^*(t) = z^{*'}(\mathcal{G}^0)^{-1}(t, s)$, one may observe that the *integrands* in functional $\Phi(l^*(t), \mathcal{L}^0(\cdot), t, t_0)$ of (7.47)–(7.51) will be independent of t. This property ensures the existence of a *recurrent* computational procedure, as indicated in the next section (see also Sect. 2.3). The modified maximum principle of this section thus allows a solution in recurrent form. This is not the case for the standard maximum principle.

We now pass to the construction of ellipsoidal approximations for the reach sets.

7.2.2 External Ellipsoids

Despite the linearity of the system, the direct calculation of reach sets is rather difficult. Among effective methods for such calculations are those that rely on ellipsoidal techniques. Indeed, although the initial set $\mathcal{E}(x^0, X^0)$ and the control set $\mathcal{E}(p(t), \mathcal{P}(t))$ are ellipsoids, the reach set $X[t] = X(t, t_0, \mathcal{E}(x^0, X^0))$ will of course *not* generally be an ellipsoid.

---

[2] In the general case, under Assumption 7.2.1, the optimal trajectory may visit the smooth boundary of the state constraint for a countable set of closed intervals, and function $L_*^0(\cdot)$ allows not more than a countable set of discontinuities of the first order.

7.2 State-Constrained Control: Computation

As shown in Chap. 3, in the absence of state constraints the reach set $X[t]$ may be approximated both externally and internally by ellipsoids $\mathcal{E}_-$ and $\mathcal{E}_+$, with $\mathcal{E}_- \subseteq X[t] \subseteq \mathcal{E}_+$. Here we deal only with external approximations, but taken under state constraints.

An approximation $\mathcal{E}(x_+, X_+)$ is said to be *tight* if there exists a vector $z \in \mathbb{R}^n$ such that $\rho(z \mid \mathcal{E}(x_+, X_+)) = \rho(z \mid X[t])$ (the ellipsoid $\mathcal{E}(x_+, X_+)$ touches $X[t]$ along direction z). We shall look for external approximations that are tight, on the one hand, and are also recursively computable, on the other.

In order to apply ellipsoidal techniques to state-constrained problems, recall Lemma 7.2.1 and Corollary 7.2.1, which indicate how the reach set $X[t] = X(t, t_0, X^0)$ may be presented as an intersection of reach sets X_L or X_L for system (7.30) *without state constraints*. We first study how to approximate sets X_L by ellipsoids.

To demonstrate the nature of the procedures suggested in this section we shall introduce the ellipsoidal technique in two stages. First, transform the system coordinates in (7.30) from x to z according to the formula $z(t) = S(t, t_0)x(t)$, with

$$dS(t, t_0)/dt = -S(t, t_0)A(t), \quad S(t_0, t_0) = I.$$

Then system (7.30) and state constraint $x(t) \in \mathcal{Y}(t)$ transform into

$$\dot{z} = S(t, t_0)B(t)u = B_N(t)u, \qquad (7.57)$$

$$z(t) \in S(t, t_0)\mathcal{E}(y(t), Y(t)), \quad z(t_0) \in S(t, t_0)\mathcal{E}(x^0, X^0),$$

while the constraint on control $u \in \mathcal{P}(t) = \mathcal{E}(p(t), P(t))$ remains the same.

Returning to the *old notation*, we will deal with system

$$\dot{x} = B(t)u, \quad x(t_0) \in \mathcal{E}(x^0, X^0), \quad y(t) \in \mathcal{Y}(t). \qquad (7.58)$$

Thus, without loss of generality, in the forthcoming formulas we assume $A(t) \equiv 0$.

Problem 7.2.1. *Given a vector function $l^*(t)$, continuously differentiable in t, find external ellipsoids $\mathcal{E}_{L+}[t] = \mathcal{E}(x_L^*(t), X_L^*(t)) \supset X_L[t]$ such that for all $t \geq t_0$, the equalities*

$$\rho(l^*(t) \mid X_L[t]) = \rho(l^*(t) \mid \mathcal{E}_{L+}[t]) = \langle l^*(t), x_L^*(t) \rangle$$

hold, so that the supporting hyperplane for $X_L[t]$ generated by $l^*(t)$, namely, the plane $\langle x - x_L^*(t), l^*(t) \rangle = 0$ which touches $X_L[t]$ at point $x^*(t)$, is also a supporting hyperplane for $\mathcal{E}_{L+}[t]$ and touches it at the same point.

Remark 7.2.3. Under Assumption 7.2.1 for the optimal trajectories $x^{(0)}(\cdot)$ of Problem 7.1.2 the boundary $\partial X[\tau]$ of the reach set $X[\tau]$ consists of three types of points, namely (I) those that are reached by $x^{(0)}(\tau)$ *without having visited the boundary of the state constraint*, then (II) those that are reached by $x^{(0)}(\tau)$ *after having visited the boundary of the state constraint*, and (III) those that lie *on the boundary of the state constraint*. Since case (I) has been investigated in detail in Chap. 3, Sects. 3.2–3.4, and case (III) is trivial, our present investigation actually deals only with case (II).

Apart from Assumptions 7.1.4 and 7.1.5, let us assume in this section that the functions $L(\cdot)$ in what follows do not have any delta function components. (The case when such components are present is treated in the next section.)

The solution to Problem 7.2.1 is given within the following statement.

Theorem 7.2.3. *With $l^*(t)$ given, the solution to Problem 7.2.1 is an ellipsoid $\mathcal{E}_{L+}^*[t] = \mathcal{E}_L(x_L^*(t), X_L^*[t])$, in which*

$$X_L^*[t] = \left(\int_{t_0}^{t} (p_u(t,s) + p_Y(t,s)) ds + p_0(t) \right) \tag{7.59}$$

$$\times \left(\int_{t_0}^{t} (p_u(t,s))^{-1} G_L(t,s) B(s) P(s) B'(s) G_L'(t,s) ds \right.$$

$$\left. + \int_{t_0}^{t} (p_Y(t,s))^{-1} G_L(t,s) L^t(s) Y(s) L^{t'}(s) G_L'(t,s) ds + p_0^{-1}(t) G_L(t,t_0) X^0 G_L'(t,t_0) \right),$$

and

$$p_u(t,s) = (l^*(t), G_L(t,s) P(s) G_L'(t,s) l^*(t))^{1/2}, \tag{7.60}$$

$$p_Y(t,s) = (l^*(t), G_L(t,s) L^t(s) Y(s) L^t G_L'(t,s) l^*(t))^{1/2},$$

$$p_0(t) = (l(t_0), G_L(t,t_0) X^0 G_L'(t,t_0) l(t_0))^{1/2},$$

with $x_L^*(t) = x_L^*[t]$.

This result follows from Chap. 3, Sect. 2.3. Since the calculations have to be made for all t, the parametrizing functions $p_u(t,s), p_Y(t,s), s \in [t_0, t], p_0(t)$ must also formally depend on t.

In other words, relations (7.59), (7.60) need to be calculated "afresh" for each t. It may be more convenient for computational purposes to have them in recursive form. As indicated in Sects. 3.3–3.4, in the absence of state constraints this could be done by selecting function $l^*(t)$ in an appropriate way. For the case of state constraints we follow Remark 7.2.2, arriving at the next assumption.

Assumption 7.2.2. *The function $l^*(t)$ is of the form $l^{*'}(t) = l_*' G_L(t_0, t)$, with $l_* \in \mathbb{R}^n$ given.*

Under Assumption 7.2.2 the function $l^*(t)$ is the solution to the differential equation

$$\dot{l}^* = L'(t) l^*, \quad l^*(t_0) = l_*. \tag{7.61}$$

7.2 State-Constrained Control: Computation

The tight external ellipsoids to $\mathcal{X}_L(t)$, which are

$$\mathcal{E}(x_L^*(t), X_L^*(t)) \supseteq \mathcal{X}_L(t) \supseteq \mathcal{X}[t],$$

may be described applying formulas of Sect. 3.3 to system (7.43). Each of these ellipsoids will be an external estimate for $\mathcal{X}[t]$.

But to describe the collection of tight ellipsoids for $\mathcal{X}[t]$ we have to ensure the following property:

$$\rho(l|\mathcal{X}[t]) = \min\{\langle l, x_L^*(t)\rangle + \langle l, X_{L+}^*(t)l\rangle^{1/2} \mid L(\cdot) \in \mathcal{C}_{00}\}. \tag{7.62}$$

Under our assumptions *the minimum over $L(\cdot)$ in (7.47) is attained for any $l = l^*(t) = G'_L(t_0, t)l_* \in \mathbb{R}^n$*, the minimizers being denoted as $L_*^t(\cdot) \in \mathcal{C}_{00}$.[3]

But prior to moving ahead we have to investigate the following. Suppose element $L_*^t(s)$ is the minimizer of functional

$$\Phi(G'_L(t_0, t)l_*, L(\cdot), t, t_0) = \Phi(l_*^0, L(\cdot), t, t_0), \quad l_*^0 = (G'_L(t_0, t)l_*,$$

The question is, if we minimize function $\Phi(l_*^0, L(\cdot), t + \sigma, t_0), \sigma > 0$, over a larger interval $[t_0, t + \sigma]$ than before, will the minimizer $L_*^{t+\sigma}(s)$, for the latter problem, taken within $s \in [t_0, t]$, be the same as the minimizer $L_*^t(s), s \in [t_0, t]$ for $\Phi(l_*^0, L(\cdot), t, t_0)$?

The answer to this question is given by the next lemma.

Lemma 7.2.4. *Taking $l = l^*(t)$ according to Assumption 7.2.2, suppose $L_*^t(s), s \in [t_0, t]$ and $L_*^{t+\sigma}(s), s \in [t_0, t + \sigma], \sigma > 0$, are the minimizers of functionals $\Phi(l_*^0, L(\cdot), t, t_0)$ and $\Phi(l_*^0, L(\cdot), t + \sigma, t_0)$, respectively. Then*

$$L_*^t(s) \equiv L_*^{t+\sigma}(s), \quad s \in [t_0, t].$$

The proof is achieved by contradiction. Note that under Assumption 7.2.2 and due to Property 7.2.1 of Sect. 7.2.5 we may always take $L(t) \equiv 0$ when $t < \tau_1, t > \tau_2$, where τ_1, τ_2 are the points of arrival and departure at the state constraint.

Following Assumption 7.2.2, we proceed further by selecting $L(\cdot)$ to be the minimizer in L of functional $\Phi(l_*^0, L(\cdot), t, t_0)$, namely, we now take $L(s) = L_*^t(s) = L_*(s), \quad s \in [t_0, t]$, which depends on l_*, s, but, as indicated in Lemma 7.2.4, does not depend on t.

Let $G_*(t, s) = G_L(t, s)$ under conditions $A(t) \equiv 0$, $L(t) \equiv L_*(t)$.
Then $p_u(t, s), p_Y(t, s), p_0(t)$ of (7.60) transform into

$$p_u(t, s) = \langle l_*, G_*(t_0, s)B(s)P(s)B'(s)G'_*(t_0, s)l_*\rangle^{1/2} = p_u(s), \tag{7.63}$$

$$p_Y(t, s) = \langle l_*, G_*(t_0, s)L_*(s)Y(s)L_*'(s)G'_*(t_0, s)l_*\rangle^{1/2} = p_Y(s),$$

$$p_0(t) = \langle l_*, X^0 l_*\rangle^{1/2} = p_0;$$

---
[3]The same minimum value is also attained here in classes of functions broader than $\mathcal{C}_{00}$.

matrix $X_L^*[t]$ transforms into

$$X_+^*[t] = \left(\int_{t_0}^t (p_u(s) + p_Y(s))ds + p_0\right) \times \quad (7.64)$$

$$\times \left(\int_{t_0}^t (p_u(s))^{-1} G_*(t_0, s) B(s) P(s) B'(s) G_*'(t_0, s) ds\right.$$

$$\left. + \int_{t_0}^t (p_Y(s))^{-1} G_*(t_0, s) L_*(s) Y(s) L_*'(s) G_*'(t_0, s) ds + p_0^{-1} X^0\right),$$

and the functional $\Phi(l^*(t), L_*(\cdot), t, t_0)$ transforms into

$$\Phi(l^*(t), L_*(\cdot), t, t_0) \quad (7.65)$$

$$= \langle l_*, x^*(t)\rangle + \langle l_*, X^0 l_*\rangle^{1/2} + \int_{t_0}^t \langle l_*, G_*(t_0, s) B(s) P(s) B'(s) G_*'(t_0, s) l_*\rangle^{1/2} ds$$

$$+ \int_{t_0}^t \langle l_*, G_*(t_0, s) L_*(s) Y(s) L_*'(s) G_*'(t_0, s) l_*\rangle^{1/2} ds,$$

with $x_L^*(t)$ transformed into

$$x_+^*[t] = x^0 + \int_{t_0}^t G_*(t_0, s)(B(s)p(s) + L_*(s)y(s))ds.$$

We may now differentiate $X_+^*[t], x_+^*[t]$. According to Sect. 7.2.5 "Specifics", below, in Sect. 7.2.5, the necessary condition (7.84) for a jump in $L_*(t)$ is not fulfilled, and therefore $L_*(t) \equiv L^0(t)$.

Denoting

$$\pi_u(t) = p_u(t) \left(\int_{t_0}^t (p_u(s) + p_Y(s))ds + p_0\right)^{-1},$$

$$\pi_Y(t) = p_Y(t) \left(\int_{t_0}^t (p_u(s) + p_Y(s))ds + p_0\right)^{-1}, \quad (7.66)$$

and differentiating $X_+^*[t], x_+^*[t]$, we arrive at

$$\dot{X}_+^*[t] = (\pi_u(t) + \pi_Y(t)) X_+^*[t] + (\pi_u(t))^{-1} G_*(t_0, t) B(t) P(t) B'(t) G_*'(t_0, t)$$
$$+ (\pi_Y(t))^{-1} G_*(t_0, t) L_*(t) Y(t) L_*'(t) G_*'(t_0, t), \quad (7.67)$$

$$\dot{x}_+^*[t] = G_*(t_0, t)(B(t)p(t) + L_*(t)y(t)), \quad (7.68)$$

$$X_+^*[t_0] = X^0, \quad x^*[t_0] = x^0. \quad (7.69)$$

7.2 State-Constrained Control: Computation

Returning to function $l^*(t) = G'_L(t_0, t)l_*$ with $G_L(t, s) = G_*(t, s)$, we have

$$X_+[t] = G_*(t, t_0) X_+^*[t] G'_*(t, t_0)$$

and

$$\dot{X}_+[t] = -L_*(t) X_+[t] - X_+[t] L_*(t)' + G_*(t, t_0) \dot{X}_+^*[t] G'_*(t, t_0), \quad (7.70)$$

so that

$$\dot{X}_+[t] = -L_*(t) X_+[t] - X_+[t] L'_*(t) \quad (7.71)$$
$$+ (\pi_u(t) + \pi_Y(t)) X_+[t] + (\pi_u(t))^{-1} B(t) P(t) B'(t) + (\pi_Y(t))^{-1} L_*(t) Y(t) L_*'(t)$$

and

$$\dot{x}_+ = -L_*(t) x_+ + B(t) p(t) + L_*(t) y(t).$$

Finally, returning to the case $A(t) \neq 0$, we have

$$\mathbf{X}_+[t] = S(t_0, t) X_+[t] S'(t_0, t), \quad \mathbf{x}(t) = S(t_0, t) x_+(t),$$

$$\dot{\mathbf{X}}_+[t] = (A(t) - L_*(t)) \mathbf{X}_+[t] + \mathbf{X}_+[t](A'(t) - L'_*(t)) \quad (7.72)$$

$$+ (\pi_u(t) + \pi_Y(t)) \mathbf{X}_+[t] + (\pi_u(t))^{-1} B(t) P(t) B'(t)$$

$$+ (\pi_Y(t))^{-1} L_*(t) Y(t) L_*'(t), \quad \mathbf{X}_+[t_0] = X^0,$$

and

$$\dot{\mathbf{x}} = (A(t) - L_*(t)) \mathbf{x} + B(t) p(t) + L_*(t) y(t), \quad \mathbf{x}(t_0) = x^0. \quad (7.73)$$

Rewriting $\pi_u(t)$, $\pi_Y(t)$ following schemes of [181], p. 189, for these items (see also below: Remark 7.2.6 of Sect. 7.2.5), we have

$$\pi_u(t) = \langle l^*(t), B(t) P(t) B'(t) l^*(t) \rangle^{1/2} / \langle l^*(t), \mathbf{X}_+(t) l^*(t) \rangle^{1/2}, \quad (7.74)$$
$$\pi_Y(t) = \langle l^*(t), L_*(t) Y(t) L'_*(t) l^*(t) \rangle^{1/2} / \langle l^*(t), \mathbf{X}_+(t) l^*(t) \rangle^{1/2}.$$

Summarizing the results we come to the following.

Theorem 7.2.4. *With shape matrix $Y(t)$ of the state constraint being nondegenerate and if, under Assumptions 7.1.4, 7.2.1, and 7.2.2, the multipliers $L_*(t)$ responsible for the state constraint have no jumps $(d\Lambda(t) = \lambda(t) dt)$, then the external ellipsoids $\mathcal{E}(\mathbf{x}[t], \mathbf{X}_+[t]) \supseteq X[t]$ have the form (7.72), (7.73) with $\mathbf{X}[t_0] = X^0$, $\mathbf{x}[t_0] = x^0$.*

Moreover, with $l^*{}'(t) = G_*{}'(t_0,t)l_*$, $l_* \in \mathbb{R}^n$, the ellipsoids $\mathcal{E}(\mathbf{x}[t], \mathbf{X}[t])$ touch the set $X[t]$ along the curve $x^*(t)$ which satisfies the condition

$$\langle l^*(t), x^*(t)\rangle = \rho(l^*(t) \mid X[t]), \ t \leq \tau_1, \ t \geq \tau_2,$$

so that

$$\langle l^*(t), x^*(t)\rangle = \rho(l^*(t) \mid X[t]) = \rho(l^*(t) \mid \mathcal{E}(\mathbf{x}[t], \mathbf{X}_+[t])). \tag{7.75}$$

On the state constraint boundary, with $t \in [\tau_1, \tau_2]$, the trajectory which satisfies the first equality of (7.75) may be uniquely determined by the maximum condition (7.55).

The overall procedure of calculation for Problem 7.1.2 starting from a point in the interior of the state constraint is as follows.

(i) Consider an array of initial vectors l_* located on the unit ball $\mathcal{B}(0)$ in $\mathbb{R}^n$. Then for each l_* proceed as follows (possibly, in parallel).

(ii) For each l_* construct the array of external ellipsoids $\mathcal{E}_+[\tau]_{l_*}$ for Problem 7.1.2 of system (7.30) *without state constraints* following the explicit formulas of Sects. 3.2–3.4.

(iii) Construct the array of ellipsoids $\mathcal{E}(\mathbf{x}[t], \mathbf{X}[t])$, namely,

(iii-a) for a given l'_* find related $l^{*'}(t) = l'_* G_*(t_0, t)$ and, further on, the related curve $x^{(0)}(t)$ for the reachability problem without state constraints. Note that on the interval $[t_0, \tau_1)$, where $L_*(t) \equiv 0$, we have

$$x^{(0)}(t) = x^*(t) + X_+(t)l^*(t)\langle l^*(t), X_+(t)l^*(t)\rangle^{1/2},$$

where explicit relations for elements $x^*(t)$, $X_+(t)$ of this formula are indicated in Remark 7.2.6 of Sect. 7.2.5. Follow curve $x^{(0)}(t)$ until the state constraint boundary is reached at time, say, τ_1;

(iii-b) for $l^*(\tau_1)$ solve the optimization problem (7.49) over $L(\cdot)$. Use Lemma 7.2.5 of the appendix for additional information on optimizer $L_*(\cdot)$, taking τ_2 as a parameter which varies within interval $[\tau_1, \tau]$;

(iii-c) given optimizer $L_*(\cdot)$, which depends on the selected l_*, solve Eqs. (7.72), (7.73), determining the related external ellipsoid $\mathcal{E}_{l_*}(\mathbf{x}[t], \mathbf{X}_+[t])$.

(iv) In view of Remark 7.2.3, the intersection

$$\bigcap\{\mathcal{E}_+[\tau]_{l_*} | l_* \in \mathcal{B}(0)\} \bigcap\{\mathcal{E}_{l_*}(\mathbf{x}[t], \mathbf{X}_+[t]) \mid l_* \in \mathcal{B}(0)\}\bigcap \mathcal{E}(y(\tau), Y(\tau)) = \mathcal{X}_a[\tau] \supseteq \mathcal{X}[\tau]$$

is an external approximation of $\mathcal{X}[\tau]$. Loosely speaking, the greater is the number of vectors l_* used as starting points of the calculations, the more accurate will be the approximation $\mathcal{X}_a$.

7.2 State-Constrained Control: Computation

Remark 7.2.4. If it is not necessary to find the tight external approximations for the exact reach set $X[\tau]$, but a conservative estimate of $X[\tau]$ using *only one* ellipsoid would suffice, then there is no need to solve optimization problem (7.46) in L, since ellipsoids $\mathcal{E}(\mathbf{x}[t], \mathbf{X}_+[t])$, defined by (7.72), (7.73), happen to be external estimates of $X[\tau]$ *for all* $L(\cdot)$. One may single out one of these through some appropriate criterion.

An obviously conservative estimate may also be obtained by intersecting the reach set without state constraints with the ellipsoid that defines the state constraint. The objective of this paper, however, is to indicate the calculation of the *exact reach set under state constraints*, especially when the state constraint produces a reach set different from the one without state constraints.

We now pass to the more general case, in which the multipliers may have delta function components.

7.2.3 Generalized Multipliers

In this section, Assumptions 7.1.4 (of Sect. 7.1.3) and 7.2.1 (of Sect. 7.2.1) and the conditions of Lemma 7.1.6 are taken to be true together with Assumption 7.1.5. This allows us to consider ellipsoidal approximations without the additional requirement of Property 7.2.2 of Sect. 7.2.5.

Now functional $\Phi(l, L(\cdot), \tau, t_0)$ has the form (7.65), and its minimizer for a given $l = l^*(\tau)$ is of the form

$$L^0(t) = L_*^0(t) + L_1^0 \delta(t - \tau_1) + L_2^0 \delta(t - \tau_2)$$

for all $l^*(\tau)$ with $L_*^0(t)$ absolutely continuous and $L_*^0(t) \equiv 0$ whenever $x^*(t) \in \text{int}\mathcal{Y}(t)$. Then the respective transition matrix is $G(t, s) = G_*^0(t, s)$ and $l^*(t) = G_*^{0'}(t, s) l^*$.

Here $L^0(\cdot)$ may be interpreted as the generalized derivative of a generalized Lagrange multiplier $\Lambda^{00}(\cdot)$ similar to $\Lambda^{(0)}(\cdot)$ of Sect. 7.2.1. $\Lambda^{00}(\cdot)$ is piecewise absolutely right-continuous, with possible jumps at points τ_1, τ_2 and possible jumps at t_0 and/or τ (when it happens that $t_0 = \tau_1$ and/or $\tau = \tau_2$). $\Lambda^{00}(t) \equiv \text{const}$, whenever $x^*(t) \in \text{int}\mathcal{Y}(t)$.

Following the reasoning of the previous section, we may derive equations for the approximating external ellipsoids similar to (7.72), (7.73). The necessary prerequisites for such a move are similar to Lemma 7.2.4 and Theorems 7.2.3 and 7.2.4.

Thus, we come to equations for tight external ellipsoids:

$$dx_+ = ((A(t) - L_*^0(t))x_+ + (B(t)p(t) + L_*^0(t)y(t))dt - \sum_{i=1}^{2} L_i^0(x_+ - y(t))d\chi(t, \tau_i), \quad (7.76)$$

$$dX_+[t] = ((A(t) - L_*^0(t))X_+[t] + X_+[t](A'(t) - L_*^{0'}(t)))dt \quad (7.77)$$

$$+ \left(\pi_u(t) + \pi_Y(t) + \sum_{i=1}^{k} \pi_i \right) X_+[t] dt + (\pi_u(t))^{-1} B(t) P(t) B'(t) dt$$

$$+ (\pi_Y(t))^{-1} L_*^{0'}(t) Y(t) L_*^{0'}(t) dt + \sum_{i=1}^{2} \pi_i^{-1} L_i^{0} Y(t) L_i^{0'} d\chi(t, \tau_i),$$

$$X_+[t_0] = X^0, \quad x_+(t_0) = x^0.$$

The $\pi_i > 0$ are additional parametrizing coefficients similar to $\pi_Y(t)$.

The last equation may be interpreted as before (see Remark 7.2.1).

Theorem 7.2.5. *Under Assumptions 7.1.5, 7.2.1, 7.2.2, and those of Lemma 7.1.6 with $l^*(t) = G_*^{0'}(t_0, t) l_*$, $l_* \in \mathbb{R}^n$, the external ellipsoids $\mathcal{E}(x_+(t), X_+[t]) \supseteq X[t]$ have the form (7.76), (7.77) with $X_+[t_0] = X^0$, $x_+(t_0) = x^0$. One may select the parametrizing coefficients $\pi_u, \pi_{Y(t)}, \pi_i$ so that ellipsoids $\mathcal{E}(x_+(t), X_+[t])$ would touch set $X[t]$ along the curve $x_+(t)$, which satisfies condition (7.75).*

If the optimal trajectories visit the boundary of the state constraint $m > 1$ times, then the reasoning is the same as in the above, and the sums in (7.76), (7.77) will have additional $2m$ terms.

Remark 7.2.5. The results of Sects. 7.2.2 and 7.2.3 remain true if the state constraint is applied to a system output $z(t) = Tx(t)$, $z \in \mathbb{R}^k$, $k < n$, rather than to the whole state vector x. The state constraint $\mathcal{Y}$ is then given by an elliptical cylinder in $\mathbb{R}^n$. The proofs may be achieved either by directly following the scheme of this Chap. 7 or by substituting the elliptical cylinder with an ellipsoid having $n - k$ axes of size r followed by limit transition $r \to \infty$.

7.2.4 An Example

The system is the double integrator:

$$\dot{x}_1 = x_2, \quad \dot{x}_2 = u, \quad x(0) = 0, \tag{7.78}$$

under constraints

$$|u| \leq \mu, \quad |x_2| \leq \nu. \tag{7.79}$$

We wish to find the reach set $X[\tau]$ at time $\tau > 0$.

The state constraint may be treated either directly or as a limit as $\epsilon \to 0$ of the ellipsoid $\epsilon^2 x_1^2 + x_2^2 \leq \nu^2$.

7.2 State-Constrained Control: Computation

The treatment of this example indicates the following *boundary types*:

1. The boundary of the reach set $X[\tau]$ at each time τ consists of three types of points as follows:

   (a) points reached without visiting the boundary of the state constraint;
   (b) points reached after visiting the boundary of the state constraint for some time;
   (c) points on the boundary of the state constraint.

2. Each optimal trajectory $x^{(0)}(t)$, for given $l \in \mathbb{R}^2$, visits the state constraint not more than once.
3. The necessary conditions for a jump of the multiplier responsible for the state constraint are not fulfilled, so there are no jumps.
4. Assumption 7.2.3 of Sect. 7.2.5 is fulfilled (see below)

Case 1(a) is treated according to Chap. 3, while case 1(c) is trivial. Therefore we concentrate on case 1(b).

The exact parametric equations for case 1(b) may be derived, by minimizing over η the functional

$$\rho(l|X[\tau]) = \min\left\{\mu\int_0^\tau |s_2(t)|d\xi + \nu\int_0^\tau |\eta(t)|d\xi \mid \eta(\cdot)\right\}, \quad (7.80)$$

where $s_2(t) = -l_1(\tau - t) + l_2 - \int_t^\tau \eta(\xi)d\xi$ is the solution to the adjoint equation

$$\dot{s}_1 = 0, \ \dot{s}_2 = -s_1 + \eta(t), s_1(\tau) = l_1, s_2(\tau) = l_2.$$

This gives parametric equations for a part of the boundary $\partial X[\tau]$, which is:

$$x_1 = -\mu(\tau - \sigma)^2/2 + c, \ x_2 = \mu\sigma + d, \ \tau_1 \le \sigma \le \tau.$$

Here $\tau_1 = \nu/\mu < \tau$ is the first instance when the trajectory reaches the boundary of the state constraint, while σ is a parameter that indicates the instance of τ_2 when the trajectory leaves this boundary. Also $c = \nu\tau - \nu^2/2\mu$; $d = \nu - \mu\tau$. Note that we have used Assumption 7.1.4 of Sect. 7.2.1, which implies that optimizer $\eta^0(t)$ satisfies relations

$$\eta^0(t) \equiv l_1, \ t \in [\tau_1, \tau_2], \ \eta^0(t) \equiv 0, \ t \notin [\tau_1, \tau_2],$$

and $s_1(t) \equiv l_1$.

A typical trajectory for case 1(b) is when it reaches the boundary of the state constraint at time τ_1 and then runs along the boundary and leaves it at time τ_2. Later it runs toward the boundary $\partial X[\tau]$ while staying in the interior of the constraint.

For the interval $[0, \tau]$, the adjoint system given above, when written in matrix form (7.42), is

$$\dot{S} = -SA(t) + M(t)\eta(t). \quad (7.81)$$

To solve the problem through a numerical procedure involving external ellipsoids, using recursive formula and following Sect. 7.2.2, transform the adjoint Eq. (7.81) of the above into the modified form

$$\dot{S}_L = -S_L(A(t) - L(t)), \; S_L(\tau) = I.$$

Taking $l_* = l_*(0)$, we will have $l'_* M(t)\eta^0(t) = (0, 1)\eta^0(t)$ when $t \in [\tau_1, \tau_2]$. This yields $M(t) = M_0(t)$ depending on η^0. Direct calculation then allows one to find for each l_* the relation
for the optimizer $L_*(t)$ of (7.53). This is

$$L_*(t) = S_{M_0}(t)^{-1} M_0(t), \; t \in [\tau_1, \tau_2],$$

with $L_*(t) \equiv 0$ when $t < \tau_1, t > \tau_2$. Here $S_{M_0}(t)$ is the solution to (7.81) with initial condition $S_{M_0}(\tau) = I$ and $M(t) \equiv 0, \; t, \tau_1, t > \tau_2$.

Recalling the ellipsoidal equations, we have

$$\dot{X} = (A(t) - L(t))X + X(A(t) - L(t))' \quad (7.82)$$
$$+ (\pi_u(t) + \pi_Y(t))X + \pi_u^{-1}(t)BP(t)B' + \pi_Y^{-1}(t)L(t)Y(t)L'(t).$$

Here one should take $L(t) = L_*(t)$ and π_u, π_Y according to (7.74).

The algorithm for this example involves the following steps:

1. For the given starting direction l_* find time $\tau_1 = \nu/\mu$ of the first exit (i.e., encounter with the boundary of the state constraint).
2. Solve extremal problem (7.80), determining $l_1 = \eta^0 = $ const and l_2, for each $\tau_2 = \sigma \in [\tau_1, \tau]$.
3. Construct the ellipsoidal approximations for the system without constraint $t \leq \tau_1$, taking $L(t) \equiv 0$. Denote these as $\mathcal{E}(0, X_l^+(t))$.
4. For each $\tau_2 = \sigma \in [\tau_1, \tau]$ construct the ellipsoidal approximation following (7.82).

The part of the boundary related to case (b) is described by the boundary of the intersection $\mathcal{E}(0, X(\tau)) \cap L(\cdot)$ where $L(\cdot)$ is actually reduced to parameter $\sigma = \tau_2$. Denote this intersection as $\mathcal{E}(0, X_\sigma^+(\tau))$.

The total reach set is the intersection of approximations for each group 1(a), 1(b), and 1(c), namely,

$$\mathcal{X}[\tau] = X(\tau, 0, 0) = \bigcap_l \mathcal{E}(0, X_l^+(\tau)) \bigcap_\sigma \mathcal{E}(0, X_\sigma^+(\tau)) \bigcap \mathcal{E}(0, Y).$$

Figure 7.1 shows the structure of the reach set boundary, given here by a thick line. The system trajectory OAG reaches its end point G without visiting the boundary of the state constraint. Point G is attained at time τ. It lies on the reach set boundary of type 1(a) (see beginning of this Sect. 7.2.4). Trajectory OBF visits

7.2 State-Constrained Control: Computation

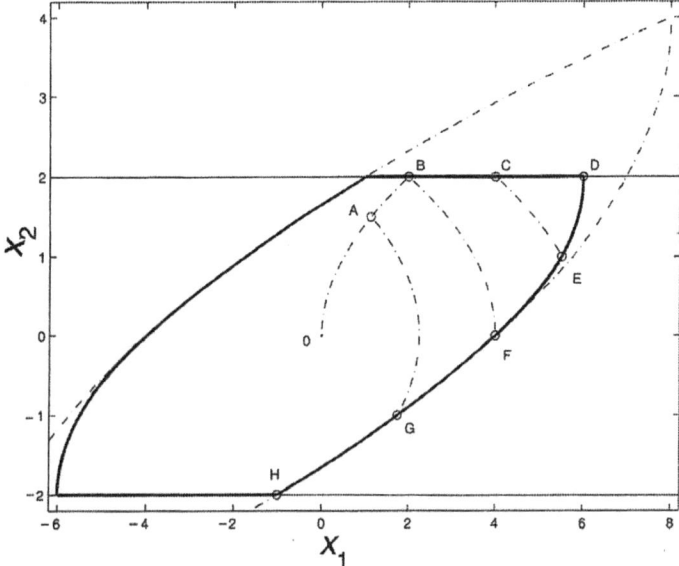

Fig. 7.1 Reach set: structure of the boundary

the boundary of the state constraint at only one point $\tau_1 = \tau_2 < \tau$. Its end point F is of both types 1(a) and 1(b). Trajectory OBCE lies on the boundary of the state constraint during the interval $[\tau_1, \tau_2)$ with $\tau_1 < \tau_2 < \tau$, and its end point E lies on the reach set boundary of type 1(b). Finally, trajectory OBD reaches the boundary of the state constraint at time $\tau_2 < \tau$ and lies on it until final time τ, so that point D is of both types 1(b) and 1(c). Thus, the part of the reach set boundary along points BDEFGH consists of segments BD [of type 1(c)], DF [of type 1(b)], and FGH [of type 1(a)].

Figure 7.2 shows the structure of the generalized Lagrange multiplier $\eta(t)$ for the case of trajectory $x_2(t)$ of type OBCE shown in the top of the figure. Here $\eta \equiv 0$ on intervals $[0, \tau_1)$ and $[\tau_2, \tau]$ and $\eta \equiv $ const on $[\tau_1, \tau_2)$. Note that the related trajectory $x_2(t)$ runs along the boundary of the state constraint during the interval $[\tau_1, \tau_2)$, where $\eta(t) \neq 0$. Points τ_1, τ_2 potentially could have δ-functions as components of η. But the necessary conditions for the existence of such components are not fulfilled in this example.

Figure 7.3 shows an external ellipsoid (thin line) for the boundary of the reach set under state constraints (thick line), indicating its difference from the reach set without state constraints (dash-dotted line). Figure 7.4 illustrates the final reach tube with and without state constraints for this example.[4]

---

[4]The computer illustrations for this subsection were made by M.N. Kirilin.

306   7 Dynamics and Control Under State Constraints

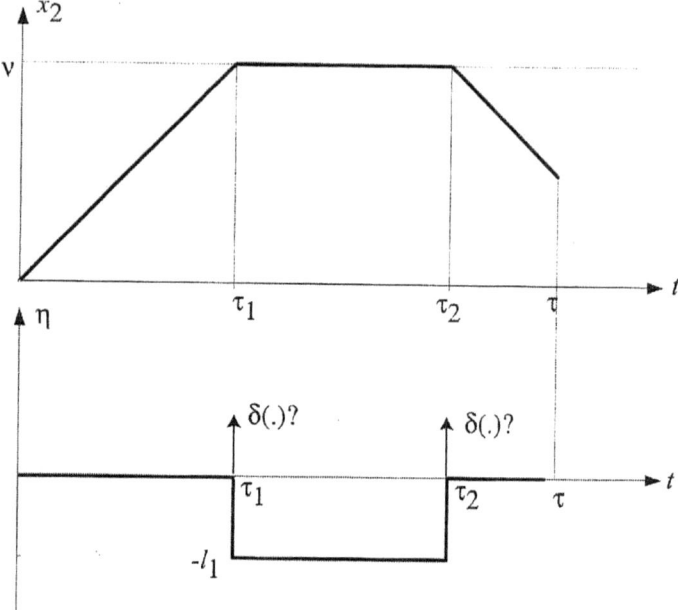

Fig. 7.2 Structure of multiplier η

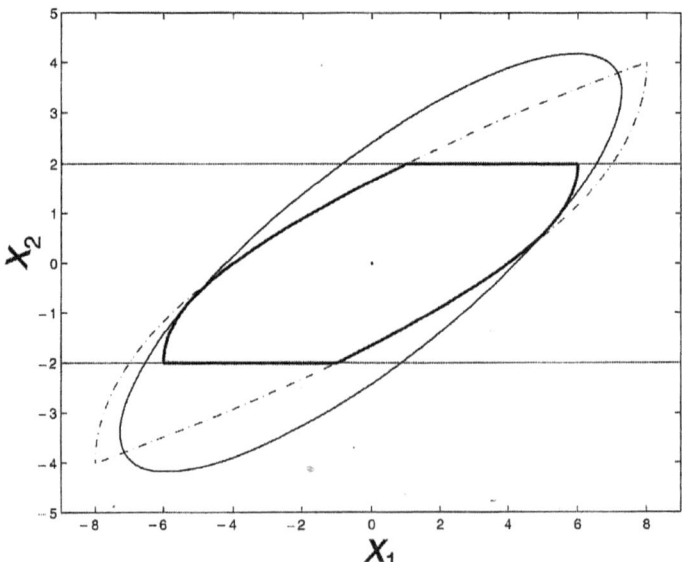

Fig. 7.3 External ellipsoid at boundary point of type (b)

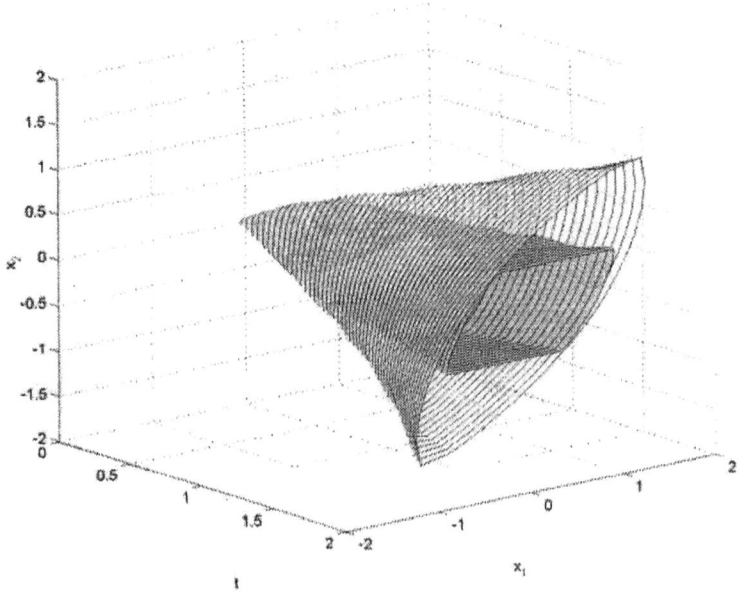

Fig. 7.4 Reach tubes with and without state constraints

7.2.5 Specifics: Helpful Facts for State-Constrained Control Design

In this subsection we present some additional facts useful for proving the main assertions of this paper.

One difficulty in solving the control problem under state constraints (Problem 7.1.2) is to determine the set of times $\{t : x(t) \in \partial \mathcal{Y}(t)\}$. This is a union of closed intervals during which the optimal trajectory is on the boundary of the state constraint.

Here are some helpful facts. For Problem 7.1.2 the Properties 7.2.1–7.2.3 are given below under Assumptions 7.1.4 and 7.2.1,

Property 7.2.1. For any $l \in \mathbb{R}^n$, the minimizer of (7.35) is $\Lambda^{(0)}(t) = \text{const}$ and in (7.39) the corresponding minimizer is $\lambda^{(0)}(t) = \text{const}$ during time intervals for which $x(t) \in \text{int}\mathcal{Y}(t)$, the interior of $\mathcal{Y}(t)$.

Since $l' M^{(0)}(t) d\lambda^{(0)}(t) = d\Lambda^{(0)}(t)$, we may track whether the trajectory is on the boundary of $\mathcal{Y}(t)$ by the multiplier $\lambda^{(0)}(t)$. Thus, we need not be interested in values of $M^{(0)}(t)$ when the trajectory is not on the boundary $\partial \mathcal{Y}(t)$.

Property 7.2.2. Suppose $l \in \mathbb{R}^n$ is given and $x^{(0)}(t)$ is the solution of Problem 7.1.2. For the function $\Lambda^{(0)}(t)$ to have a jump at t^*, under the smoothness conditions on the state constraint of Lemma 7.1.6, t^* must be a time of arrival or departure from the boundary of the tube $\mathcal{E}(y(t), Y(t))$, $t_0 \leq t \leq \tau$, and the trajectory $x^{(0)}(t)$ must be *tangent* to the tube $\mathcal{E}(y(t), Y(t))$ at t^*. Thus, $x^{(0)}(t)$ is differentiable at t^* and

$$\langle \chi, \dot{x}^{(0)}(t^* - 0) \rangle = \langle \chi, \dot{x}^{(0)}(t^* + 0) \rangle = 0. \tag{7.83}$$

Here χ is the support vector to the state constraint $\mathcal{Y}(t^*)$ at $x^{(0)}(t^*)$. In general, if the jump is $\Lambda(t^* + 0) - \Lambda(t^* - 0) = \chi$, then the necessary condition for such a jump is

$$\langle \chi, \dot{x}^{(0)}(t^* + 0) - \dot{x}^{(0)}(t^* - 0) \rangle \geq 0. \tag{7.84}$$

For example, if $x^{(0)}(t^* - \sigma)$, $\sigma > 0$, lies inside the interior of the constraint set and $x(t^* + \sigma)$ lies on the boundary, then

$$\langle \chi, \dot{x}^{(0)}(t^* + 0) - \dot{x}^{(0)}(t^* - 0) \rangle \leq 0,$$

and (7.84) will be fulfilled only if this inner product equals zero.

We also need the following assumption.

Assumption 7.2.3. *For Problem 7.1.2, with given l, there exists no control $u^*(s)$ that satisfies the maximum principle*

$$l'S(s, \tau \mid M^{(0)}(\cdot), \lambda^{(0)}(\cdot))B(s)u^*(s)$$

$$= \max\{l'S(s, \tau \mid M^{(0)}(\cdot), \lambda^{(0)}(\cdot))B(s)u \mid u \in \mathcal{E}(p(s), P(s))\}$$

for $\{s \mid l'S(s, \tau \mid M^{(0)}(\cdot), \lambda^{(0)}(\cdot))B(s) \neq 0\}$ and at the same time ensures for these values of s that the corresponding trajectory $x^*(s) \in \partial \mathcal{Y}(s)$.

Property 7.2.3. Assumption 7.2.3 holds.

This means that if the control $u^{(0)}(s)$ is determined by the maximum principle, with $h^{(0)}(\tau, s) = l'S(s, \tau \mid M^{(0)}(\cdot), \lambda^{(0)}(\cdot))B(s) \neq 0$, and therefore attains its extremal values under given hard bounds, then this control cannot also keep the corresponding trajectory $x^{(0)}(s)$ along the boundary $\partial \mathcal{Y}(s)$. In other words, in this case *the maximum principle is degenerate along the state constraint*, i.e. $h^{(0)}(\tau, s) \equiv 0$ and does not help to find the control when the trajectory lies on the boundary $\partial \mathcal{Y}(s)$.

This assumption excludes the case when solution of Problem 7.1.2 *without state constraints* already satisfies the given state constraints, which therefore turn out to be passive.

Lemma 7.2.5. *Under Assumption 7.83 applied to Problem 7.1, the function*

$$h^{(0)}(\tau, s) = l'S(s, \tau \mid M^{(0)}(\cdot), \lambda^{(0)}(\cdot))B(s) \equiv 0$$

whenever $x^{(0)}(s) \in \partial \mathcal{Y}(s)$.

7.2 State-Constrained Control: Computation

Remark 7.2.6. Recall that equations for tight external ellipsoids *without state constraints* are explained in detail in Chap. 3, Sects. 3.2–3.3, being described by equations (see (3.59), (3.60), (3.17))

$$\dot{x}^* = A(t)x^* + B(t)p(t), \quad x^*(t_0) = x^0,$$

$$\dot{X}_+^* = A(t)X_+ + X_+ A'(t) + \pi(t)X_+ + \pi^{-1}(t)P(t), \quad X_+(t_0) = X^0,$$

$$l^*(t) = -A(t)'l^*(t), \quad l^*(t_0) = l_*,$$

and

$$\pi(t) = \langle l^*(t), B(t)P(t)B'(t)l^*(t)\rangle^{1/2}/\langle l^*(t), X_+(t)l^*(t)\rangle^{1/2}.$$

Exercise 7.2.1. *Compare equations of Remark 7.2.6 with (7.72)–(7.73) derived under state constraints.*

Chapter 8
Trajectory Tubes State-Constrained Feedback Control

Abstract This chapter begins with the *theory of trajectory tubes* which are necessary elements of realistic mathematical models for controlled processes and their evolutionary dynamics. We then deal with the evolution in time of state-constrained forward and backward reachability tubes also known as "viability tubes." The backward tubes are then used to design feedback controls under state constraints that may also appear in the form of obstacles to be avoided by system trajectories.

Keywords Trajectory tubes • Viable solutions • Funnel equations • Feedback control • Obstacle problems

In this chapter we study the evolution of trajectory tubes. These tubes are needed for problems of dynamics and control, especially those that deal with incompletely described systems. Such a *theory of trajectory tubes* was introduced in [158] and is described here in appropriate form for the topics of this book. The theory is then applied to state-constrained reachability ("reach") tubes, also known as "viability tubes." We introduce the technical tools in set-valued dynamics that are used for calculating such tubes. These backward tubes are then used to define state-constrained closed-loop strategies for target control. Finally we study similar issues for obstacle problems that generate a class of more complex state constraints: the target-oriented trajectories should develop within a bounded control set while also avoiding obstacles that lie on route to the target set. Throughout the investigations we assume that the notions of continuity and measurability of single-valued and multivalued maps are taken in the sense of [41, 238].

8.1 The Theory of Trajectory Tubes: Set-Valued Evolution Equations

We start with a definition of trajectory tubes that are at the object of further discussion.

8.1.1 Trajectory Tubes and the Generalized Dynamic System

Consider the nonlinear differential inclusion derived from Sect. 7.1 , (7.1), (7.2),

$$\dot{x} \in \mathcal{F}(t,x),\ t \in T,\ x(t_0) = x^0 \in \mathcal{X}^0 \in \text{comp}\, \mathbb{R}^n, \tag{8.1}$$

with $\mathcal{F}(t,x) = f(t,x,\mathcal{P}(t))$,
Assume that $\mathcal{F}(t,x)$ satisfies the Lipschitz condition with constant $k > 0$

$$h(\mathcal{F}(t,x'), \mathcal{F}(t,x'')) \le k\|x' - x''\|, \quad \forall x', x'' \in \mathbb{R}^n. \tag{8.2}$$

Taking set $\mathcal{X}^0 \in \text{comp}\,\mathbb{R}^n$ to be given, denote, as before, $x[t] = x(t,t_0,x^0)$, $t_0 \in T = [t_0,t_1]$, to be a solution of (8.1) (an isolated trajectory) that starts at point $x[t_0] = x^0 \in \mathcal{X}^0$.[1]

We regard the solution to be a Caratheodory-type trajectory $x[\cdot]$, namely, an absolutely continuous function $x[t]$ ($t \in T$) that satisfies the inclusion

$$dx[t]/dt = \dot{x}[t] \in \mathcal{F}(t,x[t]),\ x[t_0] = x^0 \tag{8.3}$$

for almost all $t \in T$.

We require all solutions $\{x[t] = x(t,t_0,x_0)\ |\ x_0 \in X_0\}$ to be extendible up to time t_1 [7, 26, 75].

Let $\mathcal{Y}(t)$ be a continuous multivalued map ($\mathcal{Y} : T \to \text{conv}\,\mathbb{R}^n$) of (7.2), that defines the state constraint

$$x(t) \in \mathcal{Y}(t),\ t \in T. \tag{8.4}$$

The map $\mathcal{Y}(t)$ may be obtained from measurement equations (considered further, in the next chapter)

$$y(t) = G(t)x + \xi(t), \xi(t) \in \mathcal{R}(t),$$

so that

$$G(t)x(t) \in \mathcal{Y}(t) = y(t) - \mathcal{R}(t), \tag{8.5}$$

with $y(t)$ given.
A more general form could be

$$y(t) \in G(t,x) \tag{8.6}$$

---

[1] Recall that a trajectory $x(t)$ of system (8.1) generated by given initial condition $\{t_0, x^0\}$ is marked by square brackets, as $x[t]$, in contrast with an unspecified trajectory $x(t)$.

8.1 The Theory of Trajectory Tubes: Set-Valued Evolution Equations

where $G(t, x)$ is a multivalued map ($G : T \times \mathbb{R}^n \to \text{conv } \mathbb{R}^p$). Other requirements on $G(t, x)$ will be indicated below, along with specific theorems.

The problem will now consist in describing the tube $\mathcal{X}_y[\cdot]$ of solutions to system (8.1) under state constraints (8.4), (8.5), (or (8.6)) and also the equations for the evolution of tube $\mathcal{X}_y[t]$.

As mentioned in the Preface to Chap. 7, the problem of jointly solving the system of inclusions (8.4), (8.5), could be also treated as a version of the viability problem [5]. The solution set to this system will be called the *viable tube* and defined more precisely below.

As we will observe, the differential inclusion (8.1) with viability constraints (8.5) may generate a class of *generalized dynamic systems* that are relevant for describing tube-valued solutions to problems of dynamics and control.

We now proceed with a rigorous formulation and a constructive theory for problems under discussion.

Definition 8.1.1 (See [5, 158]). A trajectory $x[t] = x(t, t_0, x^0), (x^0 \in \mathcal{X}^0, t \in T)$ of the differential inclusion (8.1) is said to be viable on $[t_0, \tau] \, \tau \leq t_1$, if

$$x[t] \in \mathcal{Y}(t) \text{ for all } t \in [t_0, \tau]. \tag{8.7}$$

We proceed under assumption that there exists at least one solution $x^*[t] = x^*(t, t_0, x_0^*)$ of (8.4) (together with a starting point $x^*[t_0] = x_0^* \in \mathcal{X}^0$) that satisfies condition (8.7) with $\tau = t_1$. The trajectory $x^*[t]$ is therefore assumed to be viable on the whole segment $[t_0, t_1]$. Conditions for the existence of such trajectories may be given in terms of generalized duality concepts [137, 238]. Known theorems on viability also provide the existence of such trajectories $x^*[t]$ (see [5]).

Let $\mathcal{X}(\cdot, t_0, \mathcal{X}_0)$ be the set of all solutions to inclusion (8.4) without the state constraint (8.5) that emerge from $\mathcal{X}^0$ (the "solution tube") With $\mathcal{X}[t] = \mathcal{X}(t, t_0, \mathcal{X}^0)$ being its cross-section at time t. Define

$$\Xi = \bigcup \{\mathcal{X}(t, t_0, \mathcal{X}^0) \mid t_0 \leq t \leq t_1\}$$

to be the integral funnel for (8.1), [224]. One may observe that under our assumptions $\Xi \in \text{comp } \mathbb{R}^n$ [8, 26].

The subset of $\mathcal{X}(\cdot, t_0, \mathcal{X}_0)$ that consists of all solutions to (8.1) viable on $[t_0, \tau]$ is denoted by $\mathcal{X}_y(\cdot, \tau, t_0, \mathcal{X}^0)$ and its s-cross-sections as $\mathcal{X}_y(s, \tau, t_0, \mathcal{X}^0), s \in [t_0, \tau]$. We also use the notation

$$\mathcal{X}_y[\tau] = \mathcal{X}_y(\tau, t_0, \mathcal{X}^0) = \mathcal{X}_y(\tau, \tau, t_0, \mathcal{X}^0).$$

As indicated earlier, it is not difficult to observe that sets $\mathcal{X}_y[\tau]$ are actually the reachable sets at instant τ for the differential inclusion (8.1) with state constraint (8.7).

As in Chap. 7, the map $\mathcal{X}_y(t, t_0, \mathcal{X}^0)$ ($\mathcal{X}_y : T \times T \times \mathrm{comp}\, \mathbb{R}^n \to \mathrm{comp}\, \mathbb{R}^n$) satisfies the *semigroup property*

$$\mathcal{X}_y(t, \tau, \mathcal{X}_y(\tau, t_0, \mathcal{X}^0)) = \mathcal{X}_y(t, t_0, \mathcal{X}^0), \quad t_0 \leq \tau \leq t \leq t_1,$$

and therefore defines a generalized dynamic system with set-valued trajectories $\mathcal{X}_y[t] = \mathcal{X}_y(t, t_0, \mathcal{X}^0)$. The multivalued functions $\mathcal{X}[t]$, $\mathcal{X}_y[t]$, $t \in T$, are the *trajectory tube* and the *viable trajectory tube*, respectively.

8.1.2 Some Basic Assumptions

We will usually work under one of the two following groups of hypotheses, unless otherwise noted.

Denote the *graph* of map $\mathcal{F}(t, \cdot)$ as $\mathrm{graph}_t\, \mathcal{F}$ (t is fixed):

$$\mathrm{graph}_t\, \mathcal{F} = \{\{x, z\} \in \mathbb{R}^{n \times n} \mid z \in \mathcal{F}(t, x)\}.$$

The next assumption is related to convex compact-valued tubes.

Assumption 8.1.1. *1. For some $\mathcal{D} \in \mathrm{conv}\,\mathbb{R}^n$ such that $\Xi \subseteq \mathrm{int}\,\mathcal{D}$ the set $(\mathcal{D} \times \mathbb{R}^n) \cap \mathrm{graph}_t\, \mathcal{F}$ is convex for every $t \in T$.*
2. There exists a solution $x_*[\cdot]$ to (8.4) such that $x_*[t_0] \in \mathcal{X}^0$ and

$$x_*[t] \in \mathrm{int}\, \mathcal{Y}(t), \quad \forall t \in T.$$

3. The set $\mathcal{X}^0 \in \mathrm{conv}\,\mathbb{R}^n$.

In order to formulate the next group of assumptions, related to *star-shaped sets*, we recall the following notion.

Definition 8.1.2. *A set $\mathrm{St} \subseteq \mathbb{R}^n$ is said to be star-shaped (with center at c) if $c \in \mathrm{St}$ and $(1-\lambda)c + \lambda \mathrm{St} \subseteq \mathrm{St}$ for all $\lambda \in (0, 1]$.*

The family of all star-shaped compact sets $\mathrm{St} \subseteq \mathbb{R}^n$ with center at c is denoted as $St(c, \mathbb{R}^n)$ and with center at $0 \in \mathbb{R}^n$ as $St\mathbb{R}^n$.

Assumption 8.1.2. *1. For every $t \in T$, one has $\mathrm{graph}_t\, \mathcal{F} \in St(\mathbb{R}^{2n})$.*
2. There exists $\varepsilon > 0$ such that $\varepsilon \mathbf{B}(0) \subseteq \mathcal{Y}(t)$ for all $t \in T$, $(\mathbf{B}(0) = \{x \in \mathbb{R}^n \mid \|x\| \leq 1\})$.[2]
3. The set $\mathcal{X}^0 \in St(\mathbb{R}^n)$.

---

[2] If $\mathcal{Y}(t)$ has no interior points, being located in space $\mathbb{R}^m$, $m < n$, then $\mathcal{Y}(t)$ may be substituted by its neighborhood $\mathcal{Y}_r(t) = \mathcal{Y}(t) + r\mathbf{B}(0) > 0$, with some additional regularization.

Observe that Assumptions 8.1.1 and 8.1.2 are overlapping, but in general neither of them implies the other.

One important property of trajectory tubes and viable trajectory tubes is to preserve some geometrical characteristics of sets $X[t]$, $X_y[t]$ along the system trajectories, as in the following assertions.

Lemma 8.1.1. *Under Assumption 8.1.1 the cross-sections $X[t]$, $X_y[t]$ are convex and compact for all $t \in T$ ($X[t]$, that is, $X_y[t] \in \text{conv } \mathbb{R}^n$.*

Lemma 8.1.2. *Under Assumption 8.1.2 the cross-sections $X[t]$, $X_y[t]$ are star-shaped and compact for all $t \in T$, that is, $X[t]$, $X_y[t] \in St(\mathbb{R}^n)$.*

Loosely speaking, a differential inclusion with a convex graph$_t$ $\mathcal{F}$ generates convex-valued tubes $X[t]$, $X_y[t]$, while a star-shaped graph$_t$ $\mathcal{F}$ generates tubes with star-shaped cross-sections $X[t]$, $X_y[t]$. Convexity and the star-shape property are therefore the two simplest basic geometrical invariants for cross-sections of the trajectory tubes and viable trajectory tubes.

8.1.3 The Set-Valued Evolution Equation

Having defined the notion of viable trajectory tube and observed that mapping $X_y[t] = X_y(t, t_0, X^0)$ defines a generalized dynamic system, we come to the following natural question: does there exist some sort of evolution equation that describes the tube $X_y[t]$ as solutions to a related generalized dynamic system?

It should be emphasized here that the space $\mathcal{K} = \{\text{comp } \mathbb{R}^n\}$ of all compact subsets of $\mathbb{R}^n$ to which the "states" $X_y[t]$ belong is only a metric space with a rather complicated nonlinear structure. In particular, there does not exist even an appropriate universal definition for the difference of sets $A, B \in \text{comp } \mathbb{R}^n$. Hence the evolution equation for $X_y[t]$ is devised to avoid using such "geometric" differences. It is mainly due to this reason that the construction of an infinitesimal generator for set-valued transition maps generated by such generalized system dynamics is cumbersome.

In this subsection we rely on an approach to the evolution of trajectory tubes through *funnel equations* for set-valued functions in contrast with Hamiltonian techniques where this is done, as we have seen in Sect. 7.1, through single-valued functions described by PDEs of the HJB type.

We shall further require that one of the following assumptions concerning the mapping $\mathcal{Y}(\cdot)$ is fulfilled.

Assumption 8.1.3. *The graph* graph $\mathcal{Y} \in \text{conv } \mathbb{R}^{n+1}$.

Assumption 8.1.4. *For every $l \in \mathbb{R}^n$ the support function $f(l, t) = \rho(l \mid \mathcal{Y}(t))$ is differentiable in t and its derivative $\partial f(l, t)/\partial t$ is continuous in $\{l, t\}$.*

Theorem 8.1.1. *Suppose Assumption 8.1.1 or 8.1.2 holds and map $\mathcal{Y}$ satisfies either Assumption 8.1.3 or 8.1.4. Then the multifunction $X_y[t] = X_y(t, t_0, X^0)$ is a set-valued solution to the following evolution equation*

$$\lim_{\sigma \to +0} \sigma^{-1} h\left(X_y[t+\sigma], \bigcup\{x + \sigma \mathcal{F}(t,x) \mid x \in X_y[t]\} \cap \mathcal{Y}(t+\sigma)\right) = 0, \tag{8.8}$$

$$t_0 \leq t \leq t_1, \quad X_y[t_0] = X^0.$$

Rigorous proofs of this theorem in terms of the Hausdorff distance (see Sect. 2.3) and of the following uniqueness theorem are rather lengthy and are given in detail in [158, Sects. 6, 7]. The uniqueness of solution to Eq. (8.8) is indicated in the next theorem.

Denote $Z[t_0, t_1]$ to be the set of all multivalued functions $Z[t]$ ($Z : [t_0, t_1] \to$ comp $\mathbb{R}^n$) such that

$$\lim_{\sigma \to +0} \sigma^{-1} h\left(Z[t+\sigma], \bigcup\{x + \sigma \mathcal{F}(t,x) \mid x \in Z[t]\} \cap \mathcal{Y}(t+\sigma)\right) = 0, \tag{8.9}$$

$$Z[t_0] = X^0, \quad t_0 \leq t \leq t_1,$$

where the Hausdorff limit in (8.9) is uniform in $t \in [t_0, t_1]$.

Under assumptions of Theorem 8.1.1 we have

$$X_y[\cdot] = X_y(\cdot, t_0, X^0) \in Z[t_0, t_1].$$

Theorem 8.1.2. *Suppose the assumptions of Theorem 8.1.1 hold. Then the multivalued function $X_y[\tau] = X_y(\tau, t_0, X^0)$ is the unique solution to the funnel equation (8.8) in the class $Z[t_0, t_1]$ of all the set-valued mappings $Z(\cdot)$ that satisfy this equation uniformly in $t \in [t_0, t_1]$.*

8.1.4 The Funnel Equations: Specific Cases

In this subsection we indicate some examples of differential inclusions for which Assumptions 8.1.1 and 8.1.2 are fulfilled, presenting also the specific versions of the funnel equations.

Linear Systems

Consider the linear differential inclusion

$$\dot{x} \in A(t)x + B(t)Q(t), \quad x(t_0) = x^0 \in X^0, \quad t_0 \leq t \leq t_1, \tag{8.10}$$

8.1 The Theory of Trajectory Tubes: Set-Valued Evolution Equations   317

where $x \in \mathbb{R}^n$, $A(t)$ and $B(t)$ are continuous $n \times n$- and $n \times m$-matrices, $Q(t)$ is a continuous map $(Q : [t_0, t_1] \to \text{conv}\,\mathbb{R}^m)$, $\mathcal{X}^0 \in \text{conv}\,\mathbb{R}^n$.

Here Assumptions 8.1.1 (1), 8.1.1 (3) are fulfilled automatically. So to retain Assumption 8.1.1 (2) we introduce

Assumption 8.1.5. *There exists a solution $x_*[\cdot]$ of (8.10) such that*

$$x_*[t_0] \in \mathcal{X}^0, \quad x_*[t] \in \text{int}\,\mathcal{Y}(t), \quad \forall t \in [t_0, t_1].$$

The following result is a consequence of Theorem 8.1.1.

Theorem 8.1.3. *Suppose Assumptions 8.1.5 and 8.1.3 or 8.1.4 hold. Then the set-valued function $\mathcal{X}_y[t] = \mathcal{X}_y(t, t_0, \mathcal{X}^0)$ is the solution to the evolution equation*

$$\lim_{\sigma \to +0} \sigma^{-1} h(\mathcal{X}_y[t+\sigma], ((I + \sigma A(t))\mathcal{X}_y[t] + \sigma B(t)Q(t)) \cap \mathcal{Y}(t+\sigma)) = 0 \quad (8.11)$$

$$t_0 \le t \le t_1, \quad \mathcal{X}[t_0] = \mathcal{X}^0.$$

The solution to this equation may be obtained using ellipsoidal techniques described earlier in Chap. 3 and in Chap. 7, Sect. 7.2.

Bilinearity: Uncertainty in Coefficients of a Linear System

Consider a differential equation of type (8.1) where

$$f(t, x, v) = A(t, v)x + h(t), \quad v(t) \in Q(t),$$

with $A(t, Q(t)) = \mathcal{A}(t)$ and $h(t) \in \mathcal{P}(t)$ unknown but bounded ($\mathcal{P}(t) \in \text{conv}\,\mathbb{R}^{n \times n}$ being continuous in t). This yields

$$\dot{x} \in \mathcal{A}(t)x + \mathcal{P}(t), \quad x(t_0) = x^0 \in \mathcal{X}^0, \quad x(t) \in \mathcal{Y}(t), \quad t_0 \le t \le t_1, \quad (8.12)$$

and the right-hand side $\mathcal{F}(t, x) = \mathcal{A}(t)x + \mathcal{P}(t)$ depends bilinearly upon the state vector x and the set-valued map $\mathcal{A}$.

We assume that $\mathcal{A}(\cdot)$ is a continuous mapping from $[t_0, t_1]$ into the set conv $\mathbb{R}^{n \times n}$ of convex and compact subsets of the space $\mathbb{R}^{n \times n}$ of $n \times n$-matrices.

Equation (8.12) is an important model of an uncertain bilinear dynamic system with set-membership description of the unknown matrices $A(t) \in \mathcal{A}(t)$, and inputs $h(t) \in \mathcal{P}(t)$ and with $x^0 \in \mathcal{X}^0$, [158, Sect. 3].

It is not difficult to demonstrate that here Assumption 8.1.1 does not hold, so that sets $\mathcal{X}[t] = \mathcal{X}(t, t_0, \mathcal{X}^0)$ (and also $\mathcal{X}_y[t] = \mathcal{X}_y(t, t_0, \mathcal{X}^0)$) need not be convex.

Example 8.1.1. Indeed, consider a differential inclusion in $\mathbb{R}^2$:

$$\dot{x}_1 \in [-1, 1]x_2, \quad \dot{x}_2 = 0, \quad 0 \le t \le 1,$$

with initial condition

$$x(0) = x_0 \in X^0 = \{x = (x_1, x_2) \mid x_1 = 0, |x_2| \leq 1/2\}$$

Then

$$X[1] = X(1, 0, X^0) = X^1 \cup X^2,$$

where

$$X^1 = \{x = (x_1, x_2) \mid |x_1| \leq x_2 \leq 1/2\}, \quad X^2 = \{x = (x_1, x_2) \mid |x_1| \leq -x_2 \leq 1/2\}.$$

It is obvious that set $X[1]$ is not convex.

However one can verify that though Assumption 8.1.1 is not fulfilled for the system (8.12), the Assumption 8.1.2 (1) is indeed satisfied, provided $0 \in X^0$ and $0 \in \mathcal{P}(t)$ for all $t \in [t_0, t_1]$. Assumption 8.1.2 may hence be rewritten in the following reduced form.

Assumption 8.1.6. *1. Condition $0 \in \mathcal{P}(t)$ is true for all $t \in [t_0, t_1]$.*
2. There exists an $\varepsilon > 0$ such that $\varepsilon \mathbf{B}(0) \subseteq \mathcal{Y}(t), \forall t \in [t_0, t_1]$.
3. Set $X^0 \in St(\mathbb{R}^n)$ and $0 \in X^0$.

To formulate an analog of Theorem 8.1.1 we introduce an additional notation: $M * X = \{z = Mx \in \mathbb{R}^n \mid M \in \mathcal{M}, x \in X\}$, where $\mathcal{M} \in \text{conv}\,\mathbb{R}^{n \times n}$, $X \in \text{conv}\,\mathbb{R}^{n \times n}$.

Theorem 8.1.4. *Under Assumptions 8.1.6 and 8.1.3 or 8.1.4 the set-valued map $X_y[t] = X_y(t, t_0, X^0)$ is the solution to the following evolution equation*

$$\lim_{\sigma \to +0} \sigma^{-1} h(X_y[t+\sigma], ((I + \sigma \mathcal{A}(t)) * X_y[t] + \sigma \mathcal{P}(t)) \cap \mathcal{Y}(t+\sigma)) = 0, \quad (8.13)$$

$$t_0 \leq t \leq t_1, \quad X[t_0] = X^0.$$

A Nonlinear Example

We indicate one more example, which is given by a set-valued function $\mathcal{F}(t, x)$ with a star-shaped graph$_t$ $\mathcal{F}$.

Namely, let $\mathcal{F}(t, x)$ be of the form

$$\mathcal{F}(t, x) = G(t, x)U + \mathcal{P}(t), \quad (8.14)$$

where the $n \times n$-matrix function $G(t, x)$ is continuous in t, Lipschitz-continuous in x and $\mathcal{F}(t, x)$ is positively homogeneous in $U \in \text{conv}\,\mathbb{R}^n$. The map $\mathcal{P}(t)$ is the same as before. One may immediately verify that now Assumption 8.1.2 (1) holds, so that the cross-section $X_y[t]$ of the solution tube to Eq. (8.13) is star-shaped.

8.1.5 Evolution Equation Under Relaxed Conditions

We now mention an approach similar to the described, but with the evolution equation (8.8) written using the Hausdorff semidistance

$$h^+(A, B) = \max_a \min_b \{\|a - b\| \mid a \in A, b \in B\}, \quad A, B \in \text{comp}\, \mathbb{R}^n,$$

rather than the Hausdorff distance (a metric) $h(A, B) = \max\{h^+(A, B), h^+(B, A)\}$. Here is such an equation (see [174] and also [163]) for system (7.1), (7.2):

$$\lim_{\sigma \to +0} \frac{1}{\sigma} h_+ \left(Z[t + \sigma], \bigcup \left\{ x + \sigma(f(t, x, \mathcal{P}(t)) \mid x \in Z[t] \cap \mathcal{Y}(t) \right\} \right) = 0, \; Z(t_0) = X^0, \tag{8.15}$$

Remark 8.1.1. The solution $Z[t]$ to this evolution equation is a multivalued function with $Z[t_0] = X^0$, which satisfies (8.15) almost everywhere. As a rule, this solution is not unique. However, we may single out a solution $X[t]$ which is the **inclusion-maximal** among all solutions $Z[t]$ to (8.15). Namely, $X[t] \supset Z[t]$, where $Z[t]$ is any solution that starts from $Z(t_0) = X^0$. Note that Eq. (8.15) makes sense for any piecewise continuous function $y(t)$. As mentioned above, when dealing with information tubes, we presume these functions to be right-continuous. The h^+-techniques may be more adequate for dealing with discontinuous set-valued functions $X_y[t] = X_y(t, t_0, X_0)$.

The evolution equation (8.15) delivers a formal model for the generalized dynamic system generated by mapping $X_y[t] = X_y(t, t_0, X_y^0)$. However, if the sets $X_y[t]$ are convex, there exists an alternative presentation of this evolutionary system. Namely, each of the sets $X_y[t]$ could be described by its support function

$$\varphi(l, t) = \rho(l \mid X_y[t]) = \max\{(l, x) \mid x \in X_y[t]\}.$$

The dynamics of $X_y[t]$ would then be reflected by a *generalized partial differential equation* for $\varphi(l, t)$. The description of this equation is the next issue to be discussed.

8.2 Viability Tubes and Their Calculation

8.2.1 The Evolution Equation as a Generalized Partial Differential Equation

In this section we restrict our attention to a system for which Assumption 8.1.1 is satisfied. In this case the tube $X_y(\cdot, \tau, t_0, X^0)$ will be a convex compact subset of $C^n[t_0, t_1]$. With set $X_y[\tau] = X_y(\tau, t_0, X_y^0) \in \text{conv}\, \mathbb{R}^n$ for every $\tau \in [t_0, t_1]$.

Since there is an equivalence between an inclusion $x \in X$ and the system of inequalities $(l, x) \leq \rho(l \mid X)$, $\forall l \in \mathbb{R}^n$, we may now describe the evolution of sets $X_y[\tau]$ through differential relations for the support function $\varphi(l, t) = \rho(l \mid X_y[t])$. The primary problem on the law of evolution for multivalued "states" $X_y[t]$ will now be replaced by one of evolution in time of a *distribution* $\phi_t(l) = \varphi(l, t)$ which is defined over all $l \in \mathbb{R}^n$, and is positively homogeneous and convex in l.

We now calculate the directional derivative

$$\partial^+ \varphi(l, t)/\partial t = \lim_{\sigma \to +0} \sigma^{-1}(\varphi(l, t + \sigma) - \varphi(l, t))$$

for each instant $t \in [t_0, t_1]$ and for any fixed $l \in \mathbb{R}^n$.

The next result will be proved as a direct consequence of Theorem 8.1.1.

Theorem 8.2.1. *Suppose Assumptions 8.1.1 together with either 8.1.3 or 8.1.4 hold. Then the support function $\varphi(l, t) = \rho(l \mid X_y[t])$ is right-differentiable in t and its directional derivative in time is*

$$\partial^+ \varphi(l, t)/\partial t = \min_q \max_x \{\psi(q, x, l, t) \mid q \in Q(l, t), x \in \partial_l \varphi(l, t)\} =$$

$$= \max_x \min_q \{\psi(q, x, l, t) \mid x \in \partial_l \varphi(l, t), q \in Q(l, t)\}, \qquad (8.16)$$

where

$$\psi(q, x, l, t) = \rho(q \mid \mathcal{F}(t, x)) + \partial \rho(l - q \mid \mathcal{Y}(t))/\partial t, \qquad (8.17)$$

$$Q(l, t) = \{q \in \mathbb{R}^n \mid \varphi(l, t) - \varphi(q, t) = \rho(l - q \mid \mathcal{Y}(t))\}, \quad \partial_l \varphi(l, t)$$

$$= \{x \in X_y[t] \mid (l, x) = \varphi(l, t)\}.$$

Proof. From Theorem 8.1.1 we deduce

$$\varphi(l, t + \sigma) = \rho(l \mid R(\sigma, t, X_y[t]) \cap \mathcal{Y}(t + \sigma)) + o(\sigma) \|l\|,$$

where $\sigma^{-1} o(\sigma) \to 0$ ($\sigma \to +0$),

$$R(\sigma, t X_y[t]) = \bigcup \{x + \sigma \mathcal{F}(t, x) \mid x \in X_y[t]\}, \quad t_0 \leq t \leq t + \sigma \leq t_1,$$

and $R(\sigma, t, X_y[t])$ is convex due to Assumption 8.1.1.

Taking the infimal convolution of support functions for respective sets in the above intersection (under assumptions of Theorem 8.2.1 the inf-convolution (8.16), (8.17) is exact, [194]), we come to

8.2 Viability Tubes and Their Calculation

$$\varphi(l, t + \sigma) = \tag{8.18}$$

$$= \min_{q}\{\rho(q \mid R(\sigma, t, X_y[t])) + \rho(l - q \mid \mathcal{Y}(t + \sigma)) \mid q \in r\mathbf{B}(0) + o(\sigma), \|q\| \leq 1\}$$

for a certain $r > 0$.

Hence we have to verify only the right-differentiability in σ of the following function

$$H(\sigma, l, t) = \min_{q} \max_{x}\{h(\sigma, t, l, q, x) \mid q \in rS, \ x \in X_y[t]\}, \tag{8.19}$$

where

$$h(\sigma, t, l, q, x) = (q, x) + \sigma\rho(q \mid \mathcal{F}(t, x)) + \rho(l - q \mid \mathcal{Y}(t + \sigma)) \tag{8.20}$$

with $\{l, t\}$ fixed.

The differentiation of this minimax relation may be realized using the rules given in [60, Theorem 5.3]. Applying this theorem, all the conditions of which are satisfied, we arrive at (8.16).

Let us elaborate on this result. Due to properties of $X_y[t]$ we have $X_y[t] \subseteq \mathcal{Y}(t)$, so that (8.17) yields (with $q \in Q(l, t)$)

$$\varphi(l, t) - \varphi(q, t) = \rho(l - q \mid \mathcal{Y}(t)) \geq \varphi(l - q, t) \tag{8.21}$$

On the other hand, since the function $\varphi(l, t)$ is convex and positively homogeneous in $l \in \mathbb{R}^n$ with t fixed, we also observe

$$\varphi(l, t) \leq \varphi(q, t) + \varphi(l - q, t) \tag{8.22}$$

Comparing this with (8.21), we come to the equality

$$\varphi(l, t) - \varphi(q, t) = \varphi(l - q, t). \tag{8.23}$$

Without loss of generality we could have assumed $0 \in X_y[t]$.

Suppose we now require

$$0 \in \text{int } X_y[t], \quad \forall t \in [t_0, t_1], \tag{8.24}$$

which is an additional assumption. It is not difficult to prove the following sufficient condition for the sets $X[t]$ to be symmetric with respect to 0.[3]

---

[3] Recall that a set $\mathcal{A} \subseteq \mathbb{R}^n$ is defined to be symmetric with respect to 0 if $\mathcal{A} = -\mathcal{A}$.

Lemma 8.2.1. *Suppose assumption (8.24) holds and sets X^0, graph$_t$ $\mathcal{F}$, $Y(t)$ are symmetric for all $t \in [t_0, t_1]$. Then set $X[t]$ is also symmetric whatever be the instant $t \in [t_0, t_1]$.*

Under the assumptions of Lemma 8.2.1 the inequality (8.22) turns into an equality if and only if the vectors l, q are collinear ($l = \alpha q$ and $\alpha \geq 0$). Then (8.17) turns into

$$(1 - \alpha)(\rho(l \mid X[t]) - \rho(l \mid Y(t))) = 0$$

so that with $\rho(l \mid X[t]) < \rho(l \mid Y(t))$ we have $\alpha = 1$ and in (8.16) the set $Q(l, t) = \{l\}$. Otherwise, $q = \alpha l$, $\alpha \in [0, 1]$ and

$$\psi(q, x, l, t) = \alpha \rho(l \mid \mathcal{F}(t, x)) + (1 - \alpha)\partial \rho(l \mid Y(t))/\partial t,$$

so that the minimum in (8.16) is to be taken over $\alpha \in [0, 1]$.

Finally, we have

Theorem 8.2.2. *Under assumptions of Theorem 8.1.1 the next relations are true:*

$$\frac{\partial^+ \varphi(l, t)}{\partial t} = \begin{cases} \rho_{\max}, & \text{if } \varphi(l, t) < \rho(l \mid \mathcal{Y}(t)); \\ \min\{\rho_{\max}, \partial \rho(l \mid \mathcal{Y}(t))/\partial t\}, & \text{if } \varphi(l, t) = \rho(l \mid \mathcal{Y}(t)), \end{cases} \quad (8.25)$$

$$\rho_{\max} = \max\{\rho(l \mid \mathcal{F}(t, x)) \mid x \in \partial_l \varphi(l, t)\}.$$

For a linear system this turns into

$$\frac{\partial^+ \varphi(l, t)}{\partial t} = \begin{cases} \rho_*, & \text{if } \varphi(l, t) < \rho(l \mid \mathcal{Y}(t)); \\ \min\{\rho_*, \partial \rho(l \mid \mathcal{Y}(t))/\partial t\}, & \text{if } \varphi(l, t) = \rho(l \mid \mathcal{Y}(t)), \end{cases} \quad (8.26)$$

$$\rho_* = \rho(A'l \mid \partial_l \varphi(l, t)) + \rho(l \mid B(t)Q(t)).$$

Each of the relations (8.25), (8.26) is actually a *generalized partial differential equation* which has to be solved under boundary condition

$$\varphi(l, t_0) = \rho(l \mid X[t_0]). \quad (8.27)$$

The last result will later be used in Sect. 8.4 in describing feedback solution strategies for the problem of closed-loop control under state constraints.

8.2.2 Viability Through Parameterization

In this subsection the description viable trajectory tubes is reduced to the treatment of trajectory tubes for a variety of specially designed new differential inclusions

8.2 Viability Tubes and Their Calculation

without state constraints. These new inclusions are designed depending upon certain parameters and have a relatively simple structure. The overall solution is then presented as an intersection of parallel solution tubes for the parametrized inclusions over all the parameters.

The General Case

We start with the general nonlinear case, namely, with differential inclusion (8.1) and set-valued state constraint (8.4), repeated here as

$$\dot{x} \in \mathcal{F}(t, x), \quad x(t_0) \in \mathcal{X}^0, \quad t_0 \leq t \leq t_1, \tag{8.28}$$

with constraints

$$x(t) \in \mathcal{Y}(t), \quad t_0 \leq t \leq \tau, \tag{8.29}$$

supposing also that the basic conditions of Sect. 8.1 on maps $\mathcal{F}$, Y and set $\mathcal{X}^0$ are satisfied.

In this section we do not require however that either of the Assumptions 8.1.1, 8.1.2 or 8.1.3, 8.1.4 would hold.

We further need to introduce *the restriction* $\mathcal{F}_Y(t, x)$ *of map* $\mathcal{F}(t, x)$ *to a set* $\mathcal{Y}(t)$ (at time t). This is given by

$$\mathcal{F}_Y(t, x) = \begin{cases} \mathcal{F}(t, x), & \text{if } x \in \mathcal{Y}(t); \\ \emptyset, & \text{if } x \notin \mathcal{Y}(t). \end{cases}$$

The next property follows directly from the definition of viable trajectories.

Lemma 8.2.2. *An absolutely continuous function $x[t]$ defined on the interval $[t_0, \tau]$ with $x^0 \in \mathcal{X}^0$ is a viable trajectory to (8.28) for $t \in [t_0, \tau]$ if and only if the inclusion $\dot{x}[t] \in \mathcal{F}_Y(t, x[t])$ is true for almost all $t \in [t_0, \tau]$.*

We will now represent $\mathcal{F}_Y(t, x)$ as an intersection of certain multifunctions. The first step to achieve that objective will be to prove the following auxiliary assertion.

Lemma 8.2.3. *Suppose A is a bounded set, B a convex closed set, both in $\mathbb{R}^n$. Then*

$$\bigcap \{A + LB \mid L \in \mathbb{R}^{n \times n}\} = \begin{cases} A & \text{if } 0 \in B, \\ \emptyset & \text{if } 0 \notin B, \end{cases}$$

Proof. First assume $0 \in B$. Then

$$A \subseteq \bigcap\{A + L\{0\} \mid L \in \mathbb{R}^{n \times n}\} \subseteq \bigcap\{A + LB \mid L \in \mathbb{R}^{n \times n}\} \subseteq A + 0 \cdot B = A.$$

Hence, $A = \bigcap \{A + LB \mid L \in \mathbb{R}^{n \times n}\}$. If $0 \notin B$, then there exists a hyperplane in $\mathbb{R}^n$ that strictly separates the origin from B. By the boundedness of A, for a sufficiently large number $\lambda > 0$, we then have

$$(A + \lambda B) \cap (A - \lambda B) = \emptyset.$$

Setting $L^+ = \lambda I$ and $L^- = -\lambda I$ we conclude that

$$\bigcap \{A + LB \mid L \in \mathbb{R}^{n \times n}\} \subseteq (A + L^+ B) \cap (A + L^- B) = \emptyset.$$

The lemma is thus proved.

From the above two lemmas above we obtain the following description of viable trajectories.

Theorem 8.2.3. *An absolutely continuous function $x(\cdot)$ defined on an interval $[t_0, t_1]$ with $x^0 = x(t_0)$ is a viable trajectory of (8.28) for $[t_0, \tau]$ iff the inclusion*

$$\dot{x}(t) \in \bigcap \{(\mathcal{F}(t, x) - Lx(t) + L\mathcal{Y}(t)) \mid L \in \mathbb{R}^{n \times n}\}$$

is true for almost all $t \in [t_0, \tau]$.

We introduce a family of differential inclusions that depend on a matrix parameter $L \in \mathbb{R}^{n \times n}$. These are given by

$$\dot{z} \in \mathcal{F}(t, z) - Lz + L\mathcal{Y}(t), \quad z(t_0) \in \mathcal{X}^0, \quad t_0 \le t \le t_1. \tag{8.30}$$

By $z[\cdot] = z(\cdot, \tau, t_0, z_0, L)$ denote the trajectory to (8.30) defined on the interval $[t_0, \tau]$ with $z[t_0] = z_0 \in \mathcal{X}^0$. Also denote

$$Z(\cdot, \tau, t_0, \mathcal{X}^0, L) = \bigcup \{Z(\cdot, \tau, t_0, z^0, L) \mid z^0 \in \mathcal{X}^0\},$$

where $Z(\cdot, \tau, t_0, z^0, L)$ is the tube of all trajectories $z[\cdot] = z(\cdot, \tau, t_0, z^0, L)$ issued at time t_0 from point z^0 and defined on $[t_0, \tau]$. The cross-sections of set $Z(\cdot, \tau, t_0, \mathcal{X}^0, L)$ at time t are then denoted as $Z(\tau, t_0, \mathcal{X}^0, L)$.

Theorem 8.2.4. *For each $\tau \in [t_0, t_1]$ one has*

$$\mathcal{X}_y(\cdot, \tau, t_0, \mathcal{X}^0) = \bigcap \{Z(\cdot, \tau, t_0, \mathcal{X}^0, L) \mid L \in \mathbb{R}^{n \times n}\}.$$

Moreover, the following inclusion is true

$$\mathcal{X}_y[\tau] = \mathcal{X}_y(\tau, t_0, \mathcal{X}^0) \subseteq \bigcap \{Z(\tau, t_0, \mathcal{X}^0, L) \mid L \in \mathbb{R}^{n \times n}\}.$$

The Linear-Convex Case

For the case of linear system (8.10)

$$\dot{x} \in A(t)x + B(t)Q(t), \quad x(t_0) = x^0 \in \mathcal{X}^0, \quad t_0 \leq t \leq t_1,$$

inclusion (8.30) turns into

$$\dot{z} \in (A(t)-L(t))z + Q(t) + L(t)\mathcal{Y}(t), \quad z(t_0) = z^0 \in \mathcal{X}^0, \quad t_0 \leq t \leq \tau. \quad (8.31)$$

whose solution tubes parametrized by $L(\cdot)$ are $Z(\tau, t_0, \mathcal{X}^0, L(\cdot))$ However, the main point for a linear system is that the second inclusion of Theorem 8.2.4 actually turns into an equality, producing the next result.

Theorem 8.2.5. *The following equality is true for any* $\tau \in [t_0, t_1]$:

$$\mathcal{X}_y[\tau] = \bigcap \{Z(\tau, t_0, \mathcal{X}^0, L(\cdot)) \mid L(\cdot) \in \mathbb{R}^{n \times n}[t_0, \tau]\}. \quad (8.32)$$

The proof of this theorem is a rather cumbersome procedure which relies on direct calculations of Sects. 8.1 and 8.2, including the modified maximum principle of Sect. 7.2.1, is given in detail in [158] (see Sects. 10–13 of that paper).

Remark 8.2.1. Dealing with nonlinear systems we could look at a broader class of functions that parametrize (8.30), taking $L = L(\cdot, \cdot) = \{L(t, x) \; t \in [t_0, t_1], x \in \mathbb{R}^n\}$. Question: does there then exist a class of nonlinearities and a related class of functions $L(t, x)$ that would ensure an equality similar to (8.32)?

8.3 Control Synthesis Under State Constraints: Viable Solutions

The techniques of Chaps. 2 and 3 with those of Sects. 7.1 and 7.2 allow to approach the problem of control synthesis under state constraints (see [136]). Such issues were treated later under the term "viability problems," with some renaming of earlier terminology [5]. Therefore we may also interpret our problem as that of terminal control "under viability constraints," meaning that solutions subject to the state constraint are "viable solutions." As in the case of systems without state constraints, the synthesized solution strategy described here will be nonlinear.

8.3.1 The Problem of State-Constrained Closed-Loop Control: Backward Reachability Under Viability Constraints

We now return to system (7.30), namely,

$$\dot{x} = f(t, x, u) = A(t)x + B(t)u, \quad u \in \mathcal{P}(t) = \mathcal{E}(p(t), P(t)) \tag{8.33}$$

$$x \in \mathcal{Y}(t) = \mathcal{E}(y(t), Y(t)) \tag{8.34}$$

$$x^0 = x(t_0) \in \mathcal{X}^0, \tag{8.35}$$

where $p(t)$, $P(t)$ are continuous and $y(t)$, $Y(t)$ are absolutely continuous. The target set is also ellipsoidal: $\mathcal{M} = \mathcal{E}(m, M)$

We may present constraints (8.34) and $x \in \mathcal{E}(m, M)$ as

$$\varphi(t, x) = \langle x - y(t), Y^{-1}(t)(x - y(t)) \rangle \leq 1, \quad \varphi_M(x) = \langle x - m, M^{-1}(x - m) \rangle \leq 1.$$

The following problems are considered.

Problem 8.3.1. *Given interval $[\tau, t_1]$, system position $\{\tau, x\}$ and convex compact target set $\mathcal{M} \subset \mathbb{R}^n$, find*

(i) solvability set $W[\tau] = W(\tau, t_1, \mathcal{M})$ and
(ii) feedback control strategy $u = \mathcal{U}(t, x)$, $\mathcal{U}(\cdot, \cdot) \in \mathbf{U}_Y$ such that all the solutions to the differential inclusion

$$\dot{x} \in A(t)x + B(t)\mathcal{U}(t, x) \tag{8.36}$$

that start from a given position $\{\tau, x\}$, $x = x[\tau]$, $x[\tau] \in W(\tau, t_1, \mathcal{M})$, $\tau \in [t_0, t_1]$, would satisfy the inclusion (8.34), $\tau \leq t \leq t_1$, with $x(t_1) \in \mathcal{M}$.

Here $\mathbf{U}_Y$ is an appropriate class of functions defined below, in (8.42). The given problem is non-redundant, provided $W[\tau] = W(\tau, t_1, \mathcal{M}) \neq \emptyset$.

The value function for Problem 8.3.1 may be defined as follows

$$V^s(\tau, x) = \min_u \max \left\{ \max_t \{\varphi(t, x[t]) \mid t \in [\tau, t_1]\}, \varphi_M(x[t_1]) \mid x[\tau] = x \right\}. \tag{8.37}$$

Then

$$W[\tau] = \{x : V^s(\tau, x) \leq 1\}$$

will be the *backward reach set relative to* $\mathcal{M}$ *under state constraint* $\mathcal{Y}(t)$, which is the set of points x for which there exists *some* control $u(t)$ that steers trajectory $x[t] = x(t, \tau, x)$ to $\mathcal{M}$ with $x(t) \in \mathcal{Y}(t)$, $t \in [\tau, t_1]$.

8.3 Control Synthesis Under State Constraints: Viable Solutions

Denote $V^s(t,x) = V^s(t, x \mid V^s(t_1, \cdot))$, emphasizing the dependence of $V^s(t,x)$ on the boundary condition $V^s(t_1, x) = \max\{\varphi(\vartheta, x), \varphi_M(x)\}$.

Lemma 8.3.1. *The value function $V^s(t,x)$ satisfies the **principle of optimality**, which has the semigroup form*

$$V^s(\tau, x \mid V^s(t_1, \cdot)) = V^s(\tau, x \mid V^s(t, \cdot \mid V^s(t_1, \cdot))). \tag{8.38}$$

As in Sect. 2.1 this property is established through conventional reasoning and implies a similar property for the corresponding reachability sets. Namely, if we take $\mathcal{W}[\tau] = \mathcal{W}(\tau, t_1, \mathcal{M})$, we have $\mathcal{W}(\tau, t_1, \mathcal{M}) = \mathcal{W}(\tau, t, \mathcal{W}(t, t_1, \mathcal{M}))$.

To proceed further, denote

$$\mathcal{H}(t, x, \mathbf{p}, u) = p_0 + \langle p, f(t, x, u)\rangle, \quad \mathbf{p} = \{p_0, p\}.$$

Consider the $(n+1)$-dimensional vector $\mathbf{p} = \{V_t^s, V_x^s\}$. Then, under control u the total derivative of $V^s(t,x)$ along a trajectory $x(t)$ of (8.33) will be

$$dV^s/dt|_u = \mathcal{H}(t, x, \mathbf{p}, u), \quad p_0 = V_t^s, \, p = V_x^s.$$

This relation is true for linear systems of type (8.33), with a differentiable $V^s(t, x)$. Otherwise, the property of differentiability for V^s may be relaxed to directional differentiability which in our case always holds.

The class of solution strategies. As in Chap. 2, we shall look for the solution in the class $\mathbf{U}_Y$ of all set-valued functions with convex compact values that are measurable in t and upper semicontinuous in x, ensuring thus the solvability and extendability of solutions to (8.36) for any $x^0 \in \mathcal{X}^0 \in \text{conv}\,\mathbb{R}^n$. But the difference will be in that now $\mathbf{U}_Y$ of Problem 8.3.1 will be

$$\mathbf{U}_Y = \{\mathcal{U}(t,x)\}, \; \mathcal{U}(t,x) = \begin{cases} \mathcal{P}_w(t) = \mathcal{E}(p(t), P(t)) \cap \{u : d\varphi(t,x)/dt \mid_u \leq 0\}, \\ \qquad \qquad \text{if } \varphi(t,x) = 1, \\ \mathcal{P}_w(t) = \mathcal{E}(p(t), P(t)), \\ \qquad \qquad \text{if } \varphi(t,x) < 1. \end{cases} \tag{8.39}$$

Assume

$$V^s(t,x) > \varphi(t,x).$$

Then the HJB equation for V^s will be

$$V_t^s(t,x) + \min_u \langle V_x^s, A(t)x + B(t)u\rangle = 0, \; u \in \mathcal{P}_w(t), \tag{8.40}$$

under boundary condition $V^s(t_1, x) = \max\{\varphi(t_1, x), \varphi_M(x)\}$.

Note that under control u the total derivative along the trajectory of (8.33) is

$$dV^s/dt|_u = \mathcal{H}(t,x,\mathbf{p},u), \quad \mathbf{p} = \{V_t^s, V_x^s\},$$

and $d\varphi(t,x)/dt\,|_u = \mathcal{H}(t,x,\varphi_{t,x},u)$.

Lemma 8.3.2. *The backward reachability set, at time $\tau \leq t_1$, from target set $\mathcal{M}$, under state constraint $x(t) \in \mathcal{Y}$ is the level set*

$$\mathcal{W}(\tau,t_1,\mathcal{M}) = \mathcal{W}[\tau] = \{x : V^s(\tau,x) \leq 1\}.$$

The proof of this assertion is similar to the reasoning of Sect. 7.1.2

Exercise 8.3.1. *Prove Lemma 8.3.2.*

Now consider the second part (ii) of Problem 8.3.1 on closed-loop control synthesis. Without loss of generality we shall assume $A(t) \equiv 0$, then indicate formal changes for $A(t) \neq 0$. In order to do that we will need to know set $\mathcal{W}[\tau]$ and may therefore use the description given above, repeating ellipsoidal techniques of Sect. 7.1.2. This route relies on reaching the solutions through single-valued functions, as thoroughly discussed in Sects. 7.1 and 7.2.

However we shall now rely on the alternative vision of the same problem which involves set-valued representations.

8.3.2 State-Constrained Closed-Loop Control

We thus deal with system

$$\dot{x} = B(t)u, \tag{8.41}$$

$$u \in \mathcal{P}(t) = \mathcal{E}(p(t), P(t)), \quad x \in \mathcal{Y}(t) = \mathcal{E}(y(t), Y(t)), \quad x_\tau = x(\tau) \in \mathcal{X}_\tau = \mathcal{E}(x_\tau^*, X_\tau) \tag{8.42}$$

Following the scheme of Sect. 2.6 we will look for the closed-loop solution strategy $\mathcal{U}^*$ by using solvability (or backward reach) sets introduced in Chap. 2, Sects. 2.3 and 2.6.

We recall that apart from Lemma 8.3.2, the backward reach set $\mathcal{W}[\tau] = \mathcal{W}(\tau,t_1,\mathcal{M})$ is also given by a *funnel equation*, similar to equations derived for forward reach in Sect. 8.1 (see (8.8) for the general case and (8.11) for the linear case):

Lemma 8.3.3. *The set-valued function $\mathcal{W}[t]$ satisfies the following evolution equation*

$$\lim_{\sigma \to 0} \sigma^{-1} h\left(\mathcal{W}[t-\sigma], (\mathcal{W}[t] - \sigma B(t)\mathcal{P}(t)) \cap \mathcal{Y}(t-\sigma)\right) = 0, \tag{8.43}$$

$$t_0 \leq t \leq t_1, \quad \mathcal{W}[t_1] = \mathcal{M}.$$

8.3 Control Synthesis Under State Constraints: Viable Solutions

The tube $\mathcal{W}[t]$ will now be used to introduce a scheme for finding the synthesizing control strategy $\mathcal{U}(t, x)$.

Given $\mathcal{W}[t]$, $t \in [\tau, t_1]$, let

$$dh^+(x(t), \mathcal{W}[t])/dt\big|_{(8.41)} \tag{8.44}$$

denote the total derivative at time $t + 0$ of the distance $d(x, \mathcal{W}[t]) = h^+(x, \mathcal{W}[t])$ due to Eq. (8.41).

Define

$$\mathcal{U}^*(t, x) = \left\{ u \ \Big|\ dh^+(x, \mathcal{W}[t])/dt\big|_{(8.41)} \leq 0 \right\}.$$

We now have to prove that the set-valued strategy $\mathcal{U}^*(t, x) \neq \emptyset$ for all $\{t, x\}$ and that it solves the problem of control synthesis.

Calculating $h^+(x, \mathcal{W}[t])$, we have

$$h^+(x, \mathcal{W}[t]) = \max\{\langle l, x \rangle - \rho(l \mid \mathcal{W}[t]) \mid \|l\| \leq 1\},$$

so that the maximizer $l^0 = l^0(t, x) \neq 0$ yields

$$h^+(x, \mathcal{W}[t]) = \langle l^0, x \rangle - \rho(l^0 \mid \mathcal{W}[t])$$

if $h^+(x, \mathcal{W}[t]) > 0$ (otherwise $l^0 = 0$ and $h^+(x, \mathcal{W}[t]) = 0$).

To calculate the derivative (8.41) we need to know the partial derivative of $\rho(l \mid \mathcal{W}[t])$ at time $t + 0$, which is

$$\frac{\partial}{\partial(t + 0)} \rho(l^0 \mid \mathcal{W}[t]) = -\frac{\partial}{\partial(s - 0)} \rho(l^0 \mid \mathcal{W}[-s]), \quad s = -t.$$

Here the right-hand part can be calculated similarly to [158, Sect. 9], as done for $\partial \rho(l \mid \mathcal{X}[t])/\partial(t + 0)$, but now in backward time.

Since $\mathcal{W}[t] \subseteq \mathcal{Y}(t)$, this gives

$$\frac{\partial \rho(l^0 \mid \mathcal{W}[t])}{\partial(t - 0)}$$
$$= \begin{cases} \rho(-l^0 \mid B(t)\mathcal{P}(t)), & \text{if } \rho(l^0 \mid \mathcal{W}[t]) < \rho(l^0 \mid \mathcal{Y}(t)), \\ \min\{\rho(-l^0 \mid B(t)\mathcal{P}(t)), -\partial/\partial t \ (\rho(l^0 \mid \mathcal{Y}(t)))\}, & \text{if } \rho(l^0 \mid \mathcal{W}[t]) = \rho(l^0 \mid \mathcal{Y}(t)). \end{cases} \tag{8.45}$$

Exercise 8.3.2. *Check the formula of (8.45).*

The last relation indicates that

$$\mathcal{U}^0(t, x) = \{u : \langle -l^0, B(t)u \rangle = \rho(-l^0 \mid B(t)\mathcal{P}(t))\} \neq \emptyset,$$

and therefore $\mathcal{U}^*(t, x) \neq \emptyset$, since $\mathcal{U}^0(t, x) \subseteq \mathcal{U}^*(t, x)$. Thus we come to

Lemma 8.3.4. *The strategy* $\mathcal{U}^*(\cdot,\cdot) \in \mathbf{U}_Y$.

The necessary properties that ensure $\mathcal{U}^*(\cdot,\cdot) \in \mathbf{U}_Y$ may be checked directly. A standard type of reasoning yields the following

Theorem 8.3.1. *The strategy* $u = \mathcal{U}^*(t,x)$ *solves the Problem 8.3.1 of target control synthesis under "viability" constraint* $\mathcal{Y}(t)$ *for any position* $\{t,x\}$ *that satisfies the inclusion* $x \in \mathcal{W}[t]$.

This theorem follows from

Lemma 8.3.5. *For any solution* $x[t] = x(t,\tau,x_\tau)$ *to the inclusion*

$$\dot{x} \in B(t)\mathcal{U}^*(t,x), \quad x[\tau] = x_\tau \in \mathcal{W}[\tau], \quad \tau \le t \le t_1, \quad (8.46)$$

one has $x[t] \in \mathcal{W}[t]$, $\tau \le t \le t_1$, *and therefore* $x[t] \in \mathcal{Y}(t)$, $\tau \le t \le t_1$, $x[t_1] \in \mathcal{M}$.

Indeed, if we suppose $x[\tau] \in \mathcal{W}[\tau]$ and $h^+(x[t^*], \mathcal{W}[t]) > 0$ for some $t^* > \tau$, then there exists a point $t^{**} \in (\tau, t^*]$ where $dh^+(x[t^{**}], \mathcal{W}[t^{**}])/dt > 0$ in contradiction with the definition of $\mathcal{U}^*(t,x)$.

The theory of trajectory tubes and the results presented particularly in this section allow to give a further description of the synthesized solution to (8.36) under $\mathcal{U}(t,x) \equiv \mathcal{U}^*(t,x)$. Namely, it is now necessary to describe the tube of all solutions to (8.36) that start with $x[\tau] \in X_\tau$. The main point is that this tube may be described without knowledge of the strategy $\mathcal{U}^*(t,x)$ itself, but only on the basis of the information given just for the original Problem 8.3.1. This is due to the property of *colliding tubes*, described in Sect. 2.6, which is also true under state constraints.

Exercise 8.3.3. *Prove Corollary 2.5.1 under state constraint* $x(t) \in \mathcal{Y}(t)$, $t \in [t_0, t_1]$.

Together with (8.43) consider the following evolution equation

$$\lim_{\sigma \to 0} \sigma^{-1} h(\Gamma(t+\sigma), (\Gamma(t) + \sigma\mathcal{P}(t)) \cap \mathcal{W}(t+\sigma)) = 0, \quad (8.47)$$

$$\Gamma(\tau) = X_\tau, \quad X_\tau \in \text{conv } \mathbb{R}^n, \quad X_\tau \subseteq \mathcal{Y}(\tau), \quad \tau \le t \le t_1.$$

Assumption 8.3.1. *(i) The support function* $f(t,l) = \rho(l \mid \mathcal{Y}(t))$ *is differentiable in t for each l and the derivative* $\partial f(t,l)/\partial t$ *is continuous.*
(ii) The set graph $\mathcal{Y}(\cdot) = \{t, x : x \in \mathcal{Y}(t), t \in [t_0, t_1]\}$ *is convex and compact.*

Suppose one of the conditions (i), (ii) of Assumption 8.3.1 holds. Then it is possible to prove the following assertions (see [158], Sects. 6–8).

Theorem 8.3.2. *The solution tube* $\Gamma[t] = \Gamma(t, \tau, X_\tau)$ *to the synthesized system (8.46),* $X[\tau] = X_\tau$ *under the viability constraint (8.34) is the tube of all solutions to the evolution equation (8.47).*

8.3 Control Synthesis Under State Constraints: Viable Solutions

Corollary 8.3.1. *The tube of all synthesized trajectories from a given point $x_\tau \in W[\tau]$ may be defined as the tube of all solutions to the differential inclusion*

$$\dot{x} \in B(t)\mathcal{P}(t), \quad x(\tau) = x_\tau, \quad t \in [\tau, t_1] \tag{8.48}$$

that satisfy $x(t) \in W(t)$, $\tau \le t \le t_1$, and are therefore viable relative to $W(t)$, where $W[t]$ is the solution to (8.43).

The synthesized trajectory tube is therefore a set-valued solution to the "two-set boundary value problem" (8.43), (8.47) due to the respective evolution equations. The calculation of this solution obviously does not require the knowledge of the strategy $u = \mathcal{U}^*(t,x)$ itself.

Remark 8.3.1. Observe that set $W[\tau]$ is *weakly invariant* relative to $M, \mathcal{Y}(\cdot)$ due to differential inclusion (8.48) and *strongly invariant* relative to $M, \mathcal{Y}(\cdot)$ due to differential inclusion (8.46).

Exercise 8.3.4. *A theorem of 8.3.2 type was proved for a nonlinear differential inclusion [158] under conditions different from Assumption 8.3.1. Prove this theorem under Assumption 8.3.1 of this section which imply original equation (8.33) to be linear.*

Note that with $A(t) \ne 0$ the funnel equation for $W[t]$ is as follows

$$\lim_{\sigma \to 0} \sigma^{-1} h\left(W[t-\sigma], ((I - \sigma A(t))W[t] - (B(t)\mathcal{P}(t))\sigma) \cap \mathcal{Y}(t-\sigma)\right) = 0, \tag{8.49}$$

$$t_0 \le t \le t_1, \quad W[t_1] = M.$$

Exercise 8.3.5. *Prove the results of this section for $A(t) \ne 0$, using (8.34) and the results of Sect. 2.6.*

Exercise 8.3.6. *Calculate $W[\tau]$ using the techniques of convex analysis, similar to Sect. 2.4.*

8.3.3 Example

Consider the double integrator system of Sect. 7.2.4, taken on the time interval $t \in [0, 2]$, namely,

$$\dot{x}_1 = x_2, \quad \dot{x}_2 = u, \tag{8.50}$$

with control $|u| \le k$. Let $m = (m_1, 0)$, $\varphi_M(x) = \langle x-m, M(x-m)\rangle$, $M = M' > 0$ and $\varphi(x_2) = |x_2|^2$.

Fig. 8.1 Growth of the backward reach set subject to convex constraints for a double integrator (*solid lines*; final time is shown *thicker*). The target set (*dotted line*) and constraints (*shaded region*) are shown as well. The unconstrained backward reach set at the same final time is shown for comparison (*dashed line*)

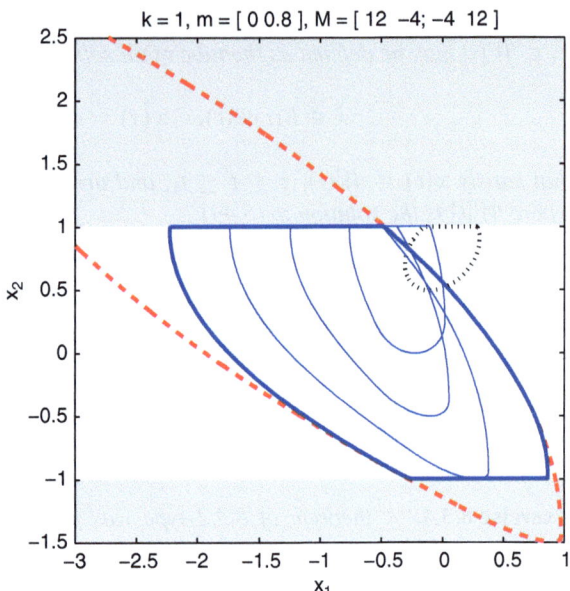

The objective is to calculate at time $\tau \in [0, 2)$ the *backward reach set* $\mathcal{W}[\tau] = \mathcal{W}(\tau, 2, \mathcal{M})$ from target set $\mathcal{M} = \{x : \langle x - m, M(x - m) \rangle^{1/2} \leq 1\}$ under state constraint $|x_2[t]| \leq 1$, $t \in [\tau, 2]$.

Such calculations may be done using schemes of Sect. 7.2.4 or following funnel equation (8.43).

Based on such a scheme Fig. 8.1 shows a particular version of this example worked out by Ian Mitchell, with solution approximated numerically, using level set methods of [221, 244]. The chosen parameters are

$$k = 1, \qquad m = (0, 0.8)', \qquad M = \begin{pmatrix} 12 & -4 \\ -4 & 12 \end{pmatrix}.$$

This figure shows final reach set at $t_1 = 2$, without constraint (dashed line) and with convex constraints $|x_2(t)| \leq 1$ (thick solid line). The constrained reach set is much smaller than the intersection of the state constraints and the unconstrained reach set.

To present system (8.50) in the form of (8.41), we apply transformation $z(t) = G(0, t)x(t)$

$$G(0, t) = \begin{pmatrix} 1 & -t \\ 0 & 1 \end{pmatrix},$$

keeping for variable z the original notation x. We then come to equations

$$\begin{cases} \dot{x}_1 = -tu \\ \dot{x}_2 = u \end{cases},$$

or

$$\dot{x} = b(t)u, \ b'(t) = (-t, 1)$$

which is of the form (8.41). Hence the related funnel equation (8.43) will be

$$\lim_{\sigma \to 0} \sigma^{-1} h \left(\mathcal{W}[t - \sigma], (\mathcal{W}[t] - \sigma B(t)\mathcal{P}(t)) \cap \mathcal{Y}(t - \sigma) \right) = 0, \qquad (8.51)$$

$$t_0 \le t \le t_1, \ \mathcal{W}[t_1] = \mathcal{M}.$$

Here the state constraint $\mathcal{Y}$ is a stripe which may be approximated internally by an ellipsoid $\mathcal{E}(0, Y_\epsilon)$, where $\rho(l \mid \mathcal{E}(0, Y_\epsilon)) = (\varepsilon^{-2} l_1^2 + l_2^2)^{1/2}$.

Exercise 8.3.7. *Present $\mathcal{W}[\tau]$ as an intersection of several ellipsoids.*

Exercise 8.3.8. *(a) Indicate the control that connects a given point on the boundary of $\mathcal{W}[\tau]$ with a point given on the boundary of $\mathcal{M}$ at time t_1.*
(b) Solve similar problem when both points need not be on the related boundaries.

8.4 Obstacle Problems

8.4.1 Complementary Convex State Constraints: The Obstacle Problem

Consider function $\psi(t, x) = \langle x - z, Z^{-1}(t)(z - x) \rangle$ with $\mathcal{Z}(t) = \{x : \psi(t, x) \ge 1\}$. Then inclusion $x(t) \in \mathcal{Z}(t)$ will be referred to as the *complementary convex state constraint*. The next problem has such type of state constraint.

Problem 8.4.1. *Given time interval $[\tau, t_1]$, and functions $\psi(t, x)$, $\varphi_M(x)$, find set*

$$\mathcal{W}_c[\tau] = \left\{ x : \exists u(\cdot), \forall t \in [\tau, \vartheta] \ \psi(t, x[t]) \ge 1, \ \varphi_M(x[\vartheta]) \le 1; x[\tau] = x \right\}.$$

$\mathcal{W}_c[\tau] = \{x : V^c(\tau, x) \le 1\}$ is the level set for the value function

$$V^c(\tau, x) = \min_u \max_t \left\{ \max\{-\psi(t, x[t]) + 2 \mid t \in [\tau, t_1]\}, \varphi_M(x[t_1]) \ \bigg| \ x[\tau] = x \right\}.$$

This is the set of points x from which *some* controlled trajectory $x[t] = x(t, \tau, x_\tau)$, starting at time τ, reaches $\mathcal{M}$ at time $t = t_1$ under state constraint $\psi(t, x[t]) \geq 1$ or, equivalently, $x[t] \in \mathcal{Z}(t)$, $\forall t \in [\tau, \vartheta]$.

Note that Problem 8.4.1 requires $x(t)$ to stay outside the interior of convex compact set $\mathcal{Z}(t)$. So function $V^c(t, x)$ and sets $\mathcal{W}[t]$ in general lack the property of convexity.

The value function $V^c(t, x)$ also satisfies a semigroup property similar to (8.43). This gives, for any $u(s) \in \mathcal{E}(p(s), P(s))$, $s \in [t, t + \sigma]$, $\sigma \geq 0$, the inequality

$$\max \left\{ \max_s \{-\psi(s, x[s]) + 2 \mid s \in [t, t + \sigma]\} - V^c(t, x), \right.$$

$$\left. V^c(t + \sigma, x[t + \sigma]) - V^c(t, x) \,\Big|\, x[t] = x \right\} \geq 0,$$

with equality reached along the optimal trajectory, and

$$-\psi(t, x) + 2 \leq V^c(t, x).$$

We assume that $V^c(t, x)$ and $\psi(t, x)$ are differentiable.

Case (a). Assuming

$$-\psi(t, x) + 2 < V^c(t, x),$$

we have

$$V^c_t(t, x) + \min_u \langle V^c_x, f(t, x, u) \rangle = 0, \ u \in \mathcal{E}(p(t), P(t)). \tag{8.52}$$

Case (b). Assuming

$$-\psi(t, x) + 2 = V^c(t, x),$$

with $u^0(t), x^0(t)$ being the optimal solution to Problem 8.4.1, we have, through reasoning similar to the above

$$\max\{\mathcal{H}(t, x[t], V^c_{tx}(t, x[t]), u), \mathcal{H}(t, x[t], -\psi_{tx}(t, x[t]), u) \geq 0,$$

$$0 = \mathcal{H}(t, x^0[t], V^c_{tx}(t, x^0[t]), u^0) \geq \mathcal{H}(t, x^0[t], -\psi_{tx}(t, x^0[t]), u^0). \tag{8.53}$$

The boundary condition is

$$V^c(t_1, x) = \max\{-\psi(t_1, x) + 2, \ \varphi_M(x)\}. \tag{8.54}$$

8.4 Obstacle Problems

Theorem 8.4.1. *The solution to Problem 8.4.1 is given by*

$$W_c[\tau] = \{x : V^c(\tau, x) \leq 1\},$$

in which $V^c(\tau, x)$ is the solution to (8.52), (8.54).

8.4.2 The Obstacle Problem and the Reach-Evasion Set

The next *obstacle problem* combines the previous two state constraints, namely those of Problems 8.3.1 and 8.4.1. We thus have

Problem 8.4.2. *Given time interval $[\tau, t_1]$ and functions $\varphi(t, x), \psi(t, x), \varphi_M(x)$, find the set*

$$W_o[\tau] = \left\{ x \in \mathbb{R}^n : \exists u(\cdot), \forall t \in [\tau, t_1], \; \varphi(t, x[t]) \leq 1, \right.$$

$$\left. \psi(t, x[t]) \geq 1, \; \varphi_M(x[t_1]) \leq 1; \; x[\tau] = x \right\}.$$

Hence $W_o[\tau] = \{x : V^o(\tau, x) \leq 1\}$ is the level set of the value function

$$V^o(\tau, x) = \min_u \max_t \left\{ \max\{\max\{\varphi(t, x[t]), -\psi(t, x[t]) + 2\} \mid t \in [\tau, t_1]\} \right.,$$

$$\left. \varphi_M(x[t_1]) \,\Big|\, x[\tau] = x \right\}.$$

This is the set of points x from which some controlled trajectory $x[t] = x(t, \tau, x)$, starting at time τ, reaches $\mathcal{M}$ at time $t = t_1$ and also satisfies the state constraints $x[t] \in \mathcal{Y}(t), \; x[t] \in \mathcal{Z}(t), \; \forall t \in [\tau, t_1]$.

Set $W[\tau]$ is known as the *reach-evasion set* [203, 204].

Finally, in Problem 8.4.2, the value function $V^o(t, x)$ satisfies an analog of Lemma 8.3.1. We have, for any $u(s) \in \mathcal{P}(s), \; s \in [t, t + \sigma], \; \sigma \geq 0$, the relations

$$\max \left\{ \max_s \{\max\{\varphi(s, x[s]), -\psi(s, x[s]) + 2\} \mid s \in [t, t + \sigma]\} - V^o(t, x), \right.$$

$$\left. V^o(t + \sigma, x[t + \sigma]) - V^o(t, x) \mid x[t] = x \right\} \geq 0, \qquad (8.55)$$

with equality along the optimal trajectory, and

$$\varphi(t, x) \leq V^o(t, x), \quad -\psi(t, x) + 2 \leq V^o(t, x).$$

Case (a-o). Assuming $V^o(t, x) > \varphi(t, x)$, $V^o(t, x) > -\psi(t, x) + 2$, we have the HJB equation

$$V_t^o(t, x) + \min_u \langle V_x^o, f(t, x, u) \rangle = 0, \ u \in \mathcal{P}(t); \tag{8.56}$$

Case (b-o). Assuming $V^o(t, x) = \varphi(t, x)$, but $V^0(t, x) > -\psi(t, x) + 2$, we have

$$\max\{\mathcal{H}(t, x, V_{tx}^o, u), \mathcal{H}(t, x, \varphi_{tx}, u)\} \geq 0,$$

$$0 = \mathcal{H}(t, x^{(0)}[t], V_{tx}^o(t, x^{(0)}[t]), u^0) \geq \mathcal{H}(t, x^{(0)}[t], \varphi_{tx}(t, x^{(0)}[t]), u^0) \tag{8.57}$$

and lastly,
Case (c-o). Assuming $V^o(t, x) = -\psi(t, x) + 2$, but $V^0(t, x) > \varphi(t, x)$, we have

$$\max\{\mathcal{H}(t, x, V_{tx}^o, u), \mathcal{H}(t, x, -\psi_{tx}, u)\} \geq 0,$$

$$0 = \mathcal{H}(t, x, V_{tx}^o, u^0) \geq \mathcal{H}(t, x, \psi_{tx}, u^0). \tag{8.58}$$

The boundary condition is

$$V^o(t_1, x) = \max\{\varphi(t_1, x), -\psi(t_1, x) + 2, \varphi_M(x)\} \tag{8.59}$$

Theorem 8.4.2. *The solution to Problem 8.4.1 is given by*

$$W_o[\tau] = \{x : V^o(\tau, x) \leq 1\},$$

in which $V^o(\tau, x)$ is the solution to (8.56)–(8.59).

Here we have presumed $\mathcal{Z}(t) \cap \mathcal{Y}(t) \neq \emptyset, \ \forall\, t \in [\tau, t_1]$.

8.4.3 Obstacle Problem: An Example

Here we consider an illustrative example for the obstacle problem worked out by I. Mitchell. We introduce a complementary convex constraint of Problem 8.4.1, then add a convex constraint coming to Problem 8.4.2. On a finite interval $t \in [0, t_1]$ we consider system

$$\dot{x}_1 = u_1, \ \dot{x}_2 = u_2, \tag{8.60}$$

with controls $|u_i| \leq k$, $i = 1, 2$, $k \geq 1$.

Now let $m = (0, m_2)$ be a terminal point, with $m_2 > 1$ and $\varphi_M(x) = \langle x - m, x - m \rangle$, $\varphi(x) = x_2$, $\psi(x) = (x_1^2 + x_2^2)$. Our objective is to calculate at time

8.4 Obstacle Problems

$\tau \in [0, \vartheta)$ the *backward reach set* $\mathcal{W}[\tau] = \mathcal{W}(\tau, t_1, \mathcal{M})$ from the terminal set $\mathcal{M} = \{x : \langle x - m, x - m \rangle^{1/2} \leq 1\}$ under state constraints $\varphi(x) \leq 2$, $\psi(x[t]) \geq 1$, $t \in [\tau, t_1]$.

For our system (8.60) we can calculate the related value function $V^o(t, x)$ as

$$V(\tau, x) = \min_u \max \left\{ \max_t \{\max\{\varphi(x[t]) - 1, -\varphi_2(x[t]) + 2\} \mid t \in [\tau, t_1]\}, \varphi_{\mathcal{M}}(t_1) \mid x[\tau] = x \right\}. \quad (8.61)$$

We then have $W_o[\tau] = \{x : V^o(\tau, x) \leq 1\}$.

A calculation of this function through methods of convex analysis is given in detail in [161].

The calculations through a level set algorithm, due to Mitchell, are indicated in Fig. 8.2.

This figure shows a particular version of this example approximated numerically in the same manner as for Fig. 8.1. The parameters are

$$k = 1, \quad m = (0, 1.3)'.$$

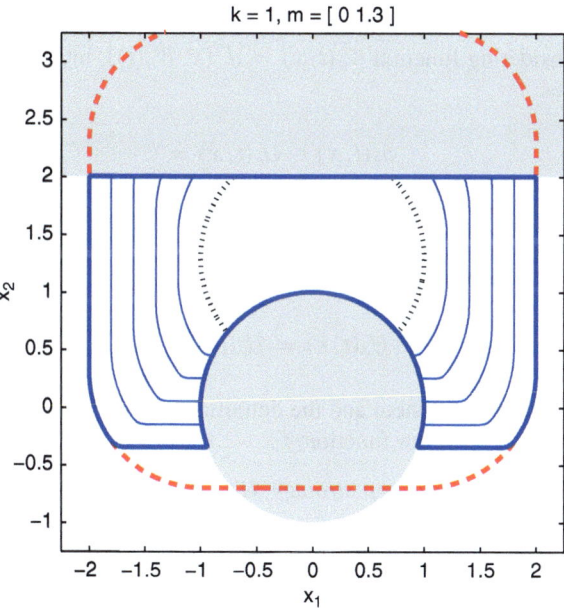

Fig. 8.2 Growth of reach $\mathcal{W}[t]$ set subject to convex and nonconvex constraints (*solid lines*; final time is shown *thicker*). The target set (*dotted circle*) and constraints (*shaded regions*) are shown as well. The unconstrained reach set at the same final time is shown for comparison (*dashed line*)

The figure shows the final reach set at $t_1 = 1$ with no constraint (dashed line) and subject to the combination of convex constraint $|x_2[t]| \leq 2$ and nonconvex constraint $\sqrt{x_1^2 + x_2^2} \geq 1$ (thick solid line).

8.4.4 Closed-Loop Control for the Obstacle Problem

Once the backward reachability sets $\mathcal{W}_o[t]$ are described, one may proceed to the solution of the control synthesis problem, to obtain the closed-loop control.

Problem 8.4.3. *Given set $\mathcal{W}_o[\tau]$, find control strategy $u_e(t, x)(U_e(t, x))$ that steers system (8.33) from any position $\{\tau, x\}$, $x \in \mathcal{W}_o(t)$ to a position $\{t_1, x(t_1)\}$ under following additional constraints:*

(i=1) $\varphi(t, x(t)) \leq 1$, $t \in [\tau, t_1]$; $\varphi_M(x(t_1)) \leq 1$;
(i=2) $\psi(t, x(t)) \geq 1$, $t \in [\tau, t_1]$; $\varphi_M(x(t_1)) \leq 1$;
(i=3) $\varphi(t, x(t)) \leq 1$, $\psi(t, x(t)) \geq 1$, $t \in [\tau, \vartheta]$; $\varphi_M(x(\vartheta)) \leq 1$.

Each of the strategies $u_e(t, x)(U_e(t, x))$ may be sought for directly, from the respective HJB-type equations or inequalities given above for calculating $V_o(t, x)$. However we will apply the generalized "aiming" scheme, used in Sect. 8.1.2 and introduced earlier, for systems without state constraints, in Sect. 2.6 (see [123, 247]).

Namely, considering function $V_o(t, x) = d^2(x, W_o(t))$, introduce either single-valued strategies

$$u_e(t, x) \in \mathcal{U}_e(t, x) =$$

$$\arg\min\{\exp(-2\lambda t)\langle (V_o)_x(t, x), f(t, x, u)\rangle \mid u \in \mathcal{P}(t)\}, \qquad (8.62)$$

$\lambda > 0$, or the set-valued strategies

$$U_e(t, x) = \mathcal{U}_e(t, x),$$

depending on the type of system and the definition of solutions used. Here λ is the Lipschitz constant in $\{t, x\}$ for function f.

Theorem 8.4.3. *The closed-loop strategy $\mathcal{U}_e(t, x)$ for Obstacle Problem 8.4.2 is given by (8.62).*

The problem is that the proposed strategy $u_e(t, x)$ or $U_e(t, x)$ must satisfy in some appropriate sense the equation

$$\dot{x} = f(t, x, u_e(t, x)), \qquad (8.63)$$

8.4 Obstacle Problems

or the differential inclusion

$$\dot{x} \in f(t, x, U_e(t, x)). \tag{8.64}$$

For a general nonlinear system of type (8.63) the solution may be defined as a "constructive motion" introduced in [123], while in the case of linear systems with convex compact constraints on the controls the solutions $U_e(t, x)$ may be taken in the class of upper semi-continuous set-valued strategies with synthesized system (8.64) treated as a differential inclusion [121, 123, 174].

Chapter 9
Guaranteed State Estimation

Abstract This chapter deals with the problem of set-membership or "guaranteed" state estimation. The problem is to estimate the state of a dynamic process from partial observations corrupted by unknown but bounded noise in the system and measurement inputs (in contrast with stochastic noise). The problem is treated in both continuous and discrete time. Comparison with stochastic filtering is also discussed.

Keywords State estimation • Bounding approach • Information states • Information tubes • Hamiltonian techniques • Ellipsoidal approximations

The problem of model and system state estimation through incomplete observations under non-probabilistic noise (the "theory of guaranteed estimation") was introduced and developed in [25, 45, 119, 135–137, 210, 241, 271]. Various versions of the problem had been worked out using different tools and serving numerous applications. The specifics of this problem are that the system operates under unknown but bounded input disturbances while the available observations, corrupted by similar set-membership noise, satisfy the *measurement equation* which yields an on-line state constraint (in contrast with such constraints given in advance). Hence the problem to be considered is actually one of finding the *reachability set* for a system subjected to *state constraints which arrive on-line*.

The solution to the state estimation problem (an "observer") is set-valued, with output given by "information sets" of states consistent with the system dynamics and the available measurements. The calculation of such sets—the observer outputs, which evolve in time as an "information tube," pose a challenging computational problem, especially in the nonlinear case.

In this chapter we proceed with a further application of Hamiltonian techniques indicated previously in Chap. 2 for systems without state constraints and in Chap. 7 for those with state constraints given in advance. This gives the solution to the guaranteed (set-membership) state estimation problem by describing information tubes using HJB equations. The idea applied here is that these information sets may be expressed as level sets of related value functions, called "information states", which are the solutions to special types of the HJB equation. Such a turn yields a deeper insight into the investigated problem and opens new routes for designing computational algorithms. It applies to both linear and nonlinear systems.

Since calculating solutions to HJB equations is not simple, we proceed, as in Sect. 7.1, with substituting the original HJB equations by variational inequalities due to a related *comparison principle*. As in Sects. 5.1 and 7.1, the comparison theorems generated now by the guaranteed state estimation (filtering) problem are applicable to both smooth and non-smooth solutions to related HJB equations. It is also shown that in case of linear systems with convex constraints on the disturbances this approach may lead to effective external and internal approximations of the information sets and tubes using ellipsoidal techniques.

Further on, a later detour is to discrete-time systems introducing the reader to a collection of formulas useful for computation. This is followed by interrelations between guaranteed state estimation theory of this book, as taken for linear systems, and the well-known stochastic approaches to state estimation based on the Kalman filtering theory. Finally mentioned is the important issue of dealing with *discontinuous measurements* which extend previous results of this chapter to a broader array of applied problems.

9.1 Set-Membership State Estimation: The Problem. The Information Tube

Consider the n-dimensional system

$$\dot{x} = f(t, x, v) \qquad (9.1)$$

where function $f(t, x, v)$ is similar to (1.1), namely, continuous in all variables and ensuring standard conditions of uniqueness and extendibility of solutions throughout a finite interval $[t_0, \vartheta]$ for any initial condition $x^0 = x(t_0) \in \mathbb{R}^n$, and also for any admissible disturbances $v(t)$, restricted by geometrical constraints ("hard bounds"), so that

$$v(t) \in Q(t), \ t \in [t_0, \vartheta], \text{ and } x(t_0) \in \mathcal{X}^0. \qquad (9.2)$$

Here $Q(t)$ is a multivalued function with values in the set $\text{comp}\mathbb{R}^q$ of all compacts of space $\mathbb{R}^q$, continuous in the Hausdorff metric, and set $\mathcal{X}^0$ is compact. The pair $\{t_0, \mathcal{X}_0\}$ is the "initial position" of system (9.1).

We also assume that set $\mathcal{F}(t, x) = f(t, x, Q(t))$ is convex and compact. Then, due to the indicated properties, the set-valued function $\mathcal{F}(t, x)$ will be Hausdorff-continuous in all the variables.

The on-line information on vector x is given through *observations* which arrive through one of the two following schemes.

9.1 Set-Membership State Estimation: The Problem. The Information Tube

Scheme I. State x is measured via continuous observations

$$y(t) = g(t, x) + \xi(t), \quad t \in [t_0, \vartheta], \tag{9.3}$$

where $y(t) \in \mathbb{R}^r$ is the measurement and $\xi(t)$—the unknown but bounded disturbance ("noise"), which is restricted to the inclusion

$$\xi(t) \in \mathcal{R}(t), \quad t \in [t_0, \vartheta], \tag{9.4}$$

while the properties of set $\mathcal{R}(t)$ are similar to $Q(t)$. Function $g(t, x)$ is taken to be continuous over both variables.

Scheme II. State x is measured via discrete observations

$$y(\tau_i) = g(\tau_i, x(\tau_i)) + \xi(\tau_i), \tag{9.5}$$

where $y(\tau_i) \in \mathbb{R}^r, x(\tau_i) \in \mathbb{R}^n$, $\tau_i < \tau_{i+1}$, $\tau_i \in [t_0, \vartheta]$, and noise $\xi(\tau_i)$ is unknown but bounded, restricted by (9.4). The measurement times τ_i are taken to be given.

Remark 9.1.1. An interesting situation is the case of communication-type constraints where the measurement signals $y(\tau_i)$ are assumed to arrive at random time instants τ_i distributed, (say) as a Poisson process, with frequency $\lambda > 0$. Such case lies beyond the scope of this book and is treated in [56].

Taking Scheme I, we assume that given are: the initial position $\{t_0, X^0\}$, functions $f(t, x, v), g(t, x)$, set-valued functions $Q(t), \mathcal{R}(t)$, and available on-line measurements $y_\tau(\sigma) = y(\tau + \sigma)$ ($\sigma \in [t_0 - \tau, 0]$).

Definition 9.1.1. *The information set* $X(\tau, y_\tau, \cdot) = X[\tau]$ *of system* (9.1)–(9.4) *is the collection of all its states* $x[\tau] = x(\tau, t_0, x(t_0)), x(t_0) \in X^0$, *consistent with its parameters and with observed measurements* $y_\tau(\sigma)$. □

Thus, the actual *on-line position* of the system may be taken as the pair $\{\tau, X[\tau]\}$.

Problem 9.1.1. *Calculate sets* $X[t], t \in [t_0, \vartheta]$, *and derive an equation for their evolution in time.*

The set-valued function $X[t]$ is called *the information tube*. Such tubes are estimates of the system dynamics due to on-line measurements. They will be calculated through two approaches—*Hamiltonian techniques and set-valued calculus*.

A similar definition and a problem are introduced within Scheme II under available on-line measurements $y[t_0, \tau] = \{y(\tau_i) : \tau_i \in [t_0, \tau]\}$. Then one may define a similar *information set* $X_d(\tau, y[t_0, \tau]) = X_d[\tau]$. The information tube $X_d[t]$ may turn out to be discontinuous. This depends on the properties of the function $y(t)$. Under Scheme I the discontinuities may be caused by discontinuous noise while under Scheme II they arrive naturally, since measurements $y(\tau_i)$ are made at isolated times.

Concentrating on Scheme I, note that the information set $X[\tau]$ is a *guaranteed estimate* of the unknown actual state $x(\tau)$ of system (9.1), so we always have

$x(\tau) \in X[\tau]$, whatever be the unknown noise. Given starting position $\{t_0, X^0\}$ and measurement $y(t)$, $t \in [t_0, \tau]$, it makes sense to construct for system (9.1), (9.2) a *forward reachability tube* $X[t] = X(t; t_0, X^0), t \geq t_0$, which consists of all those solutions to the differential inclusion $\dot{x} \in \mathcal{F}(t, x)$ that emanate from the set-valued *initial position* $\{t_0, X^0\}$ and develop in time under the *on-line state constraint*

$$g(t, x) \in y[t] - \mathcal{R}(t), \quad t \geq t_0, \tag{9.6}$$

generated by the available measurement $y(t) = y[t]$. Then function $X[t]$ is precisely the information tube that solves Problem 9.1.1 of *guaranteed filtering*.

We now proceed with solving the announced problem.

9.2 Hamiltonian Techniques for Set-Membership State Estimation

Problem 9.1.1 indicated above is not the one of optimization. But as before we shall solve it through alternative formulations of *dynamic optimization*. Following Sect. 7.1.1 these are given in two versions.

9.2.1 Calculating Information Tubes: The HJB Equation

Version A

Problem 9.2.1. *Given system (9.1)–(9.3), with available measurement $y(s) = y[s]$, $s \in [t_0, t]$, and set-valued starting position $\{t_0, X^0\}$, find function*

$$V(t, x) = \min_v \{d^2(x(t_0), X^0) + \int_{t_0}^t d^2(y[s] - g(s, x(s)), \mathcal{R}(s)) ds \mid x(t) = x\} \tag{9.7}$$

over the trajectories of system (9.1), under constraints (9.2).

The information set $X[\tau] = X(\tau, \cdot)$ may then be expressed as the level set

$$X[\tau] = \{x : V(\tau, x) \leq 0\}. \tag{9.8}$$

Here $V(t, x)$ is the *value function* related to Problem 9.2.1. This function $V(t, x)$ is henceforward referred to as the **information state** of system (9.1)–(9.3). Its level set or "cross-section" $X[\tau]$ at level 0, namely (9.8), is the *reachability set* for (9.1) under *on-line state constraints* (9.3), given $y(t)$. Relation

$$V(t_0, x) = d^2(x, X^0) \tag{9.9}$$

9.2 Hamiltonian Techniques for Set-Membership State Estimation

defines the *boundary condition* for $V(t, x)$. And as in Theorem 7.1.1 we have the "semigroup" property

$$V(t, x| V(t_0, \cdot)) = V(t, x|V(\tau, \cdot|V(t_0, \cdot))), \quad t_0 \leq \tau \leq t,$$

which implies the HJB type equation

$$V_t + \max\{\langle V_x, f(t, x, v)\rangle \mid v \in Q(t)\} - d^2(y(t) - g(t, x), \mathcal{R}(t)) = 0, \quad (9.10)$$

with boundary condition (9.9). Note that the term $d^2(y(t) - g(t, x), \mathcal{R}(t)) \neq 0$ only if $y(t) - g(t, x)$ is outside of the constraint $\mathcal{R}(t)$.

Here again $V(t, x)$ is assumed to be differentiable and if not, then (9.10) is a formal symbolic notation for an equation whose solution should be considered in a generalized "viscosity" sense (see [16, 17, 50, 80]) or equivalent "minimax" sense (see [247]). For linear systems with convex value functions $V(t, x)$ the total derivative $dV/dt = V_t + \langle V_x, f(t, x, v)\rangle$ may be substituted for a directional derivative of $V(t, x)$ along the direction $\{1, f(t, x, v)\}$. This directional derivative here exists for any direction $\{1, f\}$.

The value function V may be also defined under additional assumptions on smoothness and convexity of the constraints. Given proper continuously differentiable functions $\varphi_0(t, x), \varphi(t, \xi)$, convex in x, ξ respectively, the initial set and the constraint on the disturbance ξ in (9.3), may be presented in the form

$$\mathcal{X}^0 = \{x : \varphi_0(t_0, x) \leq 1\}, \quad \mathcal{R}(t) = \{\xi : \varphi(t, \xi) \leq 1\}. \quad (9.11)$$

The second inequality, generated by the measurement equation, defines an on-line state constraint for system (9.1). In particular, functions φ_0, φ may be defined through equalities

$$\varphi_0(t_0, x) = d^2(x, \mathcal{X}^0) + 1, \quad \varphi(t, \xi(t)) = d^2(\xi(t), \mathcal{R}(t)) + 1, \quad \xi(t) = y(t) - g(t, x),$$

with set-valued mapping $\mathcal{R}(t)$ being *Hausdorff-continuous* and function $y(t)$ being *right-continuous*. As before, here $d(y^*, Y) = \min\{\|y^* - y\| \; y \in Y\}$ is the Euclidean *distance function* for set Y. Such types of constraints are common for both linear and nonlinear systems. Using these constraints in what follows will allow to present the solutions in a more transparent form.

Version B

Assume given is the realization $y(\cdot) = y[\cdot]$ of the observation y on the interval $[t_0, t]$.

Problem 9.2.2. *Given are system (9.1) and starting position $\{t_0, \mathcal{X}^0\}$, according to (9.11). Find the value function*

$$V^{(1)}(t,x) =$$

$$= \min_{v}\{\varphi_0(t_0, x[t_0]) \mid x[t] = x, \; \varphi(s, y(s) - g(s, x[s])) \leq 1, s \in [t_0, t]\}. \quad (9.12)$$

Denoting $\langle V_x^{(1)}, f(t, x, v)\rangle = \mathcal{H}(t, x, V_x^{(1)}, v)$, consider the equation

$$V_t^{(1)} + \max_{v}\{\mathcal{H}(t, x, V_x^{(1)}, v) \mid v \in Q^\theta(t, x)\} = 0, \; V^{(1)}(t_0, x) = \varphi_0(t_0, x) \quad (9.13)$$

where

$$Q^\theta(t,x) = \begin{cases} Q(t), & \theta(t,x) < 1, \\ Q(t) \cap \{v : d\theta(t,x)/dt \mid_v \leq 0\}, & \theta(t,x) \geq 1 \end{cases}$$

and $\theta(t, x) = \varphi(t, y[t] - g(t, x))$.

Theorem 9.2.1. *Function $V^{(1)}$ is a solution to Eq. (9.13), with indicated boundary condition.*

The proof of this assertion follows the lines of (7.9)–(7.15) and the proof of Lemma 7.1.4. Similar to our previous conclusions we have

Lemma 9.2.1. *If for $y = y[s]$ ($s \in [t_0, t]$) Eq. (9.13) has a solution $V^{(1)}(t, x)$, (classical or generalized), then the following equality is true*

$$X[\tau] = \{x : V^{(1)}(\tau, x) \leq 1\}. \quad (9.14)$$

Remark 9.2.1. The suggested Versions A and B lead to different HJB equations and serve to solve the original Problem 9.1.1 under different assumptions on the problem data. Namely, Version-B requires additional assumptions on smoothness and convexity of the state constraints produced by the measurement equation. In the general nonlinear case this difference may affect the smoothness properties of respective solutions $V(t, x)$, $V^{(1)}(t, x)$. On the other hand, the Comparison Principle of the next subsection does not depend on the smoothness properties of original HJB equations for Versions A and B.

9.2.2 Comparison Principle for HJB Equations

Version A

Introduce the notation

$$\mathbf{H}(t, x, p) = \max\{\langle p, f(t, x, v)\rangle \mid v \in Q(t)\} - d^2(y(t) - g(t, x), \mathcal{R}(t)).$$

9.2 Hamiltonian Techniques for Set-Membership State Estimation

Then the related HJB equation (9.10) transforms into

$$V_t + \mathbf{H}(t, x, V_x) = 0, \quad V(t_0, x) = d^2(x, \mathcal{X}^0). \tag{9.15}$$

We will now obtain an *external approximation* of $\mathcal{X}[t]$ which in the linear case will be exact. We proceed as follows.

Assumption 9.2.1. *Given are functions $H(t, x, p)$, $w^+(t, x) \in C_1$ and $\mu(t) \in L_1$, which satisfy the inequalities*

$$\mathbf{H}(t, x, p) \leq H(t, x, p), \quad \forall \{t, x, p\}, \tag{9.16}$$

$$w_t^+ + H(t, x, w_x^+) \leq \mu(t). \tag{9.17}$$

Theorem 9.2.2. *Suppose $H(t, x, p)$, $w^+(t, x)$, $\mu(t)$ satisfy Assumption 9.2.1. Then the following estimate for the information set $\mathcal{X}[t]$ holds:*

$$\mathcal{X}[t] \subseteq \mathcal{X}_+[t], \tag{9.18}$$

where

$$\mathcal{X}_+[t] = \left\{ x : w^+(t, x) \leq \int_{t_0}^t \mu(s)ds + \max\{w^+(t_0, x) \mid x \in \mathcal{X}^0\} \right\}. \tag{9.19}$$

The proof is similar to Sect. 7.1.2, Theorem 7.1.1.

Remark 9.2.2. Condition (9.19) may be complemented by the following:

$$w^+(t_0, x) \leq V(t_0, x).$$

Then the last term in (9.19) may be substituted by $V(t_0, x)$.

Version-B

After introducing notation

$$\mathbf{H}^1(t, x, p, p_0, \lambda) = \max\{\langle p, f(t, x, v)\rangle - \lambda\langle p_0, f(t, x, v)\rangle \mid v \in Q(t)\},$$

the corresponding "forward" HJB equation (9.13) transforms into

$$V_t^{(1)} + \lambda\theta_t + \mathbf{H}^1(t, x, V_x^{(1)}, \theta_x, \lambda) = 0, \quad V^{(1)}(t_0, x) = \varphi_0(t_0, x) \tag{9.20}$$

where the multiplier λ is to be found from the condition

$$d\theta(t, x)/dt \mid_{(9.1)} = \theta_t + \langle \theta_x, f(t, x, v)\rangle \leq 0, \tag{9.21}$$

which allows one to calculate $X[t] = X(t, t_0, X^0)$ as the level set $X[\tau] = \{x : V^{(1)}(\tau, x) \leq 1\}$. This property is independent of whether $V^{(1)}$ is a classical or a generalized solution to Eq. (9.20). The comparison theorem and the respective proof are similar to Sect. 7.1.2, Theorem 7.1.1.

We now pass to the class of linear systems with convex constraints where the previous results may be developed with greater detail.

9.2.3 Linear Systems

Consider system (9.1) to be linear, so that

$$\dot{x} = A(t)x + C(t)v, \qquad (9.22)$$

under same constraint (9.2) on v. The measurement equation is also taken to be linear

$$y(t) = G(t)x + \xi(t), \qquad (9.23)$$

with constraint (9.4) on ξ. The functions $Q(t), R(t)$ are assumed ellipsoidal-valued, described by nondegenerate ellipsoids,

$$(a)\ v \in Q(t) = \mathcal{E}(q(t), Q(t)); \quad (b)\ \xi(t) \in \mathcal{R}(t) = \mathcal{E}(0, R(t)), \qquad (9.24)$$

with matrix-valued functions $Q'(t) = Q(t) > 0$, $R'(t) = R(t) > 0$ and Hausdorff-continuous. The bound on the initial vector $x(t_0)$ is also an ellipsoid:

$$x(t_0) \in \mathcal{X}^0 = \mathcal{E}(x^0, X^0),$$

so that the *starting position* is $\{t_0, \mathcal{E}(x^0, X^0)\}$.

With $y(t) = y[t]$ being the available measurement that arrives on-line, we come to the on-line state constraint

$$G(t)x(t) \in y[t] + \mathcal{E}(0, R(t)) = \mathcal{R}_y(t). \qquad (9.25)$$

To solve Problem 9.1.1 for the linear case we now approach the versions for Problems 9.2.1 and 9.2.2 of dynamic optimization.

Denote

$$k^2(t, x) = \langle y[t] - G(t)x, R^{-1}(t)(y[t] - G(t)x) \rangle,$$

and

$$\chi(t) = 1,\ \text{if}\ k^2(t, x(t)) - 1 > 0; \quad \chi(t) = 0,\ \text{if}\ k^2(t, x(t)) - 1 \leq 0.$$

Further denote $(f(x))_+ = f(x)$ if $f \geq 0$ and $f(x) = 0$ otherwise.

9.2 Hamiltonian Techniques for Set-Membership State Estimation

Version LA

The information set $X[t]$ may be expressed as the level set

$$X[t] = \{x : V(t, x) \le 0\}, \tag{9.26}$$

for value function

$$V(t, x) = \min_v \{\varphi_0(x) + \int_{t_0}^t (k^2(s, x(s)) - 1)_+ ds \mid x(t) = x\}, \ t \ge t_0. \tag{9.27}$$

where $\varphi_0(x) = (\langle x - x^0, (X^0)^{-1}(x - x^0)\rangle - 1)_+$ and $(k^2(t, x(t)) - 1)_+ = \chi(t)(k^2(t, x(t)) - 1)$. Then Eq. (9.10) transforms into

$$V_t + \langle V_x, A(t)x + C(t)q(t)\rangle + \langle V_x, C(t)Q(t)C'(t)V_x\rangle^{1/2} - \chi(t)(k^2(t, x) - 1) = 0, \tag{9.28}$$

with boundary condition

$$V(t_0, x) = \varphi_0(x) = (\langle x - x^0, (X^0)^{-1}(x - x^0)\rangle - 1)_+. \tag{9.29}$$

Version LB

To track the on-line state constraint generated by measurement $y(t)$ we have to calculate the total derivative of function $k^2(t, x) = \theta(t, x)$ along our system (9.22). So

$$d\theta(t, x)/dt = \theta_t(t, x) - 2\langle R^{-1}(t)(y[t] - G(t)x), G(t)(A(t)x + C(t)v)\rangle$$

and Eq. (9.13) transforms into

$$V_t^{(1)} + \lambda \theta_t + \max_v \{\langle V_x^{(1)} + \lambda \theta_x, A(t)x + C(t)v\rangle \mid v \in \mathcal{E}(q(t), Q(t))\} = 0,$$

or, with additional notations $G'(t)R^{-1}(t)y(t) = z(t), G'(t)R^{-1}(t)G(t) = \mathbf{G}(t)$, into

$$V_t^{(1)} + \lambda \theta_t + \lambda \langle V_x^{(1)} - 2\lambda(z(t) - \mathbf{G}(t))x, A(t)x + C(t)q(t)\rangle + H^0(t, x, \lambda) = 0, \tag{9.30}$$

where

$$H^0(t, x, \lambda) = \max_v \{\langle V_x^{(1)} - 2\lambda(z(t) - \mathbf{G}(t)x), C(t)v\rangle \mid v \in \mathcal{E}(0, Q(t))\} =$$

$$= \langle V_x^{(1)} - 2\lambda(z(t) - \mathbf{G}(t)x), C(t)Q(t)C'(t)(V_x^{(1)} - 2\lambda(z(t) - \mathbf{G}(t)x))\rangle^{1/2}.$$

The multiplier $\lambda = \lambda(t, x)$ should be calculated from relation

$$\theta_t(t, x) - 2 \langle z(t) - \mathbf{G}(t)x, A(t)x + C(t)v^* \rangle \leq 0, \tag{9.31}$$

where

$$v^* = Q(t)C'(t)(V_x^{(1)} - 2\lambda(z(t) - \mathbf{G}(t)x))H^0(t, x, \lambda)^{-1}.$$

The next move is to approximate the exact solutions given in this subsection by using the comparison principle. This is done through an application of ellipsoidal approximations. The next scheme will rely on a *deductive* approach to the derivation of ellipsoidal estimates (see also Sect. 5.1.2) in contrast with Chap. 7, where such estimates were achieved due to *inductive* procedures introduced in Chap. 3.

9.3 Ellipsoidal Approximations

These approximations are given in two versions.

9.3.1 Version AE

Recalling (9.15) with function

$$\mathbf{H}(t, x, p) = \max\{\langle p, A(t)x + C(t)v \rangle \mid v \in Q(t)\} - \chi(t)(k^2(t, x) - 1)$$

we use relation

$$\langle p, C(t)Q(t)C'(t)p \rangle^{1/2} \leq \gamma^2(t) + (4\gamma^2(t))^{-1} \langle p, C(t)Q(t)C'(t)p \rangle, \quad \forall p \in \mathbb{R}^n, \tag{9.32}$$

with equality reached for $\gamma^2(t) = (1/2)\langle p, C(t)Q(t)C'(t)p \rangle^{1/2}$. Denoting

$$H(t, x, p) = \langle p, A(t)x + C(t)q(t) \rangle + \gamma^2(t) + (4\gamma^2)^{-1} \langle p, C(t)Q(t)C'(t)p \rangle -$$
$$- \chi(t)(k^2(t, x) - 1),$$

we then have

$$H(t, x, p) \geq \mathbf{H}(t, x, p)$$

for all values of $\{t, x, p\}$.

We shall further approximate the function $V(t, x)$ in the domain $x \in \mathbb{R}^n$, $t \geq t_0$ by one of type

9.3 Ellipsoidal Approximations

$$w(t, x) = \langle x - x^*(t), P(t)(x - x^*(t)) \rangle + h^2(t) - 1, \quad (9.33)$$

with differentiable $h^2(t)$, $P(t) = P'(t) > 0$. We assume $k^2(t, x) \geq 1$, so that $\chi(t) = 1$ explaining later that here this is the main case. Then

$$w_t + \mathbf{H}(t, x, w_x) =$$
$$= w_t + \langle w_x, A(t)x + C(t)q(t) \rangle +$$
$$+ \langle w_x, C(t)Q(t)C'(t)w_x \rangle^{1/2} - \chi(t)(k^2(t,x) - 1) \leq$$
$$\leq w_t + H(t, x, w_x) = w_t + \langle w_x, A(t)x + C(t)q(t) \rangle +$$
$$+ \gamma^2(t) + (4\gamma^2(t))^{-1} \langle w_x, C(t)Q(t)C'(t)w_x \rangle -$$
$$- \chi(t)(\langle y(t) - G(t)x, R^{-1}(t)(y(t) - G(t)x) \rangle - 1),$$

and substituting $w(t, x)$, into these relations we get

$$w_t + \mathbf{H}(t, x, w_x) \leq \langle x - x^*(t), \dot{P}(t)(x - x^*(t)) \rangle -$$
$$-2 \langle \dot{x}^*(t), P(t)(x - x^*(t)) \rangle + 2 \langle P(t)(x - x^*(t)), A(t)x + C(t)q(t) \rangle +$$
$$+ (\gamma^2(t))^{-1} \langle P(t)(x - x^*(t)), C(t)Q(t)C'(t)P(t)(x - x^*(t)) \rangle + \gamma^2(t) -$$
$$- \chi(t)(\langle y(t) - G(t)x, R^{-1}(t)(y(t) - G(t)x) \rangle - 1) + \chi(t)d(h^2(t))/dt, \quad (9.34)$$

with $\chi(t) = 1$.

Continuing further with $\chi(t) = 1$, we demand that the right-hand side in (9.34) is treated as follows. We separately equate with zero the terms with multipliers of second order in $x - x^*$, then those of first order in the same variable, then free terms, excluding $\gamma^2(t) + \chi(t)$. We finally observe that the equated elements of the right-hand side of (9.34) will be zeros if and only if the following equations are true:

$$\dot{P} + P'A(t) + A'(t)P + (\gamma^2(t))^{-1}PC(t)Q(t)C'(t)P - \chi(t)G'(t)R(t)G(t) = 0, \quad (9.35)$$

$$\dot{x}^* = A(t)x^*(t) + C(t)q(t) + \chi(t)P^{-1}(t)G'(t)R(t)(y(t) - G(t)x^*(t)), \quad (9.36)$$

$$\dot{h^2} = \langle y(t) - G(t)x^*(t), R(t)(y(t) - G(t)x^*(t)) \rangle, \quad (9.37)$$

with boundary conditions

$$P(t_0) = (X^0)^{-1}, \quad x^*(t_0) = x^0, \quad h^2(t_0) = 0.$$

Relations (9.33)–(9.37) together with some further calculation bring us the inequality

$$dw/dt \leq \gamma^2(t) + \chi(t). \tag{9.38}$$

Functions $\mathbf{H}(t, x, p)$, $w(t, x)$ satisfy Assumption 9.2.1. Hence we are in the position to apply Theorem 9.2.2. We proceed as follows.

Definition 9.3.1. A measurement $y(t)$ is said to be **informative** on the interval $[t, t + \sigma]$, $\sigma > 0$, if constraint (9.25) **is active** throughout this interval.

Let $\mathcal{X}^0[s] = \mathcal{X}^0(s, t, \mathcal{X}[t])$ stand for the reachability set at time $s \geq t$, for system (9.22) *without state constraints*, emanating at time t from $\mathcal{X}[t]$ and recall that $\mathcal{X}[s] = \mathcal{X}(s, t, \mathcal{X}[t])$ is the information set gotten *under active measurement output state constraints*. Then Definition 9.3.1 means that for all $s \in [t, t + \sigma]$, it should be $\mathcal{X}[s] \subset \mathcal{X}^0[s]$. Namely, taking into account the measurement $y(s)$ must diminish set $\mathcal{X}^0[s]$ towards $\mathcal{X}[s]$.

Assumption 9.3.1. *The available measurement* $y(s) = y[s]$ *is informative on any interval* $t < s \leq t + \sigma$ *taken within* $[t_0, \vartheta]$.

Remark 9.3.1. Since we are actually approximating the reach set from above, we always deal with informative measurements which yield solutions that satisfy Assumption 9.3.1.

This means we further assume $\chi(t) \equiv 1$ throughout $[t_0, \vartheta]$.

Integrating inequality (9.38) from t_0 to t, along a trajectory $x[s] = x(s, t_0, x^0)$, $x[t] = x$, $s \in [t_0, t]$ that runs as $x[s] \in X[s]$, under disturbance $v(s) \in \mathcal{E}(q(s), Q(s))$, with measurement $y(s) = y[s]$ under disturbance $\xi(s) \in \mathcal{E}(0, R(s))$, we get

$$w(t, x) - \int_{t_0}^{t} (\gamma^2(s) + 1)ds \leq w(t_0, x[t_0]) = V(t_0, x[t_0]). \tag{9.39}$$

Further substituting $w(t, x)$ and taking into account (9.33), we have

$$\langle x - x^*(t), P(t)(x - x^*(t)) \rangle + h^2(t) - 1 \leq \int_{t_0}^{t} (\gamma^2(s) + 1)ds + V(t_0, x(t_0)).$$

Then, keeping (9.26) in mind, we come to inequality

$$\langle x - x^*(t), P(t)(x - x^*(t)) \rangle \leq 1 - h^2(t) + \int_{t_0}^{t} (\gamma^2(s) + 1)ds = \beta(t), \tag{9.40}$$

where $P(t)$ depends on parameter $\gamma^2(t)$.

This leads to the final conclusion

9.3 Ellipsoidal Approximations

Theorem 9.3.1. *An external set-valued estimate of information set $X[t]$ is the ellipsoid*

$$\mathcal{E}(x^*(t), P^{-1}(t)\beta(t)),$$

defined by (9.40), where functions $P(t), x^*(t), h(t)$ are described by Eqs. (9.35)–(9.37) under boundary conditions

$$P(t_0) = (X^0)^{-1}, \quad x^*(t_0) = x^0, \quad h^2(t_0) = 0.$$

This estimate is true for any Lebesgue-integrable function $\gamma^2(t)$.

This theorem gives us an array of ellipsoids $\mathcal{E}_\gamma[t] = \mathcal{E}(x^*(t), \beta(t)P^{-1}(t))$, parameterized by γ, each of which is an external estimate for $X[t]$. Moreover, the schemes of Chap. 3 yield the next property

Lemma 9.3.1. *The information set $X[t] = \cap\{\mathcal{E}_\gamma[t] \mid \gamma^2(\cdot)\}$.*

The exact information set is therefore presented as an intersection of all external ellipsoidal estimates of the form given above. Among these one may of course seek for an optimal one or the one most appropriate for a related problem of control.

Remark 9.3.2. (i) Condition $\chi(t) = 0$ indicates that measurement $y[t]$ brings no innovation into the estimation process and is redundant, so $X^0[s] = X[s]$. Then, instead of the information set under on-line state constraint, the formulas achieved above describe the reachability set $X^0[s]$ without state constraints. You just have to take $\chi(t) = 0, h(t) = 0$ in all these relations.

9.3.2 Version-BE

Consider Eq. (9.30). Applying inequality (9.32) to $H^0(t, x, \lambda)$, we get

$$V_t^{(1)} + \lambda\theta_t + \lambda\langle V_x - 2\lambda(z(t) - \mathbf{G}(t)x), A(t)x + C(t)q\rangle + H^0(t, x, \lambda) \leq$$

$$\leq V_t^{(1)} + \lambda\theta_t + \lambda\langle V_x - 2\lambda(z(t) - \mathbf{G}(t)x), A(t)x + C(t)q\rangle + \gamma^2(t) + (4\gamma^2(t))^{-1}(H^0(t, x, \lambda))^2. \tag{9.41}$$

Then we again take $w(t, x)$ in the form (9.33). Substituting $w(t, x)$ into the right-hand side of the last relation we further follow the previous scheme of Version-A now applying it to (9.41). This time we come to equations

$$\dot{P} + P(A(t) + \lambda \mathbf{G}(t))' + (A(t) + \lambda \mathbf{G}(t))P +$$
$$+ (\gamma^2(t))^{-1} P C(t) Q(t) C'(t) P + \lambda^2(t) G'(t) R^{-1}(t) G(t) = 0, \tag{9.42}$$

$$\dot{x}^* = A(t)x^*(t) + C(t)q(t) - \lambda(z(t) - \mathbf{G}(t)x^*(t)), \quad (9.43)$$

$$\dot{h}^2 = \langle z(t) - \mathbf{G}(t)x^*, R^{-1}(t)(z(t) - \mathbf{G}(t)x^*)\rangle, \quad (9.44)$$

with boundary conditions

$$P(t_0) = (X^0)^{-1}, \ x^*(t_0) = x^0, \ h^2(t_0) = 0.$$

These equations should be taken with any multiplier $\lambda < 0$, that satisfies (9.31). The set of all such multipliers is further denoted as Λ. Therefore now $P(t) = P_{\gamma\lambda}(t)$. The further procedures are similar to Version AE, transforming P into $P_{\gamma\lambda}(t)$ and thus allowing to formulate in the next form.

Theorem 9.3.2. *The set* $X[t] \subseteq \mathcal{E}(x^*(t), \beta(t)P_{\gamma\lambda}^{-1}(t))$, *where*

$$\mathcal{E}(x^*(t), \beta(t)P_{\gamma\lambda}^{-1}(t)) = \{x : \langle x - x^*(t), \beta^{-1}(t)P_{\gamma\lambda}(t)(x - x^*(t))\rangle \leq 1\},$$

and functions $P_{\gamma\lambda}(t), \dot{x}^*(t)$ *are described by Eqs. (9.42)–(9.44), with* $\gamma^2(t) > 0$ *and* $\lambda \in \Lambda$.

9.3.3 Example: Information Set for a Linear System

Given is the system

$$\begin{aligned} \dot{x}_1 &= x_2, \\ \dot{x}_2 &= -x_1 + v, \ |v| \leq 1 \end{aligned} \quad (9.45)$$

at $t \in [0, \vartheta]$, with observation

$$y(t) = x_1 + \xi(t), \ \xi(t) \in \mathcal{R} = [-1, 1] \quad (9.46)$$

and initial set

$$x(0) \in \mathcal{X}^0 = \{x \mid x_1^2 + x_2^2 \leq 1\}. \quad (9.47)$$

Following (4.5) the reachability set without measurement state constraint will be described as $X[t] = X(t, 0, \mathcal{X}^0) = X_0[t] + X_v[t]$ where

$$X_0[t] = G(t, 0)\mathcal{X}^0, \ G(t, s) = \begin{pmatrix} \cos(t-s) & \sin(t-s) \\ -\sin(t-s) & \cos(t-s) \end{pmatrix}$$

and

$$X_y[t] = \{x : x = \int_0^t G(t-s)Q(s)ds\}, \quad Q(s) = Q = \{q : q = bv, |v| \leq 1\}, \quad b' = \{0, 1\}.$$

The measurement state constraint is $x(t) \in \mathcal{Y}(t) = \{x : x_1(t) \in y(t) - \mathcal{R}\}$.

Thus the reachability set $X_y[t]$ under constraint $\mathcal{Y}(t)$ is the solution to the funnel equation

$$\lim_{\sigma \to 0+0} \sigma^{-1} h(X_y[t+\sigma], \{(I + \sigma A)X_y[t] + \sigma Q(t)\} \cap \mathcal{Y}(t)) = 0, \quad X_y[0] = X^0.$$

The illustrations below are done for $y(t) \equiv 0$. Then $\mathcal{Y}(t) = \mathcal{R}$ and the recurrent relation for calculating $X_y[t]$ is

$$X_y[t+\sigma] = ((I + \sigma A)X_y[t] + \sigma Q(t)) \cap \mathcal{R} + o(\sigma).$$

The boundary of the reachability set $X(t) = X(t, 0, X^0)$ at time $t = \pi$ without measurement constraint in the illustrations is green. The measurement constraint $\mathcal{Y}$ is a stripe bounded by parallel red lines. The other sets are illustrated for measurement $y(t) \equiv 0$ generated under actually realized inputs $x^0 = 0$, $|v(t)| \leq 1, |\xi(t)| \leq 1$, which we do not know. Then, at $t = \pi$ the boundary of the exact reachability set $X_y[t]$ at $t = \pi$ under such measurement consists of four parts: two circular arcs $\{(1 - 2\cos\theta, 2\sin\theta), \theta \in [0, \frac{\pi}{2}]\}, \{(2\cos\theta - 1, -2\sin\theta), \theta \in [0, \frac{\pi}{2}]\}$ and two intervals $\{-1\} \times [-2, 0], \{1\} \times [0, 2]$ located on the red lines. This boundary is shown in orange. Also shown are the ellipsoidal approximations of $X_y[\pi]$.

Exercise 9.3.1. *Find the boundaries for sets $X[t]$ (green) and $X_y[t]$ (orange) shown in Fig. 9.1 by straightforward calculation.*

We now make a detour to systems with discrete observations under continuous and discrete dynamics. Related formulas may be useful for calculations.

9.4 Discrete Observations

9.4.1 Continuous Dynamics Under Discrete Observations

Let $y(\tau_i)$, $\tau_i \in [t_0, t]$, $i = 1, \ldots, N(t)$, be the sequence of measurements (9.5) for system (9.1). Consider

Problem 9.4.1. *Given system (9.1), (9.5) with starting position $\{t_0, X^0\}$ and measurements $y(\tau_i)$, find function*

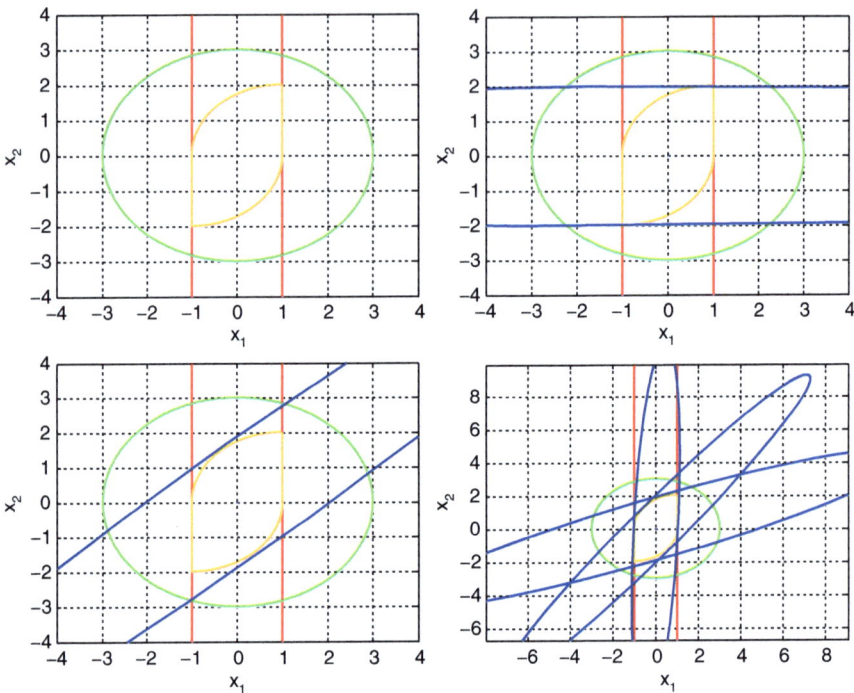

Fig. 9.1 Reachability set $X[\pi]$ without constraint $\mathcal{Y}$ in *green*. Reachability set $X_y[\pi]$ with constraint $\mathcal{Y}$ in *orange*. State constraint $\mathcal{Y}$ in *red*

$$V(t,x) = \min_v \{d^2(x(t_0), X^0) + \sum_{i=1}^{N(t)} d^2(y(\tau_i) - g(\tau_i, x(\tau_i)), \mathcal{R}(\tau_i)) \mid x(t) = x\}. \tag{9.48}$$

Following (9.26), (9.25) denote

$$X[\tau] = \{x : V(\tau, x) \leq 0\}, \quad X_y[\tau] = \{x : g(\tau, x) \in y(\tau) - \mathcal{R}(\tau)\}.$$

The solution to the last problem is to be taken as a recurrent procedure with $V(s,x)$ being the solution to equations

$$V_s + \max\{\langle V_x, f(s,x,v)\rangle \mid v \in \mathcal{Q}(s)\} = 0, \quad V(t_0, x) = d^2(x, X^0), \quad s \in [t_0, \tau_1],$$

so that $X_d[\tau_1] = X[\tau_1] \cap X_y[\tau_1]$ and with further relations

$$V_s + \max\{\langle V_x, f(s,x,v)\rangle \mid v \in \mathcal{Q}(s)\} = 0, \quad s \in [\tau_i, \tau_{i+1}], \quad V(\tau_i, x) = V(t,x)|_{t=\tau_i}, \tag{9.49}$$

9.4 Discrete Observations

and

$$X_d[\tau_{i+1}] = X[\tau_{i+1}] \cap X_y[\tau_{i+1}], \quad X[\tau_{i+1}] = \{x : V(\tau_{i+1}, x) \leq 0\}. \tag{9.50}$$

This results in

Theorem 9.4.1. *The information set for Problem 9.4.1 is*

$$X_d[t] = X(t, \tau_{N(t)}, X_d(\tau_{N(t)})), \tag{9.51}$$

where with $t = \tau_{N(t)}$ *we have* $X_d(t) = X(\tau_{N(t)})$.

For the linear case, keeping in mind the formal relation

$$\sum_{i=1}^{N(t)} d^2(y(\tau_i) - g(\tau_i, x(\tau_i)), \mathcal{R}(\tau_i)) = \int_{t_0}^{t} d^2(y(s) - g(s, x(s)), \mathcal{R}(s)) \sum_{i=1}^{N(t)} \delta(s - \tau_i) ds,$$

we may derive for $V(t, x)$ a related HJB-type equation, similar to (9.10), presenting it in symbolic form as an equation in distributions, [242],

$$V_s + \max\{\langle V_x, f(s, x, v)\rangle | v \in Q(s)\} - \sum_{i=1}^{N(t)} \delta(s - \tau_i) d^2(y(s) - g(s, x(s)), \mathcal{R}(s)) = 0, \tag{9.52}$$

seeking $V(t, x)$, according to the theory of distributions, as a weak solution to this equation.

In the last case the formal procedures for deriving tight external ellipsoids may follow those of Version-A. However, here we indicate a slightly different route by directly applying Theorem 9.4.1. Namely, for $s \in [t_0, \tau_1]$, take

$$\dot{P} + PA(s) + A'(s)P - \pi^2(s)P - (\pi^2(s))^{-1}PC(s)Q(s)C'(s)P = 0, \quad P(t_0) = (X^0)^{-1},$$

$$\dot{x}^* = A(s)x^*(s) + C(s)q(s), \quad x^*(t_0) = x^0, \quad h(s) \equiv 0.$$

Then at $s = \tau_1$ we should find an external ellipsoidal approximate $E_d^+[\tau_1] = \mathcal{E}(x_d^*(\tau_1), P_d(\tau_1))$ for

$$\mathcal{E}(x^*(\tau_1), P(\tau_1)) \cap X_y(\tau_1) \subseteq \mathcal{E}(x_d^*(\tau_1), P_d(\tau_1)),$$

continuing at $s > \tau_1$ with $P(\tau_1 + 0) = P_d(\tau_1)$, $x^*(\tau_1 + 0) = x_d^*(\tau_1)$.

Proceeding further, we have, for $s \in (\tau_i, \tau_{i+1}]$:

$$\dot{P} + PA(s) + A'(s)P - \pi^2(s)P - (\pi^2(s))^{-1}PC(s)Q(s)C'(s)P = 0, \quad P(\tau_i + 0) = P_d(\tau_i),$$

$$\dot{x}^* = A(s)x^*(s) + C(s)q(s) = 0, \quad x^*(\tau_i + 0) = x_d^*(\tau_i),$$

where

$$E_d^+[\tau_{i+1}] = \mathcal{E}(x_d^*(\tau_{i+1}), P_d(\tau_{i+1})) \supseteq \mathcal{E}(x^*(\tau_i), \mathcal{P}(\tau_i)) \cap \mathcal{X}_y(\tau_i), i = 1, \ldots, N(t) - 1,$$

and the array of external ellipsoids E_d^+ may be found through one of the schemes given in Chap. 3, Sect. 3.2.

The treatment of ellipsoidal techniques for internal approximations of information sets in linear systems may be achieved along the schemes of Sects. 3.7 and 3.12.

9.4.2 Discrete Dynamics and Observations

A frequent approach is to restrict the investigation to discrete-time models of system dynamics and observations. Here are the related equations.

Discrete-Time Processes

Consider a multistage process described by a recurrent equation

$$x(k+1) = f(k, x(k)) + C(k)v(k), \quad k \geq k_0, \ x(k_0) = x^0, \tag{9.53}$$

Here $f(k, x)$ is a given map from $\mathbb{N} \times \mathbb{R}^n$ into $\mathbb{R}^n$ ($\mathbb{N}$ is the set of integers), $C(k)$ are given matrices. The vector-valued inputs $v(k)$ are taken in $\mathbb{R}^p$, $p \leq n$. These vectors, together with x^0, are unknown but bounded, with hard bounds

$$x^0 \in \mathbf{X}^0, \quad v(k) \in \mathbf{P}(k), \tag{9.54}$$

where $\mathbf{X}^0, \mathbf{P}(k)$ are convex, compact sets in spaces $\mathbb{R}^n$ and $\mathbb{R}^p$, respectively.

The available information on $x(k)$ is confined to measurement outputs given by

$$y(k) = G(k)x(k) + \xi(k), \quad k = k_0 + 1, \ldots, N, \tag{9.55}$$

where observations $y(k) \in \mathbb{R}^m$ and matrices $G(k)$ are known while measurement disturbances $\xi(k)$ are unknown but bounded, with

$$\xi(k) \in \mathbf{Q}(k), \tag{9.56}$$

where $\mathbf{Q}(k) \subset \mathbb{R}^m$ is convex and compact.

Here $y[k, l] = \{y(k), \ldots, y(l)\}$ will stand for the sequence of measurements that arrive due to Eq. (9.55) throughout stages s, from k to l. Similarly $h[r, s] = \{h(r), \ldots, h(s)\}$ will stand for a sequence of vectors h_i, $i = r, \ldots, s$, taken, respectively, from set-valued sequence $\mathbf{F}[r, s] = \{\mathbf{F}(r), \ldots, \mathbf{F}(s)\}$. Then $h[r, s] \in \mathbf{F}[r, s]$ will stand for a sequence of inclusions $h(i) \in \mathbf{F}(i), i = r, \ldots, s$. Hence, with $h(i) \in \mathbb{R}^q$ we will have $h[1, s] \in \mathbb{R}^{q \times s} = \mathbb{R}_1^q \times \ldots \times \mathbb{R}_s^q$, where $\mathbb{R}_i^q = \mathbb{R}^q$, for $i = 1, \ldots, s$.

9.4 Discrete Observations

Symbol $x[k] = x(k, v[k_0, k-1], x^0)$ will be the end of trajectory $x(j)$ of (9.53), that develops through $[k_0, k]$ with x^0 and $v[k_0, k-1]$ given.

As before **the information set** $\mathbf{X}_y[s] = \mathbf{X}_y(s \mid t, k, \mathbf{F})$ is the collection of all points $x[s] \in \mathbb{R}^n$ that arrive at time s ($k \leq s$) due to such trajectories of system (9.53) that emanate from state $x(k) = x^*$, evolve throughout interval $[k, t]$ and also produce the measured realization $y[k, s]$ for some triplet $\{x^*, v, \xi\}$ constrained by inclusions

$$x^* \in \mathbf{F}, \quad v[k, t-1] \in \mathbf{P}[k, t-1], \quad \xi[k+1, t] \in \mathbf{Q}[k+1, t]. \qquad (9.57)$$

We also denote $\mathbf{X}_y(t, k, \mathbf{F}) = \mathbf{X}_y(t \mid t, k, \mathbf{F})$.

Lemma 9.4.1. *Whatever be s, l, τ, given realization $y[k, l]$ given, the following relations are true*

$$\mathbf{X}_y(t, k, \mathbf{F}) = \mathbf{X}_y(t, s, \mathbf{X}_y(s, k, \mathbf{F})), \quad t \geq s \geq k, \qquad (9.58)$$

$$\mathbf{X}_y(s \mid t, k, \mathbf{F}) = \mathbf{X}_y(s \mid t, l, \mathbf{X}_y(l \mid \tau, k, \mathbf{F})).$$

The given conditions ensure that the mapping $\mathbf{X}_y(t, k, \mathbf{F}) : \operatorname{comp} \mathbb{R}^n \to \operatorname{comp} \mathbb{R}^n$ defines a generalized dynamic system similar to $\mathcal{X}_y(t, t_0, X^0)$ of Sect. 8.3.1, but under discrete time.

In the general case sets $\mathbf{X}_y[t]$ defined here may turn out to be nonconvex and even disconnected. However the next property is true.

Lemma 9.4.2. *Let $\mathbf{F}$ be a closed, convex set in $\mathbb{R}^n$, and $f(k, x) = A(k)x$ be a linear map. Then sets $\mathbf{X}_y(s \mid t, k, \mathbf{F})$ are closed and convex.*

As before, the problem is to describe information sets $\mathbf{X}_y[s]$ and the related information tube $\mathbf{X}_y[k, s]$, $s \in [k_0, t]$—now a sequence of sets $\mathbf{X}_y[s]$.

Multistage Systems: Information Tubes

Returning to system (9.53)–(9.56), consider set $\mathbf{X}_y[s] = \mathbf{X}_y(s, k_0 + 1, \mathbf{X}^0)$ omitting further the lower index y.

Together with $\mathbf{X}[s]$ consider two other sets, namely,

$$\mathbf{X}_*[s] = \mathbf{X}_*(s, k_0 + 1, \mathbf{X}^0) \text{ and } \mathbf{X}^*[s] = \mathbf{X}^*(s, k_0 + 1, \mathbf{X}^0).$$

Here are their definitions. Taking

$$z(k+1) \in f(k, \operatorname{co} Z_*(k)) + C(k)\mathbf{P}(k),$$

$$y(k+1) \in G(k+1)z(k+1) + \mathbf{Q}(k+1), \quad Z_*(k_0) = \mathbf{X}^0, \quad k \geq k_0, \qquad (9.59)$$

where $Z_*(s) = Z_*(s, k_0 + 1, \mathbf{X}^0) = \{z(s)\}$ are solutions to (9.59) at stage $s \geq k_0 + 1$, we get

$$\mathbf{X}_*[s] = \text{co}\, Z_*[s] = \text{co}\, Z_*(s, k_0 + 1, \mathbf{X}^0) = X_*(s, k_0 + 1, \mathbf{X}^0).$$

Similarly, define system

$$z^*(k+1) \in \text{co}\, f(k, Z^*(k)) + C(k)\mathbf{P}(k),$$

$$y(k+1) \in G(k+1)z^*(k+1) + \mathbf{Q}(k+1), \tag{9.60}$$

with solution $Z^*(s) = \{z^*(s)\} = Z^*(s, k_0 + 1, \mathbf{X}^0)$, $s \geq k_0 + 1$, under $Z^*(k_0) = \mathbf{X}^0$. Then we get $Z^*(s) = \mathbf{X}^*[s]$.

Sets $\mathbf{X}_*[s]$, $\mathbf{X}^*[s]$ are obviously convex and the next inclusions are true;

$$\mathbf{X}[s] \subseteq \mathbf{X}_*[s] \subseteq \mathbf{X}^*[s]. \tag{9.61}$$

Remark 9.4.1. Note that apart from $\mathbf{X}_*[s]$, $\mathbf{X}^*[s]$ one may introduce other sequences which at each respective stage may be chosen as either (9.59), or (9.60).

We first describe sets $\mathbf{X}^*[s] = \mathbf{X}(s, k_0 + 1, \mathbf{X}^0)$, using notation

$$\mathbf{Y}(k) = \{x \in \mathbb{R}^n \mid y(k) - G(k)x \in \mathbf{Q}(k)\},$$

and $\mathbf{X}^\star(s, j, \mathbf{F})$ as the solution $\mathbf{X}^\star(s)$ to equation

$$\mathbf{X}^\star(k+1) = \text{co}\, f(k, \mathbf{X}^\star(k)) + C(k)\mathbf{P}(k), \quad j \leq k \leq s-1$$

with $\mathbf{X}^\star(k_0) = \mathbf{F}$. Then the recurrent evolution equation will be as follows.

Lemma 9.4.3. *Let $y[k_0 + 1, k]$ be the sequence of realized measurements y due to system (9.53)–(9.56). Then the next equalities are true:*

$$\mathbf{X}^*[k] = \mathbf{X}^*(k, k_0 + 1, \mathbf{X}^0) = \mathbf{X}^\star(k, k-1, \mathbf{X}^*[k-1]) \cap \mathbf{Y}(k), \quad k_0 \leq k, \tag{9.62}$$

The last equation indicates that the innovation introduced by each new measurement $y(k)$ arrives through an intersection of set $\mathbf{X}^\star(k, k-1, \mathbf{X}^*[k-1])$— the estimate at stage k *before* arrival of measurement $y(k)$ with $\mathbf{Y}(k)$ derived from $y(k)$.

A One-Stage System. For further calculation we need some additional details. Consider the one-stage system ($z \in \mathbb{R}^n$, $y \in \mathbb{R}^m$)

$$z \in f(\mathbf{X}) + C\mathbf{P}, \quad Gz - y \in \mathbf{Q}, \tag{9.63}$$

9.4 Discrete Observations

and

$$z^* \in \text{co } f(\mathbf{X}) + C\mathbf{P}, \quad Gz^* - y \in \mathbf{Q}, \qquad (9.64)$$

where $f(\mathbf{X}) = \bigcup \{f(x) \mid x \in \mathbf{X}\}$ and as before, co $\mathbf{F}$ is the closure of the convex hull of set $\mathbf{F}$.

Let Z and Z^* stand for the sets of all solutions $\{z\}$ and $\{z^*\}$ to the last system. Obviously we have

$$Z \subseteq \text{co} Z \subseteq Z^*$$

where set Z^* is convex.

Exercise 9.4.1. *Give an example indicating when sets Z, $\text{co}Z$, Z^* do not coincide (see [113]).*

Applying methods of convex analysis, we come to the following property.

Lemma 9.4.4. *The next relations are true:*

$$Z \subseteq R(M, f(\mathbf{X})), \quad Z^* \subseteq R(M, \text{co } f(\mathbf{X})), \forall M \in \mathbb{R}^{m \times n},$$

where

$$R(M, \mathbf{F}) = (I_n - MG)(\mathbf{F} - C\mathbf{P}) + M(y - \mathbf{Q}).$$

These relations allow to approach exact solutions due to the next assertion, given in a form similar to Chaps. 3, 7, and 8.

Theorem 9.4.2. *The exact relations for Z, Z^* are:*

$$Z = \bigcap \{R(M, f(\mathbf{X})) \mid M \in \mathbb{R}^{m \times n}\}, \qquad (9.65)$$

$$Z^* = \bigcap \{R(M, \text{co } f(\mathbf{X})) \mid M \in \mathbb{R}^{m \times n}\}. \qquad (9.66)$$

Multistage Information Tubes: Evolution Equations

These relations may now be propagated to multistage recurrent equations that describe information tubes $\mathbf{X}[k]$, $\mathbf{X}_*[k]$, $\mathbf{X}^*[k]$. Consider system

$$Z(k+1) = (E_n - M(k+1)G(k+1))[\mathbf{F}(k, \mathbf{D}(k)) + C(k)\mathbf{P}(k)] +$$

$$+ M(k+1)(y(k+1) - \mathbf{Q}(k+1)), \qquad (9.67)$$

$$\mathbf{D}(k+1) = \bigcap \{Z(k+1) \mid M(k+1) \in \mathbb{R}^{m \times n}\}, \quad \mathbf{D}(k_0) = \mathbf{X}^0, \qquad (9.68)$$

where $\mathbf{F}(k, \mathbf{D})$ is a mapping from $N \times \mathbb{R}^n$ into $\mathbb{R}^n$. Solving these equations for $k_0 \leq k \leq s$ and having in mind Theorem 9.4.2, we get

$$Z(s) = \mathbf{X}[s], \text{ if } \mathbf{F}(k, \mathbf{D}(k)) = f(k, \mathbf{D}(k)), \tag{9.69}$$

$$Z(s) = X^*[s], \text{ if } \mathbf{F}(k, \mathbf{D}(k)) = \operatorname{co} f(k, \mathbf{D}(k)). \tag{9.70}$$

If we now substitute (9.70) for

$$\mathbf{D}(k+1) = \operatorname{co} \bigcap \{Z(k+1) \mid M(k+1) \in \mathbb{R}^{m \times n}\}, \tag{9.71}$$

then solutions to (9.67), (9.71) yield the following result

$$Z(s) = X_*[s], \text{ if } \mathbf{F}(k, \mathbf{D}(k)) = f(k, \mathbf{D}(k)). \tag{9.72}$$

We finally come to the conclusion

Theorem 9.4.3. *(i) Inclusions (9.61) are true.*
(ii) Equalities (9.69), (9.70) are true due to system (9.67) and (9.68).
(iii) Equality (9.72) is true due to relations (9.67) and (9.71).

Remark 9.4.2. The last theorem indicates relations that may be used to approach exact set-valued estimates of $\mathbf{X}[k], \mathbf{X}_*[k], \mathbf{X}^*[k]$ for $k_0 \leq k \leq s$. They require intersections over M at each stage "k" of the evolution process. However the intersections may all be made at the end $k = s$ of the process, being taken over all sequences $M[k_0, s]$. Such procedures may be especially convenient for linear systems in conjunction with ellipsoidal calculus.

Exercise 9.4.2. *Compare the computational complexity for both types of intersections indicated in previous remark.*

Following the last remark, consider linear map $f(k, Z(k)) = A(k)Z(k) = \operatorname{co} f(k, Z(k))$. Now take the set-valued equation

$$Z(k+1) = (E_n - M(k+1)G(k+1))(\mathbf{F}(k, Z(k)) + C(k)\mathbf{P}(k)) +$$
$$+ M(k+1)(y(k+1) - \mathbf{Q}(k+1)), \quad Z(k_0) = \mathbf{X}^0. \tag{9.73}$$

Denote the solution tube to this equation as

$$Z(s) = Z(s, k_0, \mathbf{F}(\cdot), M(\cdot), \mathbf{X}^0), \text{ where } \mathbf{F}(\cdot) = \mathbf{F}(k, Z(k)),$$

$$M(\cdot) = M[k_0, s], \text{ where } M[k_0, s] = \{M(k_0 + 1), \ldots, M(s)\}$$

Denoting

$$Z(s, \mathbf{F}(\cdot), \mathbf{X}^0) = \bigcap \{Z(s, k_0, \mathbf{F}(\cdot), M(\cdot), \mathbf{X}^0) \mid M[k, s] \in \mathbb{R}^{m \times n}\}, \quad (9.74)$$

we come to the conclusion

Theorem 9.4.4. *For a linear mapping* $\mathbf{F}(k, Z(k)) = A(k)Z(k)$, $(k = k_0, \ldots, s)$, *with convex* $\mathbf{X}^0$, *we have*

$$Z(s, \mathbf{F}(\cdot), \mathbf{X}^0) = \mathbf{X}[s] = \mathbf{X}_*[s] = \mathbf{X}^*[s]. \quad (9.75)$$

9.5 Viability Tubes: The Linear-Quadratic Approximation

Though the problem considered further is formulated under hard bounds on the system inputs and system trajectories, the calculation of related trajectory tubes may be also achieved through quadratic approximations.

Let us return to the problem of finding viability tube $\mathcal{X}_y[t] = \mathcal{X}_y(t, t_0, \mathcal{X}^0)$ for system

$$\dot{x} \in A(t)x + Q(t), \quad x(t_0) \in \mathcal{X}^0, \quad (9.76)$$

$$G(t)x \in \mathcal{Y}(t), \quad (9.77)$$

where set-valued maps $Q(t)$, $\mathcal{Y}(t)$ and set $\mathcal{X}^0$ are the same as in Sect. 9.1.

The support function for convex compact set $\mathcal{X}_y[t]$ may be calculated directly, as a linear-quadratic problem, similar to Sects. 1.6 and 2.2. This gives

$$\rho(l \mid \mathcal{X}_y[t]) = \inf\{\Psi_t(l, \lambda(\cdot)) \mid \lambda(\cdot) \in L_2^m[t_0, \tau]\}, \quad (9.78)$$

where

$$\Psi_t(l, \lambda(\cdot)) = \rho\left(S'(t_0, t)l - \int_{t_0}^t S'(t_0, \tau)G'(\tau)\lambda(\tau)d\tau \,\bigg|\, \mathcal{X}_0\right) +$$

$$+ \int_{t_0}^t \rho\left(S'(\tau, t)l - \int_\tau^t S'(\tau, s)G'(s)\lambda(s)ds \,\bigg|\, Q(\tau)\right) d\tau +$$

$$+ \int_{t_0}^t \rho(\lambda(\tau) \mid \mathcal{Y}(\tau))d\tau.$$

We shall rewrite system (9.76), (9.77) in the form

$$\dot{x} = A(t)x + v(t), \quad x(t_0) = x^0, \quad G(t)x = w(t), \quad (9.79)$$

where measurable functions $v(t)$, $w(t)$ and vector x^0 satisfy the inclusions

$$v(t) \in Q(t), \quad w(t) \in \mathcal{Y}(t), \quad x^0 \in \mathcal{X}^0. \tag{9.80}$$

The viable information tube $\mathcal{X}_y[t]$ may then be described as a multivalued map $\mathcal{X}_y(t, t_0, \mathcal{X}^0)$ generated by the set (the bundle) of all trajectories $x(t, t_0, x^0)$ of system (9.79) that are consistent with constraints (9.80).

The calculation of $\mathcal{X}_y[t]$ for a given instant $t = \tau$ will now run along the following scheme. Let us fix a triplet $k^*(\cdot) = \{v^*(\cdot), w^*(\cdot), x_0^*\}$ where $v^*(\cdot)$, $w^*(\cdot)$, x_0^* ($t \in [t_0, \tau]$) satisfy the constraints (9.80):

$$k^*(\cdot) \in Q(\cdot) \times \mathcal{Y}(\cdot) \times \mathcal{X}^0, \quad \mathcal{Y}(t) = y(t) - \mathcal{R}(t), \tag{9.81}$$

as in (9.2), (9.3).

Instead of handling (9.80) we shall now consider a "perturbed" system, which is

$$\dot{z} = A(t)z + v^*(t) + \eta(t), \quad z(t_0) = x^{0*} + \zeta^0, \quad t_0 \leq t \leq \tau, \tag{9.82}$$

$$G(t)z = w^*(t) + \xi(t),$$

Elements $d(\cdot) = \{\zeta^0, \eta(\cdot), \xi(\cdot)\}$ represent the unknown disturbances bounded jointly by quadratic inequality

$$\zeta^{0\prime} M \zeta^0 + \int_{t_0}^{\tau} \eta'(t) R(t) \eta(t) dt + \int_{t_0}^{\tau} \xi'(t) H(t) \xi(t) dt \leq \mu^2, \tag{9.83}$$

where $\{M, R(\cdot), H(\cdot)\} \in \mathfrak{I}$ and symbol $\mathfrak{I}$ stands for the product space

$$\mathfrak{I} = \mathbb{R}_+^{n \times n} \times \mathbb{R}_+^{n \times n}[t_0, \tau] \times \mathbb{R}_+^{m \times m}[t_0, \tau],$$

with $\mathbb{R}_+^{r \times r}[t_0, \tau]$ denoting the class of all $r \times r$-matrix functions $N(\cdot) \in \mathbb{R}^{r \times r}[t_0, \tau]$ whose values $N(t)$ are symmetric and positive definite.

For every fixed $k^*(\cdot)$, $\Lambda = \{M, R(\cdot), H(\cdot)\}$ and μ denote $Z[\tau] = Z(\tau, k^*, \Lambda, \mu)$ to be the set of all states $z(\tau)$ of system (9.82) that are consistent with constraint (9.83). The support function of this set is

$$\rho(l \mid Z[\tau]) = \max_{d(\cdot)} \{\langle l, z(\tau) \rangle \mid d(\cdot) : (9.83)\} \tag{9.84}$$

It is well known that $Z[\tau] = Z(\tau, k^*, \Lambda, \mu)$ is an ellipsoid with center $z_0[\tau] = z_0(\tau, k^*, \Lambda)$ that does not depend on μ [25, 137]. This can be observed by direct calculations (see Exercise 1.6.5) which also indicate that $z_0[s]$ satisfies the linear differential equation

9.5 Viability Tubes: The Linear-Quadratic Approximation

$$\dot{z} = (A(s) - \Sigma(s)G'(s)H(s)G(s))z + \Sigma(s)G'(s)H(s)w^*(s) + v^*(s), \quad (9.85)$$

$$z(t_0) = x_0^*, \quad t_0 \leq s \leq \tau,$$

where $\Lambda = \{M, R(\cdot), H(\cdot)\} \in \mathfrak{I}$ are fixed and $\Sigma(\cdot)$ is the matrix solution to Riccati equation

$$\dot{\Sigma} = A(s)\Sigma + \Sigma A'(s) - \Sigma G'(s)H(s)G(s)\Sigma + R^{-1}(s), \quad (9.86)$$

$$\Sigma(t_0) = M^{-1}, \quad t_0 \leq s \leq \tau.$$

Let us now introduce the set

$$Z_0(\tau, \Lambda) = \bigcup \{z_0(\tau, k(\cdot), \Lambda) \mid k(\cdot) \in \mathcal{Q}(\cdot) \times \mathcal{Y}(\cdot) \times \mathcal{X}^0\}$$

which is the union of centers $z_0[\tau]$ over all triplets $k(\cdot) = k^*(\cdot)$ from (9.81).

Set $Z_0(\tau, \Lambda)$ is convex and compact, being the reachability set for system (9.85) under constraints (9.80), or in other words, of the differential inclusion

$$\dot{z} \in (A(t) - \Sigma(t)G'(t)H(t)G(t))z + \Sigma(t)G'(t)H(t)\mathcal{Y}(t) + \mathcal{Q}(t), \quad (9.87)$$

$$\mathcal{Y}(t) = y(t) - \mathcal{R}(t), \quad z(t_0) \in \mathcal{X}^0, \quad t_0 \leq t \leq \tau.$$

The support function for $Z_0(\tau, \Lambda)$, which gives a complete description of this set, is now to be calculated as in Sect. 2.4. It is a closed convex, positively homogeneous functional

$$\rho(l \mid Z_0(\tau, \Lambda)) = \Upsilon(\tau, l, \Lambda). \quad (9.88)$$

where

$$\Upsilon(\tau, l, \Lambda) = (\mathcal{D}_\Lambda l)(\tau) + \mathcal{L}_\Lambda^{-1} l$$

with linear operators $\mathcal{D}_\Lambda : \mathbb{R}^n \to L_2^m[t_0, \tau]$ and $\mathcal{L}_\Lambda : L_2^m[t_0, \tau] \to L_2^m[t_0, \tau]$ defined by relations

$$(\mathcal{D}_\Lambda l)(t) = G(t)\big(S(t_0, t)M^{-1}S'(t_0, \tau) +$$

$$+ \int_{t_0}^t S(s, t)R^{-1}(s)S'(s, \tau)ds\big)l, \quad l \in \mathbb{R}^n, \quad t_0 \leq t \leq \tau,$$

$$\mathcal{L}_\Lambda \lambda(\cdot) = (K_1 + K_2)\lambda(\cdot), \quad \lambda(\cdot) \in L_2^m[t_0, \tau],$$

$$(K_1\lambda(\cdot))(t) = \int_{t_0}^{\tau} K(t,s)\lambda(s)ds,$$

$$K(t,s) = G(t)(S(t_0,t)M^{-1}S'(t_0,s)+$$

$$+ \int_{t_0}^{\min\{t,s\}} S(\sigma,t) \times R^{-1}(\sigma)S'(\sigma,s)d\sigma)G'(s),$$

$$(K_2\lambda(\cdot))(t) = H^{-1}(t)\lambda(t), \quad t_0 \le t \le \tau. \tag{9.89}$$

Here $\mathcal{L}_\Lambda$ is a nondegenerate Fredholm operator of the second kind, so that functional $\Upsilon(\tau, l, \Lambda)$ is defined correctly for all $l \in \mathbb{R}^n$ and $\Lambda \in \mathfrak{I}$ [110].

Exercise 9.5.1. *Calculate given formulas for operators $\mathcal{D}_\lambda$ and $\mathcal{L}_\lambda$.*

Lemma 9.5.1. *The following inequality is true*

$$\rho(l \mid X_y[\tau]) \le \rho(l \mid Z_0(\tau, \Lambda)) \tag{9.90}$$

for all $l \in \mathbb{R}^n$ and $\Lambda \in \mathfrak{I}$.

From here we immediately come to the next assertion.

Corollary 9.5.1. *The viable information tube $X_y[\tau]$ may be estimated from above as follows*

$$X_y[\tau] \subseteq \bigcap \{Z_0(\tau, \Lambda) \mid \Lambda \in \mathfrak{I}\}. \tag{9.91}$$

Since here we deal with a linear system, the further objective is to emphasize, as in Theorem 8.2.5 of Sect. 8.2.2, that inclusion (9.91) is actually an equality.

Lemma 9.5.2. *Suppose the $m \times n$-matrix $G(t)$ is of full rank: $r(G(t)) = m$ for any $t \in [t_0, \tau]$. Then for every $l \in \mathbb{R}^n$ the following equalities are true*

$$\rho(l \mid X_y[\tau]) = \inf\{\Psi_\tau(l, \lambda(\cdot)) \mid \lambda(\cdot) \in L_2^m[t_0, \tau]\} = \tag{9.92}$$

$$= \operatorname{co} \inf\{\Upsilon(\tau, l, \Lambda). \mid \Lambda \in \mathfrak{I}\}.$$

Combining formula (9.78), Lemma 9.5.2 and taking into account (9.91) as an equality, we get

Theorem 9.5.1. *Let $r(G(t)) = m$ for every $t \in [t_0, \tau]$. Then*

$$X_y[\tau] = \bigcap \{Z_0(\tau, \Lambda) \mid \Lambda \in \mathfrak{I}\}. \tag{9.93}$$

Theorem 9.5.1 gives a precise description of $\mathcal{X}_y[\tau]$ through solutions $Z_0(\tau, \Lambda)$ of the linear-quadratic problem (9.82)–(9.83) by varying the matrix parameters $\Lambda = \{M, R(\cdot), H(\cdot)\}$ in the joint integral constraint (9.83).

Finalizing this section we emphasize that described above is a parameterized family of set-valued estimators $Z_0(\tau, \Lambda)$ which together, in view of Theorem 9.5.1, exactly determine the set $\mathcal{X}_y[\tau]$.

Theorem 9.5.2. *Sets $Z_0(\tau, \Lambda)$ are the "cuts" (cross-sections) at time τ of the solution tube (the integral funnel) to the differential inclusion (9.87) where $\Lambda = \{M, R(\cdot), H(\cdot)\} \in \mathfrak{I}$ and $\Sigma(\cdot)$ is the matrix solution to Riccati equation (9.86).*

Each of sets $Z_0(\tau, \Lambda)$ is a guaranteed external estimator of $\mathcal{X}_y[\tau]$. This property opens the route for parallel calculations that yield description of $\mathcal{X}_y[\tau]$ with high accuracy.

Remark 9.5.1. The last result of Theorem 9.5.2, together with Theorem 9.5.1, provides a special structure of matrix functions $L(\cdot)$ for these relations. Namely, here we may set $L(t) = \Sigma(t)G'(t)H(t)G(t)$, $t_0 \leq t \leq \tau$, where $\Sigma(\cdot)$ is defined by (9.86) with $\Lambda = \{M, R(\cdot), H(\cdot)\}$ varying within set $\mathfrak{I}$.

Note that several techniques based on using auxiliary uncertain systems under quadratic integral constraints for various classes of system models have been discussed in this context in [113, 140, 167].

One of these is the guaranteed state estimation problem of this chapter. Apart from deterministic techniques in earlier lines of this section the problem may also be solved through a stochastic scheme which ends up with results of similar type.

9.6 Information Tubes vs Stochastic Filtering Equations. Discontinuous Measurements

The two approaches to the state estimation problem, namely the stochastic and the set-membership filtering, may seem rather different. However it turns out that, apart from differences, there are useful connections between the techniques of calculating the results.

9.6.1 Set-Valued Tubes Through Stochastic Approximations

Suppose that system (9.76), (9.77) is specified as follows

$$\dot{x} \in A(t)x + \mathcal{Q}(t), \quad t \geq t_0, \quad x(t_0) \in \mathcal{X}^0, \tag{9.94}$$

$$y(t) \in G(t)x + \mathcal{R}(t), \tag{9.95}$$

where (9.95) describes the measurement (observation) equation while continuous set-valued function $\mathcal{R}(t) : T = [t_0, t_1] \to \text{conv } \mathbb{R}^m$ reflects the unknown but bounded noise w in the observations.

Given measurement $y(t) = y[t]$, $t \in [t_0, \tau]$, the guaranteed state estimation problem is to specify information sets $X_y[t]$ and their evolution as an information tube $X_y[\cdot]$ which is to be viable relative to inclusion (9.95), when $y[t]$ arrives online. The evolution of $X_y[t]$ was described by relations (9.93), (9.84), (9.86) under appropriate assumptions on $\mathcal{Y}(t)$.

It is well known, however, that a conventional stochastic filtering technique for similar linear systems subjected to stochastic Gaussian noise are given by equations of the "Kalman filter" [108]. Our next question therefore will be to see whether equations of the Kalman filter type could be also used to describe the deterministic information tubes $X_y[t]$ for the guaranteed estimation problem of the above.

This question is justified by the fact that, on the one hand, the tube $X_y[t] = X_y(t, t_0, X^0)$ may be described through the linear-quadratic approximations of Sect. 9.5, while, on the other, by the well-established connections between the Kalman filtering equations and the solutions to the linear-quadratic problem of control.

Using solutions of the previous section, fix a triplet $k(\cdot) = k^*(\cdot) = \{v^*(\cdot), w^*(\cdot), x_0^*\}$ with $k^*(\cdot) \in Q(\cdot) \times (y(\cdot) - \mathcal{R}(\cdot)) \times X^0$ and consider the stochastic differential equations

$$dz = (A(t)z + v^*(t))dt + \sigma(t)d\xi, \tag{9.96}$$

$$dq = (G(t)z + w^*(t))dt + \sigma_1(t)d\eta, \tag{9.97}$$

$$z(0) = x_0^* + \zeta, \quad q(0) = 0, \tag{9.98}$$

where ξ, η are standard normalized Brownian motions with continuous diffusion matrices $\sigma(t)$, $\sigma_1(t)$ and $\det(\sigma(t)\sigma'(t)) \neq 0$ for all $t \in T$, ζ is a Gaussian vector with zero mean and variance $M^* = \sigma_0 \sigma_0'$.

Denoting $\sigma(t)\sigma'(t) = R^*(t)$, $\sigma_1(t)\sigma_1'(t) = H^*(t)$ and treating $q = q(t)$ as the available measurement we may find equations for the minimum variance estimate $z^*(t) = \mathbf{E}(z(t) \mid q(s), t_0 \leq s \leq t)$ (the respective "Kalman filter").

These are

$$dz^*(t) = (A(t) - \Sigma(t)G'(t)H^{*-1}(t)G(t))z^*(t)dt + \tag{9.99}$$

$$+\Sigma(t)G'(t)w^*(t)dt + v^*(t)dt, \quad z^*(t_0) = x_0^*,$$

$$\dot{\Sigma}(t) = A(t)\Sigma(t) + \Sigma(t)A'(t) - \tag{9.100}$$

$$-\Sigma(t)G'(t)H^{*-1}(t)G(t)\Sigma(t) + R^*(t), \quad \Sigma(t_0) = M^*.$$

The estimate $z^*(t)$ thus depends on the triplets $k^*(\cdot)$ and $\Lambda^* = \{M^*, R^*(\cdot), H^*(\cdot)\} \in \mathfrak{I}$.

9.6 Information Tubes vs Stochastic Filtering Equations. Discontinuous...

Now let us consider the set

$$Z^*(t) = Z^*(t, \Lambda^*) = \bigcup \{z^*(t) \mid k^*(\cdot) \in Q(\cdot) \times \mathcal{Y}(\cdot) \times \mathcal{X}^0\}$$

which, with a given realization $q(t)$, is the reachability set for Eq. (9.99).

Theorem 9.6.1. *Assume equalities*

$$M^* = M^{-1}, \quad R^*(t) \equiv R^{-1}(t), \quad H^*(t) \equiv H^{-1}(t) \tag{9.101}$$

to be true and $\mathcal{Y}(t) = y(t) - \mathcal{R}(t)$, $t \in T$. Also assume

$$q(t) \equiv \int_{t_0}^{t} y(\tau) d\tau. \tag{9.102}$$

Then sets $Z_0(t, \Lambda)$ of Sect. 9.5 and $Z^*(t, \Lambda^*)$ of this section coincide, namely,

$$Z_0(t, \Lambda) \equiv Z^*(t, \Lambda^*), \quad t \in T, \quad \Lambda = \{M, R(\cdot), H(\cdot)\}.$$

Corollary 9.6.1. *Under the assumptions of Theorem 9.6.1, the following equality is true*

$$X_y[\tau] = \bigcap \{Z^*(\tau, \Lambda^*) \mid \Lambda^* \in \mathfrak{I}\}. \tag{9.103}$$

The proof of Theorem 9.6.1 follows from the fact that under the assumptions of the last theorem equation (9.99), with $k^*(\cdot) \in \{Q(\cdot) \times \mathcal{Y}(\cdot) \times \mathcal{X}^0\}$, and the system (9.79)–(9.81), have the same reachability sets under constraint (9.83). Corollary 9.6.1 then follows from Theorem 9.5.1.

Remark 9.6.1. The last results describe a clear connection between solutions to the linear-quadratic Gaussian filtering problem (the Kalman filter) and those for deterministic guaranteed state estimation under unknown but bounded "noise" which may satisfy not only soft integral quadratic bounds on the uncertain items but also the **non-quadratic instantaneous (hard) bounds** on the unknown items. A more detailed discussion of this approach may be found in [138].

One should recall that while the theorems discussed in Sect. 9.1 were proved under Assumptions 8.1.1, 8.1.3, or 8.1.4, or 8.1.6, those of Sect. 9.2.1 contained additional conditions on smoothness and convexity of the on-line continuous state constraints, whereas those of Sects. 9.2.2 and 9.2.3, allowed $\mathcal{Y}(t)$ to be *only continuous* in time t. A certain relaxation of this condition would be to demand that $y(t)$ would be piecewise continuous. We shall now recall a technique of *singular perturbations for differential inclusions* which will eventually allow to relax the requirements on $\mathcal{Y}(\cdot)$, accepting state constraints, when $\mathcal{Y}(t)$ is only measurable on T.

A nonlinear filtering version of these results could be pursued through a combination of the reasoning of Sect. 8.1 and of [14].

9.6.2 The Singular Perturbation Approach: Discontinuous Measurements in Continuous Time

In this section we consider linear-convex differential inclusions of type

$$\dot{x} \in A(t)x + Q(t), \quad y(t) - G(t)x \in \mathcal{R}(t), \tag{9.104}$$

with $x(t_0) \in X^0$, $t_0 \leq t \leq \tau$.

Here the $n \times n$- and $m \times n$-matrices $A(t)$, $G(t)$ are taken measurable and bounded almost everywhere on $[t_0, t_1]$, with set-valued maps $Q : [t_0, t_1] \to \text{conv } \mathbb{R}^n$ and $\mathcal{R} : [t_0, t_1] \to \text{conv } \mathbb{R}^m$ measurable and bounded.

Further on we consider a system of type

$$\dot{x} \in A(t)x + Q(t), \quad L(t)\dot{r} \in \mathcal{R}(t) - y(t) + G(t)x, \tag{9.105}$$

parameterized by functions $L(\cdot)$, with

$$t_0 \leq t \leq \tau, \quad x^0 = x(t_0) \in X^0, \quad r^0 = r(t_0) \in \mathcal{R}^0,$$

and

$$L(\cdot) \in \mathbb{R}_*^{m \times m}[t_0, \tau], \quad Z^0 = X^0 \times \mathcal{R}^0 \in \text{conv } \mathbb{R}^{n+m}$$

Here $\mathbb{R}_*^{m \times m}[t_0, \tau]$ is the class of all continuous invertible matrix functions $L(\cdot) \in \mathbb{R}^{m \times m}[t_0, \tau]$.

Denote $z = \{x, r\} \in \mathbb{R}^n \times \mathbb{R}^m$ and $z[t] = z(t; \tau, t_0, z_0, L)$ to be the solution to (9.105) that starts at point $z[t_0] = z^0 = \{x^0, r^0\}$. Symbol $Z(\cdot; \tau, t_0, Z^0, L)$ will now stand for the solution tube to system (9.105) with $Z[\tau] = Z(\tau, t_0, Z^0, L) = Z(\tau; \tau, t_0, Z^0, L)$ and $Z^0 \in \mathbb{R}^n \times \mathbb{R}^m$.

Let $\pi_x W$ denote the projection of set $W \subseteq \mathbb{R}^n \times \mathbb{R}^m$ into space $\mathbb{R}^n$ of x-variables. Then the next assertion is true

Theorem 9.6.2. *Under assumption $\pi_x Z^0 = X^0$ the following formula is true for any $\tau \in [t_0, t_1]$*

$$X_y[\tau] = \pi_x \left(\bigcap \{ Z(\tau; t_0, Z^0, L) \mid L \in \mathbb{R}_*^{m \times m}[t_0, \tau] \} \right).$$

Set $X_y(\tau) \subset \mathbb{R}^n$ therefore arrives as the intersection over all functions $L(\cdot) \in \mathbb{R}^{m \times m}[t_0, \tau]$ of the projections of sets $Z[\tau] \subset \mathbb{R}^{(n+m) \times (n+m)}$ on the space $\mathbb{R}^n$ of x-variables. The proof of the last theorem is given in detail in [158, Sect. 16], with examples given in [174, Sect. 4.6] (see also [159, 168]).

It is clear that conditions of smoothness or continuity of $\mathcal{Y}(t)$ in time are not required in such setting.

Chapter 10
Uncertain Systems: Output Feedback Control

Abstract This chapter finalizes some results of earlier chapters for systems that operate under set-membership uncertainty. Its aim is to emphasize a successful application of previously described techniques to such systems. The chapter thus gives a concise presentation of solution techniques for the problem of *output feedback control* based on available measurements under set-membership uncertainty.

Keywords Set-membership uncertainty · Bounding approach · Measurement output · Information space · Information tube · Feedback control strategy · Linear-convex systems · Separation theorem · Ellipsoidal techniques

This chapter gives a concise description of controlled systems that operate *under uncertainty*—incomplete information on the system dynamics and uncertainty in the system and measurement inputs due to unknown but bounded disturbances. Considered are systems with mostly *hard bounds* on the uncertain items. The issue of uncertainty is illustrated on the problem of output feedback control (OFC) for such systems. The aim of the chapter is to demonstrate that techniques of the present book may be successfully used in solving problems of feedback control under imperfect knowledge of its model. A detailed discussion of feedback control under realistic (inaccurate or incomplete) information on the system model, its on-line dynamics and accompanying on-line measurement process is a broad topic left for separate publication.

The basic problem here is of closed-loop target control under realistic information. As everywhere in this book, the forthcoming results range from theoretical issues to computational schemes. But the specifics of the derived solution schemes lie in the combination of both approaches discussed in the previous chapters, namely, of Hamiltonian methods in the form of *dynamic programming* and of techniques taken from *set-valued analysis and minmax approaches*. The overall general solution for the considered problem then appears as a combination of two parts, which deal firstly with a finite-dimensional problem of *guaranteed state estimation* and secondly—with an infinite-dimensional problem of *feedback control under set-membership uncertainty*. For the first problem new types of *set-valued observers* were introduced. For the second problem, which is especially difficult to formalize and solve, the achieved solution is reduced to one of finite-dimensions which facilitates calculation. For systems with linear structure and convex constraints the solution is more precise. Here the computational procedure is based on new developments in *ellipsoidal calculus*, which proved to be effective and allowed to design complemental software, producing complete solutions with useful examples (see [133]). Calculations for the nonlinear case are achieved by using specifics of the problem in combination with *comparison principles* that allow to relax the original equations or variational inequalities of the Hamilton–Jacobi type to simpler, finite-dimensional relations.

10.1 The Problem of OFC

One key problem in realistic control synthesis is to design closed-loop feedback control laws based on available on-line observations under noisy measurements. One must assess how the level of uncertainty and the available incomplete, but realistic information affect the values of the achievable cost, optimizing guaranteed performance. Related problems of measurement output control were well developed in a stochastic setting, [4, 18, 59, 122, 273]. However, many applications require solving this problem under other information conditions that need not rely on probabilistic models but require other approaches which are the topics of this book. Such approaches naturally require other techniques and new formalization schemes. This may include other types of observers and other models for disturbances. The solutions to such problems were investigated mostly within the H_∞ approach, [13, 18, 98], with soft-type integral costs, while problems with hard bounds on the noise, in the set-valued perspective, were less developed [147, 189].

The following text begins with a rigorous problem formulation and emphasizes the crucial role of properly selecting the on-line *generalized state* of the system in the form of *information states*, described by either value functions for related problems of optimization, or by *information sets*—the outputs of related set-valued observers. Further developed is the detailed *overall solution scheme*, which is a combination of two: a finite-dimensional problem of *guaranteed state estimation* and an infinite-dimensional problem of *feedback control under set-membership uncertainty*. Both problem solutions are described in detail. The second one is especially difficult to formalize and solve. The novelty of the developed solution is that it indicates how to reduce the second problem *to finite-dimensions*, facilitating calculation. An optimization scheme is then introduced for the overall problem.

A further important aspect is to indicate how to solve the problems completely (*"to the end"*), through appropriate computation. For systems with linear structure and convex constraints a computational procedure is indicated, based on *ellipsoidal calculus*, which proved to be effective for many problems and is used with complementary software which is applicable to systems of large dimensions (see *ellipsoidal* toolbox [132]). The procedure also allows one to apply *polyhedral techniques*, [114, 115, 117]. For the nonlinear case calculations may be facilitated by using the specifics of the problem and applying modifications of the earlier suggested *comparison principles* that allow to relax the original equations or variational inequalities of the Hamilton–Jacobi type to simpler relations (see Sects. 5.1 and 7.1 of this book and also [149]).[1]

The results of this chapter follow the lines of [189].

10.2 The System: The Generalized State and the Rigorous Problem Formulation

The problem of *OFC* under set-membership noise is formulated for the next control system model and available information.

The System Model. Given is an n-dimensional differential equation

$$\dot{x} = f_1(t, x, u) + f_2(t, x, v), \quad x(t_0) = x^0, \; t \in [t_0, \vartheta], \tag{10.1}$$

where the functions $f_1(t, x, u)$, $f_2(t, x, v)$ are continuous with respect to the triplets of independent variables and such that their sum satisfies standard conditions of uniqueness and extendability of solutions throughout the interval $[t_0, \vartheta]$, for any initial condition $x^0 \in \mathbb{R}^n$, any admissible, bounded, closed-loop control strategies

$$u = \mathcal{U}(t, \cdot) \in \mathcal{P}(t), \tag{10.2}$$

and uncertain disturbances $v(t)$ from appropriate classes indicated below. Here, as before, $\mathcal{P}(t) \in \text{comp}\mathbb{R}^p$ is a set-valued function, with compact values, continuous in the Hausdorff metric. And as before, the system trajectory is denoted $x[t] = x(t, t_0, x^0)$.

The Measurement Equation and the Noise. The observations of vector x are due to measurement equation

$$y(t) = g(t, x(t)) + \xi(t), \tag{10.3}$$

where $g(t, x)$ is continuous in both variables, $y(t) \in \mathbb{R}^r$ is the measurement, and $\xi(t)$ is the *unknown noise*.

The Uncertain Items—starting position $x(t_0) = x^0$ and right-continuous input disturbances $v(t), \xi(t)$, are taken as either

---

[1] Effective results may be reached for some classes of problems, especially those on an infinite horizon, by Lyapunov-type methods [198].

(a) *unknown but bounded, with given bounds*

$$x^0 \in X^0, v \in Q(t), \xi \in R(t), \qquad (10.4)$$

where X^0 is compact, while $Q(t)$, $R(t)$ are set-valued functions with properties similar to $P(t)$, or

(b) *unknown, with bounds also unknown.*

The present discussion further deals with case (a). However, it is also applicable to case (b) within a unified framework that combines H_∞ approaches with set-valued interpretations [14, 103]. The *aim of the control* is to steer the system to a prespecified neighborhood of a given *target set* M, whatever are the unknown but bounded uncertain items. This is to be done in fixed time by *feedback control strategy* $U(t, \cdot)$ on the basis of *available information*:

- the system and measurement model: Eqs. (10.1) and (10.3),
- the given constraints on control u and uncertain inputs $x^0, v(t), \xi(t)$: inclusions (10.2), (10.4),
- the starting position $\{t_0, X^0\}$,
- the available measurement $y(t) = y[t]$ (arrives on-line).

Sets X^0, M are assumed convex and compact. The feedback control strategy $U(t, \cdot)$ depends on time "t" and on the generalized state $X_y[t]$ or $V(t, \cdot)$ described further within a precise setting of the problem.

For a precise problem formulation we first need to define *a new (generalized) state of the system*. Namely, what should it be now, that the on-line measurement information is incomplete and corrupted by set-membership noise?

To propagate earlier schemes to OFC the new state may be taken as

(a) $\{t, y_t(\cdot, t_0)\}$, where $y_t(\cdot, t_0) = \{y(t + \sigma) \mid \sigma \in [t_0 - t, 0]\}$ is *the memorized measurement*. This memorization is done through observers and filters.
(b) $\{t, X_y[t]\}$—the *information set* of states consistent with measurements $y_t(\cdot, t_0)$ and constraints (10.4) on the uncertain items

$$\zeta_t(\cdot, t_0) = \{x^0, v_t(\cdot, t_0), \xi_t(\cdot, t_0)\}.$$

The single-valued trajectories of the classical case should then be substituted by set-valued *information tubes*, introduced in Chap. 9

(c) $\{t, V(t, \cdot)\}$—the *information state*, which is the value function $V(t, x)$—the solution to an appropriate HJB (Hamilton–Jacobi–Bellman) PDE equation, designed such that

$$X_y[t] = \{x : V(t, x) \le \alpha\}, \quad \alpha \ge 0$$

is the *level set* of $V(t, x)$.

The overall *basic problem* of OFC now reads as follows.

10.3 The Overall Solution Scheme: General Case

Problem OFC. Specify a control strategy $\mathcal{U}(t, \cdot) \subseteq \mathcal{P}(t)$ as

$$\mathcal{U}(t, \cdot) = \mathcal{U}(t, X_y[t]) \quad \text{or} \quad \mathcal{U}(t, \cdot) = \mathcal{U}(t, V(t, \cdot)),$$

dependent on the state $\{t, X_y[t]\}$ or $\{t, V(t, \cdot)\}$, which for any starting position $\{t_0, X^0\}$, $t_0 < \vartheta$, would bring $x[\vartheta] = x(\vartheta, t_0, x^0)$ to a preassigned neighborhood $\mathcal{M}_\gamma$ of the given target set $\mathcal{M}$, at given time $t = \vartheta$, whatever be the uncertain items $\zeta_t(\cdot, t_0)$, constrained by (10.4). Here $\mathcal{M}_\gamma = \mathcal{M} + \gamma \mathcal{B}_M(0)$, where $\mathcal{B}_M(0) = \{x : \langle x, Mx \rangle \leq 1\}$, $M = M' > 0$.

Note that control strategy $\mathcal{U}$ is a functional of the *information set* (or *information state*) that depends on measurement $y(t, \cdot)$ and given constraints on the uncertain items. The notion of information sets and information states, as well as the description of their properties, are the subject of a separate *theory of guaranteed estimation* treated in Chap. 9.

Depending on what we use—the information set or the information state, there are two basic interconnected approaches, both considered in Chap. 9. The first one is to use the *information state* $V(t, x)$, calculated as the *value function* for a Dynamic Programming problem of minimaximizing a functional borrowed from H_∞ theory and producing solutions in the form of control strategies, as functionals $u = \mathcal{U}(t, V(t, \cdot))$ (see [18, 27, 103, 123]). The second is to use the *information set* described through *set-valued calculus* in terms of differential inclusions, involving further the notion of *invariant sets* and calculating set-valued strategies $u = \mathcal{U}(t, X_y[t])$ through the "aiming rule" and its analogies (see related work at [6,122,136,137,153,204]). We will use both approaches within a unified framework.

10.3 The Overall Solution Scheme: General Case

Suppose that on the interval $[t_0, \tau]$ the control $u = u[t]$ and the observation $y = y[t]$ have been realized and are therefore known.[2]

Problem OFC will be to find

$$V(t_0, X_y^0) = \min_{\mathcal{U}} \max_{\zeta(\cdot)} \{ d^2(x(\vartheta), \mathcal{M}_\gamma) \mid x(t_0) \in X_y^0 \}, \quad X_y^0 = X^0 \cap \mathcal{Y}(t_0),$$

(10.5)

under constraints (hard bounds) (10.1)–(10.4) in the class of controls $u = \mathcal{U}(t, \mathcal{V}_y) = \mathcal{U}(t, X_y)$.

---

[2] Here and in the sequel square brackets in $u = u[t]$, $y = y[t]$ mean that for the interval under consideration these functions are known, otherwise we use round brackets.

376   10 Uncertain Systems: Output Feedback Control

Under such hard bounds this will be achieved through functionals

$$\mathcal{V}(t_0, \mathcal{X}^0) = \min_{u[\cdot]} \max_{\zeta[\cdot]} \left\{ \min_u \max_{\zeta(\cdot)} \left(-d^2(x(\tau), \mathcal{X}_y[\tau]) \right. \right. \tag{10.6}$$

$$\left. \left. - \int_\tau^\vartheta d^2(y(t) - g(t, x(t)), \mathcal{R}(t))dt + d^2(x(\vartheta), \mathcal{M}) \right) \right\}$$

where

$$\mathcal{X}_y[\tau] = \mathcal{X}_y(\tau \mid y_\tau[\cdot, t_0], u = u_\tau[\cdot, t_0]) = \{x : -V_-(\tau, x \mid [y, u]) \le 0\}, \tag{10.7}$$

and

$$V_-(\tau, x \mid [y, u]) = \max_\zeta \{ -d^2(x(t_0), \mathcal{X}^0) - \tag{10.8}$$

$$- \int_{t_0}^\tau d^2(y[t] - g(t, x(t)), \mathcal{R}(t))dt \mid x(\tau) = x, u = u_\tau[\cdot, t_0] \}$$

so that $-V_-(\tau, x \mid [y, u]) = d^2(x(\tau), \mathcal{X}_y[\tau])$.

Here $x(t_0), x(\vartheta)$ are the two ends of the system trajectory $x(t)$, while, as before, $\zeta_\vartheta(\cdot, t_0)$ is the triplet of uncertain items, bounded by (10.1)–(10.4).

Combining the above into one functional, we get

$$\mathcal{V}(t_0, \mathcal{X}^0) = \min_{u[\cdot]} \max_{y[\cdot]} \left\{ \min_u \max_\zeta \left(V_-(\tau, x[\tau] \mid [y, u]) \right. \right. \tag{10.9}$$

$$\left. \left. - \int_\tau^\vartheta d^2(y(t) - g(t, x(t)), \mathcal{R}(t))dt + d^2(x(\vartheta), \mathcal{M}) \right) \right\}$$

Note that on the interval $[t_0, \tau]$, where the realization of the measurement noise is $\xi[t] = y[t] - g(t, x(t))$, the triplet $\zeta = \zeta_\tau(\cdot, t_0)$ is constrained by inclusions $x^0 \in \mathcal{X}^0, v(t) \in \mathcal{Q}(t), y[t] - g(t, x(t) \in \mathcal{R}(t)$, with observation $y[t]$ and control $u[t]$ being known (such triplets are denoted as $\zeta[\cdot]$), while on the interval $(\tau, \vartheta]$, the measurements $y(t)$ are unknown, so that $\zeta = \zeta_\vartheta(\cdot, \tau)$ has to be bounded only by $\mathcal{X}_y[\tau], \mathcal{Q}(t), \mathcal{R}(t), t \ge \tau$, to satisfy all the possible future realizations of $y(t)$. Then, on the interval $(\tau, \vartheta]$, functional (10.6) has to be maximized over all such $\zeta_\vartheta(\cdot, \tau)$ and minimized over $u \in \mathcal{P}(t)$ in the class of strategies described further.

One may now observe that the overall Problem OFC may be separated into two, namely, **Problem E** on *guaranteed estimation* (calculation of the information state $V(t, x) = -V_-(t, x)$), solved within the interval $[t_0, \tau]$ and **Problem C** of *feedback control* (finding strategy $\mathcal{U}(t, V(t, \cdot))$ as a functional of the information state), solved within the interval $(\tau, \vartheta]$. Note that $V(\tau, x)$ depends on the available measurement $y_\tau[\cdot, t_0] = y(t + \sigma), \sigma \in [t_0 - \tau, 0]$, and that its role is *similar* to *sufficient statistics* in stochastic control. Also note that the control strategy $\mathcal{U}(t, V(t, \cdot))$ has to be selected within a class of functionals that ensure existence and extendability of solutions to Eq. (10.1) with $u = \mathcal{U}(t, V(t, \cdot))$. Therefore of importance will also be *a third goal* which is to achieve a *proper mathematical formalization* for the obtained results, ensuring the *existence of solutions* to the basic Eq. (10.1) under strategies $\mathcal{U}(t, V(t, \cdot))$ obtained *in the space of information states*.

In the previous paragraph the information state $V(t, x)$ may be substituted by information set $\mathcal{X}_y[t] = \mathcal{X}_y(t, \cdot)$ due to the following: with set-valued hard bounds on $\{x^0, v, \xi\}$ being given, it turns out that

$$\mathcal{X}_y[t] = \{x : V(t, x) \leq 0\}, \quad (10.10)$$

therefore *the information set $\mathcal{X}[t]$ would be the level set for the information state $V(t, \cdot)$*. This property thus justifies the fact that the generalized state is selected, as indicated above, *either as a function* (the information state, which is the value function $V(t, \cdot)$ for an appropriate problem of Dynamic Programming) *or as a set* (the information set). The dynamics of the overall OFC problem may be thus described by either PDEs of the HJBI type (which give the dynamics of $V(t, \cdot)$ and its level sets) or, for example, through a related integral funnel equation—an evolution equation with set-valued solution which gives the dynamics of $\mathcal{X}_y[t]$ (see Chaps. 8, 9, and also [158]). We further rely on either of these two definitions. The fact is that to achieve effective solutions with natural *geometrical interpretations* it appears necessary to *combine both approaches*. This allows us to work out a unified vision of the general problem and open new routes to a detailed solution of specific Problem OFC, especially with hard bounds and working within computation in *finite-dimensional space*. Conditions for the latter to suffice, as given later, also allow us to *avoid* the common restriction that u, v should satisfy certain *matching conditions*. It is thus possible to solve the linear case *completely* and to apply *new computation methods to systems with set-valued solutions*, enabling effective computer animation of solution examples. Recall that *under feedback control the original linear system becomes nonlinear*. We now describe the solution schemes for Problems **E** and **C** the subproblems of the overall Problem OFC.

10.4 Guaranteed State Estimation Revisited: Information Sets and Information States

We consider this Problem **E** in two versions: **E** and $\mathbf{E_0}$, dealing with continuous measurements.

Problem E. Given are system (10.1) with measurement Eq. (10.3). Also given are starting position $\{t_0, \mathcal{X}^0\}$, available measurement $y[t]$, and realization of control $u = u[t]$, $t \in [t_0, \tau]$. One is to specify *the information set* $\mathcal{X}_y[\tau] = \mathcal{X}_y(\tau, \cdot)$ of all states $x(\tau)$, generated by system (10.1), (10.3), that are consistent with given realizations $u[t]$, $y[t]$ and on-line state constraints (10.4).

This definition of information set $\mathcal{X}_y[\tau]$ is similar to Definition 9.1.1, but the difference lies in the system Eq. (10.1), where also present is the realization $u[t]$ known within $[t_0, \tau)$.

Therefore, the generalized state of the system may be taken as $\{t, \mathcal{X}_y[t]\}$. An equivalent definition of such a state may be also served by the triple $\{\tau, y_\tau(\cdot, t_0), u_\tau(\cdot)\}$. Then the passage from $y_\tau(\cdot, t_0)$ to $\mathcal{X}_y[\tau]$ (now with given $u_\tau(\cdot)$, is achieved through *set-valued observers* described in Chap. 9, Sects. 9.2 and 9.6 (see also [137, 151, 158]).

To solve basic problems **E** and **C** one should start by describing the evolution of information sets $\mathcal{X}_y[\tau] = \mathcal{X}_y(\tau, \cdot)$ in time, which now is also influenced by the selected control u. Here we first explain this process through *differential inclusions*, along the lines of Sects. 8.2, 9.2 and 9.6. Denote

$$\mathcal{Y}(t) = \{x : y(t) - g(t, x) \in \mathcal{R}(t)\}, \quad \mathcal{Y}_\vartheta(\cdot, \tau) = \{\mathcal{Y}(t), \ t \in [\tau, \vartheta]\}.$$

We further assume $\mathcal{Y}(t)$ to be closed and convex-valued. Let $\mathcal{D}$ be a convex compact domain in $\mathbb{R}^n$ sufficiently large to include projections on $\mathbb{R}^n$ of all possible set-valued trajectory tubes of system (10.1), (10.4) and all possible starting sets and target sets that may occur under our considerations. Then set $\mathcal{Y}_\mathcal{D}(t) = \mathcal{Y}(t) \cap \mathcal{D}$ will be convex and compact.

Assumption 10.4.1. *Set-valued function $\mathcal{Y}_\mathcal{D}(t)$ satisfies the Lipschitz condition:*

$$h(\mathcal{Y}_\mathcal{D}(t), \mathcal{Y}_\mathcal{D}(t')) \leq \lambda_y |t'' - t'|. \tag{10.11}$$

Consider system (10.1), (10.4), with functions f_1, f_2, g continuously differentiable in $\{t, x\}$. Also assume given are the realizations of the control $u = u[t]$ and measurement $y = y[t]$, $t \in [t_0, \tau]$. Denote

$$\mathcal{F}(t, x, u[t]) = f_1(t, x, u[t]) + f_2(t, x, Q(t)), \quad f_2(t, x, Q) = \cup \{f_2(t, x, v) \mid v \in Q(t)\}.$$

Then the information set $\mathcal{X}_y[\tau]$ will be the cross-section (cut) at time τ of the solution tube (the "information tube") $\mathcal{X}_y[t] = \mathcal{X}_y(t, t_0, \mathcal{X}^0)$ for the differential inclusion (DI)

$$\dot{x} \in \mathcal{F}(t, x, u[t]), \quad x(t_0) \in \mathcal{X}^0, \tag{10.12}$$

under state constraint $x \in \mathcal{Y}[t]$, $t \in [t_0, \tau]$.[3] In the coming propositions we presume $Q(t) \in \text{conv}\mathbb{R}^q$, $\mathcal{R}(t) \in \text{conv}\mathbb{R}^r$.

According to Lemma 8.2.3 and Theorem 8.2.3 (see also [158], Sect. 10) the information tube $\mathcal{X}_y[\tau, \cdot] = \{\mathcal{X}_y[t], \ t \in [t_0, \tau]\}$ will be the solution tube to the differential inclusion (DI):

$$\dot{x} \in \bigcap \{\mathcal{F}(t, x, u[t]) + L(-y[t] + g(t, x) + \mathcal{R}(t)) \mid L \in \mathbb{R}^{n \times r}\}, \ x(t_0) \in \mathcal{X}^0, \ t \in [t_0, \tau],$$

---

[3]For $\mathcal{Y}(t)$ generated by given $y[t]$ we use notation $\mathcal{Y} = \mathcal{Y}[t]$ with square brackets.

10.4 Guaranteed State Estimation Revisited: Information Sets and Information...

where L may be either constant matrices of dimension $n \times r$ or continuously differentiable $n \times r$ matrix functions $L(t,x)$, $L(\cdot,\cdot) \in \mathbf{C}_1^{n \times r}[t_0,\tau]$ or $L(t)$, $L(\cdot) \in \mathbf{C}_1^{n \times r}[t_0,\tau]$.

Taking the DI

$$\dot{z} \in \mathcal{F}(t,z,u[t]) + L(t,z)(-y[t]+g(t,z)+\mathcal{R}(t)), \quad z(t_0) \in \mathcal{X}^0, \qquad (10.13)$$

denote its solution tube, emanating from $\mathcal{X}^0$, as $Z_u[t] = Z_u(t,t_0,\mathcal{X}^0,L(\cdot,\cdot))$.

Theorem 10.4.1. *Under Assumption 10.4.1 the next inclusion will be true:*

$$X_y[\tau] = X_y(\tau,t_0,\mathcal{X}^0) \subseteq \bigcap \{Z_u(t,t_0,\mathcal{X}^0,L) \mid L = L(\cdot,\cdot) \in \mathbf{C}_1^{n\times r}[t_0,\tau]\}$$

$$\subseteq \bigcap \{Z_u(t,t_0,\mathcal{X}^0,L) \mid L = L(\cdot) \in \mathbf{C}_1^{n\times r}[t_0,\tau]\}. \qquad (10.14)$$

Corollary 10.4.1. *For a linear system with*

$$\dot{x} \in A(t)x + B(t)u[t] + C(t)Q(t), \quad y[t] \in G(t)x + \mathcal{R}(t), \quad x(t_0) \in \mathcal{X}^0,$$

and with convex, compact $\mathcal{X}^0$, $Q(t)$, $\mathcal{R}(t)$ *the DI (10.13) turns into*

$$\dot{z} \in A(t)x + B(t)u[t] + C(t)Q(t) + L(t)(-y[t]+G(t)z+\mathcal{R}(t)), \quad z^0 = z(t_0) \in \mathcal{X}^0,$$
$$(10.15)$$

and inclusion (10.14) into an equality

$$X_y[\tau] = X_y(\tau,t_0,\mathcal{X}^0) = \bigcap \{Z_u(t,t_0,\mathcal{X}^0,L) \mid L = L(\cdot) \in \mathbf{C}_1^{n\times r}[t_0,\tau]\}.$$
$$(10.16)$$

The last formula will be used below to obtain more detailed solutions under system linearity. DIs (10.13) and (10.15), taken under given $u_\tau(\cdot)$, are actually the related *set-valued observers*.

Assumption 10.4.1 implies that measurement $y(t)$ is continuous. This requirement may be relaxed to *piecewise continuous measurements* $y(t) = y[t]$ by applying another set-valued observer generated due to the *relaxed set-valued funnel equation* of Sect. 8.1.5. For solution tube $X_y[\tau,\cdot]$ of system (10.12), under $x \in \mathcal{Y}[t]$ the set-valued function $\mathcal{Y}(t)$ may then be taken as *Hausdorff upper semi-continuous*. With $u[t]$, $y[t]$ known, the tube $X_y[t]$ then satisfies the evolution equation of the funnel type, with $X[t_0] = \mathcal{X}^0 \cap \mathcal{Y}[t_0]$:

$$\lim_{\sigma\to 0+0} \sigma^{-1}h_+\left(X[t+\sigma],\bigcup\{x+\sigma\mathcal{F}(t,x,u[t]) \mid x \in X[t]\cap\mathcal{Y}[t+0]\}\right) = 0.$$
$$(10.17)$$

Passing to the Hamiltonian approach, the dynamics of information sets $X_y[\tau]$ may be now calculated by solving the alternative problem $\mathbf{E_0}$, formulated *in terms of value functions*.

Problem E_0. Given are the starting position $\{t_0, X^0\}$, the available measurements $y_\tau[\cdot, t_0]$, and the control realization $u[t]$, $t \in [t_0, \tau)$. One is to specify

$$-V(\tau, x) = \max_v \{-d^2(x[t_0], X^0) - \int_{t_0}^{\tau} d^2(y[t] - g(t, x), \mathcal{R}(t))dt \mid x[\tau] = x \in \mathbb{R}^n, \quad (10.18)$$
$$v(t) \in Q(t), \, t \in [t_0, \tau]\},$$

due to system (10.1).

Here $x[t] = x(t, \tau, x)$, $t \leq \tau$ stands for the backward trajectory, emanating from position $\{\tau, x\}$. Function $V(\tau, x) = V(\tau, x \mid V(t_0, \cdot), y_\tau[\cdot, t_0], u_\tau[\cdot, t_0])$ turns out to be precisely the *information state* of system (10.1), (10.3) introduced in the above.

Remark 10.4.1. If one presents functional (10.6) as the sum of two parts: the first defined for $[t_0, \tau]$, and the second, defined for $(\tau, \vartheta]$, then (10.18) will coincide with the first part, reflecting the problem of state estimation, namely, the calculation of the on-line information state.

Using function $V(\tau, \cdot)$ is convenient for describing *the evolution of information states* in differential form, since $V(\tau, \cdot)$ may be represented in more conventional terms of partial differential equations. Note that in (10.7) we actually have an equality $X[\tau] = \{x : V(\tau, x) \equiv 0\}$. However, we prefer to write an inequality in (10.7), since this will also allow us to describe the neighborhoods of $X_y[\tau]$ as

$$X_y^\varepsilon = \{x : V(\tau, x) \leq \varepsilon\}, \quad (10.19)$$

without introducing new notation.

Function $V(t, x)$ satisfies an *HJB-type equation* for time $t \in [t_0, \tau]$:

$$\frac{\partial V}{\partial t} = -\max_v \left\{ \left\langle \frac{\partial V}{\partial x}, f_1(t, x, u^*[t]) + f_2(t, x, v) \right\rangle \right.$$
$$\left. -d^2(y[t] - g(t, x), \mathcal{R}(t)) \, \middle| \, v(t) \in Q(t) \right\}, \quad (10.20)$$

under boundary condition

$$V(t_0, x) = d^2(x, X^0). \quad (10.21)$$

$V(t, x)$ may be non-smooth, hence the solution to Eq. (10.20) under condition (10.21) may not be classical, but must be redefined in a generalized sense, for example, as a "viscosity" or "minmax" solution [48, 50, 80, 247].

The existence of a generalized solution to Eq. (10.20), (10.21) is necessary and sufficient for the existence of solution to Problem E_0. In this case the mapping $V(t, \cdot) = V(t, \cdot \mid V(t_0, \cdot))$ satisfies the *evolution equation* of type

$$V_t(t, \cdot) = \Phi(t, V_x(t, \cdot), \cdot \mid u[t], y[t]), \quad (10.22)$$

whose right-hand side is identical to (10.20). In other words, (10.22) is an "equation of motion" for the information state $V(t, \cdot)$. The evolution of $V(t, \cdot)$'s level set $X[t]$ (see (10.18)) is therefore *implicitly* also described by this equation.

A fairly broad case is when $V(\tau, x)$ are continuous in all variables, with nonempty compact zero level-sets $X[\tau] = \{x : V(\tau, x) \leq 0\} \neq \emptyset$. These are necessarily those that solve Problem OFC for (10.1), (10.3) being *linear*, with continuous coefficients and *convex constraints*. The indicated class of functions $V(\tau, \cdot)$ will be denoted as $\mathcal{K}_V$ and treated as a metric space with metric $h_V(V'(\tau, \cdot), V''(\tau, \cdot)) = h(X'[\tau], X''[\tau])$, where $X[\tau] = \{x : V(\tau, x) \leq 0\}$ and $h(X', X'')$ is the Hausdorff metric. The variety of closed compact sets $X[\tau]$ may be also considered as a metric space $\mathcal{K}_X$, equivalent to $\mathcal{K}_V$, with metric $h(X'[\tau], X''[\tau])$. A discussion of possible related metrics is given in monograph [98, Chap. 4, Appendix C.5].

Having specified the spaces of information states and sets (also referred to as the respective *information space*), we may proceed with the problem of feedback control itself. The difficulty is that we will now have *a problem in the information space*, with states taken as either functions $V(t, \cdot)$ or sets $X(t, \cdot)$. We therefore return to Eq. (10.22), observing, that for $t > \tau$ the realization $y(t)$ is unknown in advance, arriving on-line, while for $t \geq \tau$ the control, taken, for example in the form of $u = u(t, V(t, \cdot))$, is to be selected due to this equation treated in the space of states $V(t, \cdot)$. Further on, we indicate, when possible, the routes for overcoming these scary difficulties through simpler techniques.

10.5 Feedback Control in the Information Space

As indicated in the above, the information space may be defined in either the space of information sets or the space of value functions—the information states. We first begin with the space of information sets.

10.5.1 Control of Information Tubes

The feedback control problem in the space of information sets reads as follows.

Problem C. Given starting position $\{\tau, X_y[\tau]\}$, target set $\mathcal{M}$ and number $\gamma > 0$, specify a control strategy $u = u(t, X)$, $t \in [t_0, \tau]$ that steers system (10.20), (10.21) under $u = u(t, X)$ into a γ-neighborhood $\mathcal{M}_\gamma$ of given terminal target set $\mathcal{M}$, whatever be the function

$$y_\vartheta(\cdot, \tau) \in \mathcal{Y}_\vartheta(\cdot, \tau) = \{y(t) : \zeta_\vartheta(\cdot, \tau) \in \mathcal{Z}_\vartheta(\cdot, \tau)\},$$

and the related uncertain items $\zeta_\vartheta(\cdot, \tau)$ that generated $y_\vartheta(\cdot, \tau)$.

Here $\mathcal{Z}_\vartheta(\cdot, \tau) = \bigcup\{\zeta_\vartheta(\cdot, \tau) : x \in X_y[\tau], v(t) \in Q(t), \xi(t) \in \mathcal{R}(t), t \in [\tau, \vartheta]\}$.

Problem C$_{opt}$. Among values $\gamma > 0$, for which Problem **C** is solvable, find smallest γ_0.[4]

With $\gamma > 0$ given, Problem **C** is solvable if and only if $X_y[\tau]$ belongs to $W[\tau] \neq \emptyset$, that is the set of subsets $X_\tau \subset \mathbb{R}^n$ for which this problem with starting position $\{\tau, X_y[\tau] = X_\tau\}$, does have a solution. Further we indicate that $W[\tau]$ is understood to be *weakly invariant* relative to target set $\mathcal{M}_y$.

Similarly, but for the information space of functions, we have the next problem.[5]

Problem C$_v$. Find value function

$$\mathcal{V}(\tau, V(\tau, \cdot)) = \min_u \max_y \max_{x,v} \Big\{ -V(\tau, x) - \int_\tau^\vartheta d^2(y(t) - g(t, x(t)), \mathcal{R}(t)) dt +$$

$$+ d^2(x[\vartheta], \mathcal{M}_y) \Big| u \in \mathcal{U}, y(\cdot) \in \mathcal{Y}_\vartheta(\cdot, \tau), v(\cdot) \in Q_\vartheta(\cdot, \tau), x \in X[\tau] \Big\}, \qquad (10.23)$$

with $X[\tau] = \{x : V(\tau, x) \leq 0, \ V(\tau, \cdot) \in \mathcal{K}_V\}$.

Here the minimum is taken in the class $\mathcal{U}$ of feedback strategies $u = u(t, V)$, described below. The maximum is taken over all $y(\cdot) \in \mathcal{Y}_\vartheta(\cdot, \tau) = \{\mathcal{Y}(\vartheta + \sigma) \mid \sigma \in [\tau - \vartheta, 0]\}$ described above, and further, over all $x \in X[\tau], Q_\vartheta(\cdot, \tau) = \{v(t) \in Q(t), t \in [\tau, \vartheta]\}$.

Functional (10.23) represents the second part of (10.6), which defines the overall control process. With $\tau = \vartheta$ it must ensure the boundary condition

$$\mathcal{V}(\vartheta, V(\vartheta, \cdot)) = \max_x \{d^2(x, \mathcal{M}_y) \mid V(\vartheta, x) \leq 0\}. \qquad (10.24)$$

Number $\gamma^2 = \mathcal{V}(\vartheta, V(\vartheta, \cdot))$ is the square of "size" γ of the neighborhood $\mathcal{M}_y = \mathcal{M} + \gamma \mathcal{B}(0)$ which at time ϑ entirely includes $X[\vartheta]$, so that $X[\vartheta] \subseteq \mathcal{M}_y$. Functional $\mathcal{V}(\tau, V(\tau, \cdot))$ is defined on the product space $\mathbb{R}_+ \times \mathcal{K}_V$, where $\mathbb{R}_+ = [t_0, \infty)$. Note that in (10.23) the operations max$_y$, max$_{x,v}$ are interchangeable. After this interchange one may observe that for every $y(\cdot) \in \mathcal{Y}_\vartheta(\cdot, \tau)$ the max$_{x,v}$ results in $d^2(y(t) - g(t, x(t)), \mathcal{R}(t)) \equiv 0$. This shows that in (10.23) operations min$_u$ max$_y$ max$_{x,v}$ reduce to just min$_u$ max$_{x,v}$.

Value function $\mathcal{V}(\tau, V(\tau, \cdot))$ satisfies the generalized "backward" equation of the HJBI type, which is well known in dynamic games with complete information[6] and may be formally written as

$$\min_u \max_v \Big\{ \frac{d\mathcal{V}(\tau, V(\tau, \cdot))}{d\tau} \Big| u \in \mathcal{P}(\tau), v \in Q(\tau) \Big\} = 0 \qquad (10.25)$$

---

[4] Note that if for $\gamma = 0$ Problem **C** is not solvable, then $\gamma_0 > 0$.

[5] Here and in (10.6) we use equivalent notations $\mathcal{V}(\tau, V(\tau, \cdot))$ and $\mathcal{V}(\tau, X[\tau])$, since $X[\tau]$ is the zero-level set for $V(\tau, \cdot)$.

[6] See [18] and [123] where this is done in finite-dimensional space.

10.5 Feedback Control in the Information Space

where $d\mathcal{V}(\tau, V(\tau, \cdot))/d\tau$ is the total derivative of functional $\mathcal{V}(t, V(t, \cdot))$ due to evolution equation (10.20), (10.21), or (10.22), which is the same. Equation (10.22) may have a smooth solution, when the total derivative exists in the strong or weak sense and the equation holds everywhere. Otherwise it has to be dealt with in terms of generalized solutions, as mentioned above.

The solution strategy should then be formally determined through minimization over u in (10.25), as $u^0 = u^0(t, V(t, \cdot)) \subseteq \mathcal{P}(t)$. Substituting this into (10.20), instead of $u[t]$ with $y(t)$ instead of $y[t]$, we have

$$V_t + \max_v \left\{ \left(V_x, f_1(t, x, u^0(t, V(t, \cdot))) + f_2(t, x, v) \right) \big| v \in Q(t) \right\} -$$

$$- d^2(y(t) - g(t, x), \mathcal{R}(t)) = 0 \quad (10.26)$$

The last relation may be treated as an evolution equation in the metric space $\mathcal{K}_V$ of functions $\{V(t, \cdot)\}$. The "trajectories" $V[t] = V(t, \cdot)$, issued at time t_0 from $V(t_0, \cdot) = d^2(x, \mathcal{X}^0)$, may then be interpreted as "constructive motions" $V(t, \cdot; u^0(t, \cdot))$, that arrive as a result of limit transition from infinite-dimensional analogies of Euler broken lines, constructed under all possible partitions of the interval $[t_0, \vartheta]$, by selecting piecewise-constant realizations $u[t] \in \mathcal{P}(t)$ (see [123], p. 11). Another option could be to treat $\{V(t, \cdot)\}$ as a generalized (viscosity) solution in infinite-dimensional space $\mathcal{K}_V$ (see [98]).

Being based on dynamic programming, the overall Problem OFC allows to consider linked solutions of problems **E** and **C**, where **E** is finite-dimensional, while **C** is infinite-dimensional. In the linear case, as we shall see, these solutions are even independent and may be separated. Such a scheme produces a good insight into the problem and its complete solution. But it is difficult to calculate. However, once we need to solve concrete problems, our interest is in computation schemes feasible for applications and such that would allow procedures that *do not go beyond finite-dimensional dynamics* with computational burden not greatly above the one for problems with complete information.

10.5.2 The Solution Scheme for Problem C

As indicated above, the solution to Problem OFC is given through two types of HJB, HJBI equations (see (10.20), (10.25))—one for state estimation and one for feedback control. However, what *we actually need* here are not the value functions, but only their *level sets*. Hence, instead of directly solving the infinite-dimensional equation (10.25), we will use the notion of *invariant sets* (or backward reachability sets) relative to target set $\mathcal{M}_y$. Continuing the previous subsection we deal with the space of *information sets*. But to calculate them we shall use relations derived in the space of *information states*. We follow the scheme below.

Formally, the set $\mathcal{W}_V[\tau] = \{V(\tau, \cdot) : \mathcal{V}(\tau, V(\tau, \cdot)) \leq 0\}$ that we need is the (weakly) invariant set defined as follows.

Definition 10.5.1. (a) Set $\mathcal{W}_V[\tau]$, *weakly invariant* relative to target set $\mathcal{M}_y$, is the union *of value functions* $V(\tau, \cdot)$ (in space $\mathcal{K}_V$) for each of which Problem C is solvable from initial position $\{\tau, \mathcal{X}_y[\tau]\}$, with $\mathcal{X}_y[\tau] = \{x : V(\tau, x) \leq 0\}$.

Multi-valued function $\mathcal{W}_V[t]$, $t \in [\tau, \vartheta]$ is the solvability tube in $\mathcal{K}_V$ (an infinite-dimensional analogy of the "Krasovski bridge") for Problem C.

(b) Set $\mathcal{W}[\tau]$ *weakly invariant* relative to target set $\mathcal{M}_y$, is the union (in space $\mathcal{K}_X$) of compact sets $\mathcal{X} = \mathcal{X}[\tau]$ for each of which Problem C is solvable from initial position $\{\tau, \mathcal{X}[\tau]\}$.

Set-valued function $\mathcal{W}[t]$, $t \in [\tau, \vartheta]$ is the solvability tube in $\mathcal{K}_X$ (another infinite-dimensional analogy of "Krasovski bridge," equivalent to case (a)).

Combining alternative descriptions in terms of value functions with those given through set-valued calculus, we now pass from space $\mathcal{K}_V$ to $\mathcal{K}_X$, applying to Problem C an infinite-dimensional modification of the rule of *extremal aiming* towards weakly invariant sets $\mathcal{W}[\tau]$ (see finite-dimensional version in [123]). With on-line position $\{\tau, \mathcal{X}[\tau]\}$ given, we now have to calculate the Hausdorff semi-distance $H_+(\mathcal{X}_y[\tau], \mathcal{W}[\tau])$ between the actual state $\mathcal{X}_y[\tau]$ and the solvability set $\mathcal{W}[\tau]$, taken *in the metric of space* $\mathcal{K}_X$, and arriving at

$$\mathcal{V}_H(\tau, \mathcal{X}_y[\tau]) = H_+(\mathcal{X}_y[\tau], \mathcal{W}[\tau]),$$

$$H_+(\mathcal{X}[\tau], \mathcal{W}[\tau]) = \inf_\varepsilon \{\inf\{\varepsilon : \mathcal{X}_\tau + \varepsilon \mathcal{B}(0) \supset \mathcal{X}[\tau]\} \mid \mathcal{X}_\tau \in \mathcal{W}[\tau]\}$$

where the first (external) infimum is to be taken over all $\mathcal{X}_\tau \in \mathcal{W}[\tau]$. (This Hausdorff semi-distance is similar to $h_+(\mathcal{X}', \mathcal{X}'')$ in $\mathbb{R}^n$ which was defined earlier.)

Then the solution to Problem C would be

$$U^0(\tau, \mathcal{X}_y[\tau]) = \{u : d(\exp(-2\lambda\tau)\mathcal{V}_H(\tau, \mathcal{X}_y[\tau]))/d\tau \mid_u \leq 0 \mid u \in \mathcal{P}(\tau)\}. \tag{10.27}$$

Namely the union of controls that allow a non-positive derivative in (10.27) would be the desired solution. Here λ is the Lipschitz constant for (10.1), uniform in t, x, u, v.

This is an infinite-dimensional problem, which in the general case in not simple. However, the solution turns out to be simpler than may be expected in that it may suffice to consider instead of $\mathcal{V}_H(\tau, \mathcal{X}_y[\tau])$ the value

$$\mathcal{V}_h(\tau, \mathcal{X}_y[\tau]) = h_+(\mathcal{X}_y[\tau], \mathbf{W}[\tau]) = \min\{\varepsilon > 0 : \mathcal{X}_y[\tau] \subset \mathbf{W}[\tau] + \varepsilon \mathcal{B}(0)\},$$

where

$$\mathbf{W}[\tau] = \bigcup \left\{ \bigcup \{x : x \in \mathcal{X}_\tau\} \,\middle|\, \mathcal{X}_\tau \in \mathcal{W}[\tau] \right\}$$

is a set in *finite-dimensional space* $\mathbb{R}^n$.

10.5 Feedback Control in the Information Space

The next proposition is true with set $\mathbf{W}[\tau]$ being convex.

Theorem 10.5.1. *With set* $\mathbf{W}[\tau] \in \text{conv}\mathbb{R}^n$ *the Hausdorff semi-distances given by* $H_+(X_y[\tau], \mathcal{W}[\tau])$ *and* $h_+(X_y[\tau], \mathbf{W}[\tau])$ *are equal.*

The last relation is useful in the linear case with convex compact constraints where both $X_y[\tau], \mathbf{W}[\tau]$ turn out to be convex and compact.

Then, since $\mathbf{W}[\tau]$ coincides with the backward reachability set for system (10.1), (10.4) *without additional state constraint* (10.3), this set $\mathbf{W}[\tau]$ will be the level set for the solution of a *finite-dimensional* HJBI equation.

Relation (10.27) now reduces to

$$U^0(\tau, X_y[\tau]) = \{u : d(\exp(-2\lambda\tau) h_+^2(X[\tau], \mathbf{W}[\tau]))/d\tau \mid_u \le 0 \mid u \in \mathcal{P}(\tau)\}. \tag{10.28}$$

Theorem 10.5.2. *Under conditions of Theorem 10.5.1 the control solution* $U^0(\tau, X[\tau])$ *to Problem C is given by relation (10.28).*

We thus have two types of control solutions: $u^0(t, V(t, \cdot))$ and $U(t, X_y[t])$ which have to be realized through related equations or differential inclusions similar to (10.26), according to the specific definitions of their solutions for each control type.

10.5.3 From Problem C to Problem $\mathbf{C}_{opt}$

The previous results described solution schemes that ensure inclusion $X[\vartheta] \subseteq \mathcal{M}_\gamma$, whatever be the uncertain items $\zeta(\cdot)$. Here $\mathcal{M}_\gamma$ is the *guaranteed neighborhood* of the target set. The final problem is now to minimize this neighborhood $\mathcal{M}_\gamma$ under uncertainty and output feedback. Then we need to solve the following subproblem.

Solving Problem C_{opt} of Minimizing the Guaranteed Solution

Consider minimax problem (10.6) with $\tau = t_0$ and starting position $\{t_0, X^0\}$, where $\mathcal{M}$ is substituted for $\mathcal{M}_\gamma$ and $\mathcal{W}[\tau], \mathcal{V}(\tau, \cdot)$ for $\mathcal{W}_\gamma[\tau], \mathcal{V}_\gamma(\tau, \cdot)$. Then the solution to Problem C will be to find the smallest $\gamma = \gamma^0 \ge 0$ among those that ensure $X^0 \subseteq \mathcal{W}_\gamma[t_0] = \mathbf{W}_\gamma[t_0]$.

Here $\mathcal{V}_\gamma(t_0, X^0)$ will depend on the boundary condition (10.24), so that $\mathcal{W}_\gamma[t_0] = \{W : \mathcal{V}_\gamma(t_0, W) \le 0\} = \mathbf{W}_\gamma[t_0]$. Introducing

$$\rho(\gamma, t_0) = H_\gamma(X^0, \mathcal{W}_\gamma[t_0]) = h_\gamma(X^0, \mathbf{W}_\gamma[t_0]),$$

we reduce the problem of optimization to calculating

$$\min_\gamma \rho(\gamma, t_0) = \rho(\gamma^0, t_0) = \rho_0(t_0).$$

The final step is to solve Problem C for $\gamma = \gamma^0$ by finding the closed-loop control strategy $U^0(t, X_y)$ that for the optimal neighborhood $\mathcal{M}^0 = \mathcal{M}_{\gamma^0}$ of the target set $\mathcal{M}$, ensures the inclusion $X[\vartheta] \subseteq \mathcal{M}^0$, whatever be the unknown but bounded uncertain items $\zeta(\cdot)$.

Remark 10.5.1. The uncertain items may deviate from their maximizing "worst case") values. This allows us to improve the final result by recalculating (rebooting) the solution and decreasing γ^0 by diminishing sets $\mathcal{M}_\gamma$, $\mathcal{W}_\gamma[t]$ beyond their initial sizes.

Remark 10.5.2. Problem $\mathbf{C}_{opt}$ may be also solved by integrating the HJB–HJBI equations (10.20)–(10.26), but this direct approach appears to impose a far greater computational burden than the suggested schemes. It also requires appropriate interpretations of solution classes for these equations.

The ultimate aim of treating Problem OFC is to solve it completely through adequate computation. We shall now indicate some relaxation schemes for finding $X_y[\tau]$ and $\mathbf{W}[\tau]$ through related computation methods, restricting ourselves to linear systems. The nonlinear case may be approached through the *comparison principle* of Chap. 5, Sects. 5.1–5.3.

10.6 More Detailed Solution: Linear Systems

We shall now discuss how far can the OFC problem be solved for *linear* systems of type

$$\dot{x} = A(t)x + B(t)u + C(t)v, \tag{10.29}$$

$$y(t) = G(t)x + \xi, \tag{10.30}$$

under continuous coefficients and *convex compact* constraints $u \in \mathcal{P}(t)$ and (10.4).

10.6.1 The "Linear-Convex" Solution

As indicated in the previous section, we must find the feedback control $u^0(\tau, X_y[\tau])$, having calculated $X_y[\tau]$ and $\mathbf{W}[\tau]$, where in the linear case both sets are convex and compact. These sets may be expressed explicitly, through duality methods of convex analysis, as the level sets of related value functions similarly to formulas of Chap. 2, Sect. 2.4 and Chap. 3. And it would also allow us to calculate derivatives of such value functions. Here, however, we indicate another scheme for such calculations by using related evolution equations of the funnel type introduced in

10.6 More Detailed Solution: Linear Systems

Chap. 2, Sect. 2.3.1 and Chap. 8, Sect. 8.1.3. This move brings us, using notation $h_+(\mathcal{X}_\tau, \mathbf{W}[\tau]) = \min\{\varepsilon > 0 : \mathcal{X}_\tau \in \mathbf{W}[\tau] + \varepsilon \mathcal{B}(0)\}$, to problem

$$\mathcal{V}_h(\tau, \mathcal{X}_y[\tau]) = \max\{\rho(l \mid G(\vartheta, \tau)\mathcal{X}_y[\tau]) - \rho(l \mid G(\vartheta, \tau)\mathbf{W}[\tau]) \mid <l, l > \le 1\}. \tag{10.31}$$

Then, following (10.28) for $\mathcal{V}_H = \mathcal{V}_h$, we find the overall control solution u^0.

Hence, in order to calculate u^0 we first need to calculate the total derivative of $\mathcal{V}_h(\tau, \mathcal{X}_y[\tau])$ and ensure that it is non-positive under control u^0. Related evolution equations for $\mathcal{X}_y[t]$, $\mathbf{W}[t]$ may be written as

$$\lim_{\sigma \to 0+0} \sigma^{-1} h_+(\mathcal{X}_y[t+\sigma], (I+\sigma A(t))(\mathcal{X}_y[t] \cap \mathcal{Y}(t)) + \sigma B(t)u(t) + \sigma C(t)Q(t)) = 0, \tag{10.32}$$

$$\lim_{\sigma \to 0+0} \sigma^{-1} h(\mathbf{W}[t-\sigma] + \sigma C(t)Q(t), (I-\sigma A(t))\mathbf{W}[t] + \sigma B(t)\mathcal{P}(t)) = 0, \tag{10.33}$$

with boundary conditions $\mathcal{X}_y[t_0] = \mathcal{X}^0$, $\mathbf{W}[\vartheta] = \mathcal{M}_y$. For the first equation we use the unique inclusion maximal solution, for the second we use the ordinary solution, which is always unique.

Denote the unique maximizer in (10.31) as l^0. We have

$$d\mathcal{V}_h(t, \mathcal{X}_y[t])/dt = d\rho(l^0 \mid G(\vartheta, t)\mathcal{X}_y[t])/dt - d\rho(l^0 \mid G(\vartheta, t)\mathbf{W}[t])/dt. \tag{10.34}$$

Calculating along the lines of Sect. 8.3, we get

$$d\rho(l^0 \mid G(\vartheta, t)\mathcal{X}_y[t])/dt \le \langle l^0, G(\vartheta, t)B(t)u \rangle + \rho(l^0 \mid G(\vartheta, t)C(t)Q(t)),$$

$$d\rho(l^0 \mid G(\vartheta, t)\mathbf{W}[t])/dt = -\rho(-l^0 \mid G(\vartheta, t)B(t)\mathcal{P}(t)) + \rho(l^0 \mid G(\vartheta, t)C(t)Q(t)).$$

Substituting these derivatives into (10.34) we come to the next statement.

Theorem 10.6.1. *The total derivative*

$$d\mathcal{V}_h(t, \mathcal{X}_y[t])/dt \le \langle l^0, G(\vartheta, t)B(t)u \rangle + \rho(-l^0 \mid G(\vartheta, t)B(t)\mathcal{P}(t)). \tag{10.35}$$

Finally, selecting

$$u^0(t, \mathcal{X}_y) = \arg\max\{\langle -l^0, G(\vartheta, t)B(t)u \rangle \mid u \in \mathcal{P}(t)\}, \tag{10.36}$$

and integrating $d\mathcal{V}_h(t, \mathcal{X}_y[t])/dt \mid_{u=u^0}$ from t_0 to ϑ, we ensure the next condition.

Lemma 10.6.1. *The control strategy (10.36) ensures the inequality*

$$\mathcal{V}_h(\vartheta, \mathcal{X}[\vartheta]) \le \mathcal{V}_h(t_0, \mathcal{X}_y[t_0]) = \gamma,$$

whatever are the uncertain items $\{x^0, v(\cdot), \xi(\cdot)\}$.

The next question is how to calculate the solutions numerically. Here we first recall that $\mathcal{X}_y[\tau]$, $\mathbf{W}[\tau]$ may be indicated in explicit form. Then, using this knowledge, we suggest a *computational approach* based on approximating convex sets by parametrized families of ellipsoids that evolve in time, following the approximated sets.

10.6.2 The Computable Solution: Ellipsoidal Approximations

Set $\mathcal{X}_y[\tau]$ may be effectively calculated numerically through ellipsoidal techniques. Using notation of Chap. 3, where an ellipsoid $\mathcal{E}(p, P) = \{u : \langle u-p, P^{-1}(u-p)\rangle \leq 1\}$, $P = P' > 0$, we assume that the sets

$$\mathcal{X}^0 = \mathcal{E}(x^0, X^0), \quad \mathcal{P}(t) = \mathcal{E}(p(t), P(t)), \quad \mathcal{Q}(t) = \mathcal{E}(q(t), Q(t)),$$
$$\mathcal{R}(t) = \mathcal{E}(0, R(t)), \quad \mathcal{M} = \mathcal{E}(m, M)$$

are *ellipsoidal*. Since every convex compact set may be presented as an intersection of ellipsoids, the solutions and methods for nonellipsoidal symmetric constraints, like parallelotopes and zonotopes, could be designed as a modification and extension of those for ellipsoidal constraints (see Chap. 5, Sects. 5.4 and 5.5).

Then the information set $\mathcal{X}_y[\tau]$ has an external ellipsoidal approximation $\mathcal{X}_y[\tau] \subseteq \mathcal{E}_L[\tau] = \mathcal{E}(x_L(\tau), X_L(\tau))$, where, throughout $[t_0, \vartheta]$, with $x(t_0) \in \mathcal{E}_L[t_0] = \mathcal{E}(x^0, X^0)$, we have

$$\dot{x} = A(t)x + B(t)u[t] + C(t)v(t), \quad y[t] = G(t)x + \xi(t),$$

$$\dot{x}_L = A(t)x_L + B(t)u[t] + C(t)q(t) + L(t)(y[t] - G(t)x_L), \quad x_L(t_0) = x^0, \tag{10.37}$$

$$\dot{X}_L = (A(t) - L(t)G(t))X_L + X_L(A(t) - L(t)G(t))' + (\pi(t) + \chi(t))X_L +$$
$$+ (\chi(t))^{-1}Q(t) + \pi^{-1}(t)L(t)G(t)R(t)G'(t), \quad X_L(t_0) = X^0. \tag{10.38}$$

The *parametrizing functions* $\omega(t) = \{\pi(t) > 0, \chi(t) > 0, L(t)\}$ may be taken piecewise continuous.

Theorem 10.6.2. *The next relation is true*

$$\mathcal{X}_y[\tau] = \bigcap\{\mathcal{E}_L[\tau] \mid \omega(\cdot)\}.$$

Among these parametrizers, one may select some *optimal* or some *tight* ones when one could have, for some direction $l \in \mathbb{R}^n$, a related triplet $\omega^0(\cdot)$, that ensures the equality $\rho(l \mid \mathcal{X}_y[\tau]) = \rho(l \mid \{\mathcal{E}_L[\tau] \mid \omega^0\})$.

For $\mathbf{W}[\tau]$ we may also apply ellipsoidal approximations of the types indicated in Chap. 5 (see also paper [149]), or use discretized versions for reachability under uncertainty (see [133]). Here available are internal and external ellipsoidal bounds

10.6 More Detailed Solution: Linear Systems

$$\mathcal{E}(w^*(\tau), W_-(\tau)) \subseteq \mathbf{W}[\tau] \subseteq \mathcal{E}(w^*(\tau), W_+(\tau))$$

that depend on internal and external parametrizing functions $\pi(t)$, $S(t)$ and tuning parameter $r(t) > 0$. The equations for $w^*(t)$, $W_+(t)$, $W_-(t)$ are of the type ($\mathbf{B}(t) = (B(t)P(t)B'(t))^{1/2}$, $\mathbf{C}(t) = (C(t)Q(t)C'(t))^{1/2}$):

$$\dot{w}^* = A(t)w^* + B(t)p(t) + C(t)q(t), \quad w^*(\vartheta) = m, \tag{10.39}$$

$$\dot{W}_+ = A(t)W_+ + W_+ A'(t) - \pi(t)W_+ - \pi^{-1}(t)\mathbf{B}(t)$$
$$+ r(t)(W_+ S(t)\mathbf{C}(t) + \mathbf{C}'(t)S'(t)W_+), \quad W_+(\vartheta) = M. \tag{10.40}$$

$$\dot{W}_- = A(t)W_- + W_- A'(t) + \pi(t)W_- + \pi^{-1}(t)\mathbf{C}(t)$$
$$- r(t)(W_- S(t)\mathbf{B}(t) + \mathbf{B}'(t)S'(t)W_-), \quad W_-(\vartheta) = M. \tag{10.41}$$

With $\pi(t) > 0, r(t) > 0$ and orthogonal matrices $S(t)S'(t) = I$, we have

$$\mathcal{E}(w^*(t), W_+(t)) = W_+^0[t] \supseteq \mathbf{W}[t] \supseteq W_-^0[t] = \mathcal{E}(w^*(t), W_-(t)),$$

where

$$W_+^0[t] = \{w : \langle w - w^*(t), W_+^{-1}(t)(w - w^*(t))\rangle \leq 1\},$$

$$W_-^0[t] = \{w : \langle w - w^*(t), W_-^{-1}(t)(w - w^*(t))\rangle \leq 1\},$$

whatever be the functions $\pi(t)$ and the orthogonal matrices $S(t)$. Following [180], we have the next proposition.

Theorem 10.6.3. *The next property is true*

$$\mathbf{W}[\tau] = \bigcup\{\mathcal{E}(w^*(\tau), W_-(\tau)) \mid \pi(\cdot), S(\cdot)\}.$$

The parameter $r(t)$ in (10.35) may be taken as $r(t) = \langle l_e, W_-[t]l_e\rangle^{-1/2}$.

Suppose l_e is a unique support vector for $X_y[t], \mathbf{W}[t]$. Then, along the lines of Chap. 3, Sects. 3.2 and 3.7, one may describe the values of $\pi_e(t), \chi_e(t), L_e(t)$ for (10.33) and $\pi(t), S(t), r(t)$ for (10.35) that ensure the equalities

$$\pi_e(t) = \langle l_e, \mathbf{C}(t)l_e\rangle^{1/2}\langle l_e, X_L[t]l_e\rangle^{-1/2}, \quad \chi(t) = \langle l_e, L'_e R(t)L_e l_e\rangle^{1/2}\langle l_e, X_L[t]l_e\rangle^{-1/2}, \tag{10.42}$$

$$\pi(t) = \langle l_e, \mathbf{C}(t)l_e\rangle^{1/2}\langle l_e, W_-[t]l_e\rangle^{-1/2},$$

$$\langle l_e, W_-(t)S(t)B(t)P^{1/2}(t)l_e\rangle^{1/2} = \langle l_e, \mathbf{B}(t)l_e\rangle^{1/2}\langle l_e, W_-[t]l_e\rangle^{1/2}. \tag{10.43}$$

We now pass to the computation scheme.

We return to problem (10.31), in view of relations

$$X_y[\tau] \subseteq \mathcal{E}_L[\tau] = \mathcal{E}(x_L(\tau), X_L(\tau)), \quad \mathbf{W}[\tau] \supseteq \mathcal{E}_{\pi S}^-[\tau] = \mathcal{E}_{\pi,S}(w^*(t), W_-(t)).$$

Substituting sets $X_y[\tau]$, $\mathbf{W}[\tau]$ in (10.31) by their *ellipsoidal approximations*, we take the *externals* for $X_y[\tau]$ and the *internals* for $\mathbf{W}[\tau]$. Then, instead of $\mathcal{V}_h(\tau, X_y[\tau])$ we get its approximation $V_{h\mathcal{E}}(\tau, \cdot)$, that also depends on the parametrizing functions for approximating ellipsoids. We may then "tune" the parametrizers, selecting them so as to minimize the error of approximation (which, with proper tuning, may even tend to zero). Denote $l_\mathcal{E}^0$ to be the unique optimizer for problem

$$\mathcal{V}_{h\mathcal{E}}(\tau, \mathcal{E}_L[\tau]) = \max\{\rho(l \mid G(\vartheta, \tau)\mathcal{E}_L[\tau]) - \rho(l \mid G(\vartheta, \tau)\mathcal{E}_{\pi S}^-[\tau]) \mid <l, l> \leq 1\}. \quad (10.44)$$

The ellipsoidal-valued states involved here are *finite-dimensional* elements, of type $\{x, X\}, \{w, W\}$, whose evolution is explicitly described by *ordinary differential equations*. The approximate control strategy U_L^0 may then finally be taken as the minimizer in $u^0 = U_l^0 = U(t, X)$ of the total derivative $d\mathcal{V}_{h\mathcal{E}}(t, \mathcal{E}_L[t])/dt|_u$.

The question is: will the approximate control strategy U_L^0 perform as successfully as the exact solution U^0? And what would be the suboptimal neighborhood $\mathcal{M}_L$ of set $\mathcal{M}$ reachable through strategy U_L^0 despite the uncertainties as compared to $\mathcal{M}^0 = \mathcal{M}_y, \gamma = \gamma^0$?

In order to reach the answers we shall calculate the derivatives of functions $\rho(l \mid G(\vartheta, t)\mathcal{E}_L(t))$, $\rho(l \mid G(\vartheta, t)\mathcal{E}_{\pi S}^-(t))$ involved in (10.38), using notation $l_e(t) = G'(\vartheta, t)l_\mathcal{E}^0$. Then, taking $l^0 = l_\mathcal{E}^0$, using relations (10.42), (10.43), and the Cauchy–Bunyakovsky inequality

$$\langle l, L'GX_+l \rangle \leq \langle G'Ll, X_+G'Ll \rangle^{1/2} \langle l, X_+l \rangle^{1/2},$$

after some calculations, we have

$$d\rho(l^0 \mid \mathcal{E}_L(t))/dt = -(\langle l^0, L_0 G(t) X_+(t) G'(t) L_0 l^0 \rangle^{1/2} + \langle l^0, L_0' G(t)(x - x^*) \rangle)$$
$$+ \langle l^0, L_0 \xi(t) \rangle - \langle l^0, L_0' R(t) L_0 l^0 \rangle^{1/2} + \langle l^0, C(t)v[t] \rangle + \langle l^0, B(t)u[t] \rangle. \quad (10.45)$$

Similarly, we get

$$d\rho(l^0 \mid \mathcal{E}_{\pi S}^-(t))/dt = \langle l^0, B(t)p(t) + C(t)q(t) \rangle + \langle l^0, \mathbf{C}(t)l^0 \rangle^{1/2} - \langle l^0, \mathbf{B}(t)l^0 \rangle. \quad (10.46)$$

Here x, ξ are those that generated the realized measurement $y(t) = G(t)x + \xi(t)$. Substituting this in (10.44), we find the total derivative

$$d\mathcal{V}_{h\mathcal{E}}(\tau, \mathcal{E}_L[\tau])/dt = -\langle l^0, L_0 G(t) X_+(t) G'(t) L_0' G(t) l^0 \rangle^{1/2} + \langle l^0, L_0' G(t)(x-x^*) \rangle +$$
$$+ \langle l^0, B(t)(u - p(t)) \rangle + \langle l^0, \mathbf{B}(t)l^0 \rangle^{1/2} + \langle l^0, L_0 \xi(t) \rangle - \langle l^0, L_0' R(t) L_l^0 \rangle^{1/2} +$$
$$+ \langle l^0, C(t)(v[t] - q(t)) \rangle - \langle l^0, \mathbf{C}(t)l^0 \rangle^{1/2}. \quad (10.47)$$

Selecting

$$U_L^0 = U^0(t, \mathcal{E}_+(x^*, X_+(t)) = \arg\min\{\langle l^0, B(t)u\rangle \mid u \in \mathcal{E}(p(t), P(t))\} =$$
$$= \arg\max\{-\langle l^0, B(t)u\rangle \mid u \in \mathcal{E}(p(t), P(t))\}, \quad (10.48)$$

we minimize the derivative in (10.47), ensuring

$$\left. d\mathcal{V}_{h\mathcal{E}}(t, \mathcal{E}_{L_0}[t])/dt \right| \le 0.$$

Integrating the last inequality from t to ϑ, we come to the guaranteed condition

$$\gamma^0 \le \mathcal{V}(\vartheta, \mathcal{E}_{L_0}[\vartheta]) \le \mathcal{V}(t, \mathcal{E}_{L_0}[t]). \quad (10.49)$$

The final result is given by

Theorem 10.6.4. *The solution to Problem OFC is given by control strategy (10.48), where l^0 is the optimizer in problem (10.44).*

Exercise 10.6.1. Find the error estimates for (10.49)

10.7 Separation of Estimation and Control

In order to simplify calculations for linear systems we shall treat the given Problem OFC in another coordinate system. Following Sect. 1.1 and taking $G(t, \tau)$ to be the fundamental transition matrix

$$\frac{\partial G(t, \tau)}{\partial t} = A(t)G(t, \tau), G(\tau, \tau) = I,$$

for homogeneous system $\dot{x} = A(t)x$ from (10.29), we introduce the transformation $x = G(\vartheta, t)\mathbf{x}$. which brings us to a new variable **x**. Making the necessary changes and returning after that to the original notations, we transform system (10.29), to the next one

$$\dot{x} = B(t)u + C(t)v(t), \quad (10.50)$$

$$y(t) = H(t)x + \xi(t), \quad (10.51)$$

under constraints $u \in \mathcal{P}(t)$, (10.4).

For passing to the solution of Problem OFC we present system (10.50), (10.51) as the array

$$dx^*/dt = B(t)u, \ x^*(t_0) = 0, \quad (10.52)$$

and

$$dw/dt = C(t)v(t), \quad w(t_0) \in \mathcal{X}^0, \tag{10.53}$$

with

$$z(t) = H(t)w + \xi(t), \tag{10.54}$$

where

$$x^* + w = x, \quad z(t) = y(t) - H(t)\int_{t_0}^{t} B(s)u^*[s]ds.$$

With $u = u[s]$, $s \in [t_0, t)$ given, there will be a one-to-one mapping between realizations $y[s]$ and $z[s]$. We may now define the information set of system (10.53), (10.54), denoting it as $\mathcal{W}_z^t[t] = \mathcal{W}_z^t(t, \cdot)$. Then we have $\mathcal{X}_y[t] = x^*[t] + \mathcal{W}_z^t[t]$.

Then $x^*(t)$ will define an isolated undisturbed controlled trajectory, while the set-valued observer $\mathcal{W}_z^t[t]$ will give a joint estimate of the existing disturbances in system (10.53), (10.54). Tube $\mathcal{W}_z^t[t]$ $t \in [t_0, \tau]$ may be calculated on-line, separately from the problem of specifying the control itself.

Applying such separation to procedures of Sects. 10.5 and 10.6 may be useful for solving concrete problems of OFC.

The chapter is now concluded by a few examples.

10.8 Some Examples

These examples illustrate the solution of Problem OFC under various types of unknown disturbances. The system equation is

$$\begin{cases} \dot{x}_1 = x_2 + \varepsilon_1 v \\ \dot{x}_2 = u + \varepsilon_2 v \end{cases}$$

with measurement output

$$y(t) = x_1(t) + \xi(t),$$

constraints $|u| \leq \mu$, $|v| \leq \nu \leq \mu$, $0 < \varepsilon_1 < \varepsilon_2 < 1$, $|\xi| \leq \eta$, starting set $\mathcal{X}^0 = \mathcal{E}(0, I)$, and target set $\mathcal{E}(m, M)$, $m \neq 0$, $M \neq I$.

The actual disturbances which are unknown are modelled here by the following types of functions:

in Fig. 10.1 by $v(t) = \nu \cos 5t$, $\xi(t) = \eta \sin 4t$,

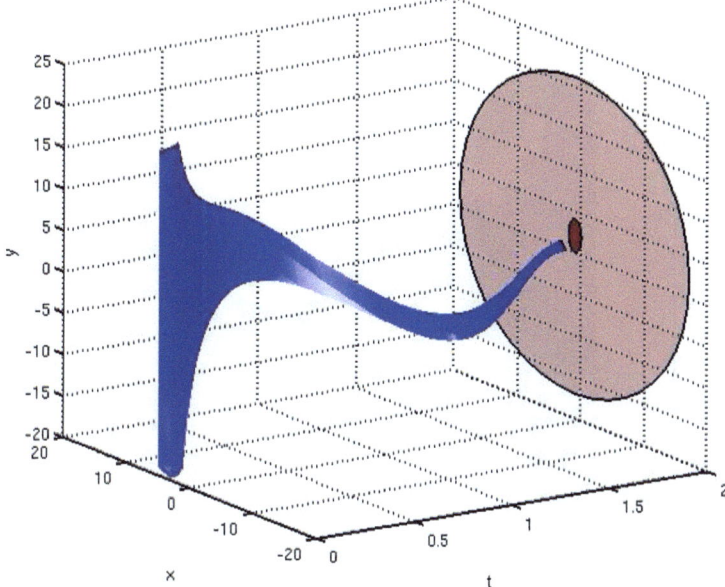

Fig. 10.1 Unknown oscillating disturbances

in Fig. 10.2 by

$$v(t) = \begin{cases} \theta_1, & \text{with } |\theta_1| \le v, \ \theta_1 \sim \mathcal{N}(0,1), \\ -v, & \text{with } \theta_1 < -v, \\ v, & \text{with } \theta_1 > v, \end{cases}$$

where $\mathcal{N}(0,1)$ stands for the normal distribution with zero mean and unit variance,
in Fig. 10.3 by $v(t) \equiv 0$, $\xi(t) \equiv 0$ which means the disturbances are zero-valued, but we do not know that. This is the worst-case disturbance.

Concluding Remarks

This chapter describes theoretical tools and computation approaches for the problem of optimizing OFC under set-membership bounds on controls and uncertain items. It indicates solution routes, which allow one to cope with uncertainty and incomplete measurements within finite-dimensional techniques, with effective computational schemes given for linear systems without imposing matching conditions for the hard bounds on controls and input disturbances. It also demonstrates the applicability of tools presented in this book to significant problems of control. The chapter is also an invitation to a broad research on challenging problems in feedback control under realistic information with increasing number of new motivations.

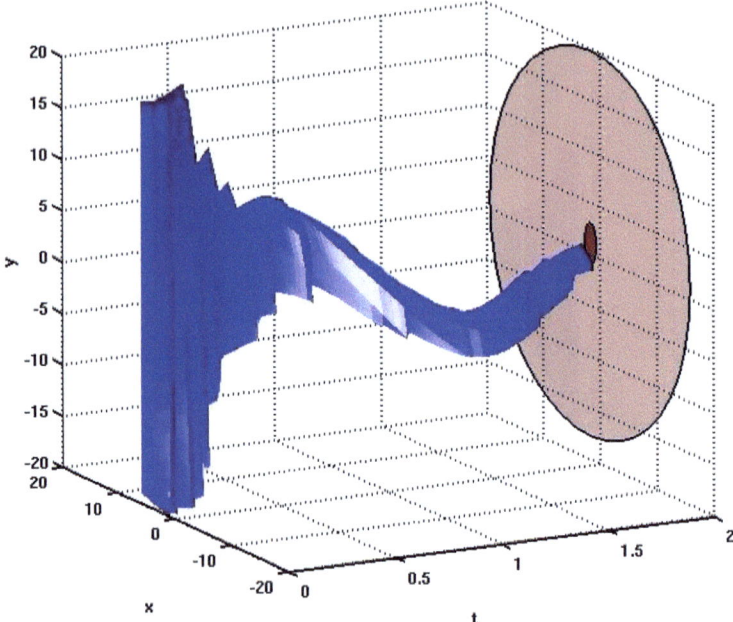

Fig. 10.2 Unknown random disturbances

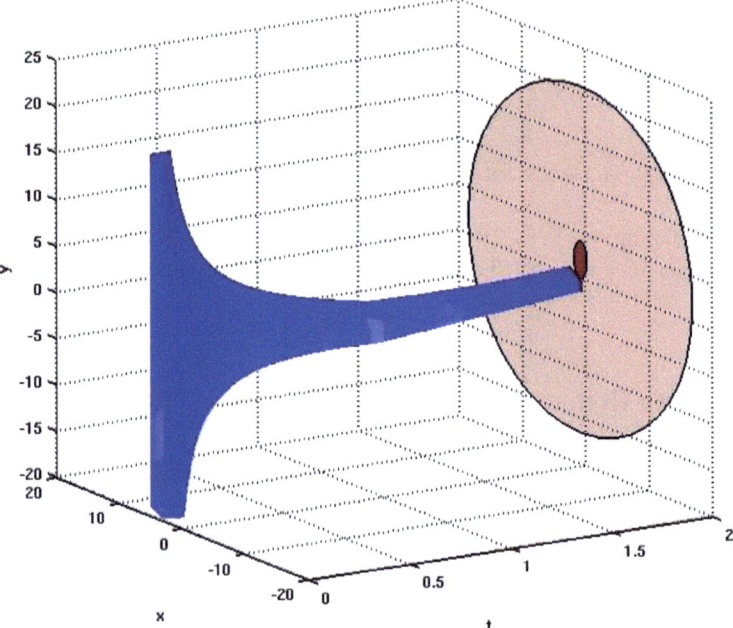

Fig. 10.3 Unknown disturbances are zero-valued

Chapter 11
Verification: Hybrid Systems

Abstract This chapter deals with a specific class of hybrid systems which combine controlled continuous dynamics through switching from one available motion to another due to discrete-time logical commands. Solutions to the reachability problem and their verification are indicated, followed by computational schemes, The application of impulse controls to the switching process is described. Examples of various difficulty are worked out. The chapter is to demonstrate applicability of methods of this book to hybrid systems.

Keywords Hybrid system • Switching • Reachability • Verification • Branching trajectory tubes • Ellipsoidal methods • Impulse feedback control

The main part of this chapter deals with a specific class of *hybrid systems* (see [39, 91, 204, 228, 258]). Their performance is due to an array of standard systems with acting motion generated by only one of them and with instantaneous switchings from one to another. The switchings are governed by a discrete time process that generates logically based commands when crossing some prespecified spatial domains ("the guards"). The reachability problem for such processes leads to branching trajectory tubes. Their description is complemented by verification problems whose solution, loosely speaking, should show which branch intersects with a given target zone or misses. Hence the chapter starts with description of the verification problem and methods of its solution, then passes to described target controlled hybrid dynamics and its verification. Ellipsoidal computation schemes for problems of reachability and verification in hybrid processes for isolated linear systems are indicated. Finally discussed is the application of impulse controls to such dynamics. Several examples of various difficulty are demonstrated. The aim of this chapter is to demonstrate the applicability of methods given in this book to hybrid processes.

11.1 Verification Problems

The problems of reachability and solvability are closely related to that of *verification* for the achieved solutions. The aim of verification is to check whether suggested algorithms do solve the intended problem.

11.1.1 The Problems and the Solution Schemes

Problem 11.1.1. *Given time ϑ, target set $M \in \text{comp}\mathbb{R}^n$ and set $X^0 = X[t_0]$,* **verify** *which of the following relations is true:*

(i) $X(\vartheta, t_0, X^0) \subseteq M$ (**all** *the reachable points are in M*),
(ii) $X(\vartheta, t_0, X^0) \cap M \neq \emptyset$ (**some** *of the reachable points are in M*),
(iii) $X(\vartheta, t_0, X^0) \cap M = \emptyset$ (*set M is **not** reachable from $\{t_0, X^0\}$ at time ϑ*).

Since $X(\vartheta, t_0, X^0)$ is the level set of the value function at time $t = \vartheta$ (see Chap. 2, Sect. 2.3):

$$V(t, x) = V(t, x \mid t_0, X^0) = \min_{u(\cdot), x(t_0)} \{d^2(x(t_0), X^0) \mid x(t) = x\},$$

$$V(t_0, x) = d^2(x, X^0),$$

we may check conditions (i)–(iii) using the following relations:

Theorem 11.1.1. *Following conditions (i)–(iii) of Problem 11.1.1 are, respectively, equivalent to the next relations, with $X(\vartheta, t_0, X^0) = \{x : V(\vartheta, x) \leq 0\}$:*

$$\max_x \{d^2(x, M) \mid V(\vartheta, x) \leq 0\} = 0, \tag{11.1}$$

$$\min_x \{d^2(x, M) \mid V(\vartheta, x) \leq 0\} = 0, \tag{11.2}$$

$$\min_x \{d^2(x, M) \mid V(\vartheta, x) \leq 0\} > 0. \tag{11.3}$$

Corollary 11.1.1. *If M is the level set for a certain convex function ϕ, such that*

$$M = \{x : \phi(x) \leq 0\},$$

then conditions (11.1)–(11.3) are, respectively, equivalent to

$$\max_x \{\phi(x) \mid V(\vartheta, x) \leq 0\} \leq 0,$$

$$\min_x \{\phi(x) \mid V(\vartheta, x) \leq 0\} \leq 0,$$

$$\min_x \{\phi(x) \mid V(\vartheta, x) \leq 0\} > 0.$$

In the previous propositions we dealt with verification of the solutions at fixed time ϑ. These may be further extended in the following way.

Theorem 11.1.2. *Recalling Problem 11.1.1, suppose*

$$\sup_t \max_x \{d^2(x, \mathcal{M})| V(t, x) \leq 0; \; t \geq t_0\} = 0$$

*then property (i) is ensured **for all** $t \geq t_0$;*

$$\text{if } \min_t \max_x \{d^2(x, \mathcal{M})|V(t, x) \leq 0; \; t \geq t_0\} = 0,$$

*then (i) is true **for some** $t \geq t_0$;*

$$\text{if } \sup_t \min_x \{d^2(x, \mathcal{M})|V(t, x) \leq 0; \; t \geq t_0\} = 0,$$

*then (ii) will be true **for all** $t \geq t_0$;*

$$\text{if } \min_t \min_x \{d^2(x, \mathcal{M})|V(t, x) \leq 0; \; t \geq t_0\} = 0,$$

*then (ii) will be true **for some** $t \geq t_0$;*

$$\text{if } \inf_t \min_x \{d^2(x, \mathcal{M})|V(t, x) \leq 0; \; t \geq t_0\} > 0,$$

*then (iii) will be true **for all** $t \geq t_0$;*

$$\text{if } \sup_t \min_x \{d^2(x, \mathcal{M})|V(t, x) \leq 0; \; t \geq t_0\} > 0,$$

*then (iii) will be true **for some** $t \geq t_0$.*

Exercise 11.1.1. *Check the validity of Theorems 11.1.1, 11.1.2 and Corollary 11.1.1*

Examples of solutions to this problem are given in Sect. 11.1.2.
An array of problems comes up in backward time. We mention some of these.

Problem 11.1.2. *Given starting set $\mathcal{F}$, and fixed time t, determine whether target set $\mathcal{M}$ is reachable from position $\{t, x\}$, at any time $\vartheta \geq t$*

(i) **for all** *states $x \in \mathcal{F}$,*
(ii) **for some** *state $x \in \mathcal{F}$,*
(iii) **is not** *reachable from any state $x \in \mathcal{F}$.*

The conditions that ensure the solutions to this Problem are now formulated through value function

$$V_0(t, x) = V_0(t, x|\vartheta, \mathcal{M}) = \min_{u(\cdot)} \{d^2(x(\vartheta), \mathcal{M}) \mid x(t) = x\}, \; \vartheta \geq t,$$

Theorem 11.1.3. *The set $\mathcal{M}$ is reachable from position $\{t, x\}$ at some ϑ*

(i) for all states $x \in \mathcal{F}$ if

$$V_0(t, \mathcal{F}) = \min_{\vartheta} \max_{x} \{V_0(t, x|\vartheta, \mathcal{M})| \ \vartheta \geq t, \ x \in \mathcal{F}\} = 0, \quad (11.4)$$

(ii) for some state $x^* \in \mathcal{F}$ if

$$V_0(t, \mathcal{F}) = \min_{\vartheta} \min_{x} \{V_0(t, x|\vartheta, \mathcal{M})|\vartheta \geq t, \ x \in \mathcal{F}\} = 0, \quad (11.5)$$

(iii) is not reachable from any state $x \in \mathcal{F}$ if

$$V_0(t, \mathcal{F}) = \min_{x} \{V_0(t, x, \mathcal{M}) \mid x \in \mathcal{F}\} = \inf_{\vartheta} \min_{x} \{V_0(t, x|\vartheta, \mathcal{M})|\vartheta \geq t, \ x \in \mathcal{F}\} > 0. \quad (11.6)$$

Remark 11.1.1. Note that if in the above and in the sequel we write $\min_\vartheta$ over domain $\vartheta \geq t_0$, then this presumes that the minimum should be attained at some finite time ϑ^*.

11.1.2 Ellipsoidal Techniques for Verification Problems

Consider system (3.1) under ellipsoidal constraints (3.4), (3.6), (3.7) and ellipsoidal target set $\mathcal{M} = \mathcal{E}(m, M)$. Dealing with formulation of the verification Problem 11.1.1 (i)–(iii) we shall solve it through ellipsoidal methods, further naming it then as a new Problem 11.1-E (i)–(iii).

Recall that due to (3.23) we have

$$\mathcal{X}[\vartheta] = \bigcap \{\mathcal{E}(x^*[\vartheta], X_+^l[\vartheta])| \ l : \langle l, l \rangle = 1\}. \quad (11.7)$$

where $E_+[l, \vartheta] = \mathcal{E}(x^*[\vartheta], X_+^l[\vartheta])$ are tight external approximations of $\mathcal{X}[\vartheta]$ along direction l.

At the same time Theorem 3.7.3 indicates

$$\mathcal{X}[\vartheta] = \bigcup \{\mathcal{E}(x^*[\vartheta], X_-^l[\vartheta])| \ l : \langle l, l \rangle = 1.\} \quad (11.8)$$

where $E_-[l, \vartheta] = \mathcal{E}(x^*[\vartheta], X_-^l[\vartheta])$ are tight internal approximations of $\mathcal{X}[\vartheta]$ along direction l.

Beginning with Problem 11.1.1-E(i) we have to check the inclusion

$$\mathcal{X}[\vartheta] \subseteq \mathcal{E}(m, M). \quad (11.9)$$

Then to ensure (11.9) we need to have either

$$\min_{l} \{\rho(l \mid \mathcal{E}(m, M)) - \rho(l \mid E_+[l, \vartheta]) \mid \langle l, l \rangle \leq 1\} \geq 0. \quad (11.10)$$

11.1 Verification Problems

or

$$\min_{l}\{\rho(l \mid \mathcal{E}(m, M)) - \rho(l \mid E_{-}[l, \vartheta]) \mid \langle l, l \rangle \leq 1\} \geq 0. \quad (11.11)$$

For Problem 11.1.1-E(ii) we have to check condition $X[\vartheta] \cap \mathcal{E}(m, M) \neq \emptyset$. This will be ensured if $E_{-}[l, \vartheta] \cap \mathcal{E}(m, M) \neq \emptyset$ for some tight internal approximation $E_{-}[l, \vartheta] = \mathcal{E}(x^*[\vartheta], X_{-}^{l}(\vartheta))$ of $X[\vartheta]$. The last requirement will be ensured iff for some such $E_{-}[l, \vartheta]$ the inequality

$$\rho(l \mid E_{-}[l, \vartheta]) \geq -\rho(-l \mid \mathcal{E}(m, M)) \geq 0,$$

for all l will be true or, in other terms, if

$$\min_{l}\{\rho(l \mid E_{-}[l, \vartheta]) + \rho(-l \mid \mathcal{E}(m, M)) \mid \langle l, l \rangle \leq 1\} \geq 0, \quad (11.12)$$

Finally, for Problem 11.1.1-E(iii) we have to check that $X[\vartheta] \cap \mathcal{E}(m, M) = \emptyset$. This will be ensured if $E_{+}[l, \vartheta] \cap \mathcal{E}(m, M) = \emptyset$ for some tight external approximation $E_{+}[l, \vartheta]$ of $X[\vartheta]$, or in other terms, if the distance $d(E_{+}[l, \vartheta], \mathcal{E}(m, M)) > 0$ for some such $E_{+}[l, \vartheta]$. The latter holds iff

$$\max_{l}\{\max_{p}\{-\rho(-p \mid E_{+}[l, \vartheta]) - \rho(p \mid \mathcal{E}(m, M)) \mid \langle p, p \rangle \leq 1\} \mid \langle l, l \rangle \leq 1\} > 0. \quad (11.13)$$

Theorem 11.1.4. *The solutions to (i)–(iii) of Problem 11.1.1, as written in terms of approximating ellipsoids, are satisfied iff the following relations are true: (11.10) or (11.11) for (i); (11.12) for (ii) and (11.13) for (iii).*

Now introduce some new notation, namely,

$$V_{+}(\vartheta, x, l) = \langle x - x^*[\vartheta], X_{+}^{l}[\vartheta](x - x^*[\vartheta]) \rangle.$$

Then

$$E_{+}[l, \vartheta] = \{x : V_{+}(\vartheta, x, l) \leq 1\}, \; E_{-}[l, \vartheta] = \{x : V_{-}(\vartheta, x, l) \leq 1\}.$$

Denote

$$V_{+}(\vartheta, x) = \max_{l}\{V_{+}(\vartheta, x, l) \mid \langle l, l \rangle = 1\}, \; V_{-}(\vartheta, x) = \min_{l}\{V_{-}(\vartheta, x, l) \mid \langle l, l \rangle = 1\}.$$

Here both functions $V_{+}(\vartheta, x, l), V_{-}(\vartheta, x, l)$ may serve as quadratic approximations of value function $V(\vartheta, x)$ that defines $X[\vartheta] = X(\vartheta, t_0, X^0)$ (see Chap. 2, Sect. 2.3). Namely,

$$V_{+}(\vartheta, x, l) \leq V_{+}(\vartheta, x) = V(\vartheta, x) = V_{-}(\vartheta, x) \leq V_{-}(\vartheta, x, l). \quad (11.14)$$

Further denote $V_m(x) = \langle x - m, M(x - m) \rangle$, so that $\mathcal{E}(m, M) = \{x : V_m(x) \leq 1\}$. Theorem 11.1.4 may then be rewritten in terms of functions V_{+}, V_{-}, V_m.

Theorem 11.1.5. *Suppose one of the conditions is true:*

(i) $V_m(x) \leq V_+(\vartheta, x) = \max_{l}\{V_+(\vartheta, x, l) \mid \langle l, l \rangle = 1\}, \forall x,$

(ii) $\{\exists\, l = \hat{l}, x = \hat{x} : V_m(\hat{x}) \leq V_-(\vartheta, \hat{x}, \hat{l})\},$

(iii) $\{\exists\, l = \hat{l} : \min\{V_m(x) \mid V_+(\vartheta, x, \hat{l}) \leq 1\} > 1\}.$

Then, respectively, the requirements (i)–(iii) of Problem 11.1.1 will be fulfilled.

This theorem implies the following conclusions (in view of $V(\vartheta, x) = V_+(\vartheta, x) = V_-(\vartheta, x)$).

Corollary 11.1.2. *The solutions to (i)–(iii) of Problem 11.1.1, as written in terms of approximating ellipsoids, are satisfied iff, respectively, the following relations are true:*

(i) $\max_{x}\{d^2(x, \mathcal{E}(m, M)) \mid V(\vartheta, x) \leq 1\} = 0,$

(ii) $\min_{x}\{d^2(x, \mathcal{E}(m, M)) \mid V(\vartheta, x) \leq 1\} = 0,$

(iii) $\min_{x}\{d^2(x, \mathcal{E}(m, M)) \mid V(\vartheta, x) \leq 1\} > 0.$

An ellipsoidal version of Theorem 11.1.2 then reads as follows.

Theorem 11.1.6. *If*

$$\sup_{t} \max_{x}\{d^2(x, \mathcal{E}(m, M)) \mid V(t, x) \leq 1,\ t \geq t_0\} = 0,$$

then property (i) will be true **for all** $t \geq t_0$;

$$\text{if } \min_{t} \max_{x}\{d^2(x, \mathcal{E}(m, M)) \mid V(t, x) \leq 1,\ t \geq t_0\} = 0,$$

then (i) will be true **for some** $t \geq t_0$;

$$\text{if } \sup_{t} \min_{x}\{d^2(x, \mathcal{E}(m, M)) \mid V(t, x) \leq 1,\ t \geq t_0\} = 0,$$

then (ii) will be true **for all** $t \geq t_0$;

$$\text{if } \min_{t} \min_{x}\{d^2(x, \mathcal{E}(m, M)) \mid V(t, x) \leq 1,\ t \geq t_0\} = 0,$$

then (ii) will be true **for some** $t \geq t_0$;

$$\text{if } \inf_{t} \min_{x}\{d^2(x, \mathcal{E}(m, M)) \mid V(t, x) \leq 1,\ t \geq t_0\} > 0,$$

then (iii) will be true **for all** $t \geq t_0$;

$$\sup_t \min_x \{d^2(x, \mathcal{E}(m, M)) \mid V(t,x) \leq 1, \ t \geq t_0\} > 0,$$

then (iii) will be true **for some** $t \geq t_0$.

Exercise 11.1.2. *Formulate and prove an ellipsoidal version of Theorem 11.1.6.*

11.2 Hybrid Dynamics and Control

The previous chapters of this book deal with various types of problems on control for systems whose mathematical models are described by standard ordinary differential equations. However, recent applications may require treating such problems in a more complicated setting, under complex dynamics. In the present chapter such systems are of the hybrid type. The notion of hybrid system has various definitions (see [36,39,204,229,258,259]). However, the main idea is that the system is defined by an array of standard systems such that the active motion is due to only one of them, with instantaneous switching from one to another. The process of switching is usually logically controlled in discrete times, in such way that a possible switch may occur only when passing through some spatial domains (the "guards"), or may not occur. The performance range of such systems is then obviously broader than that of standard systems. Hence, the overall controlled motions will develop in time as those generated by alternating isolated continuous motions due to systems whose sequence is controlled through logically based discrete commands, while their individual contribution is designed by its own controllers. The switching from one system to another may be also accompanied by an instantaneous change of some phase coordinates.

Despite the fairly complicated overall dynamics, efficient computation of control solutions for hybrid processes appears to be also available through ellipsoidal-valued approximations for systems composed of individual participants of the types described above. The discussion given below is devoted to linear controlled systems, though the overall schemes may be also applied under individual nonlinearities.

11.2.1 The Hybrid System and the Reachability Problem

The System

The considered overall system **H** is governed by an array of linear subsystems $(i = 1, \ldots, k)$

$$\dot{x} \in A^{(i)}(t)x + B^{(i)}(t)u^{(i)}(t) + C^{(i)}(t)v^{(i)}(t), \tag{11.15}$$

with continuous matrix coefficients $A^{(i)}(t), B^{(i)}(t), C^{(i)}(t)$ and $x \in \mathbb{R}^n$. Here $u^{(i)} \in \mathbb{R}^p$ are piecewise *open-loop controls* $u^{(i)}(t)$ restricted by inclusions $u^{(i)}(t) \in \mathcal{P}^{(i)}(t)$, where $\mathcal{P}^{(i)}(t)$, are set-valued functions with convex compact values, continuous in time. Functions $v^{(i)}$ are taken to be given.

In this section we consider hard bounds that are *ellipsoidal-valued*, with

$$\mathcal{P}^{(i)}(t) = \mathcal{E}(p^{(i)}(t), P^{(i)}(t)). \tag{11.16}$$

In the phase space $\mathbb{R}^n$ given are k hyperplanes (affine manifolds)

$$H_j = \{x : \langle c^{(j)}, x \rangle - \gamma_j = 0\}, \; c^{(j)} \in \mathbb{R}^n, \; \gamma_j \in \mathbb{R}, \; j = 1, \ldots, k.$$

which are *the enabling zones* (the guards).

Here is how the system operates.

At given time t_0 the motion initiates from starting set $\mathcal{X}^0 = \mathcal{E}(x^0, X^0)$ and follows one of the equations of the above (take $i = 1$, to be precise), due to one of the controls $u(t; 1)$ until at time $\tau'[1]$ it reaches some $H_j = H(j[1])$—the first of the hyperplanes along its route (we assume $j \neq 1$). Now a binary operation interferes— the motion either continues along the "old" subsystem (a "passive" crossing), with system number $i = 1$ or switches to (*is reset to*) a "new" subsystem (an "active" crossing), with system number indexed as $i = i[1] = j[1] \neq 1$. (Otherwise, if $j[1] = 1$, we presume there is a passive crossing, with no reset.)

Some Notation. Each number $[\kappa] = [1], [2] \ldots$, in square brackets is the sequential *number of crossing* a hyperplane H_j, so that the range of $\{j[\kappa], \kappa \in \mathbb{Z}_+\}$ is $j \in \{1, \ldots, k\}$. Here and further $\mathbb{Z}_+$ is *the set of positive integers*. The range of system numbers $i[\kappa]$, $\kappa \in \mathbb{Z}_+$, after each crossing is also $i \in \{1, \ldots, k\}$.

Then, before crossing $H_j[1] = H(j[1])$, the state of the system is denoted as $\{t, x, [1^+]\}$ while after the binary operation it is either $\{t, x, [1^+, i[1]^+]\}$—if the crossing was active (a reset), or $\{t, x, [1^+, i[1]^-]\}$, if it was passive (no reset).

We assume, for the sake of brevity, that *an active crossing a hyperplane H_j is a switching to system with $i = j$.*

So, after first crossing, the further motion develops along $i = 1$ or $i = i[1]$ until crossing the next hyperplane $H(j[2])$ where a similar binary operation is applied— the motion either continues along the previous subsystem or switches to (*is reset to*) subsystem with index $i[2] = j[2]$. After the second crossing the state of the system is $\{t, x, [1^+, i[1]^s, i[2]^s]\}$, where the boolean index "s" is either $s = +$ or $s = -$.

Note that *the state of the system has a part $\{t, x\}$ without memory* related to the running on-line position of the continuous-time variable and a part $[1^+, i[1]^s, i[2]^s]$ *with memory*, related to the discrete event variable $i^s[\kappa]$, which describes the sequence of switchings made earlier by the system.[1]

---

[1] The part with memory may be important for making the decision—"to switch" or "not to switch," for example, if the number of switchings is restricted.

11.2 Hybrid Dynamics and Control

At each new crossing a new term is added to this sequence. Thus the following general rules should be observed:

(i) that crossing each hyperplane H_j results either in a reset to subsystem with number j or in no reset at all,
(ii) that the crossing takes place in direction of support vectors $c^{(j)}$, and at points of crossing we have

$$\min\{\langle c^{(j)}, z\rangle | z \in F^{(i)}(t,x)\} \geq \varepsilon > 0, \ \forall i, j = 1, \ldots, k, \ \forall x \in H_j \quad (11.17)$$

where

$$F^{(i)}(t,x) = A^{(i)}(t)x + B^{(i)}(t)\mathcal{E}(p^{(i)}(t), P^{(i)}(t)) + C^{(i)}(t)v^{(i)}(t);$$

(iii) that the state space variable after κ crossings is $\{t, x, i[1]_1^s, \ldots, i[\kappa]_\kappa^s\}$ where each "boolean" index "s" is either $+$ or $-$.
(iv) that at a crossing with hyperplane H_j the sequence of type $\{i[1]^s, \ldots, i[\kappa]^+\}$, describing the "discrete event" part of the state is complemented by a new term, which is either $i[\kappa+1]^+$ if there is a switching to system, or $i[\kappa+1]^-$, if there is no switching.

Such a notation allows to trace back the array of subsystems used earlier from any current position $\{t, x\}$. Thus, if the state is $\{t, x; 1^+, i[1]^-, \ldots, i[\kappa]^-\}$ with $s = -$ for all $i[1], \ldots, i[\kappa]$, then the trajectory did not switch at all throughout any of the κ crossings, having followed one and the same subsystem with $i = 1$ throughout the whole process. Note that at each state $\{t, x, [1^+, i[1]^s, \ldots, i[\kappa]^s]\}$ the system is to follow the subsystem whose number coincides with that of the last term with index $s = +$.

We will be interested in the reachability problem for such systems.

Remark 11.2.1. The system under consideration is one of the possible types of *hybrid systems*. It differs from the so-called *switching systems* in that the time instants for crossing are not given, but are located during the spatial course of the trajectory as intervals of crossing some specified domains (the "guards") where it is also possible to reset the phase coordinates. The suggested scheme also allows to be propagated to a broad variety of options with different information requirements.

Remark 11.2.2. In this chapter the guards are taken as hyperplanes and the time of each crossing is unique for every trajectory. An example of propagating the suggested scheme is when the guards are taken as a domain bounded by two parallel hyperplanes. The switching is then assumed possible at any time within this domain. This situation with an example is indicated at the end of the present section.

The problems considered in this chapter deal with reachability under piecewise—open-loop controls with possible resets of controlled systems at given guards, taken as hyperplanes, so that between these zones the control is open-loop. For further considerations the restriction on starting set X^0 is of ellipsoidal type: $X^0 = \mathcal{E}(x^0, X^0)$.

Reachability Under Resets

The reachability problem consists of two versions.

Problem 11.2.1. *Find all vectors $\{x\}$ reachable from starting position $\{t_0, X^0\}$ at given time $t = \tau$ through all possible controls: "the reach set $X(\tau; t_0, X^0)$ at time τ from $\{t_0, X^0\}$".*

Problem 11.2.2. *Find all vectors $\{x\}$ reachable **at some time** t within interval $t \in [t', t''] = \mathcal{T}$ through all possible controls: "the reach set $X(t', t''; t_0, X^0)$ **within interval** $\mathcal{T}$ from $\{t_0, X^0\}$".*

One may observe that the problem consists in investigating *branching trajectory tubes*, in describing their *cross-sections ("cuts")* and the *unions* of such cross-sections. The reach sets may therefore turn out to be *disconnected sets*. We next discuss reach sets at given time t.

We first describe the reach set for a given sequence $[i_0, i[1]^{s_1}, \ldots, i[l]^{s_l}]$ of crossings, from position $\{t_0, X^0, i_0\}$, assuming $i_0 = 1$ to be precise. Here index s_i is either $-$ or $+$.

The Reach Set After One Crossing

(a) Suppose that before reaching H_j, $i[1] = j$, we have

$$X^{(1)}[t] = X^{(1)}(t; t_0, X^0, [1^+]) =$$

$$G^{(1)}(t, t_0)X^0 + \int_{t_0}^{t} G^{(1)}(t, s)(B(s)\mathcal{P}^{(1)}(s) + C(s)v^{(1)}(s))ds$$

where $G^{(i)}(t, s)$ is the transition function for the i-th subsystem.

(b) To be precise, suppose that before reaching H_j we have

$$\max\{\langle c^{(j)}, x \rangle \mid x \in X^{(1)}[t]\} = \rho_j^+(t) < \gamma_j.$$

The first instant of time when $X^{(1)}[t] \cap H_j \neq \emptyset$ is τ_j'. It is found as the smallest root of the equation

$$\gamma_j - \rho_j^+(t) = 0, \ t \geq t_0.$$

Introducing the function

$$\min\{\langle c^{(j)}, x \rangle \mid x \in X^{(1)}[t]\} = \rho_j^-(t),$$

we may also observe that condition $X^{(1)}[t] \cap H_j \neq \emptyset$ will hold as long as

$$\rho_j^-(t) \leq \gamma_j \leq \rho_j^+(t),$$

11.2 Hybrid Dynamics and Control

and the point of departure from H_j is the smallest positive root τ_j'' of the equation

$$\gamma_j - \rho_j^-(t) = 0, \ t \geq \tau_j'.$$

Condition (11.17) ensures that points τ_j', τ_j'' are *unique*. Note that τ_j'' is the time instant when the entire reach set $X^{(1)}[t]$ leaves H_j.

Denote $X^{(1)}[t] \cap H_j = Z_j^{(1)}(t)$.

(c) After the crossing we have to envisage two branches:

(−) with no reset—then nothing changes and

$$X(t; t_0, X^0, [1^+, j^-]) = X^{(1)}(t; t_0, X^0, [1^+])$$

(+) with reset—then we consider the union

$$\bigcup \{X^{(j)}(t; s, Z_j^{(1)}(s)) | s \in [\tau_j', \tau_j'']\} = X(t; t_0, X^0, [1^+, j^+]), \ t \geq \tau_j''.$$

Here in case (−) the reach tube develops further along the "old" subsystem (1), while in case (+) it develops along "new" subsystem (j).

A Branch of the Reachability Set

For each new crossing we may now repeat the described procedure. Thus, we may obtain the reach set $X(t, t_0, X^0; 1^+, i[1]^{s_1}, \ldots, i[l]^{s_l})$ for a branch $\{1^+, i[1]^{s_1}, \ldots, i[l]^{s_l}\}$.

We further assume the next condition

Assumption 11.2.1. *The neighboring intervals of crossing* $\left[\tau_\kappa', \tau_\kappa''\right]$, $\kappa = 1, \ldots, l$, *do not intersect. It is presumed that* $\tau_\kappa'' < \vartheta$, *where* $[t_0, \vartheta]$ *is the interval under consideration.* □

Taking an interval $[t_0, t]$, with $\tau_{i[k-1]}'' \leq t \leq \tau_{i[k]}'$ and $\tau_{i[m-1]}'' \leq t \leq \tau_{i[m]}'$, $i[l] < i[m]$, one may observe the following semigroup type property.

Lemma 11.2.1. *Each branch* $\{1^+, i[1]^{s_1}, \ldots, i[k]^{s_m}\}$ *yields the following superposition property* $(1 \leq k \leq m)$

$$X(t, t_0, X^0; 1^+, i[1]^{s_1}, \ldots, i[m]^{s_m}) =$$
$$= X(t, t^*, X(t^*, t_0, X^0; 1^+, i[1]^{s_1}, \ldots, i[k]^{s_l}); i[k]^{s_l+1}, \ldots, i[m]^{s_m}). \quad (11.18)$$
□

The computation of a branch may be done as a sequence of *one-stage crossing transformations*.

Recall that the continuous-time transition between crossings, along system $j = j_1$ from a continuous-time position $\{\tau, X\}$ with $\tau \geq \tau''_{j_1}$ to a position at time $t \leq \tau'_{j_2}$, is $X^{(j_1)}(t, \tau, X)$.

Then, for example, given position (state) $\{\tau'_j, X; [1^+]\}$, at start of crossing H_j, we may define a "one-stage crossing" transformation

$$T_j^{(s)}\{\tau'_j, X, [1^+]\} = \{\tau''_j, X^{(1)}(\tau''_j, \tau'_j, X), [1^+, j^-]\}, \text{ if } s = -,$$

and

$$T_j^{(s)}\{\tau'_j, X, [1^+]\} = \{\tau''_j, Z^{(j)}[\tau''_j], [1^+, j^+]\}, \text{ if } s = +.$$

Here

$$Z^{(j)}[\tau''_j] = \bigcup_t \{X^{(j)}(\tau''_j, t, H_j \cap X^{(1)}(t, \tau'_j, X)) \mid t \in [\tau'_j, \tau''_j]\}$$

The Branch of One-Stage Transformations

We may now represent a branch $\{1^+, i[1]^{s_1}, \ldots, i[k]^{s_k}\}$ through a sequence of interchanging operations of type $T_j^{(s)}$ and $X^{(j)}$.

For example, the reach set for branch $\{1^+, i[1]^+, i[2]^-\}$, from starting position $\{\tau, X, [1^+]\}$, $\tau \leq \tau'_{i1}$, at time $t \in [\tau''_{i2}, \tau'_{i3}]$, may be represented as the output of the following sequence of mappings. (Here and further, to clarify when necessary the notation of lower indices, we denote $i_\kappa = i\kappa$)

$$T_{i1}^+\{\tau'_{i1}, X^{(1)}(\tau'_{i1}, \tau, X), [1^+]\} = \{\tau''_{i1}, Z^{(i1)}[\tau''_{i1}], [1^+, i[1]^+]\}$$

$$X^{(i1)}[\tau'_{i2}] = X^{(i1)}(\tau'_{i2}, \tau''_{i1}, Z^{(i1)}[\tau''_{i1}]),$$

$$T_{i2}^-\{\tau'_{i2}, X^{(i1)}[\tau'_{i2}], [1^+, i[1]^+]\} = \{\tau''_{i2}, X^{(i1)}(\tau''_{i2}, \tau'_{i2}, X^{(i1)}[\tau'_{i2}]), [1^+, i[1]^+, i[2]^-]\}$$

Then, for $t > \tau''_{i2}$, the desired set of positions is given as

$$\{t; X^{(i1)}(t, \tau''_{i2}, X^{(i1)}(\tau''_{i2}, \tau''_{i1}, Z^{(i1)}[\tau''_{i1}])), [1^+, i[1]^+, i[2]^-]\}.$$

This is a branch of the overall reach set. The continuous variables for this branch at $t \in [\tau''_{i2}, \tau'_{i3})$ produce the equality

$$X(t; \tau, X) = X^{(i1)}(t; \tau''_{i2}, X^{(i1)}[\tau'_{i2}]).$$

Lemma 11.2.2. *A branch of type* $X(t; t_0, X^0) : i = 1^\pm, i[1]^{s_1}, \ldots, i[k]^{s_k}$, *may be described by a superposition of interchangings one-stage crossing transformations $T_{i\kappa}^s$ and continuous maps $X^{(\kappa)}$, $\kappa = 1, \ldots, l$.*

11.2 Hybrid Dynamics and Control

An alternative scheme for calculating reach sets is to describe them through value functions of optimization problems. Its advantage is that it is not restricted to linear systems.

11.2.2 Value Functions: Ellipsoidal Approximation

Reachability Through Value Functions

As indicated in [175], the "reach" (reachability) sets for "ordinary" (nonhybrid) systems may be calculated as level sets of solutions to HJB (Hamilton–Jacobi–Bellman) equations for some optimization problems. We will follow this scheme for the hybrid system under consideration. Consider first a one-stage crossing.

(a) *Before the crossing* $H_{i[1]} = H_j$, we assume that the system operates from position $\{t_0, \mathcal{X}^0, [1^+]\}$ and the next crossing starts at τ_1. Then, for $t < \tau_1$, we have

$$\mathcal{X}^{(1)}[t] = \{x : V^{(1)}(t, x) \le 0\},$$

where

$$V^{(1)}(t, x) = \min_{u^{(1)}} \{d^2(x^{(1)}(t_0), \mathcal{X}^0) | x^{(1)}(t) = x\},$$

and $x^{(i)}(t) = x(t)$ is the trajectory of the i-th system. We further also use the notation $V^{(1)}(t, x) = V^{(1)}(t, x_1, x_2, \ldots, x_n)$.

(b) *At the crossing* we have $\mathcal{X}^{(1)}[t] \cap H_j = \mathcal{Z}_j^{(1)}(t)$ which can be calculated as follows.

Without loss of generality we may presume $c_1^{(j)} = 1$. Then

$$\mathcal{Z}_j^{(1)}(t) = \{x : x_1 = \zeta(x), V^{(1)}(t, \zeta(x), x_2, \ldots, x_n) \le 0\} \cap H_j, \ \zeta(x) = \gamma_1 - \sum_{i=2}^{n} c_i^{(j)} x_i.$$

In particular, if the hyperplane $H_j = \{x : x_1 = \gamma_1\}$, then

$$\mathcal{Z}_j^{(1)}(t) = \{x : x_1 = \gamma_1, V^{(1)}(t, \gamma_1, x_2, \ldots, x_n) \le 0\}.$$

and the set $\mathcal{Z}_j^{(1)}(t) \ne \emptyset$ iff $\rho_j^-(t) \le \gamma_j \le \rho_j^+(t)$, where

$$\rho_j^+(t) = \max\{(c^{(j)}, x) \mid V^{(1)}(t, x) \le 0\},$$
$$\rho_j^-(t) = \min\{(c^{(j)}, x) \mid V^{(1)}(t, x) \le 0\}.$$

This happens within the time interval $[\tau_j', \tau_j'']$, $\rho_j^-(\tau_j'') = \gamma_j$, $\rho_j^+(\tau_j') = \gamma_j$.

(c) *After the crossing* we envisage two branches:

(−) with no reset—then $X(t; t_0, X^0, [1^+, j^-]) = X^{(1)}(t; t_0, X^0, [1^+])$,
(+) with reset—then we have to calculate the union

$$\bigcup_s \{X^{(j)}(t; s, Z_j^{(1)}(s)) | s \in [\tau_j', \tau_j'']\} = X(t; t_0, X^0, [1^+, j^+]), \quad t \geq \tau_j''.$$

With $t > \tau_j''$ this union may be calculated as the level set for function

$$\mathcal{V}(t, x \mid [1^+, j^+]) = \min_s \{V^{(j)}(t, s, x) \mid s \in [\tau_j', \tau_j'']\}$$

where

$$V^{(j)}(t, s, x) = \min_{u^{(j)}} \{V^{(1)}(s, \zeta(x(s)), x_2(s), \ldots, x_n(s)) \mid s \in [\tau_j', \tau_j''], \ x(t) = x\},$$

so that

$$X(t; t_0, X^0, [1^+, j^+]) = \{x : \mathcal{V}(t, x | [1^+, j^+]) \leq 0\} =$$

$$\bigcup_s \{X^{(j)}[t, s] \mid s \in [\tau_j', \tau_j'']\},$$

$$X^{(j)}[t, s] = X^{(j)}(t; s, Z_j^{(1)}(s)) = \{x : V^{(j)}(t, s, x) \leq 0\}.$$

Remark 11.2.3. Note that in general the union above is nonconvex.

(d) Repeating the procedure for each new crossing, we may obtain the reach set $X(t; t_0, X^0, [1^+, i[1]^{(s_1)}, \ldots, i[k]^{(s_k)}])$ for a branch $[1^+, i[1]^{(s_1)}, \ldots, i[k]^{(s_k)}]$.

These are the general schemes to compute reachability sets for the given class of hybrid systems. We further indicate some ellipsoidal techniques for problems of this chapter.

Reachability Through Ellipsoidal Approximations

Let us first calculate the reachability set after a one-stage crossing transformation.

(a) Starting with $X^0 = \mathcal{E}(x^0, X^0)$, $i = 1$, the reach set $X(t; t_0, X^0, [1^+])$ is due to equation

$$\dot{x} = A^{(1)}(t)x + B^{(1)}(t)\mathcal{E}(p^{(1)}(t), P^{(1)}(t)) + C^{(1)}(t)v^{(1)}(t),$$

11.2 Hybrid Dynamics and Control

which yields the following *ellipsoidal approximations* (see Chap. 3, Sects. 3.2, 3.3, 3.7 and 3.9).

$$\mathcal{E}(x^{(1)}(t), X_-^{(1)}(t)) \subseteq X(t; t_0, \mathcal{X}^0, [1^+]) \subseteq \mathcal{E}(x^{(1)}(t), X_+^{(1)}(t)),$$

where

$$\dot{X}_+^{(1)} = A^{(1)}(t) X_+^{(1)} + X_+^{(1)} A^{(1)'}(t) + \pi(t) X_+^{(1)} + (\pi(t))^{-1} B^{(1)}(t) P^{(1)}(t) B^{(1)'}(t), \tag{11.19}$$

$$\dot{X}_-^{(1)} = A^{(1)}(t) X_-^{(1)} + X_-^{(1)} A^{(1)'}(t) +$$

$$+ X_{-*}^{(1)} S(t) B^{(1)}(t) (P^{(1)})^{1/2}(t) + (P^{(1)})^{1/2}(t) B^{(1)'}(t) S'(t) X_{-*}^{(1)'},$$

$$\dot{x}^{(1)} = A^{(1)}(t) x^{(1)} + B^{(1)}(t) p^{(1)}(t) + C^{(1)}(t) v^{(1)}(t),$$

$$\pi(t) > 0, \ S(t) S'(t) = I, \ X_{-*} X'_{-*} = X_-, \ X_+^{(1)}(t_0) = X_-^{(1)}(t_0) = X^0, \ x^{(1)}(t_0) = x^0.$$

Here

$$\pi(t) = \langle l(t), B(t) P(t) B'(t) l(t) \rangle^{1/2} \langle l, X_+^{(1)}(t) l \rangle^{-1/2},$$

$$S(t) B(t) P^{1/2}(t) l(t) = \lambda(t) S_0 (X^0)^{1/2} l,$$

$$\lambda(t) = \langle l(t), B(t) P(t) B'(t) l(t) \rangle^{1/2} \langle l, X^0 l \rangle^{-1/2}, \ S'_0 S_0 = I.$$

As indicated in Chap. 3, these approximations, which depend on l, will be *tight* along a given *good curve* $l(t) = G^{(1)'}(t_0, t) l$, $l \in \mathbb{R}^n$, generated by $G^{(i)}(t_0, t)$—the fundamental transition matrix for the homogeneous system (11.15). Namely, for such $l = l(t)$ we have $X_-^{(1)}(t) = X^{(1)}(t|l)$, $X_+^{(1)}(t) = X_+^{(1)}(t|l)$ and

$$\rho(l \mid \mathcal{E}(x^{(1)}(t), X_-^{(1)}(t|l))) = \rho(l \mid X(t; t_0, \mathcal{X}^0, [1^+])) = \rho(l \mid \mathcal{E}(x^{(1)}(t), X_+^{(1)}(t|l))). \tag{11.20}$$

Moreover,

$$\bigcup_l \mathcal{E}(x^{(1)}(t), X_-^{(1)}(t|l)) = X(t; t_0, \mathcal{X}^0, [1^+]) = \bigcap_l \mathcal{E}(x^{(1)}(t), X_+^{(1)}(t|l)). \tag{11.21}$$

over all $\{l : \langle l, l \rangle \leq 1\}$.

(b) Let us now discuss the crossings $E_1^{(j)}(t|l) = \mathcal{E}(x^{(1)}(t), X_+^{(1)}(t|l)) \cap H_j$. Let $\mathbf{e}^{(i)}$ be the unit orths for the original coordinate system.

Introducing the linear map

$$\mathbf{T} c^{(j)} = \mathbf{e}^{(1)}; \ \mathbf{T} \mathbf{e}^{(i)} = \mathbf{e}^{(i)}, \ i = 1, \ldots, k, \ i \neq j, \ |\mathbf{T}| \neq 0,$$

Without loss of generality we may transform hyperplane H_j into hyperplane $x_1 = \gamma$ and take $\gamma = 0$, marking H_j as $H_1(j)$ and defining it by the equality $x_1 = 0$. Then, keeping previous notations in the new coordinates, we have[2]

$$E_1^{(j)}(t|l) = \mathcal{E}(z^{(j)}(t), Z_{1+}^{(j)}(t\mid l)) =$$

$$= \mathcal{E}(x^{(1)}(t), X_+^{(1)}(t\mid l)) \cap H_1(j) = \{x : V_+^{(1)}(t, x\mid l) \le 1,\ x_1 = 0\},$$

where

$$V_+^{(1)}(t, x\mid l) = \langle x - x^{(1)}(t), (X_+^{(1)})^{-1}(t\mid l)(x - x^{(1)}(t))\rangle. \qquad (11.22)$$

The intersection $E_1^{(j)}(t|l)$ is a degenerate ellipsoid whose support function is

$$\rho(l|E_1^{(j)}(t\mid l)) = \langle l, z^{(j)}(t)\rangle + \langle l, Z_{1+}^{(j)}(t\mid l)l\rangle^{1/2},$$

which may be calculated through standard methods of linear algebra.

Exercise 11.2.1. *Calculate the parameters of support function for ellipsoid* $E_1^{(j)}(t|l)$.

Now, in the n-dimensional space $H_1(j) = \{x : x_1 = 0\}$, we may consider an array of ellipsoids $\mathcal{E}(z^{(j)}(t), Z_{1+}^{(j)}(t\mid l))$, and

$$t \in [\tau_j', \tau_j''] = \mathcal{T} = \bigcup \left\{ \mathcal{T}_l \bigcap [\tau', \tau''] \mid \langle l, l \rangle \le 1 \right\},$$

where

$$\mathcal{T}_l = \{t : \exists x = (0, x_2, \ldots, x_n) : \langle x - x^{(1)}(t), (X_+^{(1)})^{-1}(t\mid l)(x - x^{(1)}(t))\rangle \le 1\}.$$

Then for any l we have the relations

$$\mathcal{E}(z^{(j)}(t), Z_{1+}^{(j)}(t\mid l)) \supseteq \bigcap_l \{\mathcal{E}(z^{(j)}(t), Z_{1+}^{(j)}(t\mid l)) \mid \langle l, l\rangle \le 1\} = \mathcal{Z}_1^{(j)}[t].$$

where the equality follows from (11.20).

When propagated after the reset along the new subsystem (j), with $\vartheta \ge \tau_j''$, each of the ellipsoids $\mathcal{E}(z^{(j)}(t), Z_{1+}^{(j)}(t\mid l))$ is transformed into

---

[2] From here on it is important to emphasize the dependence of ellipsoids, reach sets, and value functions on l. Therefore we further include l in the arguments of respective items.

11.2 Hybrid Dynamics and Control

$$X_+^{(j)}[\vartheta, t \mid l] = X_+^{(j)}(\vartheta, t, \mathcal{E}(z^{(j)}(t), Z_{1+}^{(j)}(t \mid l))) =$$

$$G^{(j)}(\vartheta, t)\mathcal{E}(z^{(j)}(t), Z_{1+}^{(j)}(t \mid l)) +$$

$$+ \int_t^\vartheta G^{(j)}(t, s)(B(s)\mathcal{E}(p^{(j)}(s), P^{(j)}(s)) + C(s)v^{(j)}(s))ds.$$

Note that though generated by an ellipsoid, the set $X^{(j)}[\vartheta, t]$ in general is not an ellipsoid. However, it may be externally approximated by an array of parametrized ellipsoids according to Sects. 3.2 and 3.3. Namely the following property is true.

Lemma 11.2.3. *The exact reachability set from position* $\{t, Z_{1+}^{(j)}[t]\}$ *is*

$$X_+^{(j)}[\vartheta, t] = \bigcap \{X_+^{(j)}[\vartheta, t \mid l] \mid \langle l, l \rangle \leq 1\}.$$

Here for each l there is an ellipsoidal approximation. For example, taking $t = \tau \in [\tau', \tau'']$ in the previous line gives

$$X_+^{(j)}[\vartheta, \tau \mid l] \subseteq \mathcal{E}_\tau(x^{(j)}(\vartheta), X_+^{(j)}(\vartheta \mid q, l)).$$

Elements $x^{(j)}(t), X_+^{(j)}(t), t \in [\tau, \vartheta]$ of this relation satisfy the following equations

$$\dot{X}_+^{(j)} = A^{(j)}(t)X_+^{(j)} + X_+^{(j)} A^{(j)'}(t) + \pi_q(t)X_+^{(j)} + (\pi_q(t))^{-1} B^{(j)}(t) P^{(j)}(t) B^{(j)'}(t), \quad (11.23)$$

$$\dot{x}^{(j)} = A^{(j)}(t)x^{(j)} + B^{(j)}(t)p^{(j)}(t) + C^{(j)}(t)v^{(j)}(t),$$

with starting conditions

$$x^{(j)}(\tau) = z^{(j)}(\tau), \quad X_+^{(j)}(\tau) = Z_1^{(j)}[\tau].$$

Note that the approximating sets $\mathcal{E}_\tau(x^{(j)}(\vartheta), X_+^{(j)}(\vartheta \mid q, l))$ depend on two vector parameters: $l \in \mathbb{R}^n$ (calculated with $\pi = \pi_l$, from Eq. (11.20) and responsible for finding $Z_1^{(j)}[\tau]$) and $q \in \mathbb{R}^{n-1}$ (calculated with π_q, from Eq. (11.19) and responsible for finding $X^{(j)}[\vartheta, \tau]$). Function $\pi_q(t)$ is calculated by formulas similar to π_l.

We finally come to the following conclusion

Theorem 11.2.1. *The following equality is true:*

$$X_+^{(j)}[\vartheta, t \mid Z_1^{(j)}[\tau]] = \bigcap \{\mathcal{E}_\tau(x^{(j)}(\vartheta), X_+^{(j)}(\vartheta \mid q, l)) \mid \|l\| \leq 1, \|q\| \leq 1\}.$$

The last formula indicates the possibility of using *parallel calculations* through an array of parametrized identical procedures and synchronization of their results.

The final move is now to find the union

$$X(\vartheta, t_0, X^0) = \bigcup \{X_+^{(j)}(\vartheta, \tau \mid Z_1^{(j)}[\tau]) \mid \tau \in \mathcal{T}\}. \tag{11.24}$$

which produces the final nonconvex reachability set.

Theorem 11.2.2. *The final reachability set* $X(\vartheta, t_0, X^0)$ *after one crossing with an active switching is given by the union (11.24).*

Hence the final reach set may be presented as *the union of intersections*, namely

$$X(\vartheta, t_0, X^0) =$$

$$\bigcup_{\tau} \bigcap_{l,q} \{\mathcal{E}_\tau(x^{(j)}(\vartheta), X_+^{(j)}(\vartheta \mid q, l)) \mid \|l\| \leq 1, \|q\| \leq 1, \tau \in \mathcal{T}\}. \tag{11.25}$$

The above relation is written in the form of *set-valued* functions. But it may be also written in terms of *single-valued functions*. Indeed, since

$$\mathcal{E}_\tau(x^{(j)}(\vartheta), X_+^{(j)}(\vartheta \mid q, l)) = \{x : V(\vartheta, \tau, x \mid q, l) \leq 1\},$$

and

$$V(\vartheta, \tau, x \mid q, l) = \{x : \langle x - x^{(j)}(t), (X_+^{(j)}(\vartheta, \tau \mid q, l))^{-1}(x - x^{(j)}(t)) \rangle \leq 1\},$$

define

$$\mathbf{V}(\vartheta, t_0, x) = \min_\tau \max_{l,q} \{V(\vartheta, \tau, x \mid q, l) \mid \|l\| \leq 1, \|q\| \leq 1, \tau \in \mathcal{T}\}. \tag{11.26}$$

Then

Theorem 11.2.3. *The reachability set* $X(\vartheta, t_0, X^0)$ *after one crossing with an active switching is the* level set

$$X(\vartheta, t_0, X^0) = \{x : \mathbf{V}(\vartheta, t_0, x) \leq 1\}. \tag{11.27}$$

The above value functions thus allow both external approximations and an exact description of the reach set. A similar scheme is true for internal approximations.

Recall that calculation of unions of reachability sets was discussed in Sect. 5.2 where an example was also given. For computational purposes we refer to following considerations from this subsection.

According to the theory of minimax problems we have

$$\mathbf{V}(\vartheta, t_0, x) = \min_\tau \max_{l,q} \{V(\vartheta, \tau, x \mid q, l) \mid \|l\| \leq 1, \|q\| \leq 1, \tau \in \mathcal{T}\} \geq$$

$$\max_{l,q} \min_\tau \{V(\vartheta, \tau, x \mid q, l) \mid \|l\| \leq 1, \|q\| \leq 1, \tau \in \mathcal{T}\} = \mathbf{V}^\sharp(\vartheta, t_0, x).$$

11.2 Hybrid Dynamics and Control

which in terms of sets yields

$$X(\vartheta, t_0, \mathcal{X}^0) = \bigcup_\tau \bigcap_{l,q} \{\mathcal{E}_\tau(x^{(j)}(\vartheta)), X_+^{(j)}(\vartheta \mid q, l)) \mid \|l\| \leq 1, \|q\| \leq 1, \tau \in T\} \subseteq$$

$$\bigcap_{l,q} \bigcup_\tau \{\mathcal{E}_\tau(x^{(j)}(\vartheta)), X_+^{(j)}(\vartheta \mid q, l)) \mid \|l\| \leq 1, \|q\| \leq 1, \tau \in T\} = \mathcal{X}_+(\vartheta, t_0, \mathcal{X}^0)$$

Set $\mathcal{X}_+(\vartheta, t_0, \mathcal{X}^0)$ is therefore an external approximation of the nonconvex exact set $X(\vartheta, t_0, \mathcal{X}^0)$ which captures the nonconvexity of the exact one.

Example 11.1 (Replacing Example 5.8).

$$\dot{x}_1 = x_2 + r\cos\omega^* t, \quad \dot{x}_2 = u + r\sin\omega^* t, \quad t \in [0, \vartheta], \quad |u| \leq 1,$$

we shall look for the reach set

$$X(\Theta, 0, \mathcal{X}^0) = X(\vartheta, 0, 0, \mathcal{X}^0) = \bigcup_t \{X(t, 0, \mathcal{X}^0) \mid t \in \Theta\}, \quad \Theta = [0, \vartheta],$$

where

$$X(t, 0, \mathcal{X}^0) = \bigcap \{\mathcal{E}(x^{(c)}(t), X_+(t \mid l)) \mid \langle l, l \rangle \leq 1\}.$$

is an intersection of *tight ellipsoids* produced using (11.19), with $x^{(1)} = x^{(c)}$, $X_+^{(1)} = X_+(t \mid l)$.

Denote $V_l(t, x) = \{x : \langle x - x^{(c)}, X_+(t \mid l)(x - x^{(c)}) \rangle \leq 1\}$. Then, due to relations

$$\mathbf{V}(\vartheta, 0, x) = \min_t \max_l \{V_l(t, x) \mid \langle l, l \rangle \leq 1, \ t \in [0, \vartheta]\} \geq$$

$$\geq \max_l \min_t \{V_l(t, x) \mid \langle l, l \rangle \leq 1, \ t \in [0, \vartheta]\} = \mathbf{V}^\sharp(\vartheta, 0, x),$$

we may find the exact nonconvex reachability set or its external approximation (also nonconvex), namely,

$$\{x : \mathbf{V}(\vartheta, 0, x) \leq 1\} = X(\vartheta, 0, \mathcal{X}^{(0)}) \subseteq \mathcal{X}_+(\vartheta, 0, \mathcal{X}^{(0)}) = \{x : \mathbf{V}^\sharp(\vartheta, 0, x) \leq 1\}.$$

Figure 11.1 demonstrates the intersection of nonconvex level sets over all $\{l : \langle l, l \rangle \leq 1\}$, which coincides with the external approximation $\mathcal{X}_+(\vartheta, 0, \mathcal{X}^{(0)})$.

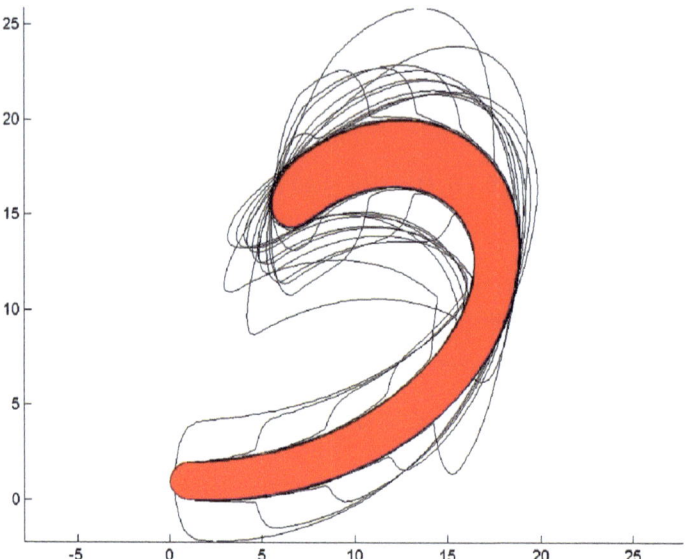

Fig. 11.1 Nonconvex level sets of function $\min\{V_l(t, x) \mid t \in \Theta\}$ for fixed values of l and their intersection

Another Type of Enabling Zone: A Gap

Consider a family of parallel hyperplanes $H_j(\gamma) = \{x : \langle c^{(j)}, x \rangle = \gamma\}$ parametrized by variable γ. The enabling zone will now be $\mathcal{D}_j$—the domain (a "gap") between hyperplanes $H_j(\gamma^{(1)})$ and $H_j(\gamma^{(2)})$, so that $\mathcal{D}_j = \mathcal{D}_j(\gamma^{(1)}, \gamma^{(2)}) = \{x \in H_j(\gamma) \mid \gamma \in [\gamma^{(1)}, \gamma^{(2)}]\}$.

We shall discuss one crossing of the gap $\mathcal{D}_j$. Then at time t, before reaching $\mathcal{D}_j$, we have

$$\max\{\langle c^{(j)}, x \rangle \mid x \in X[t]\} = \gamma^+(t) < \gamma^{(1)},$$

and the first instant of time when $X[t] \cap \mathcal{D}_j \neq \emptyset$ is τ'_j—the smallest positive root of equation $\gamma^{(1)} - \gamma^+(t) = 0$.

Now, with $t \geq \tau'$ there may be either no resets or a reset to system j while passing $\mathcal{D}_j$. If there are no resets, then introducing the function

$$\min\{\langle c^{(j)}, x \rangle \mid x \in X[t]\} = \gamma^-(t),$$

we observe that condition $X[t] \cap \mathcal{D}_j \neq \emptyset$ will hold as long as $\gamma^+(t) \geq \gamma^{(1)}$ and $\gamma^-(t) \leq \gamma^{(2)}$, and the point of departure from $\mathcal{D}_j$ will be the smallest root $\tau''_j \geq \tau'_j$ of equation $\gamma^{(2)} - \gamma^-(t) = 0$.

11.2 Hybrid Dynamics and Control

If there is a reset from $X[t-0]$ to $X^*[t]$, it may happen at any time $\tau \geq \tau'_j$, with respective $H_j = H_j(\gamma), \gamma = \gamma(\tau)$, provided $\gamma(\tau) \leq \gamma^{(2)}$. Hence, denoting $X[\tau] \cap H_j(\gamma(\tau)) = \mathcal{Z}_j(\gamma(\tau))$, we further have the union

$$X^*[t] = \bigcup \{X^*(t; s, \mathcal{Z}_j(\gamma(s))) | s \in [\tau, t]\}. \tag{11.28}$$

Then, for every $\gamma(t) \in (\gamma^{(1)}, \gamma^{(2)}]$, we introduce

$$\min\{\langle c^{(j)}, x\rangle \mid x \in X^*[t]\} = \gamma_r^-(t),$$

and the time of departure of $X^*[t]$ from $\mathcal{D}_j$ will be the smallest root τ''_{jr} of equation $\gamma^{(2)} - \gamma_r^-(t) = 0$.

Condition (11.17) ensures that the points τ'_j, τ''_{jr} are unique. What follows is now similar to what was written above for guards taken as hyperplanes.

We discuss an example.

Exercise 11.2.2. *A branching reachability tube.*

Consider the system

$$\dot{x}^{(i)} = A^{(i)}x^{(i)} + B^{(i)}u^{(i)} + f^{(i)}(t), \quad i = \overline{1, 2}, \tag{11.29}$$

with parameters

$$A^{(1)} = \begin{pmatrix} 0 & -1 \\ 1 & 0 \end{pmatrix}, \quad A^{(2)} = \begin{pmatrix} 0 & 2 \\ -2 & 0 \end{pmatrix}, \quad B^{(1)} = B^{(2)} = I,$$

$$f^{(1)}(t) \equiv \begin{pmatrix} 7 \\ -2 \end{pmatrix}, \quad f^{(2)}(t) \equiv \begin{pmatrix} 0 \\ 20 \end{pmatrix},$$

$$p^{(1)} = p^{(2)} = 0, \quad P^{(1)} = P^{(2)} = 0.005I.$$

The motion starts from position $x(0) \in \mathcal{E}(x^0, X^0)$,

$$X^0 = \begin{pmatrix} 10 & 0 \\ 0 & 0 \end{pmatrix}, x^0 = \begin{pmatrix} 1 \\ 1 \end{pmatrix}. \tag{11.30}$$

and moves throughout interval $t \in [0, \vartheta]$ due to system $i = 1$.

The reset to system $i = 2$ is possible upon intersecting with the "gap"

$$\mathcal{D} = \mathcal{D}(c, \gamma) = \{x : \langle c, x\rangle = \gamma, \gamma \in [\gamma^{(1)}, \gamma^{(2)}] :$$

$$c = \frac{1}{\sqrt{2}}(1, 1)', \gamma^{(1)} = 3, \gamma^{(2)} = 5.$$

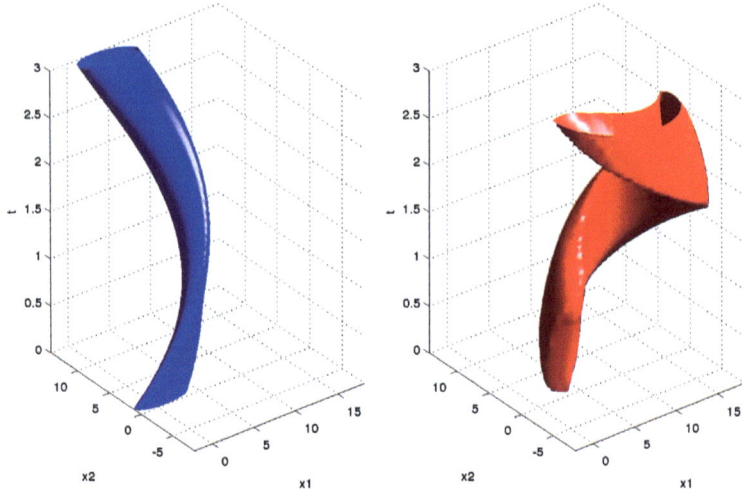

Fig. 11.2 Reachability set for each branch

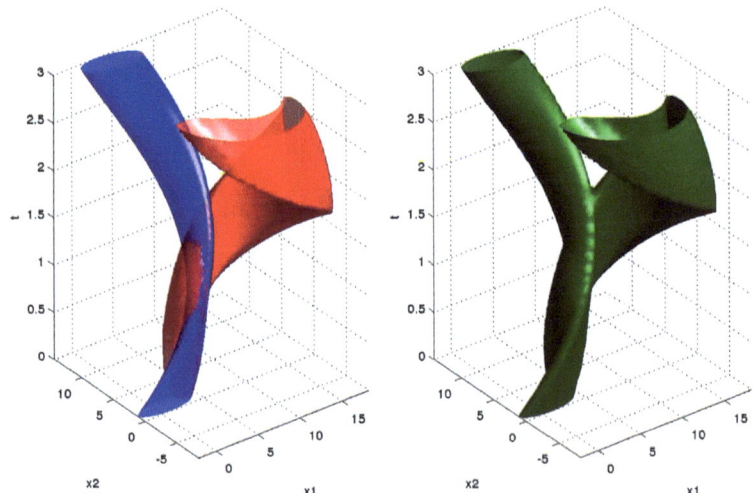

Fig. 11.3 Reachability tubes for both branches (*left*) and the overall branching reachability set (*right*)

Indicated in Figs. 11.2 and 11.3 are external approximations for the two branches of the reachability set after crossing the gap $\mathcal{D}$ at time $t = 3$. Here for $i = 1$ there is no reset and the reachability set is a convex set $X[t]$, while for $i = 2$ there is a reset from $i = 1$ to $i = 2$ and $X^*[t]$ is a nonconvex union of convex sets of type (11.28).

11.3 Verification of Hybrid Systems

We apply the results of Sect. 11.1 to hybrid systems.

11.3.1 Verification: Problems and Solutions

Suppose $\mathcal{M} = \mathcal{E}(m, M)$, $M = M' > 0$, is the given target and $\mathcal{M} \cap H_i = \emptyset, \forall i \in \{1, \ldots, k\}$.

Problem 11.3.1. *Given starting position $\{t_0, X^0, [1^+]\}$, target set $\mathcal{M}$ and time $t > t_0$, verify whether there exists a branch $I(k) = [1^+, i_1^{(s_1)}, \ldots, i_k^{(s_k)}]$ for which one of the following conditions is true (these cover all possible situations)*

$$\text{(i)} \quad X(t, t_0, X^0, I(k)) \cap \mathcal{M} = \emptyset,$$

$$\text{(ii)} \quad X(t, t_0, X^0, I(k)) \cap \mathcal{M} \neq \emptyset,$$

$$\text{(iii)} \quad X(t, t_0, X^0, I(k)) \subseteq \mathcal{M}.$$

Let us investigate the solution of this problem for a given branch $I(j) = [1^+, i_1^{(s_1)}, \ldots, i_j^{(s_j)}]$, assuming $k = j$, $t > \tau_j''$.

Theorem 11.3.1. *The following conditions are true.*

(A) *Suppose set $X^{(j)}[t] = X(t, t_0, X^0, I(j)) = \bigcup \{X^{(j)}[t, s] \mid s \in \mathcal{T}_j = [\tau', \tau'']\}$ is convex.*
Then case (i) of Problem 11.3.1 holds if and only if

$$\max\{-\rho(-l \mid \mathcal{M}) - \rho(l \mid X^{(j)}[t]) \mid \langle l, l \rangle \leq 1\} = \delta > 0. \tag{11.31}$$

Case (ii) holds if and only if

$$\max\{-\rho(-l \mid \mathcal{M}) - \rho(l \mid X^{(j)}[t]) \mid \langle l, l \rangle \leq 1\} \leq 0. \tag{11.32}$$

Case (iii) holds if and only if

$$\min\{\rho(l \mid \mathcal{M}) - \rho(l \mid X^{(j)}[t]) \mid \langle l, l \rangle \leq 1\} \geq 0. \tag{11.33}$$

(B) *Suppose the sets $X^{(j)}[t, s]$ are convex, but their union set $X^{(j)}[t] = \bigcup\{X^{(j)}[t, s] \mid s \in \mathcal{T}_j = [\tau', \tau'']\}$ is not convex. Then case (i) of Problem 11.3.1 holds if and only if*

$$\min_s \max_l \{-\rho(-l \mid \mathcal{M}) - \rho(l \mid X^{(j)}[t, s]) \mid \langle l, l \rangle \leq 1, \, s \in \mathcal{T}_j\} = \delta > 0. \tag{11.34}$$

Case (ii) holds if and only if

$$\min_s \max_l \{-\rho(-l \mid \mathcal{M}) - \rho(l \mid X^{(j)}[t,s]) \mid \langle l,l \rangle \leq 1, \ s \in \mathcal{T}_j\} \leq 0. \quad (11.35)$$

Case (iii) holds if and only if

$$\min_s \min_l \{\rho(l \mid \mathcal{M}) - \rho(l \mid X^{(j)}[t,s]) \mid \langle l,l \rangle \leq 1, \ s \in \mathcal{T}_j\} \geq 0. \quad (11.36)$$

If the same properties have to be verified for nonconvex sets of type

$$X[\vartheta', \vartheta''] = \bigcup \{X[t] \mid t \in [\vartheta', \vartheta'']\},$$

reachable within interval $[\vartheta', \vartheta'']$, $\vartheta'' \geq \tau''$, then one has to repeat operations similar to (11.34)–(11.36), but for $X[t]$ taken within this interval.

If there is more than one crossing, then each new crossing adds a new parameter $s = s_k$, with range within related interval of type $[\tau'_k, \tau''_k]$. So, after j such active intervals, it would be necessary to compute related unions of nonconvex sets involving optimization of parametrized value functions over parameters s_k, $k = 1, \ldots, j$. The realization of such computations brings us to ellipsoidal approximations.

11.3.2 Ellipsoidal Methods for Verification

We now introduce some ellipsoidal procedures for problems of the previous subsubsection.

Suppose $X^{(k)}[t,s] = \bigcap \{E[t,s,l] \ \langle l,l \rangle \leq 1\}$, where $E[t,s,l]$ is an ellipsoidal function defined for all $\{s \in [\tau', \tau''] = \mathcal{T}_k\}$ and all $\{l : \langle l,l \rangle \leq 1\}$, continuous in all the variables, and such that for all q we have

$$\rho\left(q \mid \bigcap \{E[t,s,l] \mid \langle l,l \rangle \leq 1\}\right) = \min_l \{\rho(q \mid E[t,s,l]) \mid \langle l,l \rangle \leq 1\}.$$

Due to (11.20), (11.21) this property is true for system (11.19).

Denote $X^{(k)}[t] = \bigcup_s X^{(k)}[t,s]$, $\mu = E(m, M)$.

Lemma 11.3.1. *The next conditions are true.*

(i) *For a given $s \in \mathcal{T}_k$ condition $X^{(k)}[t,s] \cap \mathcal{E}(m, M) = \emptyset$ holds if and only if*

$$d_1[t,s] = \max_l \max_q \{-\rho(-q \mid E[t,s,l]) - \rho(q \mid E(m,M)) \mid \langle q,q \rangle \leq 1\} \geq \delta > 0$$

and $X^{(k)}[t] \cap \mathcal{E}(m, M) = \emptyset$ if and only if

$$d_1[t] = \min_s \{d_1[t, s] \mid s \in \mathcal{T}\} \geq \delta > 0.$$

(ii) For a given $s \in \mathcal{T}_k$ condition $X^{(k)}[t, s] \cap \mathcal{E}(m, M) \neq \emptyset$ holds if and only if

$$d_2[t, s] = \max_l \max_q \{-\rho(-q \mid E[t, s, l]) - \rho(q \mid E(m, M)) \mid \langle q, q \rangle \leq 1\} \leq 0,$$

and $X^{(k)}[t] \cap \mathcal{E}(m, M) \neq \emptyset$ if and only if

$$d_2[t] = \min_s \{d_1[t, s] \mid s \in \mathcal{T}\} \leq 0.$$

(iii) For a given $s \in \mathcal{T}_k$ condition $X^{(k)}[t, s] \subseteq \mathcal{E}(m, M)$ holds if and only if

$$d_3[t, s] = \min_q \max_l \{\rho(q \mid E(m, M)) - \rho(-q \mid E[t, s, l]) \mid \langle q, q \rangle \leq 1\} \geq 0,$$

and $X^{(k)}[t] \subseteq \mathcal{E}(m, M)$ if and only if

$$d_3[t] = \min_s \{d_1[t, s] \mid s \in \mathcal{T}\} \geq 0.$$

Relations of this lemma allow us to compute the union $X^{(k)}[t]$ in all the considered cases.

If it is further required to compute

$$\bigcup \{X(\vartheta, t_0, X^0) \mid \vartheta \in [\vartheta', \vartheta'']\}, \ \vartheta \geq \tau'',$$

then the previous schemes have to be applied once more, involving one more parametrized array of ellipsoids.

Such overall procedures should be repeated after each active crossing, designing a branching process, whose calculation would involve effective parallelization. Computing *sequential arrays of ellipsoids* that correspond to related directions q, l, through procedures of such parallelization and increasing the number of directions, one may approach the exact solutions with any preassigned degree of accuracy.

11.4 Impulse Controls in Hybrid System Models

In Sect. 11.2 we considered resets occurring only in the system model. We now indicate how to treat resets in both system model and the state space variables. This leads to the use of *impulse controls* for describing the resets.

11.4.1 Hybrid Systems with Resets in Both Model and System States

We now discuss situations when the hybrid system involves resets both in the system number (marked as $R_s(j)$) and in the phase coordinates (marked as $R_{crd}(i)$). Upper index $s = -$ with no crossing, $s = +$ with reset in system number and $s = \dagger$ with reset in phase coordinates.

Returning to (11.19), consider hybrid system

$$\dot{x} \in A^{(i)}(t)x + B^{(i)}(t)u^{(i)} + C^{(i)}(t)v^{(i)}(t), \quad i = 1, \ldots, k, \tag{11.37}$$

with resets of system model at hyperplanes at

$$H_j = \{x \mid \langle c^{(j)}, x \rangle - \gamma_j = 0\}, \quad c^{(j)} \in \mathbb{R}^n, \quad \gamma_j \in \mathbb{R}, \quad j = 1, \ldots, m.$$

Suppose an isolated motion, starting at $\{t_0, X^0\}$, developed due to system i_1, under fixed control $u = u(t)$, reaches the first hyperplane H_{j_1} at time $\tau(j_1)$. Then the following events may occur:

- at time $\tau(j_1)$ there is a *system reset* for i_1, so it switches to system $i_2 = R_s(j_1)i_1$, due to transformation $R_s(j_1)$ from $I(k) = \{i = 1, \ldots, k\}$ to $I(k)$, and develops further due to i_2.
- at the same time $\tau(j_1)$ there is a *reset of the phase coordinates* due to transformation $x(\tau(j_1) + 0) = R_{crd}(j_1)x(\tau(j_1) - 0)$ from $\mathbb{R}^n \to \mathbb{R}^n$. If there are no resets, then the respective transformations are identities.

So, if at instant of intersecting H_j with both types of resets active, the position (state) of the system will change from $\{t, x, [i_1]\}$, $t = \tau(j_1) - 0$,—before crossing, to another after crossing, which is $\{\tau(j_1)+0, x(\tau(j_1)+1), [i_1, i_2]\}$, where $x(\tau(j_1) + 0) = R_{crd}(j_1)x$, $i_2 = R_s(j_1)i_1$, $i_2 = i_2^s$, $s = \{+\dagger\}$, then the system will further develop due to i_2, from boundary condition $\{\tau(j_1), x(\tau(j_1) + 0)\}$. After yet one more crossing, with only one active reset R_{crd} at $\tau(j_2)$, the state at $t \in (\tau(j_2), \tau(j_3 - 0))$ will be $\{t, x, [i_1, i_2, i_3]\}$, with $i_3 = i_3^s$, $s = \{-\dagger\}$. The on-line state is therefore composed of a part $\{t, x\}$ without memory, responsible for the continuous component of the hybrid system and a part $[i_1, i_2, i_3]$ with memory, responsible for the discrete part.[3]

Remark 11.4.1. The difference between the systems considered here and standard switching systems is that here the switchings depend on spatial rather than temporal parameters and also that there may be instantaneous change of coordinates which leads to discontinuity of trajectories in the phase space.

Before introducing some general schemes we begin with illustrative examples.

---

[3] In more complicated problems, for example, with complex constraints on the number of switchings or other outputs of resets, this memorized part may be logically controlled and its knowledge may be important. Such components are also important in the design of feedback controls and computation of backward reach sets for hybrid systems. Situations mentioned in this footnote mostly lie beyond the scope of this book.

11.4.2 Two Simple Examples

We give some examples of hybrid systems with reset of their *phase coordinates*. In order to formalize such situations we will introduce additional system inputs in the form of exogenous *impulse controls*. We begin with two simple examples.

Example 11.2. The two-dimensional bouncing ball.

Consider a small freely falling heavy ball which bounces when striking the floor (a plane, inclined to horizontal level), due to an elastic impact. It then moves further along a trajectory which depends on the inclination angle of the floor at each new bounce. Such angles, which act as controls, should be selected so that the ball ends up in a fixed hole on the horizontal axis—the target. The process of bouncing may cause some small losses of energy.

Following is an analytical, two-dimensional description of such bouncing motion. A more complicated three-dimensional model of a bouncing motion is indicated later.

The system equations are

$$\dot{x}_1 = y_1, \quad \dot{y}_1 = v_1(x, y, \varphi)\delta(x_2),$$

$$\dot{x}_2 = y_2, \quad \dot{y}_2 = -kv_2(x, y, \varphi)\delta(x_2) - g, \quad 0 < k < 1,$$

where $\varphi = \varphi(x, y)$ are the values of the discrete control inputs which are presented as inclination angles of the floor to horizontal line $x_2 = 0$ at impact times t_i. Such inputs are modeled as delta functions of type $v\delta(x_2(t_i))$.

The last equations are symbolic. Another representation of the same system, according to Sect. 9.3.1 is

$$dx_1 = y_1 dt, \quad dy_1 = v_1(x, y, \varphi) dU^0(x_2, 0),$$

$$dx_2 = y_2 dt, \quad dy_2 = -kv_2(x, y, \varphi) dU^0(x_2, 0) - g dt, \quad 0 < k < 1,$$

where function $U^0(x, h) = \mathbf{1}_h(x)$ (with scalar variable x) is a unit jump, namely, $\mathbf{1}_h(x) = 0$, if $x < h$ and $\mathbf{1}_h(x) = 1$, if $x \geq h$.

The solution of the last equations may be interpreted as a solution of the next vector-valued integral equation with integral taken in the Lebesgue–Stiltjes sense [234].

$$x_1(t) = x_1^0 + \int_{t_0}^{t} y_1(s) ds, \quad y_1(t) = y_1^0 + \int_{t_0}^{t} v_1(x(t), y(t), \varphi(x(t), y(t))) dU^0(x_2(t), 0),$$

$$x_2(t) = x_2^0 + \int_{t_0}^{t} y_2(s) ds, \quad y_2(t) = y_2^0 - k \int_{t_0}^{t} v_2(x(t), y(t), \varphi(x(t), y(t))) dU^0(x_2(t), 0).$$

(11.38)

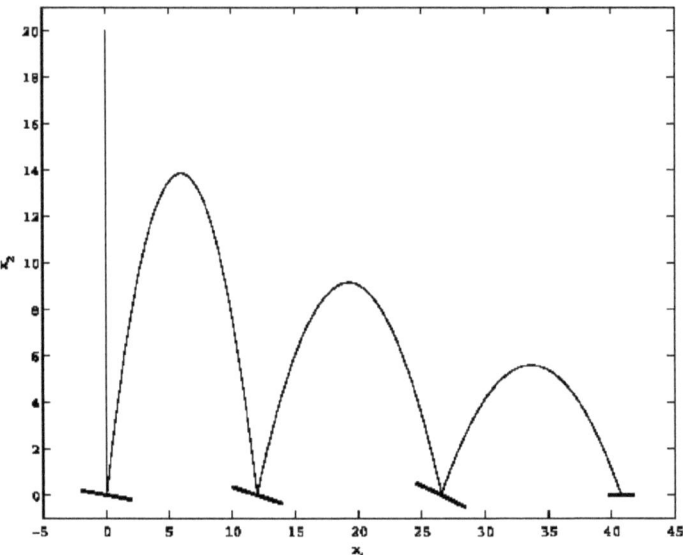

Fig. 11.4 The trajectory of a bouncing ball

Here, Fig. 11.4 shows the bouncing trajectory where at times of impact, with $x_2(t_i) = 0$, the vertical velocity y_2 of the ball instantaneously changes its sign while velocities y_1, y_2 change their values. This leads to an instantaneous jump (*a reset*) of the values of vector $\{\dot{x}_1(t), \dot{x}_2(t)\}$ as soon as $x_2(t) = 0$. The inclination angle φ of the floor at $x_2(t_i) = 0$ therefore depends on the state space position of the system and generates a feedback impulse control, thus reflecting the discrete component of the hybrid process. The reset of variables $\{\dot{x}_1(t) = y_1(t), \dot{x}_2(t) = y_2(t)\}$ is due to virtual δ-valued controls of intensities $v_1(x, y, \varphi)$ and $-kv_2(x, y, \varphi)$.

Example 11.3. One inflow for two water tanks.

Another is the well-known simple example of controlling two water tanks through one inflow [104]. The feedback dynamics of this system is now formalized using impulsive inputs.

This system is a model with two "motions": (q_1), (q_2), with alternating resets from one to the other. Here

$$(q_1): \quad \dot{x}_1(t) = w - v_1, \quad \dot{x}_2 = -v_2, \quad x_2 \geq V_2,$$

$$(q_2): \quad \dot{x}_1(t) = -v_1, \quad \dot{x}_2 = w - v_2, \quad x_1 \geq V_1.$$

These equations describe a water supply at speed w through inflow from only one pipe directed alternatingly to two water tanks from which water leaves with

11.4 Impulse Controls in Hybrid System Models

respective speeds v_1 and v_2. The water levels to be maintained at the tanks are V_1 and V_2. At each time t the pipe fills one of the tanks, then switching to the other according to the following feedback rule.

Suppose at first the water is filling tank q_2 with $x_1 > V_1, x_2 < V_2$. Then, when x_1 decreases to level V_1 the pipe switches from q_2 to q_1 (at that time it should be at $x_2 > V_2$). Then later, after x_2 had decreased to V_2, the pipe is reversed again to q_2. The speeds and the water levels have to be coordinated to ensure sustainability of the required water levels V_1, V_2.

Exercise 11.4.1. *Indicate the range of speeds w, v_1, v_2 and water levels V_1, V_2 when these levels are sustainable.*

The two systems may be combined into one, switching from one to the other through two impulsive control inputs generated due to feedback control signals. Denoting the right-hand sides of systems (q_1) and (q_2) by $F_1(x), F_2(x)$ and introducing a third variable ζ that has only two possible values ± 1, consider the following system in symbolic form

$$\begin{cases} \dot{x} = F_1(x)\left(\frac{1+\text{sign}\zeta}{2}\right) + F_2(x)\left(\frac{1-\text{sign}\zeta}{2}\right), \\ \dot{\zeta} = 2\delta(x_1 - V_1 + 0) - 2\delta(x_2 - V_2 + 0) \end{cases}$$

or as

$$\begin{cases} dx = \left(F_1(x)\left(\frac{1+\text{sign}\zeta}{2}\right) + F_2(x)\left(\frac{1-\text{sign}\zeta}{2}\right)\right) dt \\ d\zeta = 2d\,U^0(x_1 - V_1, 0) - 2d\,U^0(x_2 - V_2, 0) \end{cases}.$$

The solution of this equation is defined through integral equations taken in the sense of the Lebesgue–Stiltjes integral and similar to the type (11.37) (Fig. 11.5).

11.4.3 Impulse Controls in Hybrid Systems

We now indicate a scheme involving impulse controls to formalize the mathematical model of hybrid systems.

Consider equations

$$\dot{x} = A(t, \zeta)x + B(t, \zeta)u + Iu^{(imp)}, \tag{11.39}$$

$$\dot{\zeta} = u_\zeta, \tag{11.40}$$

$$x \in \mathbb{R}^n, \quad \zeta \in \{1, \ldots, k\}.$$

Here $u = u(t, x, \zeta)$ is the on-line p-dimensional "ordinary" control in the class of bounded functions, the one responsible for the continuous component of system

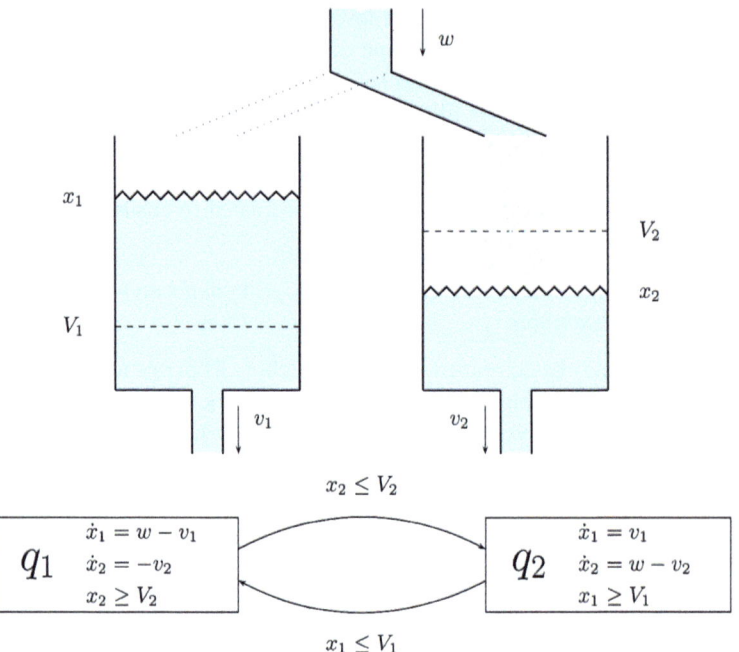

Fig. 11.5 One inflow for two water tanks

trajectory $x(t)$; the n-dimensional vector $u^{(imp)} = (u_1^{(imp)}, \ldots, u_n^{(imp)})'$ is the impulse control responsible for reset of the phase coordinates and has the form

$$u^{(imp)} = u^{(imp)}(t, x, \zeta); \quad u_j^{(imp)} = u_j^{(imp)}(t, x, \zeta) = \alpha(t, x, \zeta(t-0))\delta(f(x, \zeta)).$$

The scalar u_ζ controls the number ζ of system Eq. (11.39), and is the input in Eq. (11.40) for finding the integer $\zeta \in \{1, \ldots, k\}$ and has the form

$$u_\zeta = u_\zeta(x, \zeta) = \beta(x, \zeta(t-0))\delta(f_d(x, \zeta(t-0))).$$

The piecewise-constant, integer-valued function $\zeta(t)$ is responsible for switching the system at crossing time τ from $\zeta(\tau-0) = i$ to $\zeta(\tau+0) = j$. This is realized through Eq. (11.40) with $\beta(x, i) = j - i$.

The resets occur at instants of crossing surfaces $f(x, \zeta) = 0$ for phase coordinates and $f_\zeta(x, \zeta) = 0$ for the number $\zeta = i$ of Eq. (11.39). In system (11.37) these surfaces are presented as hyperplanes $H(j)$.

Remark 11.4.2. The application of dynamic programming techniques to hybrid systems was discussed in [23, 36, 170].

11.4.4 More Complicated Example: The Three-Dimensional Bouncing Ball [4]

Consider a three-dimensional Newtonian motion of a heavy ball under gravity "g." Starting with a free fall, the ball bounces back when reaching a plane H inclined to the horizontal level. The phase state of the ball is defined by a six-dimensional vector $x = (x_1, \ldots, x_6)' \in \mathbb{R}^6$, where $(x_1, x_2, x_3)'$ are coordinates of its center in $\mathbb{R}^3$ (x_3 stands for its vertical position), and (x_4, x_5, x_6) are coordinates of the respective velocity vector. At each time t, with no force other than gravity, it satisfies Eq. (11.37), where, with $\mathbb{O}_{i \times j}$ standing for a zero matrix of dimensions $i \times j$, we have

$$A(t) = \begin{pmatrix} \mathbb{O}_{3 \times 3} & \mathbb{I}_{3 \times 3} \\ \mathbb{O}_{3 \times 0} & \mathbb{O}_{3 \times 3} \end{pmatrix}, \; B(t) = 0, \; C(t)f(t) = f(t) = \begin{pmatrix} \mathbb{O}_{5 \times 1} \\ -g \end{pmatrix}. \qquad (11.41)$$

This system is treated within

$$t \in \mathbb{R}, \; x \in \Omega_x = \{(x_1, \ldots, x_6) : x_i \in \mathbb{R}, \; x_3 \geq 0\}.$$

If it is in horizontal position, plane H may rotate around axes l_1 and l_2, that may emanate from any point along directions of vectors $\mathbf{e}_1 = (1, 0, 0)'$ and $\mathbf{e}_2 = (0, 1, 0)'$. At each time it may be rotated around only one of these axes: to angle $\varphi_1 \in [-\alpha_1, \alpha_1]$ ($\alpha_1 \in (0, \pi/2)$) for l_1, and angle $\varphi_2 \in [-\alpha_2, \alpha_2]$ ($\alpha_2 \in (0, \pi/2)$) for l_2. Such rotations incline the plane to a new position. So, when the ball now reaches H, it bounces back causing a reset of the system trajectory in phase space $\mathbb{R}^6$, demonstrating the hybrid nature of the process. At an instant of reset H may had been turned around one of the axes l_1 and l_2 that emanate from the point of impact.

Let the state of the ball be x^- just before the reset and x^+ right after that. Introduce the additional notations

$$C_1(\varphi) = \begin{pmatrix} 1 & 0 & 0 \\ 0 & \cos(\varphi) & \sin(\varphi) \\ 0 & -\sin(\varphi) & \cos(\varphi) \end{pmatrix}, \; C_2(\varphi) = \begin{pmatrix} \cos(\varphi) & 0 & -\sin(\varphi) \\ 0 & 1 & 0 \\ \sin(\varphi) & 0 & \cos(\varphi) \end{pmatrix}.$$

If by time of reset the plane H was turned around axis l_1 to angle φ_1, and also around l_2 to angle φ_2, then vectors x^- and x^+ are linked by the next relation

$$x^+ = Cx^-, \text{ where } C = C(\varphi_1, \varphi_2) = \begin{pmatrix} \mathbb{I}_{3 \times 3} & \mathbb{O}_{3 \times 3} \\ \mathbb{O}_{3 \times 3} & M \end{pmatrix},$$

$$M = I - (1 + \gamma) C_1(\varphi_1) C_2(\varphi_2) E_{3,3} C_2'(\varphi_2) C_1'(\varphi_1), \; E_{3,3} = \begin{pmatrix} 0 & 0 & 0 \\ 0 & 0 & 0 \\ 0 & 0 & 1 \end{pmatrix}.$$

---
[4] This example was worked out by P.A. Tochilin.

Here $\gamma \in [0, 1]$ is the "recovery" coefficient: with $\gamma = 1$ we have the model for an absolutely elastic impact, and with $\gamma = 0$ for a nonelastic.

Such a model may now be also formalized involving δ-functions.

$$\dot{x} = Ax(t) + f + (C(\varphi_1, \varphi_2) - I) x \delta(x_3). \tag{11.42}$$

With $\varphi_2 = 0$ (when plane H may rotate only around axis l_1), then (11.42), taken in coordinate form, is

$$\begin{cases} \dot{x}_1 = x_4 \\ \dot{x}_2 = x_5 \\ \dot{x}_3 = x_6 \\ \dot{x}_4 = 0 \\ \dot{x}_5 = -(1+\gamma)(\sin^2(\varphi_1) x_5 + \sin(\varphi_1) \cos(\varphi_1) x_6) \delta(x_3) \\ \dot{x}_6 = -(1+\gamma)(\sin(\varphi_1) \cos(\varphi_1) x_5 + \cos^2(\varphi_1) x_6) \delta(x_3) - g \end{cases}$$

and with $\varphi_1 = 0$ (when H may rotate only around l_2), Eq. (11.42) in detail is:

$$\begin{cases} \dot{x}_1 = x_4 \\ \dot{x}_2 = x_5 \\ \dot{x}_3 = x_6 \\ \dot{x}_4 = -(1+\gamma)(\sin^2(\varphi_1) x_4 - \sin(\varphi_1) \cos(\varphi_1) x_6) \delta(x_3) \\ \dot{x}_5 = 0 \\ \dot{x}_6 = -(1+\gamma)(-\sin(\varphi_1) \cos(\varphi_1) x_4 + \cos^2(\varphi_1) x_6) \delta(x_3) - g \end{cases}$$

At each time t the on-line controls φ_1, φ_2 are determined by the running position $x(t)$ of the system.

On horizontal plane $H = H_0 = \{x : x_3 = 0\}$ consider a hole (target set)

$$\mathcal{M} = \{x \in \mathbb{R}^3 : (x_1 - m_1)^2 + (x_2 - m_2)^2 \le r^2, \ x_3 = 0\}.$$

Also specify a starting position $x^0 = x(t_0) = (x_1^0, x_2^0, x_3^0, 0, 0, 0)'$, where $x_3^0 > 0$.

The problem will be to direct the trajectory of the bouncing ball, by choice of angles φ_1, φ_2 (the exogenous controls) to the target set $\mathcal{M}$. Here the system parameters and the matrix C are independent of time.

Considered are the following *optimization problems*.

Problem 11.4.1. *Find minimal time $t_1 \ge t_0$, for which there exists a control $\varphi_1(\cdot), \varphi_2(\cdot)$, that ensures inclusion $x(t_1; t_0, x_0)|_{\varphi_1, \varphi_2} \in \mathcal{M} \times \mathbb{R}^3$ with any final velocity.*

Problem 11.4.2. *Given $\{t_0, x^0\}$, find the minimal number k^* of bounces (resets) $k \in \mathbb{Z}_+$ (the set of positive integers), such that there exist a time interval $[t_0, t_1]$, and a control $\varphi_1(\cdot), \varphi_2(\cdot)$ for which the inclusion $x[t_1] = x(t_1; t_0, x^0)|_{\varphi_1, \varphi_2} \in \mathcal{M} \times \mathbb{R}^3$ is achieved with number of bounces $k \le k^*$.*

11.4 Impulse Controls in Hybrid System Models

Consider the following value functions

$$V^{(1)}(t_1, x) = \min_{\varphi_1(\cdot), \varphi_2(\cdot)} \min_{t \in [t_0, t_1]} \Big\{ [(x_1(t; t_0, x)|_{\varphi_1, \varphi_2} - m_1)^2 + (x_2(t; t_0, x)|_{\varphi_1, \varphi_2} - m_2)^2 - r^2]_+$$

$$+ I(x_3(t; t_0, x)|_{\varphi_1, \varphi_2} \mid 0) \Big\},$$

$$V^{(2)}(x, k) = \min_{t \geq t_0} \min_{\varphi_1(\cdot), \varphi_2(\cdot)} \Big\{ [(x_1(t; t_0, x)|_{\varphi_1, \varphi_2} - m_1)^2 + (x_2(t; t_0, x)|_{\varphi_1, \varphi_2} - m_2)^2 - r^2]_+ =$$

$$+ I(x_3(t; t_0, x)|_{\varphi_1, \varphi_2} \mid 0) + \theta(x(\cdot; t_0, x)|_{\varphi_1, \varphi_2}, t, k) \Big\}.$$

Here $[h(t, x)]_+ = 0$ if $h(t, x) \leq 0$, and $[h(t, x)]_+ = h(t, x)$ if $h(t, x) > 0$; function $\theta(x(\cdot; t_0, x)|_{\varphi_1, \varphi_2}, t, k) = 0$ within interval $\tau \in [t_0, t]$, if the number of bounces does not exceed k, otherwise it is $+\infty$.[5]

To compute solutions of Problems 11.4.1 and 11.4.2, we may use the following facts.

Lemma 11.4.1. *(1) With x^0 given, the solution to Problem 11.4.1—the minimal time $t_1 = t_1^*$—is:*

$$t_1^* = \arg\min\{t_1 \geq t_0 : V^{(1)}(t_1, x^0) \leq 0\},$$

achieved through some control sequence $\varphi_1(\cdot), \varphi_2(\cdot)$.
(2) With x^0 given, the solution $k = k^*$ for which there exists a time interval $[t_0, t_1]$ together with a control sequence $\varphi_1(\cdot), \varphi_2(\cdot)$, that solve Problem 11.4.2, is:

$$k^* = \arg\min\{k \in \mathbb{Z}_+ : V^{(2)}(x^0, k) \leq 0\}.$$

For computing $V^{(1)}(t_1, x)$, $V^{(2)}(x, k)$, there exist related HJB equations with appropriate boundary conditions (see [171]).

Exercise 11.4.2. *Indicate the HJB equations with boundary conditions for computing value functions $V^{(1)}(t_1, x)$, $V^{(2)}(x, k)$.*

Remark 11.4.3. Value functions $V^{(1)}(t_1, x)$, $V^{(2)}(x, k)$ may be also computed through methods of nonlinear analysis and ellipsoidal techniques, along the lines of Chaps. 2 and 3.

With $V^{(1)}(t_1, x)$, $V^{(2)}(x, k)$ computed, one may specify the solutions to Problems 11.4.1 and 11.4.2.

---

[5] As before $I(x \mid D) = 0$, if $x \in D$ and $+\infty$ otherwise.

Denote $t_1^*(x) = \min\{t_1 \geq t_0 : V^{(1)}(t_1, x) \leq 0\} \geq 0$ and at collision time $t_1 - 0$ (just before the bounce) consider state $\{t_1 - 0, x\}$, $x = (x_1, \ldots, x_6)$, so that $x_3(t) = 0$, $x_6(t-0) < 0$.

Suppose $(x_1(t_1-0) - m_1)^2 + (x_2(t_1-0) - m_2)^2 - r^2 > 0$, $x_3 = 0$, but $V^{(1)}(t_1-0, x) \leq 0$. Then, obviously $t_1 < t_1^*$, and the control solution at this point will be of type

$$(\varphi_1^*(x), \varphi_2^*(x)) : V^{(1)}(t_1^*(x), C(\varphi_1^*, \varphi_2^*)x) \leq V^{(1)}(t_1^*(x), x),$$

until $t_1 = t_1^*$, $(x_1(t_1-0) - m_1)^2 + (x_2(t_1-0) - m_2)^2 - r^2 = 0$, $x_3 = 0$.

Similarly, denote $k^*(x) = \min\{k \in \mathbb{Z}_+ : V^{(2)}(x, k) \leq 0\}$, and suppose at collision time t_1 we have $(x_1(t_1-0) - m_1)^2 + (x_2(t_1-0) - m_2)^2 - r^2 > 0$, $x_3 = 0$. Then the number of realized bounces will be $k < k^*$ and the control solution at this point will be of type

$$(\varphi_1^*(x), \varphi_2^*(x)) : V^{(2)}(C(\varphi_1^*, \varphi_2^*)x, k) \leq V^{(2)}(x, k^*(x)),$$

until $k = k^* - 1$.

Shown in Fig. 11.6 is the trajectory of the 3-d bouncing ball (a hybrid system) which reaches the target set with minimum resets (bounces). The problem parameters are: $\alpha_1 = 0.5$, $\alpha_2 = 0.5$, $m_1 = 1$, $m_1 = 2$, $r^2 = 0.2$, $\gamma = 0.8$, and the starting position is $x^0 = (-4, -1, 4, 0, 0, 0)'$.

Shown in Fig. 11.7 is the time-optimal target-oriented trajectory of the bouncing ball with parameters $\alpha_1 = 1$, $\alpha_2 = 0.4$, $m_1 = 1$, $m_2 = 2$, $r^2 = 0.2$, $\gamma = 0.9$ and starting position $x^0 = (-2, 3, 4, 0, 0, 0)'$.

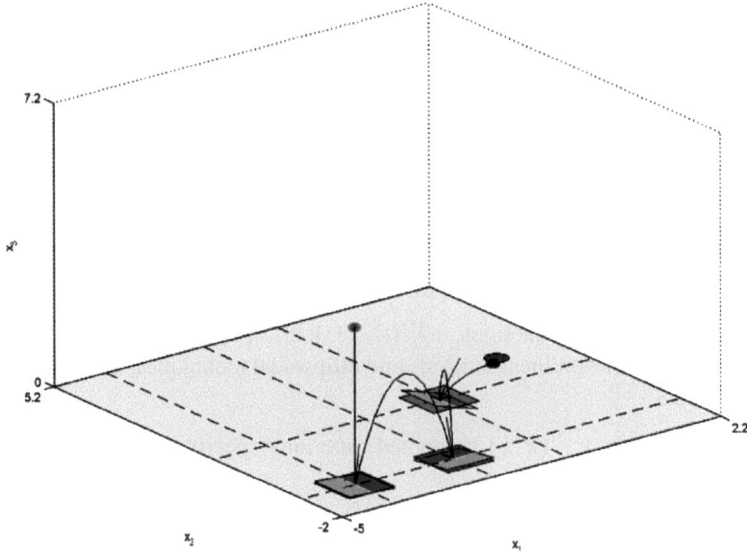

Fig. 11.6 Reaching target hole with minimal bounces

11.4 Impulse Controls in Hybrid System Models

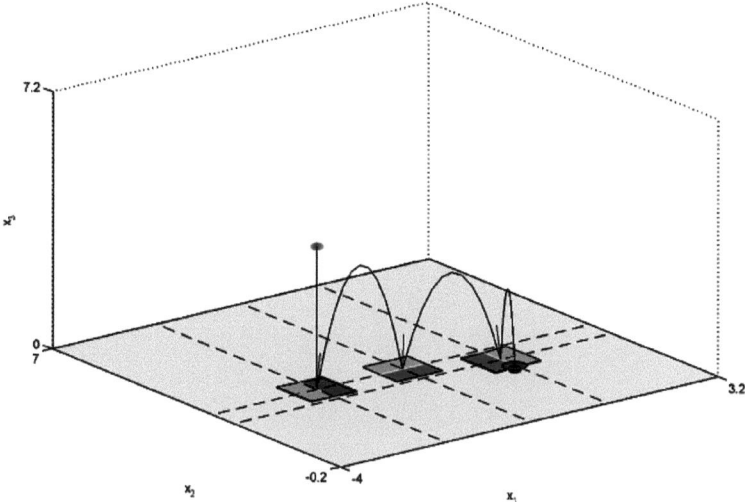

Fig. 11.7 Reaching target hole in minimal time

References

1. Akhiezer, N.I.: The Classical Moment Problem and Some Related Questions in Analysis. Oliver & Boyd, Edinburgh (1965)
2. Albert, A.: Regression and the Moor-Penrose Pseudoinverse. Mathematics in Science and Engineering, vol. 94. Academic, New York (1972)
3. Artstein, Z.: Yet another proof of the Lyapunov convexity theorem. Proc. Am. Math. Soc. **108**(1), 89–91 (1990)
4. Åström, K.J.: Introduction to Stochastic Control Theory. Academic, New York (1970)
5. Aubin, J.-P.: Viability Theory. SCFA. Birkhäuser, Boston (1991)
6. Aubin, J.-P., Bayen, A.M., St.Pierre, P.: Viability Theory: New Directions. Springer, Heidelberg (2011)
7. Aubin, J.-P., Cellina, A.: Differential Inclusions. Springer, New York (1984)
8. Aubin, J.-P., Frankowska, H.: Set-Valued Analysis. SCFA 2. Birkhäuser, Boston (1990)
9. Aumann, R.J.: Integrals of set-valued functions. J. Math. Anal. Appl. **12**, 1–12 (1965)
10. Bai, E.W., Cho, H., Tempo, R.: Optimization with few violated constraints for linear bounded error parameter estimation. IEEE Trans. Automat. Control **47**(7), 1067–1077 (2002)
11. Balakrishnan, A.V.: Applied Functional Analysis. Springer, New York (1976)
12. Balluchi, A., Benvenuti, L., Di Benedetto, M.D., Sangiovanni-Vincentelli, A.: The design of dynamical observers for hybrid systems: theory and application to an automotive control problem. Automatica **49**(4), 915–925 (2013)
13. Baras, J.S., Bensoussan, A., James, M.R.: Dynamic observers and asymptotic limits of recursive filters: special case. SIAM J. Appl. Math. **48**(5), 1147–1158 (1988) 1
14. Baras, J.S., Kurzhanski, A.B.: Nonlinear filtering: the set-membership (bounding) and the H_∞ techniques. In: Proceedings of the 3rd IFAC Symposium NOLCOS, pp. 409–418. Pergamon Press, Oxford (1995)
15. Barbashin, E.A.: On the theory of generalized dynamic systems. Sci. Notes Moscow State Univ. **135**, 110–133 (1949)
16. Bardi, M., Capuzzo Dolcetta, I.: Optimal Control and Viscosity Solutions of Hamilton–Jacobi–Bellman Equations. SCFA. Birkhäuser, Boston (1997)
17. Barron, E.N., Jensen, R.: Semicontinuous viscosity solutions for Hamilton–Jacobi equations with convex Hamiltonians. Commun. Partial Differ. Equ. **15**(12), 1713–1742 (1990)
18. Başar, T., Bernhard, P.: H_∞ Optimal Control and Related Minimax Design Problems. SCFA, 2nd edn. Birkhäuser, Basel (1995)
19. Ba sar, T., Olsder, J.: Dynamic Noncooperative Game Theory. Academic, New York (1982)

20. Basile, G., Marro, G.: Controlled and Conditioned Invariants in Linear System Theory. Prentice-Hall, Englewood Cliffs (1992)
21. Bellman, R., Dreyfus, S.: Applied Dynamic Programming. Princeton University Press, Princeton (1962)
22. Bellman, R.: Introduction to the Mathematical Theory of Controlled Processes, vol. 1/2. Academic, New York (1967/1971)
23. Bensoussan, A., Lions, J.-L.: Contrôle impulsionnel et inéquations quasi-variationnelles. Dunod, Paris (1982)
24. Bertsekas, D.P.: Dynamic Programming and Optimal Control, vol. 1/2. Athena Scientific, Belmont (1996/2012)
25. Bertsekas, D.P., Rhodes, I.: Recursive state estimation for a set-membership description of uncertainty. IEEE Trans. Automat. Control **16**(2), 117–128 (1971)
26. Blagodatskih, V.I, Filippov, A.F.: Differential inclusions and optimal control. Proc. Steklov Math. Inst. **169**, 194–252 (1985)
27. Blanchini, F., Miani, S.: Set-Theoretic Methods in Control. Springer, Boston (2008)
28. Blaquière, A., Leitmann, G. (eds.): Dynamical Systems and Microphysics: Control Theory and Mechanics. Academic, Orlando (1982)
29. Bliss, G.A.: Calculus of Variations. Mathematical Association of America, Chicago (1944)
30. Boltyanski, V.G.: Mathematical Methods of Optimal Control. Holt, New York (1971)
31. Boltyanski, V.G.: Tent methods in optimal control. ISNM International Series of Numerical Mathematics, vol. 111, pp. 3–20
32. Bohner, M., Peterson, A.: Dynamic Equations on Time Scales: An Introduction with Applications. Birkhauser, Boston (2001)
33. Bolza, O.: Lectures on Calculus of Variations. Hafer Pub. Co, New York (1946). Dover reprint
34. Bonneuil, N.: Computing the viability kernel in large state dimensions. J. Math. Anal. Appl. **323**, 1444–154 (2006)
35. Boyd, S., Ghaoui, L.E., Feron, E., Balakrishnan, V.: Linear Matrix Inequalities in System and Control Theory. Studies in Applied Mathematics, vol. 15. SIAM, Philadelphia (1994)
36. Branicky, M.S., Borkar, V.S., Mitter, S.K.: A unified framework for hybrid control: model and optimal control theory. IEEE Trans. Automat. Control **43**(1), 31–45 (1998)
37. Bressan, A., Piccoli, B.: Introduction to the Mathematical Theory of Control, vol. 2. American Institute for Mathematical Sciences, Springfield (2007)
38. Brockett, R.W.: Finite-Dimensional Linear Systems. Wiley, New York (1970)
39. Brockett, R.W.: Hybrid models for motion control systems. In: Trentelman, H., Willems, J. (eds.) Essays in Control: Perspectives in the Theory and Its Applications, pp. 29–53. Birkhäuser, Boston (1993)
40. Callier, F.M., Desoer, C.A.: Linear System Theory. Springer, New York (1991)
41. Castaing, C., Valadier, M.: Convex Analysis and Measurable Multifunctions. Lecture Notes in Mathematics, vol. 580. Springer, Berlin (1977)
42. Carter, T.E.: Optimal impulsive space trajectories based on linear equations. J. Optim. Theory Appl. **70**(2), 277–297 (1991)
43. Carter, T.E., Brient, J.: Linearized impulsive rendezvous problem. J. Optim. Theory Appl. **86**(3), 553–584 (1995)
44. Chernousko, F.L.: Ellipsoidal bounds for sets of attainability and uncertainty in control. Optim. Control Appl. Methods **3**(2), 187–202 (1982)
45. Chernousko, F.L.: State Estimation for Dynamic Systems. CRC Press, Boca Raton (1994)
46. Chutinan, A., Krogh, B.H.: Verification of polyhedral-invariant hybrid automata using polygonal flow pipe approximations. In: Hybrid Systems: Computation and Control. Lecture Notes in Control Science, vol. 1569, pp. 76–90. Springer, Berlin (1999)
47. Clarke, F.H.: Optimization and Nonsmooth Analysis. Wiley, New York (1983)
48. Clarke, F.H., Ledyaev Yu, S., Stern, R.J., Wolenski, P.R.: Nonsmooth Analysis and Control Theory. Graduate Texts in Mathematics, vol. 178. Springer, New York (1998)
49. Crandall, M.G., Evans, L.C., Lions, P.-L.: Some properties of solutions of Hamilton–Jacobi equations. Trans. Am. Math. Soc. **282**(2), 487–502 (1984)

50. Crandall, M.G., Lions, P.-L.: Viscosity solutions of Hamilton–Jacobi equations. Trans. Am. Math. Soc. **277**(1), 1–41 (1983)
51. Courant, R., Hilbert, D.: Methods of Mathematical Physics, vol. 1 & 2. Wiley, New York (1989) (Wiley GmbH & Co., KGaA, 2008)
52. Dar'in, A.N., Kurzhanskii, A.B.: Nonlinear control synthesis under two types of constraints. Differ. Equ. **37**(11), 1549–1558 (2001)
53. Daryin, A.N., Kurzhanskii, A.B.: Control under indeterminacy and double constraints. Differ. Equ. **39**(11), 1554–1567 (2003)
54. Daryin, A.N., Kurzhanskii, A.B., Seleznev, A.V.: The dynamic programming method in impulsive control synthesis. Differ. Equ. **41**(11), 1566–1576 (2005)
55. Daryin, A.N., Kurzhanski, A.B., Seleznev, A.V.: A dynamic programming approach to the impulse control synthesis problem. In: Proceedings of Joint 44th IEEE Conference on Decision and Control and European Control Conference ECC 2005, Seville, Spain, pp. 8215–8220 (2005)
56. Daryin, A.N., Digailova, I.A., Kurzhanski, A.B.: On the problem of impulse measurement feedback control. Proc. Steklov Math. Inst. **268**(1), 71–84 (2010)
57. Daryin, A.N., Kurzhanski, A.B.: Estimation of reachability sets for large-scale uncertain systems: from theory to computation. In: Proceedings of CDC-2012 (2012)
58. Daryin, A.N., Kurzhanski, A.B.: Parallel algorithm for calculating the invariant sets of high-dimensional linear systems under uncertainty. Comput. Math. Math. Phys. **53**(1), 34–43 (2013)
59. Davis, M.H.A., Varaiya, P.: Information states for linear stochastic systems. J. Math. Anal. Appl. **37**, 384–402 (1972)
60. Demyanov, V.F.: Minimax: Directional Differentiation. Leningrad University Pub. Leningrad (1974)
61. Demyanov, V.F., Malozemov, V.N.: Introduction to Minimax. Wiley, New York (1974)
62. Demyanov, V.F., Rubinov, A.M.: Quasidifferential Calculus. Optimization Software, New York (1986)
63. Digailova, I.A., Kurzhanski, A.B.: The joint model and state estimation problem under set-membership uncertainty. In: Proceedings of the 15-th IFAC Congress, Barcelona (2002)
64. Digailova, I.A., Kurzhanski, A.B.: Reachability analysis under control-dependent stochastic noise. In: Proceedings of 16th IFAC World Congress, Prague (2005)
65. Digailova, I.A., Kurzhanski, A.B., Varaiya, P.: Stochastic reachability and measurement feedback. In: Proceedings of 17th World Congress IFAC, Seoul, 6–11 July 2008, pp. 14336–14341
66. Dontchev, A.L.: Perturbations, Approximations, and Sensitivity Analysis of Optimal Control Systems. Lecture Notes in Control and Information Sciences, vol. 52. Springer, New York (1983)
67. Dunford, N., Schwartz, J.T.: Linear Operators. Interscience, New York/London (1958)
68. Dykhta, V.A., Samsonyuk, O.N.: Optimal Impulse Control and Its Applications. Fizmatlit, Moscow (2003) [in Russian]
69. Ekeland, I., Temam, R.: Analyse Convexe et Problemes Variationelles. Dunod, Paris (1973)
70. Fan, K.: Minimax theorems. Proc. Natl. Acad. Sci. USA **39**(1), 42–47 (1953)
71. Fan, K.: On systems of linear inequalities. In: Kuhn, H.W., Tucker, A.W. (eds.) Linear Inequalities and Related Systems. Annals of Mathematical Studies, vol. 38, pp. 99–156. Princeton University Press, Princeton (1956)
72. Fan, K.: Existence theorems and extreme solutions for inequalities concerning convex functions or linear transformations. Mathematische Zeitschrift **68**(1), 205–216 (1957)
73. Fenchel, W.: On conjugate convex functions. Can. J. Math. **1**, 73–77 (1949)
74. Fenchel, W.: Convex Cones, Sets and Functions. Princeton University Press, Princeton (1951)
75. Filippov, A.F.: On certain questions in the theory of optimal control. SIAM J. Control. **1**, 76–84 (1962)
76. Filippov, A.F.: Differential Equations with Discontinuous Righthand Sides. Kluwer, Dordrecht (1988)

77. Filippova, T.F.: A note on the evolution property of the assembly of viable solutions to a differential inclusion. Comput. Math. Appl. **25**(2), 115–121 (1993)
78. Filippova, T.F., Kurzhanski, A.B., Sugimoto, K., Valýi, I.: Ellipsoidal calculus, singular perturbations and the state estimation problems for uncertain systems. Working Paper WP-92-051, IIASA, Laxenburg (1992)
79. Filippova, T.: Approximation techniques in impulsive control problems for the tubes of solutions of uncertain differential systems. In: Anastassiou, G.A., Duman, O. (eds.) Advances in Applied Mathematics and Approximation Theory: Contributions from AMAT, pp. 385–396. Springer, New York (2012)
80. Fleming, W.H., Soner, H.M.: Controlled Markov Processes and Viscosity Solutions. Springer, New York (1993)
81. Gamkrelidze, R.V.: Principles of Optimal Control Theory. Mathematical Concepts and Methods in Science and Engineering, vol. 7. Plenum Press, New York (1978)
82. Gantmacher, F.R.: Matrix Theory, vol. 2. Chelsea Publishing, New York (1964)
83. Gao, Y., Lygeros, L., Quincampoix, M.: On the reachability problem for uncertain hybrid systems. IEEE Trans. Automat. Control **52**(9), 1572–1586 (2007)
84. Garcia, C.E., Prett, D.M., Morari, M.: Model predictive control: theory and practice-a survey. Automatica **25**(3), 335–348 (1989)
85. Gayek, J.E.: A survey of techniques for approximating reachable and controllable sets, Proc. of the CDC Conf. Brighton, England, 1724–1729 (1991)
86. Gelfand, I.M., Fomin, S.V.: Calculus of Variations. Academic, New York (1964)
87. Gelfand, I.M., Shilov, G.E.: Generalized Functions. Dover, New York (1991)
88. Giannessi, F., Maugeri, A. (eds.): Variational Methods and Applications. Kluwer, New York (2004)
89. Godunov, S.K.: Guaranteed Accuracy in Numerical Linear Algebra. Springer, Dordrecht (1993)
90. Godunov, S.K.: Ordinary Differential Equations with Constant Coefficients, vol. 169. AMS Bookstore, Providence (1997)
91. Goebel, R., Sanfelice, R.G., Teel, A.R.: Hybrid Dynamical Systems. Princeton University Press, Princeton (2012)
92. Golub, G.H., Van Loan, C.F.: Matrix Computations. John Hopkins Un. Press, Baltimore, MI (1996)
93. Grant, M., Boyd, S.: Graph interpretation for nonsmooth convex programs. In: Recent Advances in Learning and Control, pp. 95–110. Springer, Berlin (2008)
94. Greenstreet, M.R., Mitchell, I.: Reachability analysis using polygonal projections. In: Hybrid Systems: Computation and Control. Lecture Notes in Control Science, vol. 1569, pp. 103–116. Springer, Berlin (1999)
95. Gurman, V.I.: The Extension Principle in Problems of Control. Nauka, Moscow (1997)
96. Gusev, M.I., Kurzhanski, A.B.: Optimization of controlled systems with bounds on the controls and state coordinates, I. Differ. Equ. **7**(9), 1591–1602 (1971) [translated from Russian]
97. Gusev, M.I., Kurzhanski, A.B.: Optimization of controlled systems with bounds on the controls and state coordinates, II. Differ. Equ. **7**(10), 1789–1800 (1971) [translated from Russian]
98. Helton, J.W., James, M.R.: Extending H_∞ Control to Nonlinear Systems. SIAM, Philadelphia (1999)
99. Ioffe, A.D., Tikhomirov, V.M.: Theory of Extremal Problems. North-Holland Amsterdam (1979)
100. Isidori, A., Astolfi, A.: Disturbance attenuation and H_∞ control via measurement feedback in nonlinear systems. IEEE Trans. Automat. Control **37**, 1283–1293 (1992)
101. Isidori, A., Marconi, L.: Asymptotic analysis and observer design in the theory of nonlinear output. In: Nonlinear Observers and Applications. Lecture Notes in Control and Information Sciences, vol. 363, pp. 181–210. Springer, Berlin (2007)
102. Isaacs, R.: Differential Games. Wiley, New York (1965)

103. James, M.R., Baras, J.S.: Partially observed differential games, infinite-dimensional Hamilton–Jacobi–Isaacs equations and nonlinear H_∞ control. SIAM J. Control Optim **34**(4), 1342–1364 (1996)
104. Johansson, M.: Piecewise Linear Control Systems. Lecture Notes in Control and Information Sciences, vol. 284. Springer, Berlin (2003)
105. Kailath, T.: Linear Systems. Prentice Hall, Englewood Cliffs (1980)
106. Kalman, R.E.: Contributions to the theory of optimal control. Bol. Soc. Mat. Mexicana. **5**, 102–119 (1959)
107. Kalman, R.E.: On the general theory of control systems. In: Proceedings of the 1-st IFAC Congress, Butterworth, London, vol. 1 (1960)
108. Kalman, R.E.: A new approach to linear filtering and prediction problems. Trans. ASME **82**(Series D), 35–45 (1960)
109. Kalman, R.E., Ho, Y.-C., Narendra, K.S.: Controllability of linear dynamical systems. Contrib. Theory Differ. Equ. **1**(2), 189–213 (1963)
110. Kolmogorov, A.N., Fomin, S.V.: Elements of the Theory of Functions and Functional Analysis. Courier Dover, New York (1999)
111. Komarov, V.A.: Estimates of reachability sets for differential inclusions. Math. Notes (Matematicheskiye Zametki) **37**(6), 916–925 (1985)
112. Komarov, V.A.: Equation of reachability sets for differential inclusions under state constraints. Proc. Steklov Math. Inst. **185**, 116–125 (1989)
113. Koscheev, A.S., Kurzhanski, A.B.: On adaptive estimation of multistage systems under uncertainty. Izvestia Academy Sci. **2**, 72–93 (1983)
114. Kostousova, E.K.: State estimation for dynamic systems via parallelotopes: optimization and parallel computations. Optim. Methods Softw. **9**(4), 269–306 (1998)
115. Kostousova, E.K.: Control synthesis via parallelotopes: optimization and parallel computations. Optim. Methods Softw. **14**(4), 267–310 (2001)
116. Kostousova, E.K.: External polyhedral estimates for reachable sets of linear discrete-time systems with integral bounds on controls. Int. J. Pure Appl. Math. **50**(2), 187–194 (2009)
117. Kostousova, E.K.: On tight polyhedral estimates for reachable sets of linear differential systems. In: ICNPAA 2012 World Congress: 9th International Conference on Mathematical Problems in Engineering, Aerospace and Sciences (2012)
118. Kostousova, E.K., Kurzhanski, A.B.: Theoretical framework and approximation techniques for parallel computation in set-membership state estimation. In: Proceedings of Symposium on Modelling, Analysis and Simulation. CESA'96 IMACS Multiconference, CESA'96, Lille-France, vol. 2, pp. 849–854 (1996)
119. Krasovski, N.N.: On the theory of controllability and observability of linear dynamic systems. Appl. Math. Mech. (PMM) **28**(1), 3–14 (1964)
120. Krasovski, N.N.: Motion Control Theory. Nauka, Moscow (1968) [in Russian]
121. Krasovski, N.N.: Rendezvous Game Problems. National Technical Information Service, Springfield (1971)
122. Krasovskii, A.N., Krasovskii, N.N.: Control Under Lack of Information. Birkhäuser, Boston (1995)
123. Krasovskii, N.N., Subbotin, A.I.: Game-Theoretical Control Problems. Springer Series in Soviet Mathematics. Springer, New York (1988)
124. Krein, M., Smulian, V.: On regular convex sets in the space conjugate to a Banach space. Ann. Math. **41**, 556–583 (1940)
125. Krener, A.J.: Eulerian and Lagrangian observability of point vortex flows. Tellus A **60**, 1089–1102 (2008)
126. Krener, A.J., Ide, K.: Measures of unobservability. In: Proceedings of the IEEE Conference on Decision and Control, Shanghai (2009)
127. Krener, A.J., Wang, K.: Rational minmax filtering. In: Proceedings of IFAC-2011 World Congress, Milan (2011)
128. Krener, A.J., Kang, W.: Linear time-invariant minmax filtering. Ann. Rev. Control **35**, 166–171 (2011)

129. Kruzhkov, S.N.: Generalized solutions of nonlinear first-order PDE's with many independent variables. Math. Sb. **70**((112)3), 394–415 (1966)
130. Kuhn, H.W., Tucker, A.W. (eds.): Linear Inequalities and Related Systems. Annals of Mathematical Studies, vol. 38. Princeton University Press, Princeton (1956)
131. Kuratowski, K.: Topology, vol. 1/2. Academic, New York (1966/1968)
132. Kurzhanskiy, A.A., Varaiya, P.: Ellipsoidal Toolbox. University of California at Berkeley. http://code.google.com/p/ellipsoids/ (2005)
133. Kurzhanskiy, A.A., Varaiya, P.: Ellipsoidal techniques for reachability analysis of discrete-time linear systems. IEEE Trans. Automat. Control **52**(1), 26–38 (2007)
134. Kurzhanskiy, A.A., Varaiya, P.: Active traffic management on road networks. Philos. Trans. R. Soc. A **368**, 4607–4626 (2010)
135. Kurzhanski, A.B.: On duality of problems of optimal control and observation. Appl. Math. Mech. **34**(3), 429–439 (1970) [translated from the Russian]
136. Kurzhanski, A.B.: Differential games of observation. Russ. Math. Dokl. (Dok. Akad. Nauk SSSR) **207**(3), 527–530 (1972)
137. Kurzhanski, A.B.: Control and Observation Under Uncertainty Conditions, 392 pp. Nauka, Moscow (1977)
138. Kurzhanskii, A.B.: On stochastic filtering approximations of estimation problems for systems with uncertainty. Stochastics **23**(6), 109–130 (1982)
139. Kurzhanski, A.B.: Evolution equations for problems of control and estimation of uncertain systems. In: Proceedings of the International Congress of Mathematicians, pp. 1381–1402 (1983)
140. Kurzhanski, A.B.: On the analytical description of the set of viable trajectories of a differential system. Dokl. Akad. Nauk SSSR. **287**(5), 1047–1050 (1986)
141. Kurzhanski, A.B.: Identification: a theory of guaranteed estimates. In: Willems, J.C. (ed.) From Data to Model, pp. 135–214. Springer, Berlin (1989)
142. Kurzhanski, A.B.: The identification problem: a theory of guaranteed estimates. Autom. Remote Control (Avtomatica i Telemekhanika) **4**, 3–26 (1991)
143. Kurzhanski, A.B.: Set-Valued Analysis and Differential Inclusion. Progress in Systems and Control Theory, vol. 16. Birkhäuser, Boston (1993)
144. Kurzhanski, A.B.: Modeling Techniques for Uncertain Systems. Progress in Systems and Control Theory, vol. 18. Birkhäuser, Boston (1994)
145. Kurzhanski, A.B.: Set-valued calculus and dynamic programming in problems of feedback control. Int. Ser. Numer. Math. **124**, 163–174 (1998)
146. Kurzhanski, A.B.: The principle of optimality in measurement feedback control for linear systems. In: Ranter, A., Byrnes, C.V. (eds.) Directions in Mathematical Systems Theory and Optimization. Lecture Notes in Control Science, vol. 286, pp. 193–202. Springer, Berlin (2003)
147. Kurzhanski, A.B.: The problem of measurement feedback control. J. Appl. Math. Mech. (PMM) **68**(4), 487–501 (2004) [translated from the Russian]
148. Kurzhanski, A.B.: The diagnostics of safety zones in motion planning. Optim. Methods Softw. **20**(2/3), 231–239 (2005)
149. Kurzhanski, A.B.: Comparison principle for equations of the Hamilton–Jacobi type in control theory. Proc. Steklov Math. Inst. **253**(1), 185–195 (2006)
150. Kurzhanski, A.B.: "Selected Works of A.B. Kurzhanski", 755 pp. Moscow State University Pub., Moscow (2009)
151. Kurzhanski, A.B.: Hamiltonian techniques for the problem of set-membership state estimation. Int. J. Adapt. Control Signal Process. **25**(3), 249–263 (2010)
152. Kurzhanski, A.B.: On synthesizing impulse controls and the theory of fast controls. Proc. Steklov Math. Inst. **268**(1), 207–221 (2010)
153. Kurzhanski, A.B., Baras, J.: Nonlinear filtering: the set-membership (bounding) and the H-infinity techniques. In: Proceedings of the IFAC Symposium NOLCOS-95 at Lake Tahoe, USA, pp. 409–418. Pergamon Press, Oxford (1995)

154. Kurzhanski, A.B., Daryin, A.N.: Dynamic programming for impulse controls. Ann. Rev. Control **32**(2), 213–227 (2008)
155. Kurzhanski, A.B., Digailova, I.A.: Attainability problems under stochastic perturbations. Differ. Equ. **40**(4), 1573–1578 (2004)
156. Kurzhanski, A.B., Filippova, T.F.: On the set-valued calculus in problems of viability and control for dynamic processes: the evolution equation. Annales de l'institut Henri Poincaré. Analyse non linéaire **S6**, 339–363 (1989)
157. Kurzhanski, A.B., Filippova, T.F.: On the method of singular perturbations for differential inclusions. Dokl. Akad. Nauk SSSR **321**(3), 454–459 (1991)
158. Kurzhanski, A.B., Filippova, T.F.: On the theory of trajectory tubes: a mathematical formalism for uncertain dynamics, viability and control. In: Advances in Nonlinear Dynamics and Control. Progress in Systems and Control Theory, vol. 17, pp. 122–188. Birkhäuser, Boston (1993)
159. Kurzhanski, A.B., Filippova, T.F., Sugimoto, K., Valyi, I.: Ellipsoidal state estimation for uncertain dynamical systems. In: Milanese, M., Norton, J., Piet-Lahanier, H., Walter, E. (eds.) Bounding Approaches to System Identification, pp. 213–238. Plenum Press, New York (1996)
160. Kurzhanski, A.B., Kirilin, M.N.: Ellipsoidal techniques for reachability problems under nonellipsoidal constraints. In: Proceedings of the NOLCOS-01, St. Petersburg, pp. 768–768 (2001)
161. Kurzhanski, A.B., Mitchell, I.M., Varaiya, P.: Optimization techniques for state-constrained control and obstacle problems. J. Optim. Theory Appl. **128**(3), 499–521 (2006)
162. Kurzhanski, A.B., Mitchell, I.M., Varaiya, P.: Control synthesis for state constrained systems and obstacle problems. In: Proceedings of the IFAC Eds. Allgower F., Zeitz M., Oxford. NOLCOS 6-th Symposium at Suttgart, Elsevier, Amsterdam (2005–2014)
163. Kurzhanski, A.B., Nikonov, O.I.: On the control strategy synthesis problem. Evolution equations and set-valued integration. Dokl. Akad. Nauk SSSR. **311**(4), 788–793 (1990)
164. Kurzhanski, A.B., Nikonov, O.I.: Evolution equations for trajectory bundles of synthesized control systems. Russ. Acad. Sci. Dokl. Math. **333**(5), 578–581 (1993)
165. Kurzhanski, A.B., Osipov, Y.S.: The problem of control with bounded phase coordinates. Appl. Math. Mech. (PMM) **32**(2), 188–195 (1968)
166. Kurzhanski, A.B., Osipov, Y.S.: On optimal control under state constraints. Appl. Math. Mech. (PMM) **33**(4), 705–719 (1969)
167. Kurzhanski, A.B., Sivergina, I.F.: Method of guaranteed estimates and regularization problems for evolutionary systems. J. Comput. Math. Math. Phys. **32**(11), 1720–1733 (1992)
168. Kurzhanski, A.B., Sugimoto, K., Vályi, I.: Guaranteed state estimation for dynamical systems: ellipsoidal techniques. Int. J. Adapt. Control Signal Process. **8**(1), 85–101 (1994)
169. Kurzhanski, A.B., Tanaka, M.: Identification: deterministic versus stochastic models. Österreichische Zeitschrift für Statistik und Informatik **19**(1), 30–56 (1989)
170. Kurzhanski, A.B., Tochilin, P.A.: Weakly invariant sets of hybrid systems. Differ. Equ. **44**(11), 1585–1594 (2008)
171. Kurzhanski, A.B., Tochilin, P.A.: Impulse controls in models of hybrid systems. Differ. Equ. **45**(5), 731–742 (2009)
172. Kurzhanski, A.B., Tochilin, P.A.: On the problem of control synthesis under uncertainty from the outputs of finite observers. Differ. Equ. **47**(11), 1–9 (2011)
173. Kurzhanski, A.B., Vályi, I.: Ellipsoidal calculus for estimation and feedback control. In: Systems and Control in the Twenty-First Century, pp. 229–243. Birkhäuser, Boston (1997)
174. Kurzhanski, A.B., Vályi, I.: Ellipsoidal Calculus for Estimation and Control. SCFA. Birkhäuser, Boston (1997)
175. Kurzhanski, A.B., Varaiya, P.: Ellipsoidal techniques for reachability analysis. In: Proceedings of the Pittsburg Conference "Hybrid Systems-2000". Lecture Notes in Control Science, vol. 1790, pp. 202–214. Springer, Berlin (2000)
176. Kurzhanski, A.B., Varaiya, P.: On the reachability problem under persistent disturbances. Dokl. Math. **61**(3), 3809–3814 (2000)

177. Kurzhanski, A.B., Varaiya, P.: Ellipsoidal techniques for reachability analysis. Internal approximation. Syst. Control Lett. **41**, 201–211 (2000)
178. Kurzhanski, A.B., Varaiya, P.: Dynamic optimization for reachability problems. J. Optim. Theory Appl. **108**(2), 227–251 (2001)
179. Kurzhanski, A.B., Varaiya, P.: Reachability under state constraints - the ellipsoidal technique. In: Proceedings of the 15-th IFAC Congress, Barcelona (2002)
180. Kurzhanski, A.B., Varaiya, P.: Reachability analysis for uncertain systems—the ellipsoidal technique. Dyn. Continuous Dis. Impulsive Syst. Ser. B **9**(3), 347–367 (2002)
181. Kurzhanski, A.B., Varaiya, P.: Ellipsoidal techniques for reachability analysis. Part I: external approximations. Optim. Methods Softw. **17**(2), 177–206 (2002)
182. Kurzhanski, A.B., Varaiya, P.: Ellipsoidal techniques for reachability analysis. Part II: internal approximations. Box-valued constraints. Optim. Methods Softw. **17**(2), 207–237 (2002)
183. Kurzhanski, A.B., Varaiya, P.: On reachability under uncertainty. SIAM J. Control Optim. **41**(1), 181–216 (2002)
184. Kurzhanski, A.B., Varaiya, P.: On some nonstandard dynamic programming problems of control theory. In: Giannessi, F., Maugeri, A. (eds.) Variational Methods and Applications, pp. 613–627. Kluwer, New York (2004)
185. Kurzhanski, A.B., Varaiya, P.: Ellipsoidal techniques for reachability under state constraints. SIAM J. Control Optim. **45**(4), 1369–1394 (2006)
186. Kurzhanski, A.B., Varaiya, P.: The Hamilton–Jacobi equations for nonlinear target control and their approximation. In: Analysis and Design of Nonlinear Control Systems (in Honor of Alberto Isidori), pp. 77–90. Springer, Berlin (2007)
187. Kurzhanski, A.B., Varaiya, P.: On synthesizing team target controls under obstacles and collision avoidance. J. Franklin Inst. **347**(1), 130–145 (2010)
188. Kurzhanski, A.B., Varaiya, P.: Impulsive inputs for feedback control and hybrid system modeling. In: Sivasundaram, S., Vasundhara Devi, J., Udwadia, F.E., Lasiecka, I. (eds.) Advances in Dynamics and Control: Theory, Methods and Applications. Cambridge Scientific Publications, Cottenham (2009)
189. Kurzhanski, A.B., Varaiya, P.: Output feedback control under set-membership uncertainty. J. Optim. Theory Appl. **151**(1), 11–32 (2011)
190. Kvasnica, M., Grieder, P., Baotic, M., Morari, M.: Multiparametric toolbox (MPT). In: Hybrid Systems: Computation and Control, pp. 448–462. Springer, Berlin (2004)
191. Kwakernaak, H., Sivan, R.: Linear Optimal Control Systems. Wiley-Interscience, New York (1972)
192. Lancaster, P.: Theory of Matrices. Academic, New York (1969)
193. Lattès, R., Lions, J.-L.: Méthode de quasi-réversibilité et applications. Travaux et recherches mathématiques, vol. 15. Dunod, Paris (1967)
194. Laurent, P.J.: Approximation and Optimization. Hermann, Paris (1972)
195. Lee, E.B., Markus, L.: Foundations of Optimal Control Theory. Wiley, New York (1967)
196. Lefschetz, S.: Differential Equations: Geometric Theory. Dover, Mineola (1977)
197. Le Guernic, C., Girard, A.: Nonlinear Anal. Hybrid Syst. **4**(2), 250–262 (2010)
198. Leitmann, G.: Optimality and reachability with feedback controls. In: Blaquière, A., Leitmann, G. (eds.) Dynamical Systems and Microphysics: Control Theory and Mechanics. Academic, Orlando (1982)
199. Lempio, F., Veliov, V.: Discrete approximations of differential inclusions. Bayreuther Mathematische Schriften **54**, 149–232 (1998)
200. Lions, P.-L.: Viscosity solutions and optimal control. In: Proceedings of ICIAM 91, pp. 182–195. SIAM, Washington (1991)
201. Lions, P.-L., Souganidis, P.E.: Differential games, optimal control and directional derivatives of viscosity solutions of Bellman's and Isaac's equations. SIAM J. Control Optim. **23**, 566–583 (1995)
202. Lyapunov, A.A: On fully additive vector functions. Izvestia USSR Acad. Sci., Ser. Math., **4**(6), 465–478 (1940)

203. Lygeros, J.: On the relation of reachability to minimum cost optimal control. In: Proceedings of the 41st IEEE CDC, Las Vegas, pp. 1910–1915 (2002)
204. Lygeros, J., Tomlin, C., Sastry, S.: Controllers for reachability specifications for hybrid systems. Automatica **35**(3), 349–370 (1999)
205. Lygeros, J., Quincampoix, M., Rzezuchowski, T.: Impulse differential inclusions driven by discrete measures. In: Hybrid Systems: Computation and Control. Lecture Notes in Computer Science, vol. 4416, pp. 385–398. Springer, Berlin (2007)
206. Margellos, K., Lygeros, J.: Hamilton-Jacobi formulation for reach-avoid differential games. IEEE Trans. Automat. Control **56**(8), 1849–1861 (2011)
207. Mayne, D.Q., Rawlings, J.B., Rao, C.V., Scokaert, C.O.M.: Constrained model predictive control: stability and optimality. Automatica **36**, 789–814 (2000)
208. Mayne, D.Q., Seron, M.M., Rakovic, S.V.: Robust model predictive control of constrained linear systems with bounded disturbances. Automatica **41**, 219–224 (2005)
209. Mazurenko, S.S.: A differential equation for the gauge function of the star-shaped attainability set of a differential inclusion. Doklady Math., Moscow **86**(1), 476–479 (2012) [original Russian Text published in Doklady Akademii Nauk **445**(2), 139–142 (2012)]
210. Milanese, M., Norton, J., Piet-Lahanier, H., Walter, E. (eds.): Bounding Approach to System Identification. Plenum Press, London (1996)
211. Milanese, M., Tempo, R.: Optimal algorithms theory for robust estimation and prediction. IEEE Trans. Automat. Control **30**(8), 730–738 (1985)
212. Miller, B.M., Rubinovich, E.Ya.: Impulsive Control in Continuous and Discrete-Continuous Systems. Kluwer, New York (2003)
213. Mitchell, I.M.: Application of level set methods to control and reachability problems in continuous and hybrid systems. Ph.D. thesis, Stanford University (2002)
214. Mitchell, I.M., Tomlin, C.J.: Overapproximating reachable sets by Hamilton-Jacobi projections. J. Sci. Comput. **19**(1), 323–346 (2003)
215. Mitchell, I.M., Templeton, J.A.: A toolbox of Hamilton-Jacobi solvers for analysis of nondterministic in continuous and hybrid systems. In: Hybrid Systems: Computation and Control, pp. 480–494. Springer, Berlin (2005)
216. Morari, M., Thiele, L. (eds.): Hybrid Systems: Computation and Control. Lecture Notes in Computer Sciences, vol. 3414. Springer, Berlin/Heidelberg (2005)
217. Moreau, J.J.: Fonctionelles Convexes. College de France, Paris (1966)
218. Motta, M., Rampazzo, F.: Dynamic programming for nonlinear system driven by ordinary and impulsive control. SIAM J. Control Optim. **34**(1), 199–225 (1996)
219. Neustadt, L.W.: Optimization, a moment problem and nonlinear programming. SIAM J. Control Optim. **2**(1), 33–53 (1964)
220. Nikolski, M.S.: On the approximation of a reachability set of a differential inclusion. Herald of the Moscow Univ. (Vestnik MGU), Ser. Comput. Math. Cybern. **4**, 31–34 (1987)
221. Osher, S., Fedkiw, R.: Level Set Methods and Dynamic Implicit Surfaces. Applied Mathematical Sciences, vol. 153. Springer, New York (2003)
222. Osipov, Yu.S.: On the theory of differential games in distributed parameter systems. Sov. Math. Dokl. **223**(6), 1314–1317 (1975)
223. Osipov, Yu.S., Kryazhimski, A.V.: Inverse Problems for Ordinary Differential Equations: Dynamical Solutions. Gordon and Breach, London (1995)
224. Panasyuk, V.I., Panasyuk, V.I.: On one equation generated by a differential inclusion . Math. Notes **27**(3), 429–437 (1980)
225. Patsko, V.S., Pyatko, S.G., Fedotov, A.A.: Three-dimensional reachability set for a nonlinear control system. J. Comput. Syst. Sci. Int. **42**(3), 320–328 (2003)
226. Pontryagin, L.S., Boltyansky, V.G., Gamkrelidze, P.V., Mischenko, E.F.: The Mathematical Theory of Optimal Processes. Wiley Interscience, New York (1962)
227. Pschenichniy, B.N.: Convex Analysis and Extremal Problems. Nauka, Moscow (1980)
228. Puri, A., Varaiya, P.: Decidability of hybrid systems with rectangular differential inclusions. In: Proceedings of the 6th Workshop on Computer-Aided Verification. Lecture Notes in Control Science, vol. 818, pp. 95–104. Springer, New York (1994)

229. Puri, A., Borkar, V.S., Varaiya, P.: ε-Approximation of differential inclusions. In: Hybrid Systems. Lecture Notes in Control Science, vol. 1066, pp. 362–376. Springer, New York (1996)
230. Rakovic, S.R., Kerrigan, E., Mayne, D., Lygeros, J.: Reachability analysis of discrete-time systems with disturbances. IEEE Trans. Automat. Control **51**, 546–561 (2006)
231. Rakovic, S.R., Kouvritakis, B., Cannon, M., Panos, Ch., Findeisen, R.: Parametrized tube model predictive control. IEEE Trans. Automat. Control **57**, 2746–2761 (2012)
232. Rakovic, S.R., Kouvritakis, B., Findeisen, R., Cannon, M.: Homothetic model predictive control. Automatica **48**, 1631–1637 (2012)
233. Rakovic, S.R., Kouvritakis, B., Cannon, M.: Equi-normalization and exact scaling dynamics in homothetic tube control. Syst. Control Lett. **62**, 209–217 (2013)
234. Riesz, F., Sz-.Nagy, B.: Leçons d'analyse fonctionnelle. Akadémiai Kiadó, Budapest (1972)
235. Rockafellar, R.T.: Level sets and continuity of conjugate convex functions. Trans. Am. Math. Soc. **123**, 46–63 (1966)
236. Rockafellar, R.T.: State constraints in convex problems of Bolza. SIAM J. Control Optim **10**(4), 691–715 (1972)
237. Rockafellar, R.T.: Convex Analysis, 2nd edn. Princeton University Press, Princeton (1999)
238. Rockafellar, R.T., Wets, R.J.: Variational Analysis. Springer, Berlin (2005)
239. Roxin, E.: On generalized dynamical systems defined by contingent equations. J. Differ. Equ. **1**(2), 188–205 (1965)
240. Saint-Pierre, P.: Approximation of the viability kernel. Appl. Math. Optim. **29**(2), 187–209 (1994)
241. Schlaepfer, F.M., Schweppe, F.C.: Continuous-time state estimation under disturbances bounded by convex sets. IEEE Trans. Automat. Control **17**(2), 197–205 (1972)
242. Schwartz, L.: Théorie des Distributions. Hermann, Paris (1950)
243. Schweppe, F.C.: Uncertain Dynamic Systems. Prentice Hall, Englewood Cliffs (1973)
244. Sethian, J.A.: Level Set Methods and Fast Marching Methods. Cambridge University Press, Cambridge (1999)
245. Skelton, R.E.: Dynamic Systems and Control. Wiley, New York (1988)
246. Sorensen, A.J.: A survey of dynamic positioning control systems. Ann. Rev. Control **35**, 123–136 (2011)
247. Subbotin, A.I.: Generalized Solutions of First-Order PDE's. The Dynamic Optimization Perspective. SCFA. Birkhäuser, Boston (1995)
248. Subbotina, N.N., Kolpakova, E.A., Tokmantsev, T.B., Shagalova, L.G.: Method of Characteristics for the Hamilton-Jacobi-Bellman Equation. Ural Scientific Center/Russian Academy of Sciences, Yekaterinburg (2013)
249. Summers, S., Kamgarpour, M., Tomlin, C.J., Lygeros, J.: A stochastic reach-avoid problem with random obstacles. In: Belta, C., Ivancic, F. (eds.) Hybrid Systems: Computation and Control. ACM, New York (2013)
250. Summers, S., Kamgarpour, M., Tomlin, C.J., Lygeros, J.: Stochastic hybrid system controller synthesis for reachability specifications encoded by random sets. Automatica **49**(9), 2906–2910 (2013)
251. Tikhonov, A.N., Arsenin, V.Y.: Solutions of Ill-Posed Prolems. Winston, New York (1975)
252. Tikhonov, A.N., Goncharsky, A.V.: Ill-Posed Problems in the Natural Sciences. Oxford University Press, Oxford (1987)
253. Tolstonogov, A.A.: Differential Inclusions ina Banach Space. Kluwer Academic, Boston (2000)
254. Tomlin, C.J., Lygeros, J., Sastry, S.: A game theoretic approach to controller design for hybrid systems. Proceedings of IEEE **88**, 949–969 (2000)
255. Ushakov, V.N.: Construction of solutions in differential games of pursuit-evasion. Lecture Notes in Nonlinear Analysis, vol. 2(1), pp. 269–281. Polish Academic Publishers, Warsaw (1997)
256. Usoro, P.B., Schweppe, F., Wormley, D., Gould, L.: Ellipsoidal set-theoretic control synthesis. J. Dyn. Syst. Meas. Control **104**, 331–336 (1982)

References

257. Ustyuzhanin, A.M.: On the problem of matrix parameter identification. Probl. Control Inf. Theory **15**(4), 265–273 (1986)
258. Van der Schaft, A., Schumacher, H.: An Introduction to Hybrid Dynamical Systems. Lecture Notes in Control and Information Sciences, vol. 25. Springer, Berlin (2000)
259. Varaiya, P.: Reach set computation using optimal control. In: Proceedings of the KIT Workshop on Verification of Hybrid Systems, Grenoble, October 1998
260. Varaiya, P., Lin, J.: Existence of saddle points in differential games. SIAM J. Control Optim. **7**(1), 142–157 (1969)
261. Vazhentsev, A.Yu.: On internal ellipsoidal approximations for problems of feedback control under state constraints. Proc. Russ. Acad. Sci. Theory Syst. Control **3**, 70–77 (2000)
262. Veliov, V.: Second order discrete approximations to strongly convex differential inclusions. Syst. Control Lett. **13**(3), 263–269 (1989)
263. Veliov,V.: Stability-like properties of differential inclusions. Set Valued Anal. **5**(1), 73–88 (1997)
264. Vidal, R., Schaffert, S., Lygeros, J., Sastry, S.: Controlled invariance of discrete time systems. In: Hybrid Systems: Computation and Control. Lecture Notes in Computer Science, vol. 1790, pp. 437–450. Springer, Berlin (2000)
265. Vinter, R.B.: A characterization of the reachable set for nonlinear control systems. SIAM J. Control Optim. **18**(6), 599–610 (1980)
266. Vinter, R.B., Wolenski, P.: Hamilton–Jacobi theory for optimal control problems with data measurable in time. SIAM J. Control Optim. **28**(6), 1404–1419 (1990)
267. Vinter, R.B.: Optimal Control. Birkhäuser, Boston (2010)
268. Vostrikov, I.V., Dar'in, A.N., Kurzhanskii, A.B.: On the damping of a ladder-type vibration system subjected to uncertain perturbations. Differ. Equ. **42**(11), 1524–1535 (2006)
269. Whittaker, E.T.: Analytical Dynamics. Dover, New York (1944)
270. Willemsm, J.C. (ed.): From Data to Model. Springer, New York (1989)
271. Witsenhausen, H.: Set of possible states of linear systems given perturbed observations. IEEE Trans. Automat. Control **13**(1), 5–21 (1968)
272. Wolenski, P.: The exponential formula for the reachable set of a lipschitz differential inclusion. SIAM J. Control Optim. **28**(5), 1148–1161 (1990)
273. Wonham, W.M.: Linear Multivariable Control: A Geometrical Approach. Springer, Berlin (1985)
274. Zadeh, l.A., Desoer, C.A.: Linear System Theory: The State Space Approach. Dover, New York (2008)
275. Zakroischikov, V.N.: Reachability sets of hybrid systems under sequential switchings. Herald MSU **2**, 37–44 (2009)

Index

B

Backward reachability, 47, 61–64, 82, 83, 128–131, 144, 158–167, 170, 183, 197, 199, 326–328, 338, 383, 385

C

Closed loop target control, 75–78, 167, 371
Colliding tubes, 70–75, 330
Comparison principle
    convex reach sets, 198
    nonconvex reach sets, 208–213
    under state constraints, 281–284
Constraints (bounds)
    ellipsoidal, quadratic, 128, 369, 388
    joint (double), 7
    magnitude, integral, 5–8
Control
    closed-loop, 1, 5–10, 46, 52, 53, 54, 58, 63, 75, 78, 167, 200, 255, 322, 326–331, 338–339, 373
    feedback, 9, 47, 52, 128, 161, 170, 184, 195–196, 260, 269, 270, 272, 275, 284, 311–339, 371–394, 420, 423
    feedforward, model predictive, 46
    minimal magnitude, 12–16
    norm-minimal, minimal energy, 10
    open-loop, 1, 8–10, 24–40, 52, 63, 253, 261, 402, 403
    optimal, linear, 1–46
    state constrained, 289–309
    target, 75–78, 128, 131, 167, 170, 182, 311, 371
    time-optimal, 18–24, 83–86
Controllability, 10–18, 32, 33, 72, 90, 101, 169, 237, 256, 293

D

Differential inclusion, 10, 19, 23, 60, 61, 63, 76, 88, 89, 115, 132, 268, 270, 280, 290, 312, 313, 315–317, 322, 324, 326, 331, 339, 344, 365, 367, 369, 370, 375, 378, 385
Disturbance, 2, 9, 25, 63, 196–198, 341–343, 345, 352, 358, 364, 371–373, 392–394
Dynamic programming, 46–86, 253, 260, 371, 375, 377, 383, 424
Dynamic programming equation, 48–54

E

Ellipsoidal approximations
    box-valued constraints, 125, 229–231, 235–241, 285
    ellipsoidal constraints, 88–91, 202, 388, 398
    external, tight, 96, 98
    internal, tight, 120, 131, 161, 246
    zonotope constraints, 241–251, 388

Equation
  HJB (Hamilton–Jacobi–Bellman), 47, 51, 53, 54, 56, 58–66, 68, 69, 75, 84, 197–208, 218, 224, 253, 267, 341, 342, 344–348, 374, 383, 427
  HJB (state constrained), 277–279, 281, 282, 284, 327, 336
  HJBI (Hamilton–Jacobi–Bellman–Isaacs), 197, 205, 383, 385
  integral funnel, 102, 280, 313, 367, 377
  Riccati, 56, 214, 367

F
Function
  conjugate (Fenchel), 28, 68, 261, 268
  convex, 29, 38, 39, 66, 68, 81, 216, 254, 396
  second conjugate (Fenchel), 81
  strictly convex, 38, 39
  support, 19, 20, 22–24, 39, 47, 64, 65, 69, 80, 89, 90, 111, 119, 120, 129, 144, 150, 157, 158, 185, 188, 189, 191, 202, 241, 242, 280, 285, 287, 289–292, 315, 319, 320, 330, 363–365, 410

G
Generalized state, 372–375, 377, 378
Gram matrix, 11, 15, 16
Guaranteed state estimation, 341–370, 372, 377–381

H
Hausdorff distance, 61, 103, 105, 281, 319
Hausdorff semidistance, 61, 103, 280, 281, 319
Hybrid system
  branching trajectory tubes, 395
  ellipsoidal methods, 398–401, 418–419
  enabling zones, resets, 402, 414–415
  impulsive feedback control, 420, 423
  reachability, 395, 396, 401–416
  value functions, 396, 397, 399, 407–416, 418, 427

I
Impulse control
  approximating motion, 269–273
  closed loop, 260–266
  delta function, number of impulses, 7, 269, 421
  HJB variational inequality, 260–266
  open loop, value function, 255–260
  realistic approximation, controls, 266–269
Information space, 381–386
Information tubes, calculation
  linear-quadratic approximation, 363–367
  singular perturbations, discontinuous measurements, 370
  stochastic filtering approximation, 367–370
Integral
  set-valued, 20, 115, 144, 231, 288
  Stieltjes, Lebesgue–Stieltjes, 254, 266, 268, 286, 288, 293

L
Legendre transformation, 29
Lipschitz condition, 312, 378

M
Matching conditions, 196, 393
Maximum condition, 14, 24, 28–31, 34, 36, 44, 287, 294
Maximum principle, 1, 14, 24–40, 44, 45, 72, 74, 90, 99, 108, 109, 245, 256, 287, 289–294, 308, 325
  ellipsoidal, 107–110, 251
Measurement equation, 312, 341, 345, 346, 348, 373
Metric space, 281, 315, 381, 383
Minimum condition, 28, 35, 44
Moment problem, 10, 12, 17–18, 255

O
Obstacle problem
  closed-loop control, 338–339
  complementary convex constraint, 336
  reach-evasion set, 335–336
Optimization problem
  linear-quadratic, 46, 54–57, 363, 367, 368
  primal, dual, 40–46
Output feedback control
  ellipsoidal approximations, 388–392
  linear-convex systems, 372, 379, 381, 386–388
  overall solution, 323, 372, 375–377

P
Parallel computation, 127, 195–196
Polyhedral techniques, output feedback control, 373
Principle of optimality, 53, 59, 62, 263, 277, 278, 327

R

Reachability, 18–24, 27, 32, 47, 58–70, 78–84, 87–145, 151, 158–167, 170, 182–184, 186, 189–193, 197–199, 201, 202, 208–215, 218, 222–229, 235–239, 242, 243, 245–251, 275–289, 300, 311, 326–328, 338, 341, 344, 352–356, 365, 369, 383, 385, 388, 395, 396, 401–413, 415, 416

S

Saddle point, 26, 36, 210
Separation principle for estimation and control, 391–392
Set
    box-valued, 198
    convex, 5, 8, 38, 39, 187, 359, 416
    ellipsoidal, 115
    invariant, strong, 63, 64, 79
    invariant, weak, 61, 62, 202, 382–384
    nonconvex, 200, 211, 220, 418
    unknown but bounded, 8, 341, 358
    zonotope, 198, 242, 246
Set-valued analysis, set-membership uncertainty, 371
Set-valued observer, 378, 379, 392
Solvability set
    extremal aiming, 78, 132, 195, 384
    Krasovski bridge, 384
State constraints
    closed loop (feedback) control, 5–10, 322, 326–331
    ellipsoidal approximation, 275, 294, 301, 304
    linear-convex system, 285–289
    maximum principle (modified), 289–294, 325
    maximum principle (standard), 287, 294
    reachability set, value function, 341, 356
State estimation, bounding approach
    comparison principle, 342, 346–348, 350
    continuous dynamics, discrete observation, 355–358
    discrete dynamics, discrete observation, 358–363
    ellipsoidal approximation, 350–355, 388–391
    HJB equation, 341, 342, 344–346
    information set, information tube, 341–344, 347, 349, 352, 368, 375, 377–381
    information state, 341, 344, 377–381
    linear system, 342, 345, 348–350, 354–355, 358, 366, 368, 379
Subdifferential, 28, 60, 279
Switching system, 403, 420
System
    controlled, 1–5, 87, 141, 177, 179, 182, 185, 276, 371, 401, 403
    linear, 4, 18, 24, 54, 60, 66, 68, 88–91, 202, 203, 213, 253, 281, 284, 285, 289, 316–317, 322, 325, 327, 339, 342, 345, 348–350, 354–355, 358, 366, 368, 377, 379, 386–391, 393
    linear-convex, 24–40, 47, 64–70, 183, 229, 285–289
    quasilinear, 5

T

Trajectory tube
    evolution equation, 311–319
    funnel equations, bilinear system, 315–318
    viable tube
        calculation, 313
        linear-convex system, 313
        parametrization, 313
Transition matrix, 2, 148, 291, 292, 301, 391, 409

V

Value function, 47, 48, 50–52, 54, 56, 58–60, 62–72, 80, 81, 85, 201, 209, 217, 219, 221, 224, 255–260, 262, 264, 266, 267, 276–278, 281, 283, 286, 326, 327, 333–335, 337, 341, 344, 349, 372, 374, 375, 379, 381–384, 386, 396, 397, 399, 407–416, 418, 427
Verification problem
    ellipsoidal methods, 398–401, 418–419
    solution scheme, 396–398

MIX
Papier aus verantwortungsvollen Quellen
Paper from responsible sources
FSC® C105338

If you have any concerns about our products,
you can contact us on
ProductSafety@springernature.com

In case Publisher is established outside the EU,
the EU authorized representative is:
**Springer Nature Customer Service Center GmbH
Europaplatz 3, 69115 Heidelberg, Germany**

Printed by Libri Plureos GmbH
in Hamburg, Germany